Statistics
Concepts and Controversies

Tenth Edition

David S. Moore
Purdue University

William I. Notz
The Ohio State University

macmillan
learning

Austin • Boston • New York • Plymouth

Vice President, STEM: Daryl Fox
Program Director: Andrew Dunaway
Program Manager: Sarah Seymour
Senior Marketing Manager: Nancy Bradshaw
Marketing Assistant: Madeleine Inskeep
Executive Development Editor: Katrina Mangold
Development Editor: Leslie Lahr
Executive Media Editor: Catriona Kaplan
Associate Editor: Andy Newton
Assistant Editor: Justin Jones
Director of Digital Production: Keri deManigold
Media Project Manager: Hanna Squire
Director, Content Management Enhancement: Tracey Kuehn
Senior Managing Editor: Lisa Kinne
Senior Workflow Project Supervisor: Susan Wein
Senior Content Project Manager: Edward Dionne
Project Managers: Tristann Jones and Vanavan Jayaraman, Lumina Datamatics, Inc.
Director of Design, Content Management: Diana Blume
Design Services Manager: Natasha Wolfe
Cover Design Manager: John Callahan
Art Manager: Matthew McAdams
Executive Permissions Editor: Cecilia Varas
Rights and Billing Associate: Alexis Gargin
Production Supervisor: Lawrence Guerra
Composition: Lumina Datamatics, Inc.
Printing and Binding: LSC Communications
Cover Images: Free material from gapminder.org. Creative Commons Attribution 4.0.
https://creativecommons.org/licenses/by/4.0/

Library of Congress Control Number: 2019947188

Student Edition Paperback:
ISBN-13: 978-1-319-10902-8
ISBN-10: 1-319-10902-0
Student Edition Loose-Leaf:
ISBN-13: 978-1-319-27262-3
ISBN-10: 1-319-27262-2
Instructor Complimentary Copy:
ISBN-13: 978-1-319-27247-0
ISBN-10: 1-319-27247-9

1 2 3 4 5 6   24 23 22 21 20 19

Macmillan Learning
One New York Plaza
Suite 4600
New York, NY 10004-1562
www.macmillanlearning.com

In 1946, William Freeman founded W. H. Freeman and Company and published Linus Pauling's *General Chemistry*, which revolutionized the chemistry curriculum and established the prototype for a Freeman text. W. H. Freeman quickly became a publishing house where leading researchers can make significant contributions to mathematics and science. In 1996, W. H. Freeman joined Macmillan and we have since proudly continued the legacy of providing revolutionary, quality educational tools for teaching and learning in STEM.

Brief Contents

*This material is optional

Contents

*This material is optional

*This material is optional

Available online at www.macmillanlearning .com/scc10e:
Notes and Data Sources
Exploring the Web Exercises

*This material is optional

To the Teacher

Statistics as a Liberal Discipline

Statistics: Concepts and Controversies (*SCC*) is a book on statistics as a liberal discipline, that is, as part of the general education of "nonmathematical" students. The book grew out of one of the author's experiences in developing and teaching a course for freshmen and sophomores from Purdue University's School of Liberal Arts. We are pleased that other teachers have found *SCC* useful for unusually diverse audiences, extending as far as students of philosophy and medicine. This tenth edition is a revision of the text, with several new features. It retains, however, the goals of the original: to present statistics not as a technical tool but as part of the intellectual culture that educated people share.

Statistics among the liberal arts

Statistics has a widespread reputation as the least liberal of subjects. When statistics is praised, it is most often for its usefulness. Health professionals need statistics to read accounts of medical research; managers need statistics because efficient crunching of numbers will find its way to the bottom line; citizens need statistics to understand opinion polls and government statistics such as the unemployment rate and the Consumer Price Index. Because data and chance are omnipresent, as our propaganda line goes, everyone will find statistics useful, and perhaps even profitable.

This is true. We would even argue that for most students, the conceptual and verbal approach in *SCC* is better preparation for future encounters with statistical studies than the usual methods-oriented introduction. The joint curriculum committee of the American Statistical Association and the Mathematical Association of America recommends that any first course in statistics "emphasize the elements of statistical thinking" and feature "more data and concepts, fewer recipes and derivations." *SCC* does this, with the flavor appropriate to a liberal education: more concepts, more thinking, only simple data, fewer recipes, and no formal derivations. There is, however, another justification for learning about statistical ideas: statistics belongs among the liberal arts. A liberal education emphasizes fundamental intellectual skills, that is, general methods of inquiry that apply in a wide variety of settings. The traditional liberal arts present such methods: literary and historical studies, the political and social analysis of human societies, the probing of nature by experimental science, the power of abstraction and deduction in mathematics. The case that statistics belongs among the liberal arts rests on the fact that reasoning from uncertain empirical data is a similarly general intellectual method. *Data* and *chance,* the topics of this book, are pervasive aspects of our experience. Though we employ the tools of mathematics to work with data and chance, the mathematics implements ideas that are not strictly mathematical. In fact, psychologists argue convincingly that mastering formal mathematics does little to improve our ability to reason effectively about data and chance in everyday life.

SCC is shaped, as far as the limitations of the authors and the intended readers allow, by the view that statistics is an independent and fundamental intellectual method. The focus is on statistical thinking, on what others might call *quantitative literacy* or *numeracy*.

The nature of this book

There are books on statistical theory and books on statistical methods. This is neither. It is a book on statistical ideas and statistical reasoning and on their relevance to public policy and to the human sciences from medicine to sociology. We have included many elementary graphical and numerical techniques to give flesh to the ideas and muscle to the reasoning. Students learn to think about data by working with data. We have not, however, allowed technique to dominate concepts. Our intention is to teach verbally rather than algebraically, to invite discussion and even argument rather than mere computation, though some computation remains essential. The coverage is considerably broader than one might traditionally cover in a one-term course, as the table of contents reveals. In the spirit of general education, we have preferred breadth to detail.

Despite its informal nature, SCC is a textbook. It is organized for systematic study and has abundant exercises, many of which ask students to offer a discussion or make a judgment. Even those admirable individuals who seek pleasure in uncompelled reading should look at the exercises as well as the text. Teachers should be aware that the book is more serious than its low mathematical level suggests. The emphasis on ideas and reasoning asks more of the reader than many recipe-laden methods texts.

New in this edition

This new version of a classic text fits the current teaching environment while continuing to present statistics to "nonmathematical" readers as an aid to clear thinking in personal and professional life. The following new features and enhancements build on *SCC*'s strong pedagogical foundation:

Chapter Goals. Each chapter now opens with a brief list of learning goals to prepare the student for what lies ahead. At the end of each chapter, we list **Chapter Achievements,** which reinforce these main concepts the student has now mastered.

Multiple Variables. The revised Guidelines for Assessment and Instruction in Statistics Education (GAISE) report encourages introductory classes to expose students to multiple variable thinking. As a result, we decided to include a short subsection in Chapter 14 that gives students some insight into how relationships between several variables can be investigated graphically.

Examples and exercises. Over one-third of the examples and exercises are revised to reflect current data and a variety of topics. They cover a wide range of application areas, adding interest and relevance for students. New example and exercise topics include Brexit; determining the mood of the nation; digital media use and ADHD; Android vs. iOS; psychology and the 2016 U.S. election; New Year's Eve in Times Square; one-bedroom apartment rents; political views of college students; three-point shooting; and graduation plans.

Case Studies. Beginning each chapter, Case Studies engage students in real-life scenarios related to the chapter concepts. The Case Study is also used as an opportunity to introduce the chapter's learning objectives (in nontechnical language) as the tools that students are given to assist in their evaluation of the study. A **Case Study Evaluated** section at the end of each chapter revisits the chapter opening Case Study with follow-up questions, asking students to evaluate what they have learned from the chapter and to apply their knowledge to the Case Study. At least half the case studies have been updated or changed to include more-relevant topics. New topics include a discussion of estimating crowd size (Chapter 9), how the number of Starbucks in a town is related to home prices (Chapter 15), use of social media (Chapter 21), and political party affiliation and beliefs about freedom of speech (Chapter 24).

Design. The lively and contemporary design integrates colorful figures, vibrant photos, and a dynamic layout to engage students and enhance their understanding of text material.

What's the Verdict? This new feature describes an interesting scenario followed by a series of questions that challenge students to apply what they have just learned to evaluate some real issue. In some ways, students will mimic the sort of discussion that occurs in many of the chapter examples. These are placed at the ends of selected chapters, but icons appear in the body of a chapter indicating those questions that students are now equipped to answer. An ongoing "What's the Verdict?" issue might appear in several places within a given chapter and even span several different chapters. Real statistical examples involve multiple concepts in *SCC*, and as students learn new concepts, these questions can help them come to a fuller understanding of how to assess real examples.

In addition to the new tenth edition enhancements, *SCC* has retained the successful pedagogical features from previous editions:

- **Statistical Controversies.** These boxed features explore controversial topics and relate them to the chapter material. There is follow-up discussion and a proposed resolution to each of these topics found online at **www.macmillanlearning.com/scc10e**, in the Resolving the Controversy section.
- **Now It's Your Turn exercises.** These appear after a worked example, allowing students to test their understanding. Full solutions to these exercises are provided in the back of the text.
- **Macmillan Learning Online Resources.** At the end of each chapter, we provide a list of resources available from Macmillan Learning that provide additional support for understanding topics in the chapter.
- **Chapter summaries.** The "Statistics in Summary" sections at the end of each chapter now consist of two sections. The first is a bulleted summary of the material presented in the chapter. The second section, titled "Link It," relates the chapter content to material in previous and upcoming chapters. The goal of this format is to help students understand how individual chapters relate to each other and to the overall practice of statistics.
- **Check the Basics exercises.** Each chapter ends with a series of straight forward multiple-choice problems that test students' understanding of basic concepts. If students have difficulty with these problems, we recommend they review the basic concepts in the chapter before tackling the chapter exercises.

- **In the News exercises.** From popular news media outlets, these exercises use current events and cite recent data sources.

- **Applets.** An applet icon signals where related, interactive statistical applets can be found on the book's website.

- **Exploring the Web exercises.** These exercises direct students to investigate topics on the Web and think critically about statistical data and concepts. Available on the text's website (**www.macmillanlearning.com/scc10e**).

- **Technology output screenshots.** Most statistical analyses rely heavily on statistical software. In this book, we specifically discuss the use of JMP 12 in some parts. Other software for conducting statistical analysis includes CrunchIt!®, Minitab, SPSS (an IBM Company),[1] R, and a TI-83/-84 calculator. As specialized statistical packages, JMP, Minitab, and SPSS are the most popular software choices both in industry and in colleges and schools of business. As an all-purpose spreadsheet program, Excel provides a limited set of statistical analysis options in comparison. However, given its pervasiveness and wide acceptance in industry and the computer world at large, we believe it is important to give Excel proper attention. It should be noted that for users who want more statistical capabilities but want to work in an Excel environment, there are a number of commercially available add-on packages (if you have JMP, for instance, it can be invoked from within Excel). TI-83/-84 calculators are generally sufficient for an introductory course, although most statistical analysis is beyond the capabilities of even the best calculator, so those seeking to continue their learning of statistics should consider learning one of the specialized statistical packages.

Even though basic guidance for JMP is provided in parts of this book, it should be emphasized that *SCC* is not bound to any program. Computer output from statistical packages is very similar, so you can feel quite comfortable using any one of these packages.

---

[1] SPSS was acquired by IBM in October 2009.

Assessment—The "Office Hours" experience, while doing Homework

Tutorial-Style Formative Assessment

For select questions in our formative assessment environment, students' incorrect answers receive full solutions and feedback to guide their study. Many exercises are also designed to deliver error-specific feedback based on their common misconceptions about topics/learning objectives. This socratic feedback mechanism emulates the office hours experience, encouraging students to think critically about their identified misconception. Students are also provided with the fully-worked solutions to reinforce concepts and provide an in-product study guide for every problem.

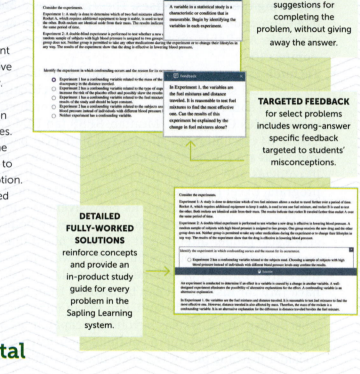

HINTS attached to select problems encourage critical thinking by providing suggestions for completing the problem, without giving away the answer.

TARGETED FEEDBACK for select problems includes wrong-answer specific feedback targeted to students' misconceptions.

DETAILED FULLY-WORKED SOLUTIONS reinforce concepts and provide an in-product study guide for every problem in the Sapling Learning system.

Adaptive Assessment Focused on Fundamental Concepts

LEARNINGCURVE

LearningCurve's adaptive quizzing encourages students to learn through practice. Through gamified elements, students are challenged to gain points by submitting correct answers to meet or exceed an instructor-determined score. Faculty and students are then provided detailed analytics on their performance via their own personal study plan with pathways to additional learning tools and insights. This tool focuses on getting students prepared for their upcoming class.

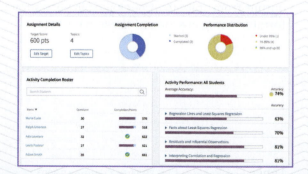

Analytics

STUDENT ANALYSIS

Through our heatmap dashboard instructors can get a quick visual representation of student performance on a given question. They can then dive deeper into each individual student to view and evaluate every submission they've made on a problem, getting to see a student's "work" on the question.

ITEM ANALYSIS

Instructors can also evaluate an aggregate view of their class's performance on a problem (item) through our **"Item Analysis"** tab. They can quickly see what percentage of students got this question correct, incorrect, or unanswered through the progress bar at the top. They can also see all of the student responses rolled up; giving them immediate insight into what their students' most common misconceptions are on this problem.

Interactive e-book

The **e-book** provides powerful study tools for students, multimedia content, and easy customization for instructors. Students can search, highlight, and bookmark specific information, making it easier to study and access key content. Media assets such as data sets, glossary terms, select exercise answers, and videos are linked within the e-book in **SaplingPlus,** allowing ease-of-use for students.

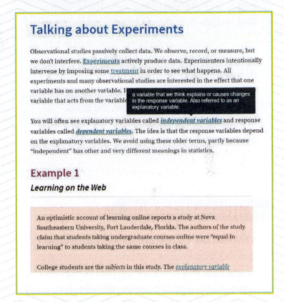

Integration

With **Deep Asset Linking,** instructors can organize individual pieces of Macmillan content into LMS Content folders. Students complete single-sign-on through the link to connect their Macmillan and LMS accounts. Once this occurs, students will have seamless access to the Macmillan content through their campus LMS using only the LMS username and password.

- **Multiple gradebook columns** generate automatically for each Macmillan assignment deployed to the LMS, allowing for assignments to be combined into gradebook categories within the LMS for gradebook calculations like dropping lowest scores or weighted percentages.

- **Automatic grade sync** occurs with Deep integration.

Media/Learning Objects

CONCEPT VIDEOS are brief videos that make important and sometimes complex concepts more digestible for beginners in statistics, presented in a manner that combines video, audio, and interactive features. These videos also include built-in, assignable assessments.

In the **CONTROVERSY VIDEOS**, students exchange views on the controversial topics introduced in the chapter. Their dialog adds detail to the main points from the book. These videos also include built-in, assignable assessments.

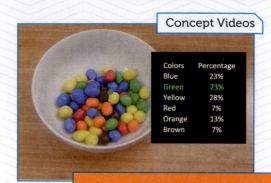

Concept Videos

The Harris Poll Online

Factors to Consider
- Sample size
- Self-selection bias
- Accuracy of results
- Survey administration

Key Terms
- nonresponse
- probability sample
- sampling frame
- frame error

Controversy Videos

TRIED AND TRUE

STATTUTOR TUTORIALS offer multimedia tutorials that explore important concepts and procedures in a presentation that combines video, audio, and interactive features. The newly revised format includes built-in, assignable assessments and a bright new interface.

STATISTICAL APPLETS give students hands-on opportunities to familiarize themselves with important statistical concepts and procedures, in an interactive setting that allows them to manipulate variables and see the results graphically. Icons in the textbook indicate when an applet is available for the material being covered.

STATBOARDS VIDEOS are brief whiteboard videos that illustrate difficult topics through additional examples, written and explained by a select group of statistics educators.

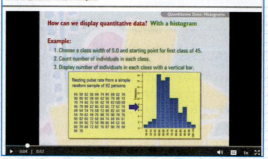

StatTutor Tutorials

StatBoards Videos

Paired with Assessment:

TRIED AND TRUE

STATISTICAL VIDEO SERIES consists of StatClips, StatClips Examples, and Statistically Speaking "Snapshots." View animated lecture videos, whiteboard lessons, and documentary-style footage that illustrate key statistical concepts and help students visualize statistics in real-world scenarios.

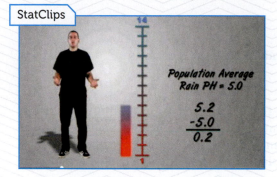

StatClips

EESEE CASE STUDIES (Electronic Encyclopedia of Statistical Examples and Exercises), developed by The Ohio State University Statistics Department, teach students to apply their statistical skills by exploring actual case studies using real data.

SCC-SPECIFIC ASSESSMENT

WHAT'S THE VERDICT? Assessment Modules offer machine-gradable activities that contain all questions from each of the new *What's the Verdict?* features from the book.

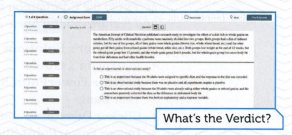

What's the Verdict?

PRE-BUILT CHAPTER HOMEWORK ASSIGNMENTS are curated assignments containing questions directly from the book, with select questions containing error specific feedback to guide students to concept mastery.

Pre-Built Chapter Homework Assignments

Data Tools

DATA FILES are available in JMP, ASCII, Excel, TI, Minitab, SPSS, R, and CSV formats.

VIDEO TECHNOLOGY MANUALS available for TI-83/84 calculators, Minitab, Excel, JMP, SPSS, R, RStudio, Rcmdr, and CrunchIt!® provide brief instructions for using specific statistical software.

CRUNCHIT!® is Macmillan Learning's own web-based statistical software that allows users to perform all the statistical operations and graphing needed for an introductory statistics course

and more. It saves users time by automatically loading data from SCC, and it provides the flexibility to edit and import additional data.

jmp **STUDENT EDITION** (developed by SAS) is easy to learn and contains all the capabilities required for introductory statistics. JMP is the leading commercial data analysis software of choice for scientists, engineers and analysts at companies throughout the globe (for Windows and Mac).

Additional resources available with *SCC*:

Companion Website www.macmillan learning.com/scc10e This open-access website includes statistical applets and data set files. The website also offers access to the *Exploring the Web* exercises for each chapter. Instructor access to the Companion Website requires user registration as an instructor and features all the open-access student web materials, plus:

- **Instructor's Solutions Manual**
- **Instructor's Guide**
- **Lecture PowerPoint Slides**
- **Test Bank**
- **Clicker Questions** and **Expanded Clicker Questions**
- **Image PowerPoint Slides**

▷ iClicker is a two-way radio-frequency classroom response solution developed by educators for educators. Each step of iClicker's development has been informed by teaching and learning. To learn more about packaging iClicker with this textbook, please contact your local sales rep or visit **www.iclicker.com**.

Acknowledgments

A very special thanks to authors Jack Miller, University of Michigan, and Michelle Everson, The Ohio State University, who both bring so much to this edition—passion, energy, time, scholarship, classroom experience, and creativity. Their fresh perspectives on all aspects of the book are invaluable and helped improve this edition substantially. Jack revised Chapters 3, 4, 7, 8, 9, 13, 16, and 24. Michelle revised Chapters 10, 11, 14, 15, 21, 22, and 23. Both contributed to the Part reviews. They tackled this work in the true spirit of collaboration and teamwork.

A hearty thanks to Ellen Gundlach for contributing the "What's the Verdict?" stories and questions, which will be sure to prove interesting and relevant to instructors and students alike.

The staff of W. H. Freeman, especially Andrew Dunaway, Leslie Lahr, Catriona Kaplan, Andy Newton, Justin Jones, Aaron Gladish, Ava Cas, Hanna Squire, Ed Dionne, Natasha Wolfe, John Callahan, and Susan Wein, have done their usual excellent job in editing, designing, and producing the book. Special thanks also to Tristann Jones and Vanavan Jayaraman. Thank you to Rose Martinez-Dawson for carefully accuracy checking the book.

A special thanks to Karen Carson for all her help with this edition. Karen provided encouragement, guidance, helpful suggestions, insights, and an unfailing positive attitude. Most of all, she has been a good friend. Her contributions improved the book and we owe her a debt of gratitude.

Our sincere appreciation and thanks go to the following instructors who created much of the ancillary package that accompanies the 10th edition: Michelle Duda and Karen Starin, both at Columbus State Community College, created the Instructor's Solution Manual, Student Solution Manual, back-of-book answers, Clicker questions, Test Bank, Lecture Slides, and Instructor's Guide. Sharona Byrnes and Ciro Ramirez, both at California State University, Los Angeles, developed the expanded Clicker questions.

Our copy editor, Christine Sabooni, our proofreaders, Lori Lewis and the team in Lumina Datamatics Inc., have done a terrific job overseeing word usage and punctuation; thank you.

We are grateful to many colleagues who commented on successive drafts of the manuscript.

Worku Aberra, Dawson College

Shemsi Alhaddad, University of South Carolina, Lancaster

Mark Angelo, The College of New Jersey

Robert W. Avakian, Oklahoma State University Institute of Technology

Marc Belanger, Vanier College

GBP Bootheway, St. Bonaventure University

Eric L. Bruns, Campbellsville University

Jaya Casukhela, The Ohio State University

Stacye Chen, The College of New Jersey

Jerry J. Chen, Suffolk County Community College

Julie M. Clark, Hollins University

Keith Coates, Drury University

Lisa McKinney Crawford, Piedmont Technical College

Alan Dabney, Texas A&M University

Timothy P. Dienes, Central Piedmont Community College

Sue Ann Jones Dobbyn, Pellissippi State Community College

Jimmy Doi, California Polytechnic State University San Luis Obispo

Kathleen Donoghue, Robert Morris University

Rob-Roy Douglas, Northern Lights College

Kimberley J. Duff, Cerritos College

Brandi Falley, Texas Woman's University

Jayne Gackenbach, Athabasca University

Mark A. Gebert, University of Kentucky

Ross Gosky, Appalachian State University

Michael Granaas, University of South Dakota

Billie-Jo Grant, California Polytechnic State University

Annie Hager, Plymouth State University

Melodie J. Hallett, San Diego State University

Shahryar Heydari, Piedmont College

William Huepenbecker, Bowling Green State University Firelands

Tim Husband, Siena Heights University

Grethe Hystad, Purdue University Northwest

Yvette Janecek, Blinn College

Rasitha Jayasekare, Butler University

Tony Jenkins, Northwestern Michigan College

Xiaoyi Ji, Utah Valley University

Lisa W. Kay, Eastern Kentucky University

Yoon G. Kim, Humboldt State University

Brian Knaeble, Utah Valley University

Palmer Kocher, State University of New York New Paltz

John W. Kulig, Plymouth State University

Subrata Kundu, George Washington University

Michael D. Larsen, Saint Michael's College

Scott Lewis, Utah Valley University

Yin Lin, University of Texas at El Paso

George Litman, National Louis University

Brent Marinan, University of Arizona

Randall Martinez, Cypress College

Eric Matsuoka, Leeward Community College

Ana Clare Mello, Shasta College

David Miller, William Paterson University

Juliann Moore, Oregon State University

Linda Myers, Harrisburg Area Community College

Jacquelyn O'Donohoe, Plymouth State University

Krzysztof Plotka, University of Scranton

Gary Popoli, Stevenson University

Herman E. Ray, Kennesaw State University

Betty C. Rogers, Piedmont College

Alicia Romero, University of Michigan

Anita A. Ross, North Park University

Soma Roy, California Polytechnic State University

James C. Scott, Colby College

Caleb Southworth, University of Oregon

Karen Jean Starin, Columbus State Community College

Gary S. Stoudt, Indiana University of Pennsylvania

Dana Sylvan, Hunter College City College of New York

Sherry Taylor, Piedmont Technical College

Brent Wessel, Saint Louis University

Lesley W. Wiglesworth, Centre College

Norbert Youmbi, Saint Francis University

Jerrold H. Zar, Northern Illinois University

Nicholas Zoller, Southern Nazarene University

Prelude

Making Sense of Statistics

Statistics is about data. Data are numbers, but they are not "just numbers." *Data are numbers with a context*. The number 10.5, for example, carries no information by itself. But if we hear that a friend's new baby weighed 10.5 pounds at birth, we congratulate her on the healthy size of the child. The context engages our background knowledge and allows us to make judgments. We know that a baby weighing 10.5 pounds is quite large, and that a human baby is unlikely to weigh 10.5 ounces or 10.5 kilograms. The context makes the number informative.

Statistics uses data to gain insight and to draw conclusions. The tools are graphs and calculations, but the tools are guided by ways of thinking that amount to educated common sense. Let's begin our study of statistics with a rapid and informal guide to coping with data and statistical studies in the news media and in the heat of political and social controversy. We will examine the examples introduced in this prelude in more detail later.

Data beats anecdotes

Belief is no substitute for arithmetic.
HENRY SPENCER

An anecdote is a striking story that sticks in our minds exactly because it is striking. Anecdotes humanize an issue, so news reports usually start (and often stop) with anecdotes. But anecdotes are weak ground for making up your mind—they are often misleading exactly because they are striking. Always ask if a claim is backed by data, not just by an appealing personal story.

Does living near power lines cause leukemia in children? The National Cancer Institute spent 5 years and $5 million gathering data on the question. Result: no connection between leukemia and exposure to magnetic fields of the kind produced by power lines. The editorial that accompanied the study report in the *New England Journal of Medicine* thundered, "It is time to stop wasting our research resources" on the question.

Now compare the impact of a television news report of a 5-year, $5 million investigation with that of a televised interview with an articulate mother whose child has leukemia and who happens to live near a power line. In the public mind, the anecdote wins every time. Be skeptical. Data are more reliable than anecdotes because they systematically describe an overall picture rather than focus on a few incidents.

We are tempted to add, "Data beat self-proclaimed experts." The idea of balance held by much of the news industry is to present a quick statement by an "expert" on either side. We never learn that one expert expresses the consensus of an entire field of science and the other is a quack with a special-interest axe to grind. As a result of the media's taste for conflict, the public now thinks that for every expert, there is an equal and opposite expert. If you really care about an issue, try to find out what the data say and how good the data are. Many issues do remain unsettled, but many others are unsettled only in the minds of people who don't care about evidence. You can start by looking at the credentials of the "experts" and at whether the studies they cite have appeared in journals that require careful outside review before they publish a claim.

Where the data come from is important

Figures won't lie but liars will figure.
CHARLES GROSVENOR

Data are numbers, and numbers always seem solid. Some are and some are not. Where the data come from is the single most important fact about any statistical study. When Ann Landers asked readers of her advice column whether they would have chosen to have children again and 70% of those who replied shouted "No," readers should have just amused themselves with Ann's excerpts from tear-stained letters describing what beasts the writers' children are. Ann Landers was in the entertainment business. Her invitation attracted parents who regretted having their children. Most parents don't regret having children. We know this because opinion polls have asked large numbers of parents, chosen at random to avoid attracting one opinion or another. Opinion polls have their problems, as we will see, but they beat just asking upset people to write in their views.

Even the most reputable publications have not been immune to bad data. The *Journal of the American Medical Association* once printed an article claiming that pumping refrigerated liquid through tubes in the stomach relieves ulcers. The patients did respond, but only because patients often respond to *any* treatment given with the authority of a trusted doctor. That is, placebos (dummy treatments) work. When a skeptic finally tried a properly controlled study in which some patients got the tube and some got a placebo, the placebo actually did a bit better. "No comparison, no conclusion" is a good starting point for judging medical studies. We would be skeptical about the ongoing interest in "natural remedies," for example. Few of these have passed a comparative trial to show that they are more than just placebos sold in bottles bearing pretty pictures of plants.

Beware the lurking variable

I have enough money to last me the rest of my life, unless I buy something.
JACKIE MASON

You read that crime is higher in counties with gambling casinos. A college teacher says that students who took a course online did better than the students in the classroom. Government reports emphasize that well-educated people earn a lot more than people with less education. Don't jump to conclusions. Ask first, "What is there that they didn't tell me that might explain this?"

Crime is higher in counties with casinos, but it is also higher in urban counties and in poor counties. What kind of counties are casinos in? Did these counties have high crime rates before the casino arrived? The online students did better, but they were older and better prepared than the in-class students. No wonder they did better. Well-educated people do earn a lot. But educated people have (on the average) parents with more education and more money than the parents of poorly educated people have. They grew up in nicer places and went to better schools. These advantages help them get more education and would help them earn more even without that education.

All these studies report a connection between two variables and invite us to conclude that one of these variables influences the other. "Casinos increase crime" and "Stay in school if you want to be rich" are the messages we hear. Perhaps these messages are true. But perhaps much of the connection is explained by other variables lurking in the background, such as the nature of counties that accept casinos and the advantages that highly educated people were born with. Good statistical studies look at lots of background variables. This is tricky, but you can at least find out if it was done.

Variation is everywhere

When the facts change, I change my mind. What do you do, sir?
JOHN MAYNARD KEYNES

If a thermometer under your tongue reads higher than 98.6°F, do you have a fever? Maybe not. People vary in their "normal" temperature. Your own temperature also varies—it is lower around 6 A.M. and higher around 6 P.M. The government announces that the unemployment rate rose a tenth of a percent last month and that new home starts fell by 3%. The stock market promptly jumps (or sinks). Stocks are jumpier than is sensible. The government data come from samples that give good estimates but not the exact truth. Another run of the same samples would give slightly different answers. And economic facts jump around anyway, due to weather, strikes, holidays, and all sorts of other reasons.

Many people join the stock market in overreacting to minor changes in data that are really nothing but background noise. Here is Arthur Nielsen, head of the country's largest market research firm, describing his experience:

> Too many business people assign equal validity to all numbers printed on paper. They accept numbers as representing Truth and find it difficult to work with the concept of probability. They do not see a number as a kind of shorthand for a range that describes our actual knowledge of the underlying condition.

Variation is everywhere. Individuals vary; repeated measurements on the same individual vary; almost everything varies over time. Ignore the pundits who try to explain the deep reasons behind each day's stock market moves, or who condemn a team's ability and character after a game decided by a last-second shot that did or didn't go in.

Conclusions are not certain

As far as the laws of mathematics refer to reality they are not certain, and as far as they are certain they do not refer to reality.
ALBERT EINSTEIN

Because variation is everywhere, statistical conclusions are not certain. Most women who reach middle age have regular mammograms to detect breast cancer. Do mammograms really reduce the risk of dying of breast cancer? Statistical studies of high quality find that mammograms reduce the risk of death in women aged 50 to 64 years by 26%. That's an average over all women in the age group. Because variation is everywhere, the results are different for different women. Some women who have mammograms every year die of breast cancer, and some who never have mammograms live to 100 and die when they crash their motorcycles.

What the summary study actually said was "mammography reduces the risk of dying of breast cancer by 26 percent (95 percent confidence interval, 17 to 34 percent)." That 26% is, in Arthur Nielsen's words, "shorthand for a range that describes our actual knowledge of the underlying condition." The range is 17% to 34%, and we are 95% confident that the truth lies in that range. We're pretty sure, in other words, but not certain. Once you get beyond news reports, you can look for phrases like "95% confident" and "statistically significant" that tell us that a study did produce findings that, while not certain, are pretty sure.

Data reflect social values

It's easy to lie with statistics. But it is easier to lie without them.
FREDERICK MOSTELLER

Good data do beat anecdotes. Data are more objective than anecdotes or loud arguments about what might happen. Statistics certainly lies on the factual, scientific, rational side of public discourse. Statistical studies deserve more weight than most other evidence about controversial issues. There is, however, no such thing as perfect objectivity. Statistics shares a social context that influences what we decide to measure and how we measure it.

Suicide rates, for example, vary greatly among nations. It appears that much of the difference in the reported rates is due to social attitudes rather than to actual differences in suicide rates. Counts of suicides come from death certificates. The officials who complete the certificates (details vary depending on the state or nation) can choose to look more or less closely at, for example, drownings and falls that lack witnesses. Where suicide is stigmatized, deaths are more often reported as accidents. Countries that are predominantly Catholic have lower reported suicide rates than others, for example. Japanese culture has a tradition of honorable suicide as a response to shame. This tradition leads to better reporting of suicide in Japan because it reduces the stigma attached to suicide. In other nations, changes in social values may lead to higher suicide counts. It is becoming more common to view depression as a medical problem rather than a weakness of character and suicide as a tragic end to the illness rather than a moral flaw. Families and doctors then become more willing to report suicide as the cause of death.

Social values influence data on matters less sensitive than suicide. The percentage of people who are unemployed in the United States is measured each month by the Bureau of Labor Statistics, using a large and very professionally chosen sample of people across the country. But what does it mean to be "unemployed"? It means that you don't have a job even though you want a job and have *actively looked for work in the last four weeks*. If you went four weeks without seeking work, you are not unemployed; you are "out of the labor force." This definition of unemployment reflects the value we attach to working. A different definition might give a very different unemployment rate.

Our point is not that you should mistrust the unemployment rate. The definition of "unemployment" has been stable over time, so that we can see trends. The definition is reasonably consistent across nations, so that we can make international comparisons. The data are produced by professionals free of political interference. The unemployment rate is important and useful information. Our point is that not everything important can be reduced to numbers and that reducing things to numbers is done by people influenced by many pressures, conscious and unconscious.

Statistics and You

What Lies Ahead in This Book

This isn't a book about the tools of statistics. It is a book about statistical ideas and their impact on everyday life, public policy, and many different fields of study. You will learn some tools, of course. Life will be easier if you have in hand *a calculator with built-in statistical functions*. Specifically, you need a calculator that will find means, standard deviations, and correlations. Look for a calculator that claims to do "two-variable statistics" or mentions "correlation." If you have access to a computer with statistical software, so much the better. On the other hand, you need little formal mathematics. If you can read and use simple equations, you are in good shape. Be warned, however, that you will be asked to think. Thinking exercises the mind more deeply than following mathematical recipes. *Statistics: Concepts and Controversies* presents statistical ideas in four parts:

I. Data production describes methods for producing data that can give clear answers to specific questions. Where the data come from really is important—basic concepts about how to select samples and design experiments are the most influential ideas in statistics.

II. Data analysis concerns methods and strategies for exploring, organizing, and describing data using graphs and numerical summaries. You can learn to look at data intelligently even with quite simple tools.

III. Probability is the language we use to describe chance, variation, and risk. Because variation is everywhere, probabilistic thinking helps separate reality from background noise.

IV. Statistical inference moves beyond the data in hand to draw conclusions about some wider universe, taking into account that variation is everywhere and that conclusions are uncertain.

Ultimately, data are used to draw conclusions or make decisions. The process of reasoning from data consists of several steps that yield a case for the validity of the final conclusion. Each part of this book discusses issues that affect the quality of the steps in this process. It is easy to focus on mastering the details in each chapter and lose track of how these details contribute to the overall argument. To help you see how the individual chapters fit into the overall argument, we end each chapter with a section that we call Link It, which briefly describes how the contents of the chapter fit into the overall reasoning process. You will find this section within the Statistics in Summary subsection.

Statistical ideas and tools emerged only slowly from the struggle to work with data. Two centuries ago, astronomers and surveyors faced the problem of combining many observations that, despite the greatest care, did not exactly match. Their efforts to deal with variation in their data produced some of the first statistical tools. As the social sciences emerged in the nineteenth century, old statistical ideas were transformed and new ones were invented to describe the variation in individuals and societies. The study of heredity and of variable populations in biology brought more advances. The first half of the twentieth century gave birth to statistical designs for producing data and to statistical inference based on probability. By midcentury, it was clear that a new discipline had been born. As all fields of study place more emphasis on data and increasingly recognize that variability in data is unavoidable, statistics has become a central intellectual method. Every educated person should be acquainted with statistical reasoning. Reading this book will enable you to make that acquaintance.

About the Authors

David S. Moore is Shanti S. Gupta Distinguished Professor of Statistics, Emeritus, at Purdue University and was 1998 president of the American Statistical Association. He received his AB from Princeton University and his PhD from Cornell University, both in mathematics. He has written many research papers in statistical theory and served on the editorial boards of several major journals. Professor Moore is an elected fellow of the American Statistical Association and of the Institute of Mathematical Statistics and an elected member of the International Statistical Institute. He has served as program director for statistics and probability at the National Science Foundation.

Professor Moore has devoted much of his career to the teaching of statistics. He was the content developer for the Annenburg/Corporation for Public Broadcasting college-level telecourse *Against All Odds: Inside Statistics* and for the series of video modules *Statistics: Decisions through Data,* intended to aid the teaching of statistics in schools. He is the author of influential articles on statistical education and of several leading textbooks. Professor Moore has served as president of the International Association for Statistical Education and has received the Mathematical Association of America's national award for distinguished college or university teaching of mathematics.

William I. Notz is professor emeritus of statistics at the Ohio State University. He received his BS in physics from Johns Hopkins University and his PhD in mathematics from Cornell University. His first academic job was as an assistant professor in the Department of Statistics at Purdue University. While there, he taught the introductory concepts course with Professor Moore, using the first edition of *Statistics: Concepts and Controversies*. As a result of this experience, he developed an interest in statistical education. Professor Notz is a coauthor of *The Electronic Encyclopedia of Statistical Examples and Exercises* (EESEE) and has coauthored several textbooks.

Professor Notz's research interests have focused on experimental design and computer experiments. He is the author of several research papers and a book on the design and analysis of computer experiments.

He is an elected fellow of the American Statistical Association and an elected member of the International Statistical Institute. He has served as the editor of the journals *Technometrics* and *Journal of Statistics Education,* as well as on the editorial boards of several journals. At The Ohio State University, he has served as the director of the Statistical Consulting Service, as acting chair of the Department of Statistics, and as associate dean in the College of Mathematical and Physical Sciences. He is a winner of The Ohio State University's Alumni Distinguished Teaching Award.

Michelle Everson is a program specialist and associated faculty member in the Department of Statistics at The Ohio State University. She received her BS in industrial psychology from California State University, East Bay, her MA in psychology from San Jose State University, and her PhD in educational psychology from the University of Minnesota. From 2002 until 2014, she was a senior lecturer in the Department of Educational Psychology at the University of Minnesota, where she worked specifically with the Statistics Education graduate program. In 2009, Dr. Everson received a teaching award from the College of Education and Human Development at the University of Minnesota, and in 2011, she received the Waller Education Award from the American Statistical Association. She started her work at The Ohio State University in 2014. Dr. Everson is passionate about involving students in classroom discussion and getting them excited about the ways in which statistics is a part of their everyday lives. She enjoys teaching in both the traditional classroom setting and the online environment, and she has developed and taught multiple online courses since 2004. She was the editor of the *Journal of Statistics Education* from 2013 through 2015, and served as co-chair of an American Statistical Association committee that updated and revised the 2005 Guidelines for Assessment and Instruction in Statistics Education (GAISE) College Report. The revised GAISE College Report was endorsed by the American Statistical Association in 2016.

Jackie Bryce Miller is a Lecturer IV in the Department of Statistics at the University of Michigan. Dr. Miller earned both a BA and BS in mathematics and statistics from Miami University and was an actuarial associate for four years before going to graduate school. Dr. Miller earned an MS in statistics and a one-of-a-kind PhD in statistics education (one of the very first statistics education doctorates!) from The Ohio State University. They were an assistant professor at Drury University for three years before returning to

Ohio State in 2003. For the next ten years, they were a Statistics Education Specialist in Ohio State's Department of Statistics. In 2013, Dr. Miller moved to the Department of Statistics at Michigan, a move loved by their current students and hated by former Ohio State students. Always passionate about statistics education, Dr. Miller was awarded the inaugural Robert V. Hogg Award for excellence in teaching introductory statistics from the Mathematical Association of America's SIGMAA on Statistics Education. They are a member of the Board of Directors for the Consortium for the Advancement of Undergraduate Statistics Education (CAUSE) and have been actively involved in CAUSE since its inception. They also currently serve the American Statistical Association on the Leadership Support Council and served in several other national statistics education leadership positions. Dr. Miller is known for introducing the HyFlex (hybrid-flexible) method of instruction at Ohio State (through a Departmental Impact Grant) and at Michigan, and recently completed an NSF grant that studied the use of HyFlex in undergraduate statistics courses.

PART I

Producing Data

You and your friends are not typical. What you listen to on the radio, for example, is probably not what we listen to. Of course, we and our friends are also not typical. To get a true picture of the country as a whole (or even of college students), we must recognize that the picture may not resemble us or what we see around us. We need *data*. Data from Nielsen (a consumer research firm) for September 11, 2018, show that the most popular radio formats during the summer of 2018 were news/talk (9.6% of listeners), adult contemporary hit radio (7.7% of listeners), and country (7.4% of listeners). If you like hot adult contemporary (5.5% of listeners) and we like classic rock (5.0% of listeners), we may have no clue about the tastes of radio audiences as a whole. If we are in the broadcasting business, or even if we are interested in pop culture, we must put our own tastes aside and look at the data.

Sam Edwards/Getty Images

You can find data in the library or on the Internet (that's where we found the radio format data). But how can we know whether data can be trusted? Good data are as much a human product as wool sweaters and tablet PCs. Sloppily produced data will frustrate you as much as a sloppily made sweater. You examine a sweater before you buy, and you don't buy it if it is not well made. Neither should you use data that are not well made. The first part of this book shows how to tell if data are well made.

Where Do Data Come From?

Case Study: An Online Poll

In this chapter you will:

- Learn the different types of studies used to collect data.
- Understand the purpose of these different types of studies used to collect data.
- Examine the types of data a study might collect.

You can read the newspaper and watch TV news for months without seeing an algebraic formula. No wonder algebra seems unconnected to life. You can't go a day, however, without meeting data and statistical studies. You hear that last month's unemployment rate was 3.9%. A news article says that 87% of AAAS (the American Association for the Advancement of Science) scientists agree that climate change is mostly due to human activity, while only 50% of all U.S. residents 18 years of age and over agree. A longer article says that low-income children who received high-quality day care did better on academic tests given years later and were more likely to go to college and hold good jobs than other similar children.

Where do these data come from? Why can we trust them? Or maybe we can't trust them. Good data are the fruit of intelligent human effort. Bad data result from laziness or lack of understanding, or even the desire to mislead others. "Where do the data come from?" is the first question you should ask when someone throws a number at you.

In 2012, Colorado voters legalized marijuana. Subsequently, voters in several other states have considered legalizing marijuana. One of these is Michigan. In February 2014, the Michigan online news site MLive ran the story "Take our online poll: Should Michigan legalize marijuana?" Of 9684 respondents,

7906 (81.64%) said Yes, 1190 (12.29%) said No, and 588 (6.07%) said decriminalize but not legalize. These results would seem to indicate overwhelming support for legalizing marijuana in Michigan.

What can we say about data from this poll? By the end of this chapter you will have learned some basic questions to ask about data from the MLive online poll. The answers to these questions will help us assess whether the data from the poll are good or bad, as we will explore further in Chapter 2.

Talking about Data: Individuals and Variables

Statistics is the science of data. We could almost say "the art of data" because good judgment and even good taste, along with good math, make good statistics. A big part of good judgment lies in deciding what you must measure in order to produce data that will shed light on your concerns. We begin with some vocabulary to describe the raw materials that go into data.

For example, here are the first lines of a professor's data set at the end of a statistics course:

NAME	MAJOR	POINTS	GRADE
ADVANI, SURA	COMM	397	B
BARTON, DAVID	HIST	323	C
BROWN, ANNETTE	LIT	446	A
CHIU, SUN	PSYC	405	B
CORTEZ, MARIA	PSYC	461	A

Key Terms

Individuals or units are the objects described by a set of data. Individuals may be people, but they may also be animals or things.

A **variable** is any characteristic of an individual. A variable can take different values for different individuals.

A **categorical variable** simply places an individual into one of several groups or categories.

A **numerical variable** takes numerical values for which arithmetic operations such as adding and averaging make sense. A numerical variable is sometimes referred to as a **quantitative variable**.

The *individuals* are students enrolled in the course. In addition to each student's name, there are three *variables*. The first says what major a student has chosen. The second variable gives the student's total points out of 500 for the course, and the third records the grade received.

Statistics deals with numbers, but not all variables are numerical. Some are "categorical" and simply place an individual into one of several groups or categories. Of the three variables in the professor's data set, only total points has numbers as its values. Major and grade are categorical, and to do statistics with these variables, we use *counts* or *percentages*. We might give the percentage of students who got an A, for example, or the percentage who are psychology majors.

Bad judgment in choosing variables can lead to data that cost lots of time and money but don't shed light on the world. What constitutes good judgment can be controversial. Here are examples of the challenges in deciding what data to collect.

Example 1

Who recycles?

Who takes the trouble to recycle? Researchers spent lots of time and money weighing the stuff put out for recycling in two neighborhoods in a California city; call them Upper Crust and Lower Mid. The *individuals* here are households because trash and recycling pickup are done for residences, not for people one at a time. The *variable* measured was the weight in pounds of the curb-side recycling basket each week.

monticello/Deposit Photos

The Upper Crust households contributed more pounds per week on the average than did the folks in Lower Mid. Can we say that the rich are more serious about recycling? No. Someone noticed that Upper Crust recycling baskets contained lots of heavy glass wine bottles. In Lower Mid, they put out lots of light plastic soda bottles and light metal beer and soda cans. The conclusion: weight tells us little about commitment to recycling.

Example 2

What's your race?

The U.S. Census asks, "What is this person's race?" for every person in every household. "Race" is a *variable,* and the Census Bureau must say exactly how to measure it. The census form does this by giving a list of races. Years of political squabbling lie behind this list.

How many races shall we list, and what names shall we use for them? Shall we have a category for people of mixed race? Asians wanted more national categories, such as Filipino and Vietnamese, for the growing Asian population. Pacific Islanders wanted to be separated from the larger Asian group. Black leaders did not want a mixed-race category, fearing that many blacks would choose it and so reduce the official count of the black population.

The 2010 census form (see Figure 1.1) ended up with six Asian groups (plus "Other Asian") and three Pacific Island groups (plus "Other Pacific Islander"). There is no "mixed-race" group, but you can mark more than one race. That is, people claiming mixed race can count as both so that the total of the racial group counts in 2010 is larger than the population count. Unable to decide what the proper term for blacks should be, the Census Bureau settled on "Black, African American, or Negro." What about Hispanics? That's a separate question because Hispanics can be of any race.

United States Census 2010

U.S. DEPARTMENT OF COMMERCE
Economics and Statistics Administration
U.S. CENSUS BUREAU

This is the official form for all the people at this address.
It is quick and easy, and your answers are protected by law.

Use a blue or black pen.

Start here

The Census must count every person living in the United States on April 1, 2010.

Before you answer Question 1, count the people living in this house, apartment, or mobile home using our guidelines.

- Count all people, including babies, who live and sleep here most of the time.

The Census Bureau also conducts counts in institutions and other places, so:

- Do not count anyone living away either at college or in the Armed Forces.
- Do not count anyone in a nursing home, jail, prison, detention facility, etc., on April 1, 2010.
- Leave these people off your form, even if they will return to live here after they leave college, the nursing home, the military, jail, etc. Otherwise, they may be counted twice.

The Census must also include people without a permanent place to stay, so:

- If someone who has no permanent place to stay is staying here on April 1, 2010, count that person. Otherwise, he or she may be missed in the census.

1. How many people were living or staying in this house, apartment, or mobile home on April 1, 2010?

Number of people =

2. Were there any additional people staying here April 1, 2010 that you did not include in Question 1? Mark X all that apply.

- Children, such as newborn babies or foster children
- Relatives, such as adult children, cousins, or in-laws
- Nonrelatives, such as roommates or live-in baby sitters
- People staying here temporarily
- No additional people

3. Is this house, apartment, or mobile home — Mark X ONE box.

- Owned by you or someone in this household with a mortgage or loan? Include home equity loans.
- Owned by you or someone in this household free and clear (without a mortgage or loan)?
- Rented?
- Occupied without payment of rent?

4. What is your telephone number? We may call if we don't understand an answer.

Area Code + Number

OMB No. 0607-0919-C: Approval Expires 12/31/2011.

Form **D-61** (9-25-2008)

5. Please provide information for each person living here. Start with a person living here who owns or rents this house, apartment, or mobile home. If the owner or renter lives somewhere else, start with any adult living here. This will be Person 1.

What is Person 1's name? Print name below.

Last Name

First Name   MI

6. What is Person 1's sex? Mark X ONE box.

- Male
- Female

7. What is Person 1's age and what is Person 1's date of birth? Please report babies as age 0 when the child is less than 1 year old.
Print numbers in boxes.

Age on April 1, 2010   Month   Day   Year of birth

→ **NOTE: Please answer BOTH Question 8 about Hispanic origin and Question 9 about race. For this census, Hispanic origins are not races.**

8. Is Person 1 of Hispanic, Latino, or Spanish origin?

- No, not of Hispanic, Latino, or Spanish origin
- Yes, Mexican, Mexican Am., Chicano
- Yes, Puerto Rican
- Yes, Cuban
- Yes, another Hispanic, Latino, or Spanish origin — Print origin, for example, Argentinean, Colombian, Dominican, Nicaraguan, Salvadoran, Spaniard, and so on.

9. What is Person 1's race? Mark X one or more boxes.

- White
- Black, African Am., or Negro
- American Indian or Alaska Native — Print name of enrolled or principal tribe.

- Asian Indian
- Chinese
- Filipino
- Japanese
- Korean
- Vietnamese
- Native Hawaiian
- Guamanian or Chamorro
- Samoan
- Other Asian — Print race, for example, Hmong, Laotian, Thai, Pakistani, Cambodian, and so on.
- Other Pacific Islander — Print race, for example, Fijian, Tongan, and so on.

- Some other race — Print race.

10. Does Person 1 sometimes live or stay somewhere else?

- No
- Yes — Mark X all that apply.
  - In college housing
  - In the military
  - At a seasonal or second residence
  - For child custody
  - In jail or prison
  - In a nursing home
  - For another reason

→ **If more people were counted in Question 1, continue with Person 2.**

U S C E N S U S B U R E A U

Source: Census.gov

Figure 1.1 The first page of the 2010 census form, mailed to all households in the country. The 2010 census form can be found online at www.census.gov/2010census/about/interactive-form.php.

Again unable to choose a short name that would satisfy everyone, the Census Bureau decided to ask if you are of "Hispanic, Latino, or Spanish origin."

The fight over "race" reminds us that data reflect society. Race is a social idea, not a biological fact. In the census, you say what race you consider yourself to be. Race is a sensitive issue in the United States, so the fight is no surprise and the Census Bureau's diplomacy seems a good compromise.

Observational Studies

As Yogi Berra, the former catcher and manager of the New York Yankees who is renowned for his humorous quotes, said, "You can observe a lot by watching." Sometimes all you can do is watch. To learn how chimpanzees in the wild behave, watch. To study how a teacher and young children interact in a schoolroom, watch. It helps if the watcher knows what to look for. The chimpanzee expert may be interested in how males and females interact, in whether some chimps in the troop are dominant, in whether the chimps hunt and eat meat. Indeed, chimps were thought to be vegetarians until Jane Goodall watched them carefully in Gombe National Park, Tanzania. Now it is clear that meat is a natural part of the chimpanzee diet.

At first, the observer may not know what to record. Eventually, patterns seem to emerge, and we can decide what variables we want to measure. How often do chimpanzees hunt? Alone or in groups? How large are hunting groups? Males alone, or both males and females? How much of the diet is meat? Observation that is organized and measures clearly defined variables is more convincing than just watching. Here is an example of highly organized (and expensive) observation.

Example 3

Do power lines cause leukemia in children?

Electric currents generate magnetic fields. So, living with electricity exposes people to magnetic fields. Living near power lines increases exposure to these fields. Really strong fields can disturb living cells in laboratory studies. What about the weaker fields we experience if we live near power lines? Some data suggested that more children in these locations might develop leukemia, a cancer of the blood cells.

We can't do experiments that deliberately expose children to magnetic fields for weeks and months at a time. It's hard to compare cancer rates among children who happen to live in more and less exposed locations because leukemia is quite rare and locations vary a lot in many ways other than magnetic fields. It is easier to start with children who have leukemia and compare them with children

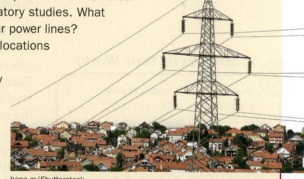
bane.m/Shutterstock

who don't. We can look at lots of possible causes—diet, pesticides, drinking water, magnetic fields, and others—to see where children with leukemia differ from those without. Some of these broad studies suggested a closer look at magnetic fields.

One really careful look at magnetic fields took five years and cost $5 million. The researchers compared 638 children who had leukemia and 620 who did not. They went into the homes and actually measured the magnetic fields in the children's bedrooms, in other rooms, and at the front door. They recorded facts about nearby power lines for the family home and also for the mother's residence when she was pregnant. Result: no evidence of more than a chance connection between magnetic fields and childhood leukemia. Similar conclusions were reached by researchers at the University of Oxford in England who reviewed data from 1962–2008 and by researchers in California who reviewed data in California from 1986–2008.

Key Terms

A **response variable** is a variable that measures an outcome or result of a study.

An **observational study** observes individuals and measures variables of interest but does not intervene to influence the responses. The purpose of an observational study is to describe some group or situation.

Statistics in Your World

You just don't understand
A sample survey of journalists and scientists found quite a communications gap. Journalists think that scientists are arrogant, while scientists think that journalists are ignorant. We won't take sides, but here is one interesting result from the survey: 82% of the scientists agree that the "media do not understand statistics well enough to explain new findings" in medicine and other fields.

"No evidence" that magnetic fields are connected with childhood leukemia doesn't prove that there is no risk. It says only that a very careful study could not find any risk that stands out from the play of chance that distributes leukemia cases across the landscape. In other words, the study could not rule out chance as a plausible explanation for what was observed. Critics continue to argue that the study failed to measure some important variables or that the children studied don't fairly represent all children. Nonetheless, a carefully designed observational study is a great advance over haphazard and sometimes emotional counting of cancer cases.

Sample Surveys

You don't have to eat the entire pot of soup to know it needs more salt. That is the idea of sampling: to gain information about the whole by examining only a part. **Sample surveys** are an important kind of observational study. They survey some group of individuals by studying only some of its members, selected not because they are of special interest but because they represent the larger group. Here is the vocabulary we use to discuss sampling.

Notice that the *population* is the group we want to study. If we want information about all U.S. college students, that is our population—even if students at only one college are available for sampling. To make sense of any sample result, you must know what population the sample represents. Did a preelection poll, for example, ask the opinions of all adults? Or citizens only? Registered voters only? Democrats only? The *sample* consists of the people we actually have information about. If the poll can't contact some of the people it selected, those people aren't in the sample.

The distinction between population and sample is basic to statistics. The following examples illustrate this distinction and also introduce some major uses of sampling. These brief descriptions also indicate the variables measured for each individual in the sample.

Key Terms

The **population** in a statistical study is the entire group of individuals about which we want information.

A **sample** is the part of the population from which we actually collect information and is used to draw conclusions about the whole.

Example 4

Public opinion polls

Polls such as those conducted by Gallup and many news organizations ask people's opinions on a variety of issues. The *variables* measured are responses to questions about public issues. Though most noticed at election time, these polls are conducted on a regular basis throughout the year. For a typical opinion poll:

Population: U.S. residents 18 years of age and over. Noncitizens and even undocumented immigrants are included.

Sample: Between 1000 and 1500 people interviewed by telephone.

Example 5

The Current Population Survey

Government economic and social data come from large sample surveys of a nation's individuals, households, or businesses. The monthly Current Population Survey (CPS) is the most important government sample survey in the United States. Many of the *variables* recorded by the CPS concern the employment or unemployment of everyone over 16 years old in a household. The government's monthly unemployment rate comes from the CPS. The CPS also records many other economic and social variables. For the CPS:

Population: The more than 126 million U.S. households. Notice that the individuals are households rather than people or families. A household consists of all people who share the same living quarters, regardless of how they are related to each other.

Sample: About 60,000 households interviewed each month.

Example 6

TV ratings

Market research is designed to discover what consumers want and what products they use. One example of market research is the television-rating service of Nielsen Media Research. The Nielsen ratings influence how much advertisers will pay to sponsor a program and whether or not the program stays on the air. For the Nielsen national TV ratings:

Population: The more than 119 million U.S. households that have a television set.

Sample: About 75,000 households that agree to use a "people meter" to record the TV viewing of all people in the household.

The *variables* recorded include the number of people in the household and their age and sex, whether the TV set is in use at each time period, and, if so, what program is being watched and who is watching it.

Example 7

The General Social Survey

Social science research makes heavy use of sampling. The General Social Survey (GSS), carried out every second year by the National Opinion Research Center at the University of Chicago, is the most important social science sample survey. The *variables* cover the subject's personal and family background, experiences and habits, and attitudes and opinions on subjects from abortion to war.

Population: Adults (aged 18 and over) living in households in the United States. The population does not include adults in institutions such as prisons and college dormitories. It also does not include persons who cannot be interviewed in English.

Sample: About 3000 adults interviewed in person in their homes.

Now it's your turn

1.1 Legalizing marijuana. The Pew Research Center conducted a poll October 25–30, 2017. They asked:

Do you think the use of marijuana should be made legal or not?

The Pew Research Center reported that the poll consisted of telephone interviews with 1504 randomly selected adult Americans. What do you think the population is? What is the sample?

Most statistical studies use samples in the broad sense. For example, the 638 children with leukemia in Example 3 are supposed to represent all children with leukemia. We usually reserve the dignified term "sample survey" for studies that use an organized plan to choose a sample that represents some specific population. The children with leukemia were patients at centers that specialize in treating children's cancer. Expert judgment says they are typical of all leukemia patients, even though they come only from special types of hospitals. A sample survey doesn't rely on judgment: it starts with an entire population and uses specific, quantifiable methods to choose a sample to represent the population. Chapters 2, 3, and 4 discuss the art and science of sample surveys.

Census

A sample survey looks at only a part of the population. Why not look at the entire population? A *census* tries to do this.

The U.S. Constitution requires a census of the American population every 10 years. A census of so large a population is expensive and takes a long time. Even the federal government, which can afford a census, uses samples such as the Current Population Survey to produce timely data on employment and many other variables. If the government asked every adult in the country about his or her employment, this month's unemployment rate would be available next year rather than next month. In fact, to save money, the 2010 census consisted of only 10 questions. Five of these were general questions, and five required answers for every person living at the address the form was sent to.

So time and money favor samples over a census. Samples can have other advantages as well. If you are testing fireworks or fuses, the sampled items are destroyed. Moreover, a sample can produce more accurate data than a census. A careful sample of an inventory of spare parts will almost certainly give more accurate results than asking the clerks to count all 500,000 parts in the warehouse. Bored people do not count accurately.

The experience of the Census Bureau reminds us that a census can only *attempt* to sample the entire population. The bureau estimated that the 2010 census overcounted the American population by 0.01% but undercounted the black population by 2.1%. A census is not foolproof, even with the resources of the government

Key Terms

A **census** is a sample survey that attempts to include the entire population in the sample.

Statistics in Your World

Is a census old-fashioned?
The United States has taken a census every 10 years since 1790. Technology marches on, however, and replacements for a national census look promising. Denmark has no census, and France plans to eliminate its census. Denmark has a national register of all its residents, who carry identification cards and change their register entry whenever they move. France will replace its census by a large sample survey that rotates among the nation's regions. The U.S. Census Bureau has a similar idea: the American Community Survey has already started and has eliminated the census "long form." In 2011, Canada eliminated its long-form census and replaced it with a controversial voluntary National Household Survey. But in 2016, Canada reintroduced the long form.

behind it. Why take a census at all? The government needs block-by-block population figures to create election districts with equal populations. The main function of the U.S. census is to provide this local information.

Experiments

Our goal in choosing a sample is a picture of the population, disturbed as little as possible by the act of gathering information. All observational studies share the principle "observe but don't disturb." When Jane Goodall first began observing chimpanzees in Tanzania, she set up a feeding station where the chimps could eat bananas. She later said that was a mistake because it might have changed the apes' behavior.

In *experiments,* on the other hand, we want to change behavior. In doing an experiment, we don't just observe individuals or ask them questions. We actively impose some treatment in order to observe the response. Experiments can answer questions such as "Does aspirin reduce the chance of a heart attack?" and "Do a majority of college students prefer Pepsi to Coke when they taste both without knowing which they are drinking?"

Key Terms

An **experiment** deliberately imposes some treatment on individuals in order to observe their responses. The purpose of an experiment is to study whether the treatment causes a change in the response.

Example 8

Helping welfare mothers find jobs

The Urban Institute in Washington DC reports that most adult welfare recipients are single mothers in their 20s and 30s with one or two children. Observational studies of welfare mothers show that many are able to increase their earnings and leave the welfare system. Some take advantage of voluntary job-training programs to improve their skills. Should participation in job-training and job-search programs be required of all able-bodied welfare mothers? Observational studies of the current system cannot tell us what the effects of such a policy would be. Even if the mothers studied are a properly chosen sample of all welfare recipients, those who seek out training and find jobs may differ in many ways from those who do not. They are observed to have more education, for example, but they may also differ in values and motivation, things that cannot be observed.

To see if a required jobs program will help mothers escape welfare, such a program must actually be tried. Choose two similar groups of mothers when they apply for welfare. Require one group to participate in a job-training program, but do not offer the program to the other group. This is an experiment. Comparing the income and work record of the two groups after several years will show whether requiring training has the desired effect.

Now it's your turn

1.2 Diet soda and weight. Many people drink diet soda to control their weight. To explore the relationship between drinking diet soda and weight, medical researchers recruited 400 adults who identified themselves as diet soda drinkers. These 400 participants recorded how many bottles or cans of diet soda they drank per week over a period of two years. Each participant was weighed at the beginning and end of this two-year period. The researchers found that the more diet soda one drank, the more likely one was to have gained weight during the two-year period. The results were published in a well-known scientific journal. Was this an observational study or an experiment?

The welfare example illustrates the big advantage of experiments over observational studies: *in principle, experiments can give good evidence for cause and effect*. If we design the experiment properly, we start with two very similar groups of welfare mothers. The *individual* women of course differ from each other in age, education, number of children, and other respects. But the two *groups* resemble each other when we look at the ages, years of education, and number of children for all women in each group. During the experiment, the women's lives differ, but there is only one systematic difference between the two groups: whether or not they are in the jobs program. All live through the same good or bad economic times, the same changes in public attitudes, and so on. If the training group does much better than the untrained group in holding jobs and earning money, we can say that the training program actually causes this happy outcome.

One of the big ideas of statistics is that experiments can give good evidence that a treatment causes a response. A big idea needs a big caution: statistical conclusions hold "on the average" for groups of individuals. They don't tell us much about one individual. If *on the average* the women in the training program earned more than those who were left out, that says that the program achieved its goal. It doesn't say that every woman in such a program will be helped. And a big idea may also raise big questions: if we hope the training will raise earnings, is it ethical to offer it to some women and not to others? Chapters 5 and 6 explain how to design good experiments, and Chapter 7 looks at ethical issues.

Chapter 1: Statistics in Summary

- Any statistical study records data about some **individuals** (people, animals, or things) by giving the value of one or more **variables** for each individual.

- Some variables, such as age and income, take numerical values. Others, such as occupation and sex, do not. Be sure the variables in a study really do tell you what you want to know.

- The most important fact about any statistical study is how the data were produced. **Observational studies** try to gather information without disturbing the scene they are observing.

- **Sample surveys** are an important kind of observational study. A sample survey chooses a **sample** from a specific **population** and uses the sample to get information about the entire population.

- A **census** attempts to measure every individual in a population.

- **Experiments** actually do something to individuals in order to see how they respond. The goal of an experiment is usually to learn whether some treatment actually causes a certain response.

This chapter summary will help you evaluate the Case Study.

Link It

In reasoning from data to a conclusion, we start with the data. Where the data come from is the first step in the argument. The nature and validity of the conclusion are affected by this first step. Two sources of data are observational studies and experiments. Observational studies are best suited for a conclusion that involves describing some group or situation without disturbing the scene we observe. Sample surveys are a type of observational study in which we draw conclusions about a population by observing only a part of the population (the sample). Experiments are best suited for a conclusion that involves determining if a treatment causes a change in a response.

In the next several chapters, we discuss these sources of data in more detail. We will see what makes for a good observational study and for a good experiment. And we will see how a bad observational study or experiment undermines the validity of the conclusions we wish to make.

Case Study Evaluated

Use what you have learned in this chapter to answer some basic questions about the data collected in the MLive poll described in the Case Study that opened the chapter. Start by reviewing the Chapter Summary. Then answer each of the following questions in complete sentences. Be sure to communicate clearly enough for any of your classmates to understand what you are saying.

To participate in the poll, you had to go online to the MLive website and click on one of the possible responses.

1. Is the poll a sample survey, census, or experiment?

2. What is the population of interest?

3. What are the individuals in the poll?

4. For each individual, what variable is measured?

5. Does this variable take numerical values?

In this chapter you have:

- Learned the different types of studies used to collect data.
- Understood the purpose of these different types of studies used to collect data.
- Examined the types of data a study might collect.

macmillan learning **Online Resources**

■ The Snapshots video, *Introduction to Statistics*, describes real-world situations for which knowledge of statistical ideas are important.

■ The Snapshots video, *Types of Studies*, provides a nice introduction to the ideas of this section.

Check the Basics

For Exercise 1.1, see page 10; for Exercise 1.2, see page 13.

1.3 Individuals and variables. A national survey by the Pew Research Center and *USA Today,* conducted August 16–Septermber 12, 2016, was based on web-based and mail responses of a national sample of 4538 adults, 18 years of age or older, living in all 50 U.S. states and the District of Columbia. Those interviewed were asked to rate the job performance of police forces across the country for holding officers accountable when misconduct occurs. Possible ratings were "Excellent," "Good," "Only fair," and "Poor." Seventy percent of white respondents gave a rating of "Excellent" or "Good," while only 31% of black respondents gave a rating of "Excellent" or "Good." For this study,

(a) the individuals are the sample of 4538 adults interviewed.

(b) the variable is the rating a respondent selected.

(c) both (a) and (b).

(d) neither (a) nor (b).

1.4 Population and sample. For the survey described in the previous exercise,

(a) the population is the 4538 adults interviewed and the sample are those who gave a rating of excellent or good.

(b) the population is all adults, 18 years of age or older, living in all 50 U.S. states and the District of Columbia, and the sample is those who gave a rating of poor.

(c) the population is all adults, 18 years of age or older, living in all 50 U.S. states and the District of Columbia, and the sample is the black and white respondents.

(d) the population is all adults, 18 years of age or older, living in all 50 U.S. states and the District of Columbia, and the sample is the 4538 adults who provided responses.

1.5 Observational studies and experiments. Researchers at UNLV studied whether the attractiveness of an instructor affected the performance of students. The researchers randomly assigned 131 UNLV students to one of several instructors. Students listened to a recording of an introductory physics lecture by their instructor, and each instructor delivered the same lecture. During the lecture, the students were shown different sets of photographs that they were told were images of the instructor. Some of the photographed people were good-looking, others less so. Students were quizzed after the lecture, and those with instructors deemed attractive performed better than those with unattractive instructors. Which of the following is true?

(a) This is an observational study, and participants were volunteers.

(b) This is an observational study, and participants were selected at random.

(c) This is an experiment, but participants themselves decided which instructor to listen to.

(d) This is an experiment, and participants were randomly assigned to treatments (instructors).

1.6 Response variable. For the study described in the previous exercise, the response variable is which of the following?

(a) Scores on a quiz.

(b) Whether a subject was assigned to a group with an attractive or unattractive instructor.

(c) The recorded lecture.

(d) The 131 students.

1.7 A census? A study is considered to be a census if

(a) the population of interest is very large.

(b) the study attempts to measure every individual in a population.

(c) all units in the study receive some treatment.

(d) it attempts to answer questions about the opinions of all citizens of a particular country.

Chapter 1 Exercises

1.8 Miles per gallon. Here is a small part of a data set that describes the fuel economy (in miles per gallon) of 2019 model motor vehicles:

Make and model	Vehicle type	Transmission type	Number of cylinders	City mpg	Highway mpg
⋮					
BMW 430i	Compact car	Manual	4	21	33
Ford Flex AWD	Sport utility vehicle (4WD)	Automatic	6	16	22
Genesis G70 RWD	Compact car	Manual	4	18	28
Toyota Avalon	Midsize car	Automatic	6	22	31
⋮					

(a) What are the individuals in this data set?

(b) For each individual, what variables are given? Which of these variables take numerical values?

1.9 Athletes' salaries. Here is a small part of a data set that describes Major League Baseball players as of opening day of the 2018 season:

Player	Team	Position	Age	Salary
⋮				
Trout, Mike	Angels	Outfielder	26	34,083
Headley, Chase	Padres	Third base	33	13,000
Axford, John	Blue Jays	Pitcher	35	1,500
Sabathia, C. C.	Yankees	Pitcher	38	10,000
⋮				

(a) What individuals does this data set describe?

(b) In addition to the player's name, how many variables does the data set contain? Which of these variables take numerical values?

(c) What do you think are the *units* in which each of the numerical variables is expressed? For example, what does it mean to give Mike Trout's annual salary as 34,083? (*Hint:* The average annual salary of a Major League Baseball player on opening day, 2018, was, with roundoff, $4,520,000.)

1.10 Who recycles? In Example 1, weight is not a good measure of the participation of households in different neighborhoods in a city recycling program. What variables would you measure in its place?

1.11 Sampling moms. Pregnant and breast-feeding women should eat at least 12 ounces of fish and seafood per week to ensure their babies' optimal brain development, according to a coalition of top scientists from private groups and federal agencies. A nutritionist wants to know whether pregnant women are eating at least 12 ounces of fish per week. To do so, she obtains a list of the 340 members of a local chain of prenatal fitness clubs and mails a questionnaire to 60 of these women selected at random. Only 21 questionnaires are returned. What is the population in this study? What is the sample from which information is actually obtained? What percentage of the women whom the nutritionist tried to contact responded?

1.12 The death penalty. A press release by the Gallup News Service says that, based on a poll conducted on October 5–11, 2017, it found that 55% of Americans respond Yes when asked this question: "Are you in favor of the death penalty for a person convicted of murder?" Toward the end of the article, you read: "Results for this Gallup poll are based on telephone interviews conducted October 5–11, 2017, with a random sample of 1,028 adults, aged 18 and older, living in all 50 U.S. states and the District of Columbia." What variable did this poll measure? What population do you think Gallup wants information about? What was the sample?

1.13 The political gender gap. There may be a "gender gap" in political party preference in the United States, with women more likely than men to prefer Democratic candidates. A political scientist interviews a large sample of registered voters, both men and women. She asks each voter whether he or she voted for the Democratic or the Republican candidate in the last presidential election. Is this study an experiment? Why or why not? What variables does the study measure?

1.14 What is the population? For each of the following sampling situations, identify the population as exactly as possible. That is, say what kind of individuals the population consists of and say exactly which individuals fall in the population. If the information given is not sufficient, complete the description of the population by making reasonable assumptions about any missing information.

(a) An opinion poll contacts 972 American adults and asks them, "Would you rather have a job working for the government or working for business?"

(b) Video adapter cables have pins that plug into slots in a computer monitor. The adapter will not work if pins are bent or broken. A computer store buys video adapter cables in large lots from a supplier. The store chooses five cables from each lot and inspects the pins. If any of the cables have bent or broken pins, the entire lot is sent back.

(c) The American Community Survey contacts 3.5 million households, including some in every county in the United States. This Census Bureau survey asks each household questions about their housing, economic, and social status.

1.15 What is the population? For each of the following sampling situations, identify the population as exactly as possible. That is, say what kind of individuals the population consists of and say exactly which individuals fall in the population. If the information given is not sufficient, complete the description of the population in a reasonable way.

(a) A sociologist is interested in determining what proportion of teens believe the drinking age should be lowered to 18 in all the states. She selects a sample of five high schools in a large city and interviews all 12th graders in each of the schools.

(b) A medical researcher is interested in the rate of dementia among former NFL football players. From a list of living, former players he selects a sample of 20 and interviews them to determine if signs of dementia are present.

(c) The host of a local radio talk show wonders if people who are actively religious are more likely to trust their neighbors than those who are not. The station receives calls from 51 listeners who voice their opinions.

1.16 Teens sleep needs. A *Washington Post* article

reported on a study about the sleep needs of teenagers. In the study, researchers measured the presence of the sleep-promoting hormone melatonin in teenagers' saliva at different times of the day. They learned that the melatonin levels rise later at night than they do in children and adults and remain at a higher level later in the morning. The teenagers who took part in the study were volunteers. Higher levels of melatonin indicate sleepiness. The researchers recommended that high schools start later in the day to accommodate the sleep needs of teens. Is this study an experiment, a sample survey, or an observational study that is not a sample survey? Explain your answer.

1.17 Power lines and leukemia. The study of power lines and leukemia in Example 3 compared two groups of individuals and measured many variables that might influence differences between the groups. Explain carefully why this study is *not* an experiment.

1.18 Treating prostate disease. A large study used records from Canada's national health care system to compare the effectiveness of two ways to treat prostate disease. The two treatments are traditional surgery and a new method that does not require surgery. The records described many patients whose doctors had chosen one or the other method. The study found that patients treated by the new method were more likely to die within eight years.

(a) Explain why this is an observational study, not an experiment.

(b) Briefly describe the nature of an experiment to compare the two ways to treat prostate disease.

1.19 Walnuts and cholesterol. Does eating walnuts increase the level of good cholesterol (HDL) and reduce the level of bad cholesterol (LDL)? Here are two ways to study this question.

1. Researchers in Australia recruited 58 adults with diabetes for a research study. These subjects were randomly assigned to two treatment groups: a low-modified-fat diet group and a low-modified-fat diet group that included a handful of walnuts each day. After six months, researchers compared changes in HDL and LDL levels for the two groups.

2. Another team of researchers recruited 58 adults with diabetes who regularly eat walnuts as part of their diet. The researchers match each with a similar adult with diabetes who does not regularly eat walnuts. The researchers measured the HDL and LDL for each adult and compared the results for both groups.

(a) Explain why the first is an experiment and the second is an observational study.

(b) Why does the experiment give more useful information about whether walnuts increase HDL and reduce LDL?

1.20 Alcohol and cancer in women. A *Washington Post* article reported on a study about alcohol consumption and cancer in women. Since 1996, a team of British researchers has been gathering detailed information from 1.28 million women aged 50 to 64. The researchers recorded how much alcohol the women reported consuming when they volunteered for the study and again three years later. The researchers then examined whether there was any link with the 68,775 cancers the women developed over an average of the next seven years. They found that even among women who consumed as little as 10 grams of alcohol a day on average (the equivalent of about one drink), the risk for cancer of the breast, liver, and rectum was elevated.

(a) Is this an experiment? Explain your answer.

(b) We would prefer a sample survey to using women who volunteer for a study. What population does it appear that the researchers were interested in? What variables did they measure?

1.21 Alcohol and longevity. A *Chicago Tribune* article reported on a study about alcohol consumption and how long you live. The research tracked 1700 nonagenarians (adults in their 90s) enrolled in the study that began in 2003 to explore impacts of daily habits on longevity. The researchers discovered that subjects who drank about two glasses of beer or wine a day were 18% less likely to experience a premature death. Meanwhile, participants who exercised 15 to 45 minutes a day cut the same risk by 11%.

(a) Is this an experiment? Explain your answer.

(b) We would prefer a sample survey to using nonagenarians who volunteer for a study. What population does it appear that the researchers were interested in? What variables did they measure?

1.22 Bullying. Researchers tracked 2668 people from early childhood through adulthood and found that 13-year-olds who are frequent targets of bullies were three times more likely than their nonvictimized peers to be depressed later as adults. What is the population in this study? What is the sample? What variables do the researchers measure?

1.23 Choose your study type. What is the best way to answer each of the questions below: an experiment, a sample survey, or an observational study that is not a sample survey? Explain your choices.

(a) Is your school's football team called for fewer penalties in home games than in away games?

(b) Are college students satisfied with the cost of textbooks that they are required to purchase?

(c) Do college students who have access to video recordings of course lectures perform better in the course than those who don't?

1.24 Choose your study purpose. Give an example of a question about pet owners, their behavior, or their opinions that would best be answered by

(a) a sample survey.

(b) an observational study that is not a sample survey.

(c) an experiment.

Exploring the Web

Access these exercises on the text website: macmillanlearning.com/scc10e.

Samples, Good and Bad

Case Study: Assessing Polls

In this chapter you will:

- Understand why some polls can be misleading while others can provide reliable information.

- Learn a simple method for selecting a sample that can provide trustworthy information about a population.

As discussed in Chapter 1, in February 2014, the Michigan online news site MLive ran the story, "Take our online poll: Should Michigan legalize marijuana." Of 9684 respondents, 7906 (81.64%) said Yes, 1190 (12.29%) said No, and 588 (6.07%) said Decriminalize but not legalize. These results would seem to indicate overwhelming support for legalizing marijuana in Michigan. However, the Pew Research Center conducted a poll on March 25–29, 2015, in which they asked, "Do you think the use of marijuana should be made legal or not?" The Pew Research Center reported that the poll consisted of telephone interviews with 1500 randomly selected adult Americans and that 53% of those surveyed favor the legal use of marijuana. This is a majority of those surveyed but not the overwhelming majority that MLive found. There is a large discrepancy in the findings of these two polls. This may be because the polls were conducted at different times, the populations sampled were different (Michigan versus all adult Americans), the MLive poll had a much larger sample than the Pew poll, the questions asked were not identical, or perhaps the data from one or both polls is simply bad. By the end of this chapter you will be able to assess whether the data from the polls in this case study are good or bad.

Statistics in Your World

Big data "Big data" is a vague term, but it is often used by some to emphasize the sheer size of data sets that now exist. Big data is often "found data," not a random sample. Proponents for big data claim that because every single data point can now be captured, old statistical sampling techniques are obsolete. In essence, they say, we have data on the entire population. Is this true? Many statisticians disagree. They challenge the notion that we could ever have all the data. The claim that we have the entire population is often an assumption, rather than a fact about the data. Although sample sizes are enormous, proponents for big data often ignore the bias that accompanies nonrandom sampling.

Key Terms

The design of a statistical study is **biased** if it systematically favors certain outcomes.

Selection of whichever individuals are easiest to reach is called **convenience sampling**.

A **voluntary response sample** chooses itself by responding to a general appeal. Write-in or call-in opinion polls are examples of voluntary response samples.

Convenience samples and voluntary response samples are often biased.

How to Sample Badly

For many years in Rapides Parish, Louisiana, only one company had been allowed to provide ambulance service. In 1999, the local paper, the *Town Talk*, asked readers to call in to offer their opinion on whether the company should keep its monopoly. Call-in polls are generally automated: call one telephone number to vote Yes and call another number to vote No. Telephone companies often charged callers to use these numbers.

The *Town Talk* got 3763 calls, which suggests unusual interest in ambulance service. Investigation showed that 638 calls came from the ambulance company office or from the homes of its executives. Many more, no doubt, came from lower-level employees. "We've got employees who are concerned about this situation, their job stability, and their families and maybe called more than they should have," said a company vice president. Other sources said employees were told to, as they say in Chicago, "vote early and often."

As the *Town Talk* learned, it is easier to sample badly than to sample well. The paper relied on *voluntary response,* allowing people to call in rather than actively selecting its own sample. The result was *biased*—the sample was overweighted with people favoring the ambulance monopoly. Voluntary response samples attract people who feel strongly about the issue in question. These people, like the employees of the ambulance company, may not fairly represent the opinions of the entire population.

There are other ways to sample badly. Suppose that we sell your company several crates of oranges each week. You examine a sample of oranges from each crate to determine the quality of our oranges. It is easy to inspect a few oranges from the top of each crate, but these oranges may not be representative of the entire crate. Those on the bottom are more often damaged in shipment. If we were less than honest, we might make sure that the rotten oranges are packed on the bottom, with some good ones on top for you to inspect. If you sample from the top, your sample results are again *biased*—the sample oranges are systematically better than the population they are supposed to represent.

Example 1

Interviewing at the mall

Squeezing the oranges on the top of the crate is one example of convenience sampling. Mall interviews are another. Manufacturers and advertising agencies often use interviews at shopping malls to gather information about the habits of consumers and the effectiveness of ads. A sample of mall shoppers is fast and cheap. But people contacted at shopping malls are not representative of the entire U.S. population. They are richer, for example, and more likely to be teenagers or retired. Moreover, the interviewers tend to select neat, safe-looking individuals from the stream of customers. Mall samples are biased: they systematically overrepresent some parts of the pop-

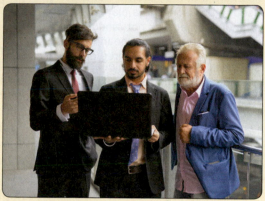
Ranta Images/Shutterstock

ulation (prosperous people, teenagers, and retired people) and underrepresent others. The opinions of such a convenience sample may be very different from those of the population as a whole.

Example 2

Write-in opinion polls

Ann Landers was a popular advice columnist, who wrote a daily column from 1955–2002. She once asked the readers of her advice column, "If you had it to do over again, would you have children?" She received nearly 10,000 responses, almost 70% saying, "NO!" Can it be true that 70% of parents regret having children? Not at all. This is a voluntary response sample. People who feel strongly about an issue, particularly people with strong negative feelings, are more likely to take the trouble to respond. Ann Landers's results are strongly biased—the percentage of parents who would not have children again is much higher in her sample than in the population of all parents.

On August 24, 2011, Abigail Van Buren (the niece of Ann Landers) revisited this question in her column "Dear Abby." A reader asked, "I'm wondering . . . if you asked the same question today, what the majority of your readers would answer."

Ms. Van Buren responded, "The results were considered shocking at the time because the majority of responders said they would NOT have children if they had it to do over again."

In October 2011, Ms. Van Buren wrote that when she again asked readers the same question that this time the majority of respondents would have children again. That is encouraging, but this was, again, a write-in poll.

Statistics in Your World

When is random too random?

The streaming music service *Spotify* received complaints from listeners that the "shuffle" feature, used to play songs in a playlist in a random order, was not random enough. Listeners were hearing the same tracks two or three days in a row or the same artists back to back. The problem was not a lack of randomness, but too much randomness. *Spotify* developer Mattias Petter Johansson explained that "to humans, truly random does not feel random."

Write-in and call-in opinion polls are almost sure to lead to strong bias. In fact, only about 15% of the public have ever responded to a call-in poll, and these tend to be the same people who call radio talk shows. That's not a representative sample of the population as a whole.

Simple Random Samples

In a voluntary response sample, people choose whether to respond. In a convenience sample, the interviewer makes the choice. In both cases, personal choice produces bias. The statistician's remedy is to allow impersonal chance to choose the sample. A sample chosen by chance allows neither favoritism by the sampler nor self-selection by respondents. Choosing a sample by chance attacks bias by giving all individuals an equal chance to be chosen. Rich and poor, young and old, black and white, all have the same chance to be in the sample.

The simplest way to use chance to select a sample is to place names in a hat (the population) and draw out a handful (the sample). This is the idea of *simple random sampling*.

Key Terms

A **simple random sample (SRS)** of size n consists of n individuals from the population chosen in such a way that every set of n individuals has an equal chance to be the sample actually selected.

An SRS not only gives each individual an equal chance to be chosen (thus avoiding bias in the choice) but also gives every possible sample an equal chance to be chosen. Drawing names from a hat does this. Write 100 names on identical slips of paper and mix them in a hat. This is a population. Now draw 10 slips, one after the other. This is an SRS because any 10 slips have the same chance as any other 10.

Now it's your turn

2.1 Sampling my class. There are 40 students in my class. My classroom has eight rows of seats with five seats per row. On the first day of class, all 40 students attend and all 40 seats are filled. I decide to take a sample of eight students. To do this, I write the numbers 1 through 5 on slips of paper, shuffle the slips of paper thoroughly, and select a number. The student sitting that number of seats in from the left in the first row is selected to be part of my sample. I repeat the process of shuffling and drawing a number once for each row. In this way I obtain my sample of eight students. Every student in the class has a 1-in-5 chance of being selected when I come to their row. Thus, every student has the same chance of being selected. Is the sample a simple random sample? Explain.

Drawing names from a hat makes clear what it means to give each individual and each possible set of *n* individuals the same chance to be chosen. That's the idea of an SRS. Of course, drawing slips from a hat would be a bit awkward for a sample of the country's 117 million households. In practice, real sample surveys use computer-generated *random digits* to choose samples. Many statistical software packages have random number generators that generate random digits. Some also allow one to choose an SRS.

Example 3

How to choose an SRS using software

A college president wants to appoint five undergraduate students to serve on a committee to make recommendations about the design of new classrooms that will meet students' perceived needs. To avoid bias, she chooses an SRS of 5 of the 30 departments in the college, and asks the department chair of each of the chosen departments to select one of their undergraduate majors to serve on the committee. Here are the steps for selecting the SRS of 5 departments.

Step 1: Label. Give each department a numerical label between 1 and 30. Here is the list of departments, with labels attached, using 1 to 30:

1	African American and African Studies	16	Geology
2	Agriculture	17	History
3	Architecture	18	Mathematics
4	Art	19	Music
5	Biology	20	Pharmacy
6	Business	21	Philosophy
7	Chemistry	22	Physics
8	Communications	23	Political Science
9	Comparative Studies	24	Psychology
10	Computer Science	25	Religion
11	Economics	26	Sociology
12	Education	27	Social Work
13	Engineering	28	Statistics
14	English	29	Theater
15	Foreign Languages	30	Women's Studies

Step 2: Software. Use statistical software to generate a random integer between 1 and 30. Repeat this process, ignoring any values that were previously generated, until you obtain five different integers between 1 and 30. The president used software and generated the numbers 18, 9, 10, 3, 9, and 1. The five different integers are 18, 9, 10, 3, and 1, so the sample is the departments of Mathematics, Comparative Studies, Computer Science, Architecture, and African American and African Studies.

Statistics in Your World

Are these random digits really random? Not a chance. The random digits in Table A were produced by a computer program. Computer programs do exactly what you tell them to do. Give the program the same input, and it will produce exactly the same "random" digits. Of course, clever people have devised computer programs that produce output that *looks* like random digits. These are called "pseudo-random numbers," and that's what Table A contains. Pseudo-random numbers work fine for statistical randomizing, but they have hidden nonrandom patterns that can mess up more refined uses.

Generating a random integer with a value between 1 and 30 is equivalent to writing the numbers 1 to 30 on identical slips of paper, placing them in a hat, mixing the slips well, and drawing one at random. The computer does the mixing and drawing.

Some statistical software may allow you to generate unique labels. A tool that is available on the Web is the Research Randomizer at www.randomizer.org. Click on the link *Randomize Now* and fill in the boxes. You can even ask the Randomizer to arrange your sample in numerical order (see Figure 2.1).

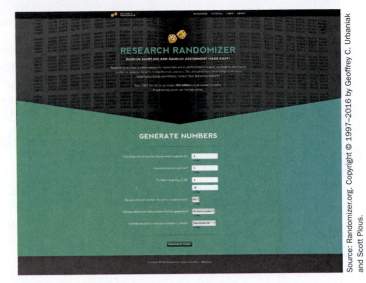

Figure 2.1 Using the Research Randomizer at www.randomizer.org.

We encourage you to use software to select an SRS, but if you are unable to use software, you can use a *table of random digits* to choose small samples by hand.

Table A at the back of the book is a table of random digits. You can think of Table A as the result of asking an assistant (or a computer) to mix the digits 0 to 9 in a hat, draw one, then replace the digit drawn, mix again, draw a second digit, and so on. The assistant's (or computer's) mixing and drawing save us the work of mixing and drawing when we need to randomize. Table A begins with the digits 19223950340575628713. To make the table easier to read, the digits appear in groups of five and in numbered rows. The groups and rows have no meaning— the table is just a long list of randomly chosen digits. Here's how to use the table to choose an SRS.

Key Terms

A **table of random digits** is a long string of the digits 0, 1, 2, 3, 4, 5, 6, 7, 8, 9 with these two properties:

1. Each entry in the table is equally likely to be any of the 10 digits 0 through 9.

2. The entries are independent of each other. That is, knowledge of one part of the table gives no information about any other part.

Example 4

How to choose an SRS using a table of random digits

To repeat Example 3, we begin by assigning numerical labels to the 30 departments.

Step 1: Label. Give each department a numerical label, using as few digits as possible. Two digits are needed to label 30 departments, so we use labels

$$01, 02, 03, \ldots, 28, 29, 30$$

It is also correct to use labels 00 to 29 or even another choice of 30 two-digit labels. Here is the list of departments, with labels attached, using 01 to 30:

01	African American and African Studies	16	Geology
02	Agriculture	17	History
03	Architecture	18	Mathematics
04	Art	19	Music
05	Biology	20	Pharmacy
06	Business	21	Philosophy
07	Chemistry	22	Physics
08	Communications	23	Political Science
09	Comparative Studies	24	Psychology
10	Computer Science	25	Religion
11	Economics	26	Sociology
12	Education	27	Social Work
13	Engineering	28	Statistics
14	English	29	Theater
15	Foreign Languages	30	Women's Studies

Step 2: Table. Enter Table A anywhere and read two-digit groups. Suppose we enter at line 130, which is

69051   64817   87174   09517   84534   06489   87201   97245

The first 10 two-digit groups in this line are

69   05   16   48   17   87   17   40   95   17

Each two-digit group in Table A is equally likely to be any of the 100 possible groups, 00, 01, 02, . . . , 99. So two-digit groups choose two-digit labels at random. That's just what we want.

The college president used only labels 01 to 30, so we ignore all other two-digit groups. The first five labels between 01 and 30 that we encounter in the table choose our sample. Of the first

10 labels in line 130, we ignore five because they are too high (over 30). The others are 05, 16, 17, 17, and 17. The clients labeled 05, 16, and 17 go into the sample. Ignore the second and third 17s because that department is already in the sample. Now run your finger across line 130 (and continue to line 131 if needed) until five departments are chosen.

The sample is the departments labeled 05, 16, 17, 20, 19. These are Biology, Geology, History, Pharmacy, and Music.

Choose an SRS in two steps

Step 1: Label. Assign a numerical label to every individual in the population. Be sure that all labels have the same number of digits if you plan to use a table of random digits.

Step 2: Software or table. Use random digits to select labels at random.

When using a table of random digits, as long as all labels have the same number of digits, all individuals will have the same chance to be chosen. Use the shortest possible labels: one digit for a population of up to 10 members, two digits for 11 to 100 members, three digits for 101 to 1000 members, and so on. As standard practice, we recommend that you begin with label 1 (or 01 or 001, as needed). You can read digits from Table A in any order—across a row, down a column, and so on—because the table has no order. As standard practice, we recommend reading across rows.

Using software or a table of random digits is much quicker than drawing names from a hat. As Examples 3 and 4 show, choosing an SRS has two steps.

Now it's your turn

2.2 Visiting Canada. Over the next few years, you want to visit the highest point in each of the 13 Canadian provinces and territories, This summer you can only visit three and you decide to select which to visit by simple random sampling. The list of 13 Canadian provinces and territories is given here. Use software, an online tool (for example, the Research Randomizer), or Table A at line 116 to choose three to be visited this year. Remember to begin by labeling the provinces and territories from 01 to 13.

Alberta	Nunavut
British Columbia	Ontario
Manitoba	Prince Edward Island
New Brunswick	Quebec
Newfoundland and Labrador	Saskatchewan
Northwest Territories	Yukon
Nova Scotia	

Can You Trust a Sample?

The *Town Talk,* Ann Landers, and mall interviews produce samples. We can't trust results from these samples because they are chosen in ways that invite bias. We have more confidence in results from an SRS because it uses impersonal chance to avoid bias. The first question to ask of any sample is whether it was chosen at random. Opinion polls and other sample surveys carried out by people who know what they are doing use random sampling.

Statistics in Your World

Golfing at random Random drawings give all the same chance to be chosen, so they offer a fair way to decide who gets a scarce good—like a round of golf. Lots of golfers want to play the famous Old Course at St. Andrews, Scotland. A few can reserve in advance. Most must hope that chance favors them in the daily random drawing for tee times. At the height of the summer season, only one in six wins the right to pay £180 (about $250) for a round.

Example 5
A Gallup Poll

A June 2018 Gallup Poll on immigration asked the question, "On the whole, do you think immigration is a good thing or a bad thing for this country today?" Gallup reported that the poll found that 75% of respondents think immigration is a good thing for the United States. Is it actually the case that a majority of Americans believe immigration is a good thing? Ask first how Gallup selected its sample. Later in the article, we read this: "Results . . . are based on telephone interviews . . . with a random sample of 1,520 adults, aged 18 and older, living in all 50 U.S. states and the District of Columbia . . ." Gallup goes on to clarify that the sample included 30% landline and 70% cellular phone numbers selected using random-digit dialing.

This is a good start toward gaining our confidence. Gallup tells us what population it has in mind (people at least 18 years old living in the continental United States, Alaska, and Hawaii). We know that the sample from this population was of size 1520 and, most important, it was chosen at random. There is more to consider in assessing a poll, and we will soon discuss this, but we have at least heard the comforting words "random sample."

Chapter 2: Statistics in Summary

We select a **sample** in order to get information about some **population**.

- How can we choose a sample that fairly represents the population? **Convenience samples** and **voluntary response samples** are common but do not produce trustworthy data. These sampling methods are usually **biased**. That is, they systematically favor some parts of the population over others in choosing the sample.

- The deliberate use of chance in producing data is one of the big ideas of statistics. Random samples use chance to choose a sample, thus avoiding bias due to personal choice.

- The basic type of random sample is the **simple random sample**, which gives all samples of the same size the same chance to be the sample we actually choose.

- To choose an SRS by hand, use a **table of random digits** such as Table A in the back of the book, or use software.

This chapter summary will help you evaluate the Case Study.

Link It

The first step in reasoning from data to a conclusion is obtaining data. In Chapter 1, we discussed sample surveys as one way to collect data in an observational study. The method of selecting the sample in a sample survey affects how well the sample represents the population. Biased sampling methods, such as convenience sampling and voluntary response samples, produce data that can be misleading, resulting in incorrect conclusions. Simple random sampling avoids bias and produces data that give us confidence that the first step in our argument is sound.

In the next chapter, we look more closely at what a simple random sample tells us about the population from which it is selected. And in Chapter 4, we discuss some of the problems faced by people who take surveys in the real world.

Case Study Evaluated

To participate in the MLive poll described in the Case Study that opened the chapter, you had to choose to go online to the MLive website and click on one of the possible responses. Use what you have learned in this chapter to answer some basic questions about the data collected in the MLive poll described in the Case Study that opened the chapter. Start by reviewing the chapter summary. Then, in complete sentences, assess whether the data collected in such an online poll are good or bad. Be sure to communicate clearly enough for any of your classmates to understand what you are saying.

In this chapter you have:

- Understood why some polls can be misleading while others can provide reliable information.
- Learned a simple method for selecting a sample that can provide trustworthy information about a population.

---

macmillan learning **Online Resources**

- The Snapshots video, *Sampling*, discusses the importance of sampling, the basics of sampling, simple random sampling, and some practical issues in selecting random samples.

- The video technology manuals explain how to select an SRS using JMP, Excel, R, Minitab, CrunchIt!, SPSS, and the TI 83/84.

Check the Basics

For Exercise 2.1, see page 24; for Exercise 2.2, see page 28.

2.3 Biased sampling methods? A *method* for selecting a sample is said to be biased if

  (a) the race or gender of respondents is taken into account.

  (b) it systematically favors certain outcomes.

  (c) the political affiliation of the person asking the questions is known by respondents.

  (d) any of the above are true.

2.4 An online survey. You go to a website to access a news story. In order to access the story, you are asked to answer a brief survey. If you choose not to answer the survey, you can only access the article for a fee.

    This method for obtaining a sample is an example of

  (a) simple random sampling.

  (b) random sampling, but not simple random sampling because people visit the website at random.

  (c) a convenience sample.

  (d) a write-in opinion poll.

2.5 Simple random sample. I plan to take a sample of 10 students in my introductory statistics class. Which of the following is a simple random sample?

  (a) I choose the 10 students sitting in the front row. Students select seats at random, so this would be a simple random sample.

  (b) I choose the first 10 students who enter the classroom. Students arrive at random, so this would be a simple random sample.

  (c) I write the names of all the students on similar slips of paper, put the slips of paper in a box, mix them well, and draw 10 slips from the box. The 10 names drawn are my sample.

  (d) All of the above are simple random samples.

2.6 An SRS? The leader of a student organization wishes to form a committee that will consist of 5 of the 50 members. He decides to take an SRS of size five from the members and those selected will comprise the committee. He chooses an SRS of size five and all are men. He would like to have some women on the task force, so he decides to take another SRS. This time there are three women and two men in the sample, so he decides these five will form the committee. We can conclude the sample obtained is

(a) an SRS because the final sample was obtained by simple random sampling.

(b) an SRS because all members had the same chance of being in the final sample.

(c) not an SRS because a sample consisting entirely of men is not allowed.

(d) not an SRS because the first sample was not balanced between men and women.

2.7 **Choosing a simple random sample.** Angela, Juan, Kevin, Lucinda, and Tanya are students in my capstone course. I wish to select a simple random sample of two of them to work on a project. I label Angela as 1, Juan as 2, Kevin as 3, Lucinda as 4, and Tanya as 5. I generate the following sequence of four random digits using statistical software: 1, 7, 8, 4. Based on these digits, my simple random sample is

(a) Juan and Tanya.

(b) Angela and Lucinda.

(c) any pair of students because all are equally likely.

(d) impossible to determine. I need to generate additional digits.

Chapter 2 Exercises

2.8 **Emails to the civic association.** You are the president of the neighborhood civic association. A brewpub is thinking of opening a restaurant and bar in the neighborhood. You report to the members of the civic association that 91 emails have been received on the issue, of which 71 oppose opening the restaurant and bar. "I'm surprised that most of our residents oppose the restaurant and bar. I thought it would be quite popular," says a member of the association. Are you convinced that a majority of the residents oppose the restaurant and bar? How would you explain the statistical issue to the member?

2.9 **Instant opinion.** In July 2018, *Green Car Reports* conducted a Twitter poll and asked their readers "Which modern electric car will be the first classic?" Of the 466 people who chose to take the poll, 48% chose the Tesla Roadster.

(a) What is the sample size for this poll?

(b) Explain why the poll may give unreliable information.

2.10 **More instant opinion.** In July 2018, the *Washington Examiner* reported on an online poll conducted by the *Drudge Report* asking participants whether Hilary Clinton should run for president ever again. Sixty-five percent of poll participants said they did not want the former secretary of state and failed 2016 Democratic presidential nominee to launch "one last" campaign for the White House, while 34 percent were in favor of a third bid. The article goes on to say that more than 318,600 people had taken part in the online survey.

(a) What is the sample size for this poll?

(b) The sample size for this poll is much larger than is typical for polls such as the Gallup Poll. Explain why the poll may give unreliable information, even with such a large sample size.

2.11 **The *Drudge Report* takes a poll.** Beginning in 2017, there was considerable interest in the Special Counsel investigation, conducted by Robert Mueller, of possible Russian interference in the 2016 presidential election. In 2018, the *Drudge Report* asked its readers, "Should President Trump fire Mueller?" As of August 9, 2018, 818,628 votes had been received on the poll with 76% answering Yes to the question. Explain why this sample is certainly biased, even though the sample size is large. What is the likely direction of the bias? That is, is 76% probably higher or lower than the truth about the opinions of the population of all adults in the United States?

2.12 We don't like one-way streets. Highway planners decided to make a main street in West Lafayette, Indiana, a one-way street. The *Lafayette Journal and Courier* took a one-day poll by inviting readers to call a telephone number to record their comments. The next day, the paper reported:

Journal and Courier *readers overwhelmingly prefer two-way traffic flow in West Lafayette's Village area to one-way streets. By nearly a 7–1 margin, callers on Wednesday complained about the one-way streets that have been in place since May. Of the 98 comments received, all but 14 said no to one-way.*

(a) What population do you think the newspaper wants information about?

(b) Is the proportion of this population who favor one-way streets almost certainly larger or smaller than the proportion of 14/98 in the sample? Why?

2.13 Design your own bad sample. Your college wants to gather student opinion about parking for students on campus. It isn't practical to contact all students.

(a) Give an example of a way to choose a sample of students that is poor practice because it depends on voluntary response.

(b) Give an example of a bad way to choose a sample that doesn't use voluntary response.

2.14 A call-in opinion poll. In 2005, the *San Francisco Bay Times* reported on a poll in New Zealand that found that New Zealanders opposed the nation's new gay-inclusive civil-unions law by a 3–1 ratio. This poll was a call-in poll that cost $1 to participate in. The *San Francisco Bay Times* article also reported that a scientific polling organization found that New Zealanders favor the law by a margin of 56.4% to 39.3%. Explain to someone who knows no statistics why the two polls can give such widely differing results and which poll is likely to be more reliable.

2.15 Call-in versus random sample polls. A national survey of TV network news viewers found that 48% said they would believe a phone-in poll of 300,000 persons rather than a random sample of 1000 persons.

Of the viewers, 42% said they would believe the random sample poll. Explain to someone who knows no statistics why the opinions of only 1000 randomly chosen respondents are a better guide to what all people think than the opinions of 300,000 callers.

2.16 Choose an SRS. A firm wants to understand the attitudes of its managers toward its system for assessing employee performance. Following is a list of all 32 of the firm's managers. Use software or Table A to choose five to be interviewed in detail about the performance appraisal system. If you use Table A, begin at line 132 to choose the five to be interviewed. Use labels 00 to 31.

Bailly	Gomes	Lukaku	Rashford
Bohui	Grant	Martia	Rojo
Chong	Hamilton	Mata	Romero
Dalot	Herrera	Matic	Sanchez
Darmian	Jones	McTominay	Shaw
De Gea	Kovar	O'Hara	Smalling
Fellaini	Lindelof	Pereira	Young
Fred	Lingard	Pogba	Valencia

2.17 Choose an SRS. Your class in ancient Ugaritic religion is poorly taught and the class members have decided to complain to the dean. The class decides to choose 6 of its 27 members at random to carry the complaint. The class list appears here. Choose an SRS of six using either software or the table of random digits, beginning at line 112. Use labels 00 to 26.

Agostinous	Domi	Ouellet
Alzner	Drouin	Pacioretty
Armia fi	Froese	Peca
Benn	Gallagher	Petry
Byron	Hudon	Plekanec
Chaputca	Juulsen	Price
Danault	Lehkonen	Reilly
De La Rose	Metei	Scherbakru
Deslauriers	Niemi	Schlemko

2.18 An election day sample. You want to choose an SRS of 20 of Indiana's 5341 voting precincts for special voting-fraud surveillance on election day.

(a) Explain clearly how you would label the 5341 precincts. If you will use Table A to choose an SRS, be sure to explain how many digits make up each of your labels.

(b) Use either software or Table A to choose the SRS, and list the labels of the precincts you selected. If you use Table A, enter Table A at line 107.

2.19 Is this an SRS? A university has 30,000 undergraduate and 10,000 graduate students. A survey of student opinion concerning health care benefits for domestic partners of students selects 300 of the 30,000 undergraduate students at random and then separately selects 100 of the 10,000 graduate students at random. The 400 students chosen make up the sample.

(a) Explain why this sampling method gives each student an equal chance to be chosen.

(b) Nonetheless, this is not an SRS. Why not?

2.20 How much do students pay for rent? A university's housing and residence office wants to know how much students pay per month for rent in off-campus housing. The university does not have enough on-campus housing for students, and this information will be used in a brochure about student housing. The population contains 12,304 students who live in off-campus housing and have not yet graduated. The university will send a questionnaire to an SRS of 200 of these students, drawn from an alphabetized list.

(a) Describe how you would label the students in order to select the sample. If you will use Table A to choose an SRS, be sure to explain how many digits make up each of your labels.

(b) Use software or Table A to select the first five students in the sample. If you use Table A, begin at line 125.

2.21 Apartment living. You are planning a report on apartment living in a college town. You decide to select three apartment complexes at random for in-depth interviews with residents. Use software or Table A, starting at line 112, to select a simple random sample of 3 of the following 36 apartment complexes. Use labels 00 to 35.

Amandari	La Reserve	Sheen Falls
Ballyfin	La Residence	Six Senses
Cala de Mar	Las Ventanas	St. Regis
Cavas Lodge	Le Meurice	Taj Exotica
Eolo	Mandarin	Taj Lake Palace
Four Seasons	Montage	Temple House
Frangipani	Nayara Springs	The Lodge at Edgewood
Gibb's Farm	Nihi Sumba	The Lowell
Huka Lodge	Rambagh Palace	The Mulia
Inkaterra	Rancho Santana	The Oberoi
Inverlochin	Rosewood	Twin Farms
Jade Mountain	Saxon	Viceroy Riviera

2.22 How do random digits behave? Which of the following statements are true of a table of random digits, and which are false? Explain your answers.

(a) Each pair of digits has a chance of 1/100 of being 00.

(b) There are exactly four 4s in each row of 40 digits.

(c) The digits 12345 can never appear as a group because this pattern is not random.

2.23 Twitter Poll. During the 2018 NBA playoffs, the San Antonio Spurs were down 3 games to 1 against the Golden State Warriors (the eventual NBA champions for the second year in a row). After game 4 of the series, KSAT in San Antonio ran a Twitter Poll asking followers "Do you think the Spurs will win game 5 and force the series back to San Antonio?" Fifty-seven percent of those responding said Yes. Explain why this opinion poll is almost certainly biased in terms of representing the chances of San Antonio beating the Warriors.

2.24 More randomization. Most sample surveys call residential telephone numbers at random. They do not, however, always ask their questions of the person who picks up the phone.

Instead, they ask about the adults who live in the residence and choose one at random to be in the sample. Why is this a good idea?

2.25 Racial profiling and traffic stops. The Denver Police Department wants to know if Hispanic residents of Denver believe that the police use racial profiling when making traffic stops. A sociologist prepares several questions about the police. The police department chooses an SRS of 200 mailing addresses in predominantly Hispanic neighborhoods and sends a uniformed Hispanic police officer to each address to ask the questions of an adult living there.

(a) What are the population and the sample?

(b) Why are the results likely to be biased even though the sample is an SRS?

2.26 Random selection? Choosing at random is a "fair" way to decide who gets some scarce good, in the sense that everyone has the same chance to win. But random choice isn't always a good idea—sometimes we don't want to treat everyone the same because some people have a better claim. In each of the following situations, would you support choosing at random? Give your reasons in each case.

(a) The basketball arena has 4000 student seats, and 7000 students want tickets. Shall we choose 4000 of the 7000 at random?

(b) The list of people waiting for liver transplants is much larger than the number of available livers. Shall we let impersonal chance decide who gets a transplant?

(c) During the Vietnam War, young men were chosen for army service at random, by a "draft lottery." Is this the best way to decide who goes and who stays home?

Exploring the Web

Access these exercises on the text website: macmillanlearning.com/scc10e.

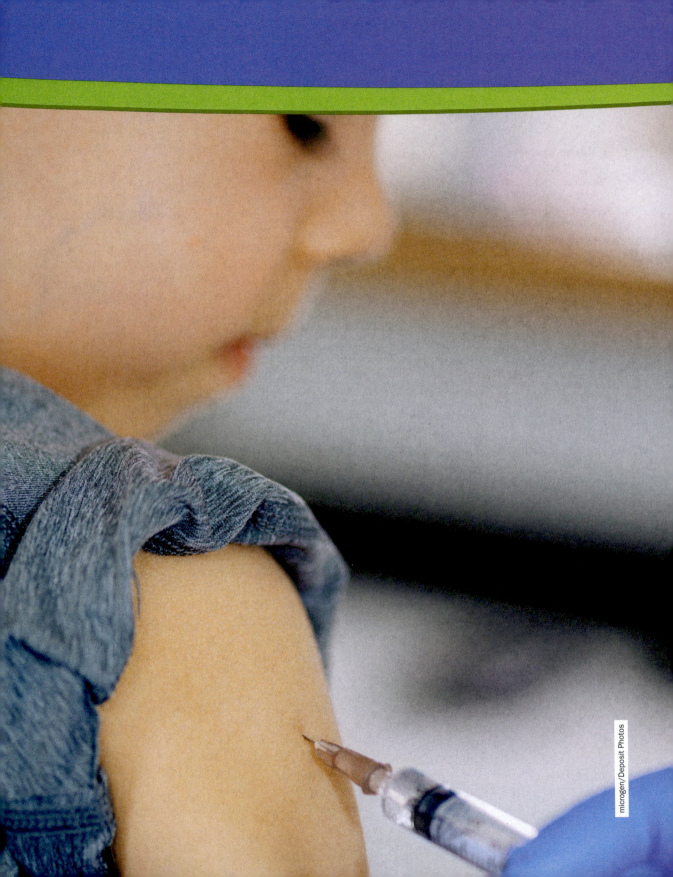

What Do Samples Tell Us?

Case Study: Childhood Vaccinations

In this chapter you will:

- Learn how and why we can generalize results from a sample to the population of interest.

- Understand that results from different samples vary from each other, but that we can quantify this variability.

- Understand what the statement "margin of error" means in poll results.

According to the Centers for Disease Control and Prevention (CDC), there were 349 confirmed cases of measles in 26 states and the District of Columbia. The 349 cases were the second highest to be reported in one year since measles was eliminated in the United States in 2000. (The 667 cases reported in 2014 were the highest.) As in previous years, the majority of people who got measles were unvaccinated. Vaccinating children against diseases like measles is controversial.

The debate about childhood vaccinations became a major news issue in late 2014 and early 2015 and continues to this day. As of 2019, California, Mississippi, and West Virginia are the only three states that do not allow exemptions for children to not be vaccinated for anything other than medical reasons. Other states allow exemptions for personal and religious beliefs. A Gallup Poll conducted from February 28–March 1, 2015, asked the following question: "How important is it that parents get their children vaccinated—extremely important, very important, somewhat important, or not at all important?" Gallup found that 54% of respondents said "extremely important" (down from 64% who responded to a similar 2001 Gallup Poll). Can we trust this conclusion?

Reading further, we find that Gallup talked with 1015 randomly selected adults to reach these conclusions. We're happy Gallup chooses at random—we wouldn't get unbiased information about the importance of childhood vaccinations by asking people attending a conference of the American Medical Association. However, the U.S. Census Bureau said that there were about 247 million adults in the United States in 2015. How can 1015 people, even a random sample of 1015 people, tell us about the opinions of 247 million people? Is 54% who feel that it is extremely important for parents to get their children vaccinated evidence that, in fact, the majority of Americans feel this way? By the end of this chapter, you will learn the answers to these questions.

The material in this chapter will provide you with the tools to address the Case Study and other statistical poll results you encounter.

From Sample to Population

Gallup's 2015 finding that "a slight majority of Americans, 54%, say it is extremely important that parents get their children vaccinated" makes a claim about the population of 247 million American adults. But Gallup doesn't know the truth about the entire population because we cannot obtain the opinions of all American adults. The poll contacted 1015 people and found that 54% of those contacted said it is extremely important for parents to vaccinate their children. Because the sample of 1015 people was chosen at random, it's reasonable to think that they represent the entire population pretty well. So Gallup turns the *fact* that 54% of the *sample* feel childhood vaccinations are extremely important into an *estimate* that about 54% of *all American adults* feel childhood vaccinations are extremely important. This is a basic move in Statistics: use a fact about a sample to estimate the truth about the whole population. To think about such moves, we must be clear whether a number describes a sample or a population. See the key terms box for the vocabulary we use.

So parameter is to population as statistic is to sample. Want to estimate an unknown population parameter? Choose a sample from the population and use a statistic as your estimate. That's what Gallup did.

Key Terms

A **parameter** is a number that describes the **population.** A parameter is a fixed number, but in practice we don't know the actual value of this number because we cannot access the entire population.

A **statistic** is a number that describes a **sample.** The value of a statistic can be determined and is known once we have taken a sample, but its value can change from sample to sample. We often use a statistic to estimate an unknown parameter.

Example 1   *Should children be vaccinated?*

The proportion of all American adults who feel childhood vaccinations are extremely important is a *parameter* describing the population of 247 million adults. Call it p, for "proportion." Alas, we do not know the numerical value of p. To estimate p, Gallup took a sample of 1015 adults.

The proportion of the sample who favor such an amendment is a *statistic*. Call it $\hat{p}$, read as "p-hat." It happens that 548 of this sample of size 1015 said that they feel childhood vaccines are extremely important, so for this sample,

$$\hat{p} = \frac{548}{1015} = 0.5399 \text{ (This decimal rounds to 0.54 and can be expressed as 54\%.)}$$

Because all American adults had the same chance to be among the chosen 1015, it seems reasonable to use the statistic $\hat{p} = 0.54$ as an estimate of the unknown parameter p. It's a *fact* that 54% of the sample feel childhood vaccines are extremely important—we know because we asked them. We don't know what percentage of all American adults feel this way, but we *estimate* that about 54% do.

Sampling Variability

If Gallup took a second random sample of 1015 adults, the new sample would have different people in it. It is almost certain that there would not be exactly 548 respondents who feel that childhood vaccines are extremely important. That is, the value of the statistic $\hat{p}$ will *vary* from sample to sample. Could it happen that one random sample finds that 54% of American adults feel childhood vaccines are extremely important and a second random sample finds that only 41% feel the same? Random samples eliminate *bias* from the act of choosing a sample, but they can still be wrong because of the variability that results when we choose at random. If the variation when we take repeated samples from the same population is too great, we can't trust the results of any one sample.

We are saved by the second great advantage of random samples. The first advantage is that choosing at random eliminates favoritism. That is, random sampling attacks bias. The second advantage is that if we took lots of random samples of the same size from the same population, the variation from sample to sample would follow a predictable pattern. This predictable pattern shows that results of bigger samples are less variable than the results of smaller samples.

Example 2   *Lots and lots of samples*

Here's another big idea of Statistics: to see how trustworthy one sample is likely to be, ask what would happen if we took many samples from the same population. Let's try it and see. Suppose that, in fact (unknown to Gallup), exactly 50% of all American adults feel childhood vaccines are extremely important. That is, the truth about the population is that $p = 0.5$. What if Gallup used the sample proportion $\hat{p}$ from a simple random sample (SRS) of size 100 to estimate the unknown value of the population proportion p?

Figure 3.1 illustrates the process of choosing many samples of the same size and finding $\hat{p}$ for each one. In the first sample, 56 of the 100 people felt childhood vaccines are extremely important so $\hat{p} = 56/100 = 0.56$. Only 36 in the next sample felt childhood vaccines are extremely important, so for that sample $\hat{p} = 0.36$. Choose 1000 samples and make a plot of the 1000 values of $\hat{p}$ like the graph (called a histogram) at the right of Figure 3.1. The different values of the sample proportion $\hat{p}$ run along the horizontal axis. The height of each bar shows how many of our 1000 samples gave the group of values on the horizontal axis covered by the bar. For example, in Figure 3.1 the bar covering the values between 0.40 and 0.42 has a height of slightly over 50. Thus, over 50 of our 1000 samples had values between 0.40 and 0.42.

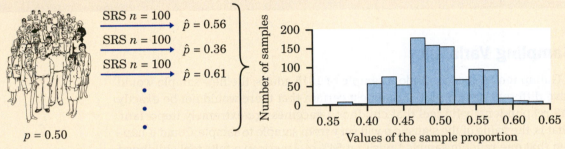

Figure 3.1 The results of many SRSs have a regular pattern. Here, we draw 1000 SRSs of size 100 from the same population. The population proportion is $p = 0.5$. The sample proportions vary from sample to sample, but their values center at the truth about the population.

Of course, Gallup interviewed 1015 people, not just 100. Figure 3.2 shows the results of 1000 SRSs, each of size 1015, drawn from a population in which the true (population) proportion is $p = 0.5$. Figures 3.1 and 3.2 are drawn on the same scale. Comparing them shows what happens when we increase the size of our samples from 100 to 1015.

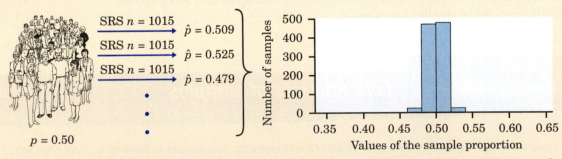

Figure 3.2 Draw 1000 SRSs of size 1015 from the same population as in Figure 3.1. The 1000 values of the sample proportion are much less variable (spread out) than was the case for smaller samples.

Look carefully at Figures 3.1 and 3.2. We flow from the population, to many samples from the population, to the many values of $\hat{p}$ from these many samples. Gather these values together and study the histograms that display them.

- In both cases, the values of the sample proportion $\hat{p}$ vary from sample to sample, but the values are centered at 0.5. Recall we are assuming that $p = 0.5$ is the true population parameter. Some samples have a $\hat{p}$ less than 0.5 and some greater, but there is no tendency to be always low or always high. That is, $\hat{p}$ has no **bias** as an estimator of p. This is true for both large and small samples.

- The values of $\hat{p}$ from samples of size 100 are much more variable than the values from samples of size 1015. In fact, 95% of our 1000 samples of size 1015 have a $\hat{p}$ lying between 0.4692 and 0.5308. That's within 0.0308 on either side of the population truth 0.5. Our samples of size 100, on the other hand, spread the middle 95% of their values between 0.40 and 0.60. That goes out 0.10 from the truth, about three times as far as the larger samples. So larger random samples have less **variability** than smaller samples.

The result is that we can rely on a sample of size 1015 to almost always give an estimate $\hat{p}$ that is close to the truth about the population. Figure 3.2 illustrates this fact for just one value of the population proportion, but it is true for any population proportion. Samples of size 100, on the other hand, might give an estimate of 40% or 60% when the truth is 50%.

Thinking about Figures 3.1 and 3.2 helps us restate the idea of bias when we use a statistic like $\hat{p}$ to estimate a parameter like p. It also reminds us that variability matters as much as bias.

We can think of the true value of the population parameter as the bull's-eye on a target and of the sample statistic as an arrow fired at the bull's-eye. Bias and variability describe what happens when an archer fires many arrows at the target. *Bias* means that the aim is off, and the arrows land consistently off the bull's-eye in the same direction. The sample values do not center about the population value. Large *variability* means that repeated shots are widely scattered on the target. Repeated samples do not give similar results but differ widely among themselves. Figure 3.3 shows this target illustration of the two types of error.

Notice that small variability (repeated shots are close together) can accompany large bias (the arrows are consistently away from the bull's-eye in one direction). And small bias (the arrows center on the bull's-eye) can accompany large variability (repeated shots are widely scattered). A good sampling scheme, like a good archer, must have both small bias and small variability.

Two types of error in estimation

Bias is consistent, repeated deviation of the sample statistic from the population parameter in the same direction when we take many samples. In other words, bias is a systematic overestimate or underestimate of the population parameter.

Variability describes how the values of the sample statistic will vary when we take many samples. Large variability means that the result of sampling is not repeatable.

A good sampling method has both small bias and small variability.

(a) Large bias, small variability

(b) Small bias, large variability

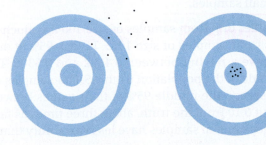
(c) Large bias, large variability   (d) Small bias, small variability

Figure 3.3 Bias and variability in shooting arrows at a target. Bias means the archer systematically misses in the same direction. More variability means that the arrows are more scattered.

Managing bias and variability

To reduce bias, use random sampling. When we start with a list of the entire population, simple random sampling produces *unbiased* estimates: the values of a statistic computed from a simple random sample (SRS) neither consistently overestimate nor consistently underestimate the value of the population parameter.

To reduce the variability of an SRS, use a larger sample. You can make the variability as small as you want by taking a large enough sample.

In practice, Gallup takes only one sample. We don't know how close to the truth an estimate from this one sample is because we don't know what the truth about the population is. But *large random samples almost always give an estimate that is close to the truth*. Looking at the pattern of many samples shows how much we can trust the result of one sample.

Margin of Error and All That

The "margin of error" that sample surveys announce translates sampling variability of the kind pictured in Figures 3.1 and 3.2 into a statement of how much confidence we can have in the results of a survey. Let's start with the kind of language we hear so often in the news.

What margin of error means

"Margin of error plus or minus 4 percentage points" is shorthand for this statement:

If we took many samples using the same method we used to get this one sample, 95% of the samples would give a result within plus or minus 4 percentage points of the truth about the population.

Note: "Margin of error" does not mean we made a mistake in our sampling methods; rather, it represents the natural sampling variability like that seen in Figures 3.1 and 3.2.

Take this step-by-step. A sample chosen at random will usually not estimate the truth about the population exactly. We need a margin of error to tell us how close our estimate comes to the truth. But we can't be *certain* that the truth differs from the estimate by no more than the margin of error. Although 95% of all samples come this close to the truth, 5% miss by more than the margin of error. We don't know the truth about the population, so we don't know if our sample is one of the 95% that hit or one of the 5% that miss. We say we are **95% confident** that the truth lies within the margin of error.

Example 3

Understanding the news

Here's what the TV news announcer says: "A new Gallup Poll finds that a slight majority of 54% of American adults feel it is extremely important that parents vaccinate their children. The margin of error for the poll was 4 percentage points." Plus or minus 4% starting at 54% is 50% to 58%. Most people think Gallup claims that the truth about the entire population lies in that range.

This is what the results from Gallup actually mean: "For results based on a sample of this size, one can say with 95% confidence that the error attributable to sampling and other random effects could be plus or minus 4 percentage points for American adults." That is, Gallup tells us that the margin of error includes the truth about the entire population for only 95% of all its samples. "95% confidence" is shorthand for that. The news report left out the "95% confidence."

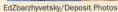
EdZbarzhyvetsky/Deposit Photos

Finding the margin of error exactly is a job for statisticians. You can, however, use a simple formula to get a rough idea of the size of a sample survey's margin of error. The reasoning behind this formula and an exact calculation of the margin of error are discussed in Chapter 21. For now, we introduce a quick and approximate calculation for the margin of error.

A quick and approximate method for the margin of error

Use the sample proportion $\hat{p}$ from a simple random sample of size n to estimate an unknown population proportion p. The margin of error for 95% confidence is approximately equal to $1/\sqrt{n}$.

Example 4

Calculating the margin of error

The Gallup Poll in Example 1 interviewed 1015 people. The margin of error for 95% confidence will be about

$$\frac{1}{\sqrt{1015}} = \frac{1}{31.8591} = 0.031 \text{ (that is, 3.1\%)}$$

Gallup announced a margin of error of 4% and our quick and approximate method gave us 3.1%. Our quick and approximate method can disagree a bit with Gallup's for two reasons. First, polls usually round their announced margin of error to a whole percent to keep their press releases simple. Second, our rough formula only works for an SRS. We will see in the next chapter that most national samples are more complicated than an SRS in ways that tend to slightly increase the margin of error. In fact, Gallup's survey methods section for this particular poll included the statement, "Each sample of national adults includes a minimum quota of 50% cellphone respondents and 50% landline respondents, with additional minimum quotas by time zone within region. Landline and cellular telephone numbers are selected using random-digit-dial methods." While these methods go beyond what we will study in the next chapter, the complexity of collecting a national sample increases the value of the margin of error.

Our quick and approximate method also reveals an important fact about how margins of error behave. Because the sample size n appears in the denominator of the fraction, larger samples have smaller margins of error. We knew that. Because the formula uses the square root of the sample size, however, *to cut the margin of error in half, we must use a sample four times as large*.

Example 5

Margin of error: Impact of sample size

In Example 2, we compared the results of taking many SRSs of size $n = 100$ and many SRSs of size $n = 1015$ from the same population. We found that the variability of the middle 95% of the sample results was about three times larger for the smaller samples.

Our quick formula estimates the margin of error for SRSs of size 1015 to be about 3.1%. The margin of error for SRSs of size 100 is about

$$\frac{1}{\sqrt{100}} = \frac{1}{10} = 0.10 \text{ (that is, 10\%)}$$

Because 1015 is roughly 10 times 100 and the square root of 10 is 3.1, the margin of error is about three times larger for samples of 100 people than for samples of 1015 people.

Now it's your turn

3.1 Drug abuse and families. In July 2018, the Gallup Poll asked a random sample of 1033 American adults, "Has drug abuse ever been a cause of trouble in your family?" The poll found that 30% of respondents said "yes," a record high percentage since the question started being asked in 1995. What is the approximate margin of error for 95% confidence?

Confidence Statements

Here is Gallup's conclusion about the views of American adults about vaccinating children in short form: "The poll found that a slight majority of Americans, 54%, feel it is extremely important to vaccinate children. We are 95% confident that the truth about all American adults is within plus or minus 4 percentage points of this sample result."
Here is an even shorter form: "We are 95% confident that between 50% and 58% of all American adults feel it is extremely important to vaccinate children." These are *confidence statements*.

A confidence statement is a fact about what happens in all possible samples and is used to say how much we can trust the result of one sample. The phrase "95% confidence" means "We used a sampling method that gives a result this close to the truth 95% of the time." Here are some hints for interpreting confidence statements:

Key Terms

A **confidence statement** has two parts: a **margin of error** and a **level of confidence.** The margin of error says how close the sample statistic lies to the population parameter. The level of confidence says what percentage of all possible samples satisfy the margin of error.

- *The conclusion of a confidence statement always applies to the population, not to the sample.* We know exactly the opinions of the 1015 people in the sample, because Gallup interviewed them. The confidence statement uses the sample result to predict something about the population of all American adults.

- *Our conclusion about the population is never completely certain.* Gallup's sample *might* be one of the 5% that miss by more than 4 percentage points.

- *A sample survey can choose to use a confidence level other than 95%.* The cost of higher confidence is a larger margin of error. For the same sample, a 99% confidence statement requires a larger margin of error than 95% confidence. If you are content with 90% confidence, you get in return a smaller margin of error. Remember that our quick and approximate method gives the margin of error only for 95% confidence.

Statistics in Your World

The telemarketer's pause
People who do sample surveys hate telemarketing. We all get so many unwanted sales pitches by phone that many people hang up before learning that the caller is conducting a survey rather than selling vinyl siding. Here's a tip. Both sample surveys and telemarketers dial telephone numbers at random. Telemarketers automatically dial many numbers, and their sellers come on the line only after you pick up the phone. Once you know this, the telltale "telemarketer's pause" gives you a chance to hang up before the seller arrives. Sample surveys have a live interviewer on the line when you answer.

- *It is usual to report the margin of error for 95% confidence.* If a news report gives a margin of error but leaves out the confidence level, it's pretty safe to assume 95% confidence.

- *Want a smaller margin of error with the same confidence? Take a larger sample.* Remember that larger samples have less variability. You can get as small a margin of error as you want and still have high confidence by paying for a large enough sample.

Example 6   *2016 election polls*

In 2016, shortly before the presidential election, SurveyUSA, a polling organization, asked voters in several states who they would vote for. In Kansas they asked a random sample of 581 likely voters, and 47% said they would vote for Donald Trump and 36% said Hillary Clinton. SurveyUSA reported the margin of error to be plus or minus 4.1%. In Oregon they sampled 654 likely voters, and 38% said they would vote for Trump and 48% said Clinton. The margin of error was reported to be plus or minus 3.9%.

There you have it: the sample of likely voters in Oregon was slightly larger, so the margin of error for conclusions about voters in Oregon is slightly smaller (3.9% compared to 4.1%). We are 95% confident that between 34.1% (that's 38% minus 3.9%) and 41.9% (that's 38% plus 3.9%) of likely voters in Oregon would vote for Trump. Note that the actual 2016 election results for Oregon were 39.1% for Trump, which is within the margin of error.

Now it's your turn

3.2 Drug abuse and families. In July 2018, the Gallup Poll asked a random sample of 1033 American adults, "Has drug abuse ever been a cause of trouble in your family?" The poll found that 30% of respondents said "yes," a record high percentage since the question started being asked in 1995. Suppose that the sample size had been 4000 rather than 1033. Find the margin of error for 95% confidence for the larger sample. How does it compare with the margin of error for a sample of size 1033?

Sampling from Large Populations

Gallup's sample of 1015 American adults was only 1 out of every 243,300 adults in the United States. Does it matter whether 1015 is 1-in-100 individuals in the population or 1-in-243,300? Read the box below to find out.

Population size doesn't matter

The variability of a statistic from a random sample is essentially unaffected by the size of the population as long as the population is at least 20 times larger than the sample.

Why does the size of the population have little influence on the behavior of statistics from random samples? Imagine sampling soup by taking a spoonful from a pot. The spoon doesn't know whether it is surrounded by a small pot or a large pot. As long as the pot of soup is well mixed (so that the spoon selects a "random sample" of the soup) and the spoonful is a small fraction of the total, the variability of the result depends only on the size of the spoon.

This is good news for national sample surveys like the Gallup Poll. A random sample of size 1000 or 2500 has small variability because the sample size is large. But remember that even a very large voluntary response sample or convenience sample is worthless because of bias. *Taking a larger sample can never fix biased sampling methods.*

However, the fact that the variability of a sample statistic depends on the size of the sample and not on the size of the population is bad news for anyone planning a sample survey in a university or a small city. For example, it takes just as large an SRS to estimate the proportion of Arizona State University undergraduate students who call themselves political conservatives as to estimate with the same margin of error the proportion of all adult U.S. residents who are conservatives. We can't use a smaller SRS at Arizona State just because there are 59,200 Arizona State undergraduate students and over 252 million adults in the United States in 2017.

STATISTICAL CONTROVERSIES

Should Election Polls Be Banned?

Preelection polls tell us that Senator So-and-So is the choice of 55% of Michigan voters. The media love these polls. Statisticians don't love them because elections often don't go as forecasted, even when the polls use all the right statistical methods. Many people who respond to the polls change their minds before the election. Others say they are undecided. Still others say which candidate they favor but won't bother to vote when the election arrives. Election

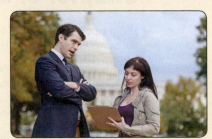

Camrocker/Deposit Photos

forecasting is one of the less satisfactory uses of sample surveys because we must ask people now how they will vote in the future.

Exit polls, which interview voters as they leave the voting place, don't have these problems. The people in the sample have just voted. A *good* exit poll, based on a national sample of election precincts, can often call a presidential election correctly long before the polls close. But, as was the case in the 2016 presidential election, exit polls can also fail to call the results correctly. These facts sharpen the debate over the political effects of election forecasts.

Can you think of good arguments *against* public preelection polls? Think about how these polls might influence voter turnout. Remember that voting ends in the East several hours before it ends in the West. What about arguments against exit polls?

What are arguments *for* preelection polls? Consider freedom of speech, for example. What about arguments for exit polls?

For some thought-provoking articles on polls, especially in light of the exit poll failures in the 2016 presidential election, see the following websites:

https://www.nytimes.com/2017/05/31/upshot/a-2016-review-why-key-state-polls-were-wrong-about-trump.html

http://www.pewresearch.org/fact-tank/2016/11/09/why-2016-election-polls-missed-their-mark/

https://www.cnn.com/election/2016/results/exit-polls

You may also go to www.google.com and search using key words "exit poll failures in the 2016 presidential election."

Chapter 3: Statistics in Summary

The purpose of sampling is to use a sample to predict information about a population. We often use a sample **statistic** to estimate the value of a population **parameter.**

■ This chapter has one big idea: to describe how trustworthy a sample is, ask, "What would happen if we took a large number of samples from the same population?" If almost all samples would give a result close to the truth, we can trust our one sample even though we can't be certain that it is close to the truth.

■ In planning a sample survey, first aim for small **bias** by using random sampling and avoiding bad sampling methods such as voluntary response. Next, choose a large enough random sample to reduce the **variability** of the result. Using a large random sample guarantees that almost all samples will give accurate results.

■ To say how accurate our conclusions about the population are, make a **confidence statement.** News reports often mention only the **margin of error.** Most often this margin of error is for **95% confidence.** That is, if we chose many samples, the truth about the population would be within the margin of error 95% of the time. Remember that margin of error does not mean we made a mistake; it's just due to variability from sample to sample.

■ We can roughly approximate the margin of error for 95% confidence based on a simple random sample of size n by the formula $1/\sqrt{n}$. As this formula suggests, only the size of the sample, not the size of the population, matters. This is true as long as the population is much larger (at least 20 times larger) than the sample.

This chapter summary will help you evaluate the Case Study.

Link It

In Chapter 1 we introduced sample surveys as an important kind of observational study. In Chapter 2 we discussed both good and bad methods for taking a sample survey. Simple random sampling was introduced as a method that deliberately uses chance to produce unbiased data. This deliberate use of chance to produce data is one of the big ideas of statistics.

In this chapter we looked more carefully at how sample information is used to gain information about the population from which it is selected. The big idea is to ask what would happen if we used our method for selecting a sample to take many samples from the same population. If almost all would give results that are close to the truth, then we have a basis for trusting our sample.

In practice, how easy is it to take a simple random sample? What problems do we encounter when we attempt to take samples in the real world? This is the topic of the next chapter.

Case Study Evaluated

Use what you have learned in this chapter to evaluate the Case Study that opened the chapter. Start by reviewing the Chapter Summary, then communicate clearly enough for any of your classmates to understand what you are saying.

In the Case Study at the beginning of the chapter, 54% of those surveyed in 2015 felt that it is extremely important to vaccinate children. The Gallup Poll stated that childhood vaccinations were considered to be extremely important by a slight majority (54%) of American adults. Is 54% evidence that, in fact, the majority of American adults in 2015 felt that childhood vaccinations are extremely important? Use what you have learned about how and why we can generalize results from a sample to the population of interest, that results from different samples vary from each other (but that we can quantify this variability), and what the statement "margin of error" means to answer this question. Your answer should be written so that someone who knows no Statistics will understand your reasoning.

> ### In this chapter you:
>
> - Learned how and why we can generalize results from a sample to the population of interest.
> - Understood that results from different samples vary from each other, but that we can quantify this variability.
> - Understood what the statement "margin of error" means in poll results that you see.

macmillan learning Online Resources

■ The video technology manuals explain how to select an SRS using JMP, Excel, R, Minitab, CrunchIt!, SPSS, and the TI 83/84.

■ The Statistical Applet, *Simple Random Sample*, can be used to select a simple random sample when the number of labels is 144 or less.

Check the Basics

For Exercise 3.1, see page 44; for Exercise 3.2, see page 46.

3.3 What's the parameter? An online store contacts 1500 customers from its list of customers who have purchased in the last year and asks the customers if they are very satisfied with the store's website. One thousand (1000) customers respond, and 696 of the 1000 say that they are very satisfied with the store's website. The parameter is

(a) the 69.6% of respondents who replied they are very satisfied with the store's website.

(b) the percentage of the 1500 customers contacted who would have replied they are very satisfied with the store's website.

(c) the percentage of all customers who purchased in the last year who would have replied they are very satisfied with the store's website.

(d) None of the above is the parameter.

3.4 What's the statistic? A state representative wants to know how voters in his district feel about enacting a statewide smoking ban in all enclosed public places, including bars and restaurants. His staff mails a questionnaire to a simple random sample of 800 voters in his district. Of the 800 questionnaires mailed, 152 were returned. Of the 152 returned questionnaires, 101 support the enactment of a statewide smoking ban in all enclosed public places. The statistic is

(a) $152/800 = 19.0\%$.

(b) $101/800 = 12.6\%$.

(c) $101/152 = 66.4\%$.

(d) Unable to be determined based on the information provided.

3.5 Decreasing sampling variability. You plan to take a sample of size 500 to estimate the proportion of students at your school who support having no classes and special presentations on Veteran's Day. To be twice as accurate with your results, you should plan to sample how many students? (*Hint*: Use the quick and approximate method for the margin of error.)

(a) 125 students

(b) 250 students

(c) 1000 students

(d) 2000 students

3.6 What can we be confident about? A May 2015 University of Michigan C.S. Mott Children's Hospital National Poll on Children's Health reported that 90% of adults are concerned that powdered alcohol will be misused by people under age 21. The margin of error was reported to be 2 percentage points, and the level of confidence was reported to be 95%. This means that

(a) we can be 95% confident that between 88% and 92% of adults are concerned that powdered alcohol will be misused by people under age 21.

(b) we can be 95% confident that between 88% and 92% of adults who were surveyed are concerned that powdered alcohol will be misused by people under age 21.

(c) we know that 90% of adults are concerned that powdered alcohol will be misused by people under age 21.

(d) Both (a) and (b) are true.

3.7 Which is more accurate? The Pew Research Center Report entitled "How Americans value public libraries in their communities," released December 11, 2013, asked a random sample of 6224 Americans aged 16 and over, "Have you used a Public Library in the last 12 months?" In the entire sample, 30% said "yes." But only 17% of those in the sample over 65 years of age said "yes." Which of these two sample percents will be more accurate as an estimate of the truth about the population?

(a) The result for those over 65 is more accurate because it is easier to estimate a proportion for a small group of people.

(b) The result for the entire sample is more accurate because it comes from a larger sample.

(c) Both are equally accurate because both come from the same sample.

(d) We cannot determine this because we do not know the percentage of those in the sample 65 years of age or younger who said "yes."

Chapter 3 Exercises

3.8 The boldface number in the next paragraph is the value of either a **parameter** or a **statistic**. State which it is.

IN THE NEWS
The Bureau of Labor Statistics announces that in July 2018 it interviewed all members of the labor force in a sample of 60,000 households; **3.9%** of the people interviewed were unemployed.

3.9 Each boldface number in the next paragraph is the value of either a **parameter** or a **statistic**. In each case, state which it is.

According to *The Independent*, the final polls conducted in Great Britain prior to "Brexit" (Britain's proposed exit from the European Union [EU]) indicated that the majority of voters would vote for Britain to stay part of the EU (and thus a minority would vote to leave). For example, the final online poll conducted by Populus indicated that only **45%** of voters in Britain would vote to leave the EU. When all votes were tallied for the referendum, **52%** of voters in Britain voted to leave the EU.

3.10 Each boldface number in the next paragraph is the value of either a **parameter** or a **statistic**. In each case, state which it is.

Voter registration records show that **25%** of all voters in the United States are registered as Republicans. However, a national radio talk show host found that of 20 Americans who called the show recently, **60%** were registered Republicans.

3.11 Each boldface number in the next paragraph is the value of either a **parameter** or a **statistic**. In each case, state which it is.

The National Health Interview Survey (NHIS) telephone survey is conducted annually in the United States. Of the first 100 numbers dialed, **55** numbers were for wireless telephones. This is not surprising because, as of the second half of 2016, **50.8%** of all U.S. households had only wireless telephones.

3.12 A sampling experiment. Figures 3.1 and 3.2 show how the sample proportion $\hat{p}$ behaves when we take many samples from the same population. You can follow the steps in this process on a small scale.

Figure 3.4 represents a small population. Each circle represents an adult. The white circles are people who favor a state constitutional amendment that would ban single-use plastic bags, and the colored circles are people who are opposed. You can check that 50 of the 100 circles are white, so in this population the proportion who favor an amendment is $p = 50/100 = 0.5$.

(a) The circles are labeled 00, 01, . . . , 99. Use line 101 of Table A to draw an SRS of size 4. What is the proportion $\hat{p}$ of the people in your sample who favor a state constitutional amendment?

(b) Take 9 more SRSs of size 4 (10 in all), using lines 102 to 110 of Table A, a different line for each sample. You now have 10 values of the sample proportion $\hat{p}$. Write down the 10 values you should now have of the sample proportion $\hat{p}$.

(c) Because your samples have only 4 people, the only values $\hat{p}$ can take are 0/4, 1/4, 2/4, 3/4, and 4/4. That is, $\hat{p}$ is always 0, 0.25, 0.5, 0.75, or 1. Mark these numbers on a line and make a histogram of your 10 results by putting a bar above each number to show how many samples had that outcome.

(d) Taking samples of size 4 from a population of size 100 is not a practical setting, but let's look at your results anyway. How

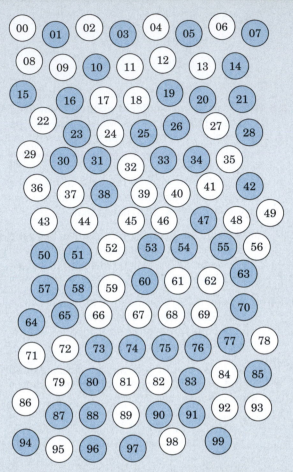

Figure 3.4 A population of 100 individuals for Exercise 3.12. Some individuals (white circles) favor a constitutional amendment and the others do not.

many of your 10 samples estimated the population proportion $p = 0.5$ exactly correctly? Is the true value 0.5 in the center of your sample values? Explain why 0.5 would be in the center of the sample values if you took a large number of samples.

3.13 A sampling experiment. Let us illustrate sampling variability in a small sample from a small population. Ten of the 25 club members listed below are female. Their names are marked with asterisks in the list. The club chooses 5 members at random to receive free trips to the national convention.

Alonso	Darwin	Herrnstein	Myrdal	Vogt*
Binet*	Epstein	Jimenez*	Perez*	Went
Blumenbach	Ferri	Luo	Spencer*	Wilson
Chase*	Gonzales*	Moll*	Thomson	Yerkes
Chen*	Gupta	Morales*	Toulmin	Zimmer

(a) Use the *Simple Random Sample* applet, other software, or a different part of Table A to draw 20 SRSs of size 5. Record the number of females in each of your samples. Make a histogram like that in Figure 3.1 to display your results. What is the average number of females in your 20 samples?

(b) Do you think the club members should suspect discrimination if none of the 5 tickets go to women?

3.14 Another sampling experiment. Let us illustrate sampling variability in a small sample from a small population. Seven of the 20 college softball players listed below are in-state students. Their names are marked with asterisks in the list. The coach chooses 5 players at random to receive a new scholarship funded by alumni.

Betsa	Blanco	Christner	Connell	Driesenga*
Falk	Garfinkel*	Lawrence	Montemarano	Ramirez
Richvalsky*	Romero	Sbonek*	Susalla*	Swearingen
Sweet	Swift*	Vargas	Wagner	Wald*

(a) Use the *Simple Random Sample* applet, other software, or a different part of Table A to draw 20 SRSs of size 5. Record the number of in-state students in each of your samples. Make a histogram like that in Figure 3.1 to display your results. What is the average number of in-state players in your 20 samples?

(b) Do you think the college should suspect discrimination if none of the 5 scholarships go to in-state players?

3.15 Canada's national health care. The Ministry of Health in the Canadian province of Ontario wants to know whether the national health care system is achieving its goals in the province. Much information about health care comes from patient records, but that source doesn't allow us to compare people who use health services with those who don't. So the Ministry of Health conducted the Ontario Health Survey, which interviewed a random sample of 61,239 people who live in the province of Ontario.

(a) What is the population for this sample survey? What is the sample?

(b) The survey found that 76% of males and 86% of females in the sample had visited a general practitioner at least once in the past year. Do you think these estimates are close to the truth about the entire population? Why?

3.16 Environmental problems. A Gallup Poll found that 62% of American adults "believe the U.S. government is not doing enough to protect the environment." Gallup's report said, "Results for this Gallup poll are based on telephone interviews conducted March 1–8, 2018, with a random sample of 1041 adults, aged 18 and older, living in all 50 U.S. states and the District of Columbia."

(a) What is the population for this sample survey? What is the sample?

(b) The survey found that 57% of Republicans and those who lean Republican and 89% of Democrats and those who lean Democratic in the sample think the U.S. government should spend more money on developing solar and wind power. Do you think these estimates are close to the truth about the entire population? Explain your answer.

3.17 Bigger samples, please. Explain in your own words the advantages of bigger random samples in a sample survey.

3.18 Sampling variability. In thinking about Gallup's sample of size 1015, we asked, "Could it happen that one random sample finds that 54% of adults feel that childhood vaccination is extremely important and a second random sample finds that only 42% favor one?" Look at Figure 3.2, which shows the results of 1000 samples of this size when the population truth is $p = 0.5$, or 50%. Would you be surprised if a sample from this population gave 54%? Would you be surprised if a sample gave 42%? Use Figure 3.2 to support your reasoning.

3.19 Health care cost. A November 2017 Gallup Poll of 1028 U.S. adults found that 627 are satisfied with the total cost they pay for their health care. The announced margin of error is ±4 percentage points. The announced confidence level is 95%.

(a) What is the value of the sample proportion $\hat{p}$ who say they are satisfied with the total cost they pay for their health care? Explain in words what the population parameter p is in this setting.

(b) Make a confidence statement about the parameter p.

3.20 Bias and variability. Figure 3.5 shows the behavior of a sample statistic in many samples in four situations. These graphs are like those in Figures 3.1 and 3.2. That is, the heights of the bars show how often the sample statistic took

various values in many samples from the same population. The true value of the population parameter is marked on each graph. Label each of the graphs in Figure 3.5 as showing high or low bias and as showing high or low variability.

3.21 Is a larger sample size always better? A politician asked his constituents via a Facebook poll to help him decide how he should vote on a bill that would establish 150m Protester Exclusion Zones around abortion clinics in New South Wales, Australia. On his Facebook page, Philip Donato stated "I believe this is a matter of conscience and the views of my electorate should supersede my own."

(a) According to *The Sydney Morning Herald*, as of 1:00 P.M. on June 6, 2018, (the day before Donato's Facebook poll closed), 74% of the 7700 poll voters supported the Protestor Exclusion Zones. Use our quick and approximate method to determine the margin of error for the poll results as of 1:00 P.M. on June 6, 2018.

(b) After the poll closed, the final Facebook poll results had 78% of the 19,600 voters indicating support for the Protester Exclusion Zones. Use our quick and approximate method to determine the margin of error for the final Facebook poll results.

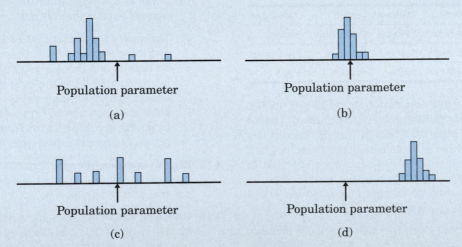

Figure 3.5 Take many samples from the same population and make a histogram of the values taken by a sample statistic. Here are the results for four different sampling methods for Exercise 3.20.

(c) Using the information you calculated in parts (a) and (b), which snapshot of the poll do you think most accurately reflects the views of Donato's constituents? Explain your answer.

(d) Does your answer to part (c) change in light of the following comment that appeared in the *Herald* article by Labor MP Penny Sharpe: "I would encourage everyone in the Central West who thinks that women should be able to go to the doctor with privacy and without interference to support the poll and encourage Mr Donato to vote for the bill"? Explain why or why not.

(e) Is it correct to say that, based on the final Facebook poll, we are 95% confident that 78% $\pm 0.7\%$ of all of Philip Donato's constituents believe that Protester Exclusion Zones should be established in New South Wales? Explain your answer. Be careful not to confuse your personal opinion with the statistical issues.

3.22 Predict the election. Just before a presidential election, a national opinion poll increases the size of its weekly sample from the usual 1000 people to 4000 people. Does the larger random sample reduce the bias of the poll result? Does it reduce the variability of the result?

3.23 Take a bigger sample. A management student is planning a project on student attitudes toward part-time work while attending college. She develops a questionnaire and plans to ask 25 randomly selected students to fill it out. Her faculty adviser approves the questionnaire but suggests that the sample size be increased to at least 100 students. Why is the larger sample helpful? Back up your statement by using the quick and approximate method to estimate the margin of error for samples of size 25 and for samples of size 100.

3.24 Sampling in the states. An agency of the federal government plans to take an SRS of residents in each state to estimate the proportion of owners of real estate in each state's population. The populations of the states range from

about 563,600 people in Wyoming to about 37.3 million in California, according to the 2010 U.S. Census.

(a) Will the variability of the sample proportion change from state to state if an SRS of size 2000 is taken in each state? Explain your answer.

(b) Will the variability of the sample proportion change from state to state if an SRS of 1/10 of 1% (0.001) of the state's population is taken in each state? Explain your answer.

3.25 Immigration poll. A June 2018 Gallup poll asked 1520 randomly selected American adults whether they felt immigration is a good thing or a bad thing for the United States. The poll found that 75% of the respondents said that immigration is a good thing for the United States.

(a) The poll announced a margin of error of ± 3 percentage points for 95% confidence in its conclusions. Make a 95% confidence statement about the percentage of all American adults who think immigration is a good thing for the United States.

(b) Explain to someone who knows no Statistics why we can't just say that 75% of all American adults think immigration is a good thing for the United States.

(c) Now explain clearly what "95% confidence" means.

3.26 Legal immigration poll. The sample survey described in Exercise 3.25 asked 1520 randomly selected American adults about immigration in general and 755 randomly selected American adults about *legal* immigration. The poll announced a margin of error of ± 3 percentage points for 95% confidence in conclusions about women. The margin of error for results concerning legal immigration was ± 4 percentage points. Why is this larger than the margin of error for those asked about immigration in general?

3.27 Explaining confidence. A student reads that we are 90% confident that the average score of those of two or more races on the 2017 SAT is between 800 and 1410. Asked to explain the meaning of this statement, the student says, "90% of all those of two or more races have 2017 SAT scores between 800 and 1410." Is the student right? Explain your answer.

3.28 The death penalty. In October 2017, the Gallup Poll asked a sample of 1028 U.S. adults, "Are you in favor of the death penalty for a person convicted of murder?" The proportion who said they were in favor was 55%.

(a) Approximately how many of the 1028 people interviewed said they were in favor of the death penalty for a person convicted of murder?

(b) Gallup says that the margin of error for this poll is ±4 percentage points. Explain to someone who knows no Statistics what "margin of error ±4 percentage points" means.

3.29 Find the margin of error. Example 6 tells us that a SurveyUSA Poll asked 581 likely voters in Kansas which presidential candidate they would vote for; 47% said they would vote for Donald Trump. Use the quick and approximate method to estimate the margin of error for conclusions about all likely voters in Kansas. How does your result compare with SurveyUSA's margin of error given in Example 6?

3.30 Find the margin of error. Exercise 3.28 concerns a Gallup Poll sample of 1028 people. Use the quick and approximate method to estimate the margin of error for statements about the population of all U.S. adults. Is your result close to the ±4% margin of error announced by Gallup?

3.31 Find the margin of error. Exercise 3.15 describes a sample survey of 61,239 adults living in Ontario. Estimate the margin of error for conclusions having 95% confidence about the entire adult population of Ontario.

3.32 Belief in a higher power. A Pew Research Center Survey conducted in December 2017 reports that 32% of a sample of 4729 adults said they believe in a higher power or spiritual force in the universe other than the God of the Bible.

(a) Use the quick and approximate method to estimate the margin of error for a random sample of this size.

(b) Assuming that this was a random sample, make a confidence statement about the percentage of all adults who believe in a higher power or spiritual force in the universe other than the God of the Bible.

3.33 Legalizing marijuana. An April 2018 Quinnipiac University Poll surveyed 1193 American voters and found that 63% support legalizing marijuana in the United States, an increase of 2% from an August 2017 survey. Make a confidence statement about the percentage of all American voters who thought marijuana should be legalized in the United States, at the time the poll was taken. (Assume that this is an SRS, and use the quick and approximate method to find the margin of error.)

3.34 Moral uncertainty versus statistical uncertainty. In Exercise 3.33 and in the Case Study, we examined polls involving controversial issues, either from a moral or personal liberty perspective (we will call this "moral uncertainty"). In both polls, national opinion was divided, suggesting that there is considerable moral uncertainty regarding both issues. What was the margin of error (the "statistical uncertainty") in both polls? Is it possible for issues with a high degree of moral uncertainty to have very little statistical uncertainty? Discuss.

3.35 Smaller margin of error. Exercise 3.28 describes an opinion poll that interviewed 1028 people. Suppose that you want a margin of error half as large as the one you found in that exercise. How many people must you plan to interview?

3.36 Satisfying Congress. Exercise 3.19 describes a sample survey of 1028 adults, with margin of error ±4% for 95% confidence.

(a) A member of Congress thinks that 95% confidence is not enough. He wants to be 99% confident. How would the margin of error for 99% confidence based on the same sample compare with the margin of error for 95% confidence?

(b) Another member of Congress is satisfied with 95% confidence, but she wants a smaller margin of error than ±4 percentage points. How can we get a smaller margin of error, still with 95% confidence?

3.37 The Current Population Survey. Though opinion polls usually make 95% confidence statements, some sample surveys use other confidence levels. The monthly unemployment rate, for example, is based on the Current Population Survey of about 60,000 households. The margin of error in the unemployment rate is announced as about two-tenths of 1 percentage point with 90% confidence. Would the margin of error for 95% confidence be smaller or larger? Why?

3.38 Detoxing from social media. The Harris Poll recently asked a sample of 2043 adults in the United States which social media app they found the hardest to stay away from. Forty-nine percent of respondents found Facebook the most difficult social media app to ignore. The last bit of information the Harris Poll shared in the same article mentioned that 31% of the poll respondents "grew anxious when they didn't check social media regularly." Write a short report of this finding, as if you were writing for a newspaper. Be sure to include a margin of error. Be careful not to confuse your personal opinion with the statistical findings.

3.39 Who is to blame? A May 2018 POLITICO/ Morning Consult poll surveyed 1993 registered voters and asked who is to blame for mass shootings in the United States. Of the 1993 respondents, 1220 blame illegal gun dealers for mass shootings in the United States. Of the subset of 945 registered voters who were asked about treatment for mental illness, 489 said that they blame lack of access to treatment for mental illness for mass shootings in the United States.

The POLITICO/Morning Consult poll reported a margin of error of ±2 percentage points. Which of the two questions resulted in the margin of error of ±2 percentage points? What is the margin of error for the other question? Explain your answers.

3.40 Simulation. Random digits can be used to *simulate* the results of random sampling. Suppose that you are drawing simple random samples of size 25 from a large number of college students and that 20% of the students are unemployed during the summer. To simulate this SRS, generate 25 random digits using the *Simple Random Sample* applet or let 25 consecutive digits in Table A stand for the 25 students in your sample. The digits 0 and 1 stand for unemployed students, and other digits stand for employed students. This is an accurate imitation of the SRS because 0 and 1 make up 20% of the 10 equally likely digits.

Simulate the results of 50 samples by counting the number of 0s and 1s in the first 25 entries in each of the 50 repetitions of the *Simple Random Sample* applet or in each of the 50 rows of Table A. Make a histogram like that in Figure 3.1 to display the results of your 50 samples. Is the truth about the population (20% unemployed, or 5 in a sample of 25) near the center of your graph? What are the smallest and largest counts of unemployed students that you obtained in your 50 samples? What percentage of your samples had either 4, 5, or 6 unemployed?

Exploring the Web

Access these exercises on the text website: macmillanlearning.com/scc10e.

4

Sample Surveys in the Real World

Case Study: Obtaining a Sample for a Survey

In this chapter you will:

- Discover that sampling errors occur in even the best samples.
- Learn that nonsampling errors may occur.

- See how question wording may impact survey responses.
- Understand that the results that are reported from samples are typically not from a simple random sample.

Consider an opinion poll that reports the results, along with a margin of error, from a recent question asked of 1000 randomly selected American adults. Should we be happy? Perhaps not. For example, how many people did the polling company have to contact to get the 1000 to participate?

Many polls don't tell the whole truth about their samples. However, the Pew Research Center for the People and the Press did tell the whole truth about its sampling methods to get the 1507 American adults for a particular survey. Here it is.

	Landline	Cell
Noncontacts	2464	3114
Not eligible	18,427	6084
Other	104	56
Unknown eligibility	3305	361
Refusals, breakoffs, and partials	4719	4836
Complete interview	902	605
Total called	29,921	15,056

Most polls are taken by telephone (both landline and cell), dialing numbers at random (to get both listed and unlisted telephone numbers) to get a random sample of households. It may be reasonable to consider a landline telephone number to represent a household, but a cell number typically represents one person and may represent a person whose residence has a landline telephone. According to Pew, "landline and cell phone numbers were sampled to yield a ratio of approximately two completed landline interviews to each cell phone interview." In addition to the balance of landline and cell numbers, Pew made sure to contact telephone numbers that would be representative of the United States geographically.

Once the collection of telephone numbers was identified, each sampled landline and cell phone number was called a minimum of seven times (assuming the phone was not answered the first six times). Calls were "staggered over times of day and days of the week (including at least one daytime call) to maximize the chances of making contact with a potential respondent." The Pew Research Center interviewers had to call a total of 44,977 telephone numbers (29,921 landline and 15,056 cell) to get its sample of 1507 (902 landline and 605 cell) respondents.

The calculation to get the response rate is more complicated than taking the number of completed interviews for landline and cell phone numbers divided by the total number of calls. The combined response rate for the poll netting 1507 respondents was 9% (10% of landline numbers and 7% of cell numbers). A higher-effort survey conducted by the Pew Research Center during the same time period resulted in a 22% combined response rate.

We know from Chapter 3 that a sample of 1507 people will give us an approximate margin of error of ±2.6 percentage points. Although Pew did obtain a sample of reasonable size, can we trust the results of this poll to make conclusions about all American adults? Can the results of the poll be extended to people who only have cell phones?

How Sample Surveys Go Wrong

Random sampling eliminates bias in choosing a sample and allows control of variability. So once we see the magic words "randomly selected" and "margin of error," do we know we have trustworthy information before us? It certainly beats voluntary response, but not always by as much as we might hope. Sampling in the real world is more complex and less reliable than choosing a simple random sample (SRS) from a list of names in a textbook exercise. Confidence statements do not reflect all the sources of error that are present in practical sampling.

Most sample surveys are afflicted by errors other than random sampling errors. These errors can introduce bias that makes a confidence statement meaningless. Good sampling technique includes the art of reducing all sources of error. Part of this art is the science of Statistics, with its random samples and confidence statements. In practice, however, good statistics isn't all there is to good sampling. Let's look at sources of errors in sample surveys and at how samplers combat them.

Sampling Errors

Random sampling error is one kind of sampling error. The margin of error tells us how serious random sampling error is, and we can control it by choosing the size of our random sample. Another source of sampling error is the use of *bad sampling methods,* such as voluntary response. We can avoid bad methods. Unfortunately, nonsampling errors are not so easy to handle.

Nonsampling Errors

Nonsampling errors are those that can plague even a census. Sampling begins with a list of individuals from which we will draw our sample. This list is called the **sampling frame.** Ideally, the sampling frame should list every individual in the population of interest. Because a list of the entire population is rarely available, the sampling frame is often not an accurate or complete representation of the population. This leads to errors known as **frame errors.** A common frame error is *undercoverage.*

If the sampling frame leaves out certain groups of people, even random samples from that frame will be biased. Using telephone directories as the frame for a telephone survey, for example, would miss everyone with an unlisted landline telephone number, everyone who cannot afford a landline phone, everyone who has a voice over Internet Protocol (VoIP) phone, and everyone who has only a cell phone. Recall from the Case Study that the Pew Research Center tries to balance its calls geographically, but both cell and VoIP numbers may betray the region the person with the number is actually in. For example, consider a household in Michigan that has no landline but has two cell phone numbers associated with its residents, one a 614 (Columbus, Ohio) area code and one a 602 (Phoenix, Arizona) area code.

> ## Errors in sampling
>
> **Sampling errors** are errors caused by the act of taking a sample. They cause sample results to be different from the results of a census of the population.
>
> **Random sampling error** is the deviation between the statistic and the parameter caused by chance in selecting a random sample. The margin of error in a confidence statement includes *only* random sampling error.
>
> **Nonsampling errors** are errors not related to the act of selecting a sample from the population. They can be present even in a census.
>
> **Undercoverage** occurs when some groups in the population have no chance of being included the sample.

Example 1   *Did we miss anyone?*

Most opinion polls can't afford to even attempt full coverage of the population of all adult residents of the United States. The interviews are done by telephone, thus missing the 2% of households without phones, either landline, VoIP, or cell. Additionally, those without *regular* access to phones, like prison inmates, or to reliable cell service, like many deployed members of the armed forces, are left out.

In addition, many polls in the United States interview only in English and Spanish, which leaves some immigrants with telephone numbers out of their samples. Those left out of these polls include a growing number of people who speak Arabic or Asian languages.

The kinds of undercoverage found in most sample surveys are most likely to leave out people who are young or poor or who move often. Nonetheless, random digit dialing comes close to producing a random sample of households with phones. Sampling errors in careful sample surveys are usually quite small. The real problems start when someone picks up (or doesn't pick up) the phone.

Frame errors can also arise from **erroneous inclusions** and **multiple inclusions.** Erroneous inclusions can occur if the frame includes units that are not in the population of interest so that invalid units have a chance of being in the sample. Multiple inclusions occur if some population members appear multiple times in the sampling frame so that they have a higher chance of being sampled. For example, if the sampling frame is a telephone book, some members of the population may have multiple listings because they have multiple telephone lines.

Nonsampling errors also include **processing errors**—mistakes in mechanical tasks such as doing arithmetic or entering responses into a computer. Fortunately, the wide availability and adoption of computers has made processing errors less common than in the past.

Example 2   *Computer-assisted interviewing*

The days of the interviewer with a clipboard are gone. Contemporary interviewers carry a laptop computer for face-to-face interviews or watch a computer screen as they carry out a telephone interview. Computer software manages the interview. The interviewer reads questions from the computer screen and uses the keyboard to enter the responses. The computer skips irrelevant items—once a respondent says that she has no children, further questions about her children never appear. The computer can check that answers to related questions are consistent with each other. It can even present questions in random order to avoid any bias due to always asking questions in the same order.

Computer software also manages the record keeping. It keeps records of who has responded and prepares a file of data from the responses. The tedious process of transferring responses from paper to computer, once a source of processing errors, has disappeared. The computer even schedules the calls in telephone surveys, taking account of the respondent's time zone and honoring appointments made by people who were willing to respond but did not have time when first called.

Another type of nonsampling error is **response error,** which occurs when a subject gives an incorrect response. A subject may lie about her age or income or about whether she has used illegal drugs. She may remember incorrectly when asked how many packs of cigarettes she smoked last week. A subject who does not understand a question may guess at an answer rather than appear ignorant. Questions that ask subjects about their behavior during a fixed time period are notoriously prone to response errors due to faulty memory. For example, the National Health Survey asks people how many times they have visited a doctor in the past year. Checking their responses against health records found that they failed to remember 60% of their visits to a doctor. A survey that asks about sensitive issues can also expect response errors, as the next example illustrates.

Example 3

The effect of race

In 1989, New York City elected its first black mayor and the state of Virginia elected its first black governor. In both cases, samples of voters interviewed as they left their polling places predicted larger margins of victory than the official vote counts. The polling organizations were certain that some voters lied when interviewed because they felt uncomfortable admitting that they had not voted for the black candidate. This phenomenon is known as "social desirability bias" and "the Bradley effect," after Tom Bradley, the former black mayor of Los Angeles who lost the 1982 California gubernatorial election despite leading in final-day preelection polls.

This effect attracted media attention during the 2008 presidential election. A few weeks before the election, polls predicted a victory, possibly a big one, for Barack Obama. Even so, Democrats worried that these polls might be overly optimistic because of the Bradley effect. In this case their fears were unfounded, but some political scientists claimed to detect the Bradley effect in polls predicting outcomes in primary races between Barack Obama and Hillary Clinton (for example, in the New Hampshire primary, polls predicted an 8 percentage point Obama victory but Clinton won by 3 percentage points).

Even the 2016 presidential election included effects due to race, despite both major party candidates being white. According to *The Washington Post*, "whites were more divided based on their perceptions of discrimination against whites than they were in 2012." As early as the primary election of 2016, *The Washington Post* reported "both white racial identity and beliefs that whites are treated unfairly are powerful predictors of support for Donald Trump in the Republican primaries." Unlike the Bradley effect, this racial divide was less about social desirability and more about white voters fearing for their rights.

Technology and attention to detail can minimize processing errors. Skilled interviewers greatly reduce response errors, especially in face-to-face interviews. There is no simple cure, however, for the most serious kind of nonsampling error, *nonresponse*.

Nonresponse is the most serious problem facing sample surveys. People are increasingly reluctant to answer questions, particularly over the phone. The rise of telemarketing, voicemail, caller ID, and spoofed phone numbers drives down response to telephone surveys. Gated communities and buildings guarded by doormen prevent face-to-face interviews. Nonresponse can bias sample survey results because different groups have different rates of nonresponse. Refusals are higher in large cities and among the elderly, for example. Bias due to nonresponse can easily overwhelm the random sampling error described by a survey's margin of error.

Example 4

How bad is nonresponse?

The U.S. Census Bureau's American Community Survey (ACS) is a monthly survey of almost 300,000 housing units and replaced the U.S. Census Bureau's "long form" that in the past was sent to some households in the decennial national census. Participation in the ACS is mandatory, and the U.S. Census Bureau follows up by telephone and then in person if a household fails to return the mail questionnaire.

The ACS has the lowest nonresponse rate of any poll we know: in 2016, only about 2.1% of the households in the sample refused to respond; the overall nonresponse rate, including "never at home" and other causes, was only 5.3%. This is a stark contrast in nonresponse from 2013 when overall response rate was 10.1%. What happened in October 2013? There was a government shutdown. During that time, there were no follow-ups by mail, phone, or in person, and the overall panel response rate was about 7 percentage points lower than is usual for the ACS. If October 2013 is excluded from the ACS when considering nonresponse, the nonresponse rate for 2013 was only 2.9%, similar to previous years.

Another survey that has a remarkable response rate is the University of Chicago's General Social Survey (GSS), the nation's most important social survey. The GSS (Example 7 in Chapter 1) contacts its sample in person, and it is run by a university. Despite these advantages, its recent surveys have a 30% rate of nonresponse.

What about polls done by the media and by market research and opinion-polling firms? We often don't know their rates of nonresponse because they won't say. That itself is a bad sign. The Pew Poll we looked at in the Case Study suggests how bad things are. Pew got 1507 responses and 15,293 who were never at home, refused, or would not finish the interview. That's a nonresponse rate of 15,293 out of 16,800, or 91%. We should keep in mind that Pew researchers are more thorough than many pollsters.

Sample surveyors know some tricks to reduce nonresponse. Carefully trained interviewers can keep people on the line if they answer at all. Calling back over longer time periods helps. So do letters sent in advance. Letters and many callbacks slow down the survey, so opinion polls that want fast answers to satisfy the media don't use them. Even the most careful surveys find that nonresponse is a problem that no amount of expertise can fully overcome. That makes the reminder in the box to the right even more important.

Careful sample surveys warn us about the other kinds of error. The Pew Research Center, for example, says, "In addition to sampling error, one should bear in mind that question wording and practical difficulties in conducting surveys can introduce error or bias into the findings of opinion polls." How true it is.

Does nonresponse make many sample surveys useless? Maybe not. We began this chapter with an account of a "standard" telephone survey done by the Pew Research Center. The Pew researchers also carried out a "rigorous" survey, with letters sent in advance, unlimited calls over eight weeks, letters by priority mail to people who refused, and so on. All this drove the rate of nonresponse down to 78%, compared with 91% for the standard survey. Pew then compared the answers to the same questions from the two surveys. The two samples were quite similar in age, sex, and race, though the rigorous sample was a bit more prosperous. The two samples also held similar opinions on all issues except one: race. People who at first refused to respond were less sympathetic toward the plights of blacks and other minorities than those who were willing to respond when contacted the first time. Overall, it appears that standard polls give reasonably accurate results. But, as in Example 3, race is again an exception.

> ## What the margin of error doesn't say
>
> The announced margin of error for a sample survey covers only random sampling error. Undercoverage, nonresponse, and other practical difficulties can cause large bias that is not covered by the margin of error.

Wording of Questions

A final influence on the results of a sample survey is the exact wording of questions. It is surprisingly difficult to word the questions so that they are completely clear. A survey that asks "Do you enjoy watching football?" will generate different answers based on the respondent's understanding of "football" (American football or soccer).

> ## Statistics in Your World
>
> **He started it!** A study of deaths in bar fights showed that in 90% of the cases, the person who died started the fight. You shouldn't believe this. If you killed someone in a fight, what would you say when the police asked you who started the fight? After all, dead men tell no tales. Now that's nonresponse!

Example 5

Words make a big difference

In May 2013, the Pew Research Center and the *Washington Post*/ABC News conducted polls asking whether people felt the U.S. Department of Justice had the right to subpoena Associated Press reporters' phone records. Each survey phrased the question differently, and each survey found different results.

When the Pew Research Center asked, "Do you approve or disapprove of the Justice Department's decision to subpoena the phone records of AP journalists as part of an investigation into the disclosure of classified information," 36% of respondents approved. The *Washington Post*/ABC News survey said, "The AP reported classified information about U.S. anti-terrorism efforts and prosecutors have obtained AP's phone records through a court order. Do you think this action by federal prosecutors is or is not justified?" Fifty-two percent (52%) of respondents to this survey said that the action of federal prosecutors was justified.

Before asking about approval or disapproval of the Justice Department's actions, Pew asked how closely respondents had followed the issues that led up to the subpoena, and 64% reported not following the news story too closely. Additionally, the difference in wording—"decision to subpoena" in the Pew Research survey versus "obtained... through a court order" used by the *Washington Post*/ABC News survey—could have led to legitimizing the actions of the Department of Justice and thus a higher "approval" for the Justice Department's actions. Or, was the higher justification based on the inclusion of the mention of U.S. anti-terrorism efforts? Or, perhaps the difference is due to Pew using "the Justice Department" and the *Washington Post*/ABC News using "federal prosecutors." We cannot begin to determine which part of the wording impacted the responses. As the Pew Research Center said in its article comparing the results of these two surveys, "each polling organization made good-faith efforts to describe the facts of the situation as accurately as possible, but the word choices and context make it impossible to identify one particular phrase or concept that tipped the public's thinking."

The wording of questions always influences the answers. If the questions are slanted to favor one response over others, we have another source of nonsampling error. A favorite trick is to ask if the subject favors some policy as a means to a desirable end: "Do you favor banning private ownership of handguns in order to reduce the rate of violent crime?" and "Do you favor imposing the death penalty in order to reduce the rate of violent crime?" are loaded questions that draw positive responses from people who are worried about crime. Here is another example of the influence of question wording.

Example 6

Paying taxes

In April 2018, a Gallup Poll asked two questions about the amount one pays in federal income taxes. Here are the two questions:

> *Do you consider the amount of federal income tax you have to pay as too high, about right, or too low?*

> *Do you regard the income tax which you will have to pay this year as fair?*

The first question had 48% of respondents say "about right" (45% said "too high") while the second question resulted in 61% of respondents saying that the taxes they paid were "fair." There appears to be a definite difference in opinions about taxes when people are asked about the amount they have to pay or whether it is fair.

Now it's your turn

4.1 Should we recycle? Is the following question slanted toward a desired response? If so, how?

> *In view of escalating environmental degradation and incipient resource depletion, would you favor economic incentives for recycling of resource-intensive consumer goods?*

How to Live with Nonsampling Errors

Nonsampling errors, especially nonresponse, are always with us. What should a careful sample survey do about this? First, **substitute other households** for the nonresponders. Because nonresponse is higher in cities, replacing nonresponders with other households in the same neighborhood may reduce bias. Once the data are in, all professional surveys use statistical methods to **weight the responses** in an attempt to correct sources of bias. If many urban households did not respond, the survey gives more weight to those that did respond. If too many women are in the sample, the survey gives more weight to the men. Here, for example, is part of a statement in the *New York Times* describing one of its sample surveys:

> *The results have been weighted to take account of household size and number of telephone lines into the residence and to adjust for variations in the sample relating to geographic region, sex, race, age and education.*

The goal is to get results "as if" the sample matched the population in age, gender, place of residence, and other variables.

The practice of weighting creates job opportunities for statisticians. It also means that the results announced by a sample survey are rarely as simple as they seem to be. In April 2018, Gallup announced that it interviewed 1509 adults and found that 56% of them have a Facebook account. It would seem that because 56% of 1509 is 845, Gallup found that 845 people in its sample had a Facebook account. Not so. Gallup no doubt used some quite fancy statistics to weight the actual responses: 56% is Gallup's best estimate of what it would have found in the absence of nonresponse. Weighting does help correct bias. It usually also increases variability. The announced margin of error must take this into account, creating more work for statisticians.

Sample Design in the Real World

The basic idea of sampling is straightforward: take an SRS from the population and use a statistic from your sample to estimate a parameter of the population. We now know that the statistic is altered behind the scenes to partly correct for nonresponse. The statisticians also have their hands on our beloved SRS. In the real world, most sample surveys use more complex designs.

Example 7 — *The Current Population Survey*

The population that the Current Population Survey (CPS) is interested in consists of all households in the United States. The scientifically chosen sample of about 60,000 occupied households is **chosen in stages.** The U.S. Census Bureau divides the nation into 2025 geographic areas called Primary

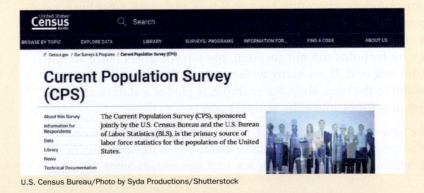

U.S. Census Bureau/Photo by Syda Productions/Shutterstock

Sampling Units (PSUs). These are generally groups of neighboring counties. At the first stage, 824 PSUs are chosen. This isn't an SRS. If all PSUs had the same chance to be chosen, the sample might miss Chicago and Los Angeles. So 446 highly populated PSUs are automatically in the sample. The other 1579 are divided into 378 groups, called **strata,** by combining PSUs that are similar in various ways. One PSU is chosen at random to represent each stratum.

Each of the 824 PSUs in the first-stage sample is divided into census blocks (smaller geographic areas). The blocks are also grouped into strata, based on such things as housing types and minority population. The households in each block are arranged in order of their location and divided into groups, called **clusters,** of about four households each. The final sample consists of samples of clusters (not of individual households) from each stratum of blocks. Interviewers go to all households in the chosen clusters. The samples of clusters within each stratum of blocks are also not SRSs. To be sure that the clusters spread out geographically, the sample starts at a random cluster and then takes, for example, every 10th cluster in the list.

The design of the CPS illustrates several ideas that are common in real-world samples that use face-to-face interviews. Taking the sample **in several stages** with **clusters** at the final stage saves travel time for interviewers by grouping the sample households first in PSUs and then in clusters. Note that clustering is not an aspect of *all* sampling strategies but can be quite helpful in situations like the CPS.

The most important refinement mentioned in Example 7 is *stratified sampling,* which is described in the box to the right.

We must of course choose the strata using facts about the population that are known before we take the sample. You might group a university's students into undergraduate and graduate students or into those who live on campus and those who commute. Stratified samples have some advantages over an SRS. First, by taking a separate SRS in each stratum, we can set sample sizes to allow separate conclusions about each stratum. Second, a stratified sample usually has a smaller margin of error than an SRS of the same size. The reason is that the individuals in each stratum are more alike than the population as a whole, so working stratum-by-stratum eliminates some variability in the sample.

It may surprise you that stratified samples can violate one of the most appealing properties of the SRS— stratified samples need not give all individuals in the population the same chance to be chosen. Some strata may be deliberately overrepresented in the sample.

Stratified sample

To choose a **stratified random sample:**

Step 1. Divide the sampling frame into distinct groups of individuals, called **strata.** Choose the strata according to any special interest you have in certain groups within the population or because the individuals in each stratum resemble each other.

Step 2. Take a separate SRS in each stratum and combine these to make up the complete sample.

Example 8   *Stratifying a sample of students*

A large university has 30,000 students, of whom 3000 are graduate students. An SRS of 500 students gives every student the same chance to be in the sample. That chance is

$$\frac{500}{30,000} = \frac{1}{60}$$

We expect an SRS of 500 to contain only about 50 grad students—because grad students make up 10% of the population, we expect them to make up about 10% of an SRS. A sample of size 50 isn't large enough to estimate grad student opinion with reasonable accuracy. We might prefer a stratified random sample of 200 grad students and 300 undergraduates.

You know how to select such a stratified sample. Label the graduate students 0001 to 3000 and use Table A to select an SRS of 200. Then label the undergraduates 00001 to 27000 and use Table A a second time to select an SRS of 300 of them. These two SRSs together form the stratified sample.

In the stratified sample, each grad student has chance

$$\frac{200}{3000} = \frac{1}{15}$$

to be chosen. Each of the undergraduates has a smaller chance,

$$\frac{300}{27,000} = \frac{1}{90}$$

Because we have two SRSs, it is easy to estimate opinions in the two groups separately. The quick and approximate method (page 43) tells us that the margin of error for a sample proportion will be about

$$\frac{1}{\sqrt{200}} = 0.07 \quad (that\ is,\ 7\%)$$

for grad students and about

$$\frac{1}{\sqrt{300}} = 0.058 \quad (that\ is,\ 5.8\%)$$

for undergraduates.

Because the sample in Example 8 deliberately overrepresents graduate students, the final analysis must adjust for this to get unbiased estimates of overall student opinion. Remember that our quick method works only for an SRS. In fact, a professional analysis would also take account of the fact that the population contains "only" 30,000 individuals—more job opportunities for statisticians.

Now it's your turn

4.2 A stratified sample. The statistics department at Cal Poly, San Luis Obispo, has 18 faculty members and 80 undergraduate majors. Use the *Simple Random Sample* applet, other software, or Table A, starting at line 111, to choose a stratified sample of 1 faculty member and 1 student to attend a reception being held by the university president.

Example 9
The woes of telephone samples

In principle, it would seem that a telephone survey that dials numbers at random could be based on an SRS. Telephone surveys have little need for clustering. Stratifying can still reduce variability, however, and so telephone surveys often take samples in two stages: a stratified sample of telephone number prefixes (area code plus first three digits) followed by individual numbers (last four digits) dialed at random in each prefix.

The real problem with an SRS of telephone numbers is that too few numbers lead to households. Blame technology. Fax numbers, cell phones, and Voice over Internet Protocols (VoIP) demand new phone numbers. Between 1997 and 2017, the number of households in the United States grew by 25%, but the number of possible phone numbers continues to grow as more individuals have access to cell phones. Some analysts believe that in the near future we may have to increase the number of digits for telephone numbers from 10 (including the area code) to 12. This will further exacerbate this problem. Telephone surveys now use "list-assisted samples" that check electronic telephone directories to eliminate prefixes that have no listed numbers before random sampling begins. Fewer calls are wasted, but anyone living where all numbers are unlisted is missed. Prefixes with no listed numbers are therefore separately sampled (stratification again), perhaps with a smaller sample size than if included in the list-assisted sample, to fill the gap.

The proliferation of cell phones has created additional problems for telephone samples. As of December 2017, about 53.9% of households had cell phones only. Random digit dialing using a machine is not allowed for cell phone numbers. Phone numbers assigned to cell phones are determined by the location of the cell phone company providing the service and need not coincide with the actual residence of the user. This makes it difficult to implement sophisticated methods of sampling, such as stratified sampling by geographic location.

It may be that the woes of telephone sampling prompted the Gallup Organization, in recent years, to drop the phrase "random sampling" from the description of their survey methods at the end of most of their polls. This presumably prevents misinterpreting the results as coming from simple random samples. In their detailed description of survey methods used for the Gallup World Poll and the Gallup Well-Being Index (available online at the Gallup website), the samples are described as involving random sampling.

This might be a good place to read the "What's the verdict?" story on page 85, and to answer the questions. These questions involve material from this chapter and Chapter 3 to assess some puzzling results from the National **WTV** Longitudinal Study of Adolescent Health.

The Challenge of Internet Surveys

The Internet is having a profound effect on many things people do, and this includes surveys. Using the Internet to conduct "Web surveys" is becoming increasingly popular. Web surveys have several advantages over more traditional survey methods. It is possible to collect large amounts of survey data at lower costs than traditional methods allow. Anyone can put survey questions on dedicated sites offering free services; thus, large-scale data collection is available to almost every person with access to the Internet. Furthermore, Web surveys allow one to deliver multimedia survey content to respondents, opening up new realms of survey possibilities that would be extremely difficult to implement using traditional methods. Some argue that eventually Web surveys will replace traditional survey methods.

Although Web surveys are easy to do, they are not easy to do well. The reasons include many of the issues we have discussed in this chapter. Three major problems are voluntary response, undercoverage, and nonresponse. Voluntary response appears in several forms. Some Web surveys invite visitors to a particular website to participate in a poll. Misterpoll.com is one such example. Visitors to this site can participate in several ongoing polls, create their own poll, and respond multiple times to the same poll. Other Web surveys solicit participation through announcements in newsgroups, email invitations, and banner ads on high-traffic sites. An example is a series of 10 polls conducted by Georgia Tech University's Graphic, Visualization, and Usability Center (GVU) in the 1990s.

Although misterpoll.com indicates that the surveys on the site are primarily intended for entertainment, the GVU polls appear to claim some measure of legitimacy. The website www.cc.gatech.edu/gvu/user_surveys/ states that the information from these surveys "is valued as an independent, objective view of developing Web demographics, culture, user attitudes, and usage patterns."

A third and more sophisticated example of voluntary response occurs when the polling organization creates what it believes to be a representative panel consisting of volunteers and uses panel members as

Statistics in Your World

New York, New York New York City, they say, is bigger, richer, faster, and ruder. Maybe there's something to that. The sample survey firm Zogby International says that as a national average, it takes 5 calls to reach a live person. When trying to reach respondents in New York City, it takes 12 calls. Survey firms assign their best interviewers to make calls to New York City and often pay them bonuses to cope with the stress.

a sampling frame. A random sample is selected from this panel, and those selected are invited to participate in the poll. A very sophisticated version of this approach is used by the Harris Poll Online.

Web surveys, such as the Harris Poll Online, in which a random sample is selected from a well-defined sampling frame, are reasonable when the sampling frame clearly represents some larger population or when interest is only in the members of the sampling frame. An example are Web surveys that use systematic sampling to select every nth visitor to a site and the target population is narrowly defined as visitors to the site. Another example are some Web surveys on college campuses. All students may be assigned email addresses and have Internet access. A list of these email addresses serves as the sampling frame, and a random sample is selected from this list. If the population of interest is all students at this particular college, these surveys can potentially yield very good results. Here is an example of this type of Web survey.

Example 10

Doctors and placebos

A placebo is a dummy treatment like a salt pill that has no direct effect on a patient but may bring about a response because patients expect it to. Do academic physicians who maintain private practices sometimes give their patients placebos? A Web survey of doctors in internal medicine departments at Chicago-area medical schools was possible because almost all doctors had listed email addresses.

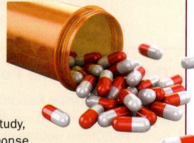

Maxx-Studio/Shutterstock

Send an email to each doctor explaining the purpose of the study, promising anonymity, and giving an individual a Web link for response. Result: 45% of respondents said they sometimes use placebos in their clinical practice.

Several other Web survey methods have been employed to eliminate problems arising from voluntary response. One is to use the Web as one of many alternative ways to participate in the survey. The Bureau of Labor Statistics and the U.S. Census Bureau have used this method. Another method is to select random samples from panels, but instead of relying on volunteers to form the panels, members are recruited using random sampling (for example, random digit dialing). Telephone interviews can be used to collect background information, identify those with Internet access, and recruit eligible persons to the panel. If the target population is current users of the Internet, this method should also potentially yield reliable results. The Pew Research Center has employed this method.

Perhaps the most ambitious approach, and one that attempts to obtain a random sample from a more general population, is the following. Take a probability sample (defined on the next page) from the population of interest. Provide all those selected with the necessary equipment and tools to participate in subsequent Web surveys. This methodology is similar in spirit to that used for the Nielsen TV ratings. It was employed by one company, InterSurvey, several years ago, although InterSurvey is no longer in business.

Several challenges remain for those who employ Web surveys. Even though Internet and email use is growing (according to a February 2018 Pew Research Center report, 89% of American adults aged 18 and older have Internet access), there is still the problem of undercoverage if Web surveys are used to draw conclusions about all American adults aged 18 and older. Weighting responses to correct for possible biases does not solve the problem because studies indicate that Internet users differ in many ways that traditional methods of weighting do not account for.

In addition, even if 100% of Americans had Internet access, there is no list of Internet users that we can use as a sampling frame, nor is there anything comparable to random digit dialing that can be used to draw random samples from the collection of all Internet users.

Finally, Web surveys often have very high rates of nonresponse. Methods that are used in phone and mail surveys to improve response rates can help, but they make Web surveys more expensive and difficult, offsetting some of their advantages.

STATISTICAL CONTROVERSIES

The Harris Online Poll

The Harris Poll Online has created an online research panel of over 6 million volunteers. According to the Harris Poll Online website, the "panel consists of a diverse cross-section of people residing in the United States, as well as in over 200 countries around the world," and "this multimillion member panel consists of potential respondents who have been recruited through online, telephone, mail, and in-person approaches to increase population coverage

belushi/Shutterstock

and enhance representativeness." One can join the panel at https://www.harrispollonline.com/#homepage.

When the Harris Poll Online conducts a survey, this panel serves as the sampling frame. A probability sample is selected from it, and statistical methods are used to weight the responses. In particular, the Harris Poll Online uses propensity score weighting, a proprietary Harris Interactive technique, which is also applied (when applicable) to adjust for respondents' likelihood of being online. They claim that "this procedure provides added assurance of accuracy and representativeness."

Are you convinced that the Harris Poll Online provides accurate information about well-defined populations such as all American adults? Why or why not?

For more information about the Harris Poll Online and Web surveys in general, see https://theharrispoll.com/ and in a special issue of *Public Opinion Quarterly* available online at https://academic.oup.com/poq/issue/72/5.

Probability Samples

It's clear from Examples 7, 8, and 9, and from the challenges of using the Internet to conduct surveys, that designing samples is a business for experts. Even most statisticians don't qualify. We won't worry about such details. The big idea is that good sample designs use chance to select individuals from the population. That is, all good samples are *probability samples*.

A stratified sample of 300 undergraduate students and 200 graduate students, for example, allows only samples with exactly that makeup. An SRS would allow any 500 students. Both are probability samples. We need only know that estimates from any probability sample share the nice properties of estimates from an SRS. Confidence statements can be made without bias and have smaller margins of error as the size of the sample increases. Nonprobability samples such as voluntary response samples do not share these advantages and cannot give trustworthy information about a population. Now that we know that most nationwide samples are more complicated than an SRS, we will usually go back to acting as if good samples were SRSs. That keeps the big idea and hides the messy details.

> ### Key Terms
>
> A **probability sample** is a sample chosen by chance. We must know what samples are possible and what chance, or probability, each possible sample has.
>
> Some probability samples, such as stratified samples, don't allow all possible samples from the population and may not give an equal chance to all the samples they do allow. As such, not all probability samples are random samples.

Questions to Ask before You Believe a Poll

Opinion polls and other sample surveys can produce accurate and useful information if the pollster uses good statistical techniques and also works hard at preparing a sampling frame, wording questions, and reducing nonresponse. Many surveys, however, especially those designed to influence public opinion rather than just record it, do not produce accurate or useful information. Here are some questions to ask before you pay much attention to poll results.

- **Who carried out the survey?** Even a political party should hire a professional sample survey firm whose reputation demands that they follow good survey practices.

- **What was the population?** That is, whose opinions were being sought?

- **How was the sample selected?** Look for mention of random sampling.

- **How large was the sample?** Even better, find out both the sample size and the **margin of error** within which the results of 95% of all samples drawn as this one was would fall.

- **What was the response rate?** That is, what percentage of the original subjects actually provided information?

- **How were the subjects contacted?** By telephone? Mail? Face-to-face interview?

- **When was the survey conducted?** Was it just after some event that might have influenced opinion?

- **What were the exact questions asked?**

Academic survey centers and government statistical offices answer these questions when they announce the results of a sample survey. National opinion polls usually don't announce their response rate (which is often low) but do give us the other information. Editors and newscasters have the bad habit of cutting out these dull facts and reporting only the sample results. Many sample surveys by interest groups and local newspapers and TV stations don't answer these questions because their polling methods are in fact unreliable. If a politician, an advertiser, or your local TV station announces the results of a poll without complete information, be skeptical.

Chapter 4: Statistics in Summary

- Sampling in the real world is complex. Even professional sample surveys don't give exactly correct information about the population.

- There are many potential sources of error in sampling. The margin of error announced by a sample survey covers only **random sampling error,** the variation due to chance in choosing a random sample.

- **Sampling errors** come from the act of choosing a sample. Random sampling error or the use of bad sampling methods are common types of sampling error.

- The most serious errors in most careful surveys, however, are **nonsampling errors.** These have nothing to do with choosing a sample—they are present even in a census.

- **Frame errors** can occur because the **sampling frame,** the list from which the sample is actually chosen, is not an accurate representation of the population. One such error is **undercoverage.** Undercoverage occurs when some members of the population are left out of the sampling frame. Other frame errors occur if the frame includes units not in the population of interest and if the frame lists units multiple times.

- The most challenging problem for sample surveys is **nonresponse:** subjects can't be contacted or refuse to answer.

- Mistakes in handling the data (**processing errors**) and incorrect answers by respondents (**response errors**) are other examples of nonsampling errors.

- Finally, the exact **wording of questions** has a big influence on the answers.

- People who design sample surveys use statistical techniques that help correct nonsampling errors, and they also use **probability samples** more complex than simple random samples, such as **stratified samples.**

- You can assess the quality of a sample survey quite well by just looking at the basics: use of random samples, sample size and margin of error, the rate of nonresponse, and the wording of the questions.

This chapter summary will help you evaluate the Case Study.

Link It

In Chapter 3 we saw that random samples can provide a sound basis for drawing conclusions about a parameter. In this chapter, we learn that even when we take a random sample, our conclusions can be weakened by undercoverage, processing errors, response error, nonresponse, and wording of questions. We must pay careful attention to every aspect of how we collect data to ensure that the conclusions we make are valid. In some cases, more complex probability samples, such as stratified samples, can help correct nonsampling errors. This chapter provides a list of questions you can ask to help you assess the quality of the results of samples collected by someone else.

Case Study Evaluated

Use what you have learned in this chapter to evaluate the Case Study that opened the chapter. Start by reviewing the Chapter Summary. Then communicate clearly enough for any of your classmates to understand what you are saying.

Use what you have learned about what sampling errors are, how nonsampling errors and question wording may impact survey results, and how the results reported from samples are typically not from an SRS to evaluate the Case Study that opened the chapter. In particular, answer the questions given in the section "Questions to Ask before You Believe a Poll" on page 75. Are the results of the Pew poll useless? You may want to refer to the discussion on pages 59–60.

In this chapter you:

- Discovered that sampling errors occur in even the best samples.
- Learned that nonsampling errors may occur.
- Saw how question wording may impact survey responses.
- Understood that the results that are reported from samples are typically not from a simple random sample.

macmillan learning **Online Resources**

- The video technology manuals explain how to select an SRS using JMP, Excel, R, Minitab, CrunchIt!, SPSS, and the TI 83/84.

- The Statistical Applet, *Simple Random Sample*, can be used to select a simple random sample when the number of labels is 144 or less.

Check the Basics

For Exercise 4.1, see page 67; for Exercise 4.2, see page 71.

4.3 What does the margin of error include? When a margin of error is reported for a survey, it includes

(a) random sampling error and other practical difficulties like undercoverage and nonresponse.

(b) random sampling error, but not other practical difficulties like undercoverage and nonresponse.

(c) practical difficulties like undercoverage and nonresponse, but not random sampling error.

(d) None of the above is correct.

4.4 What kind of sample? Archaeologists plan to examine a sample of 2-meter square plots near an ancient Greek city for artifacts visible in the ground. They choose separate samples of plots from floodplain, coast, foothills, and high hills. What kind of sample is this?

(a) A simple random sample

(b) A voluntary response sample

(c) A stratified sample

(d) A cluster sample

4.5 Sampling issues. A sample of households in a community is selected at random from the telephone directory. In this community, 4% of households have no telephone, 10% have only cell phones, and another 25% have unlisted telephone numbers. The sample will certainly suffer from

(a) nonresponse.

(b) undercoverage.

(c) false responses.

(d) all of the above.

4.6 Question wording. Which of the following represents wording that will most likely *not* influence the answers?

(a) Do you think that all instances of academic misconduct should be reported to the dean?

(b) Academic misconduct undermines the integrity of the university and education in general. Do you believe that all instances of academic misconduct should be reported to the dean?

(c) Academic misconduct can range from something as minor as using one's own work in two courses to major issues like cheating on exams and plagiarizing. Do you believe that all instances of academic misconduct should be reported to the dean?

(d) None of the above will influence the answers.

4.7 Sampling considerations. A Statistics class has 10 graduate students and 40 undergraduate students. You want to randomly sample 10% of the students in the class. One graduate student and four undergraduate students are selected at random. Which of the following is *not* correct?

(a) Because each student has a 10% chance of being selected, this is a simple random sample.

(b) Because each sample includes exactly one graduate student and four undergraduate students, this is not a random sample.

(c) It is possible to get a sample that contains only graduate students.

(d) It is possible to get a sample that contains only undergraduate students.

Chapter 4 Exercises

4.8 What kind of error? Which of the following are sources of *sampling error* and which are sources of *nonsampling error*? Explain your answers.

(a) The subject lies about past illegal drug use.

(b) A typing error is made in recording the data.

(c) Data are gathered by asking people to go to a website and answer questions online.

4.9 What kind of error? Each of the following is a source of error in a sample survey. Label each as *sampling error* or *nonsampling error,* and explain your answers.

(a) The telephone directory is used as a sampling frame.

(b) The subject cannot be contacted in five calls.

(c) Interviewers choose people on the street to interview.

4.10 Not in the margin of error. According to a December 2017 Gallup Poll, 7% of American adults report soccer as their favorite sport, up from 4% in June 2013 and just 2% in April 1997. This may seem low to you, but the United States is catching up to the rest of the world in its interest in soccer. The survey methods section of the poll states:

For results based on the total sample of national adults, the margin of sampling error is ±4 percentage points at the 95% confidence level. All reported margins of sampling error include computed design effects for weighting.

Give one example of a source of error in the poll result that is *not* included in this margin of error.

4.11 Not in the margin of error. According to an April 2018 survey, a majority of employed American adults (59%) are confident about their job security, stating it is not at all likely for them to lose their job or be laid off in the next 12 months. The survey methods section of the poll states:

For results based on the total sample of employed adults, the margin of sampling error is ±5

percentage points at the 95% confidence level. All reported margins of sampling error include computed design effects for weighting.

Give one example of a source of error in the poll result that is *not* included in this margin of error.

4.12 College parents. An online survey of college parents was conducted during February and March 2007. Emails were sent to 41,000 parents who were listed in either the College Parents of America database or the Student Advantage database. Parents were invited to participate in the online survey. Out of those invited, 1727 completed the online survey. The survey protected the anonymity of those participating in the survey but did not allow more than one response from an individual IP address.

One of the survey results was that 33% of mothers communicate at least once a day with their child while at school.

(a) What was the *response rate* for this survey? (The response rate is the percentage of the planned sample—that is, those invited to participate—who responded.)

(b) Use the quick method (page 43) to estimate the margin of error for a random sample of size 1727.

(c) Do you think that the margin of error is a good measure of the accuracy of the survey's results? Explain your answer.

4.13 Polling customers. An online store chooses an SRS of 100 customers from its list of all people who have bought something from the store in the last year. It asks those selected how satisfied they are with the store's website. If it selected two SRSs of 100 customers at the same time, the two samples would give somewhat different results. Is this variation a source of sampling error or of nonsampling error? Would the survey's announced margin of error take this source of error into account?

4.14 Ring-no-answer. A common form of non-response in telephone surveys is "ring-no-answer." That is, a call is made to an active number but no one answers. The Italian National Statistical Institute looked at nonresponse to a government survey of households in Italy during the periods January 1 to Easter and July 1 to August 31. All calls were made between 7 and 10 P.M., but 21.4% gave "ring-no-answer" in one period versus 41.5% "ring-no-answer" in the other period. Which period do you think had the higher rate of no answers? Why? Explain why a high rate of nonresponse makes sample results less reliable.

4.15 Race relations. Here are two recent opinion poll questions asked about race relations in the United States.

Would you say relations between whites and blacks are very good, somewhat good, somewhat bad, or very bad?

Do you think race relations in the United States are generally good or generally bad?

In response to the first question, 72% of non-Hispanic whites and 66% of blacks answered that relations between blacks and whites are "very good" or "somewhat good." Sixty-one (61%) of those answering the second question responded "generally bad."

The first question came from a poll that was conducted in March 2015. The second question came from a poll that was conducted between April 30 and May 3 of that same year. Between the two polls, a man named Freddie Gray died after being in the custody of the Baltimore police. In what ways do you think this event may have impacted the responses to the two different polls? Do you think the results would be different if the question from the second poll had been worded like the question from the first poll?

4.16 The environment and the economy. Here are two opinion poll questions asked about protecting the environment versus protecting the economy.

Often there are trade-offs or sacrifices people must make in deciding what is important to them. Generally speaking, when a trade-off has to be made, which is more important to you: stimulating the economy or protecting the environment?

Which worries you more: that the U.S. will NOT take the actions necessary to prevent the catastrophic effects of global warming because of fears those actions would harm the economy, or that the U.S. WILL take actions to protect against global warming and those actions will cripple the U.S. economy?

In response to the first question, 61% said stimulating the economy was more important. But only 46% of those asked the second question said they were afraid that the United States will take actions to protect against global warming and that those actions will cripple the U.S. economy. Why do you think the second wording discouraged more people from expressing more concern about the economy than about the environment?

4.17 Amending the Constitution. You are writing an opinion poll question about a proposed amendment to the Constitution. You can ask if people are in favor of "changing the Constitution" or "adding to the Constitution" by approving the amendment.

(a) Why do you think the responses to these two questions will produce different percentages in favor?

(b) One of these choices of wording will produce a much higher percentage in favor. Which one? Why?

4.18 Legal marijuana use. Recently, the issue of the legalization of marijuana has been appearing on more state ballots. In April 2018, a Quinnipiac University poll asked two questions about legal marijuana use. Here are the two questions:

Do you think that the use of marijuana should be made legal in the United States, or not?

Do you support or oppose allowing adults to legally use marijuana for medical purposes if their doctor prescribes it?

One of these questions drew 63% saying that marijuana use is okay; the other drew 93% with the same response. Which wording produced the higher percentage? Why?

4.19 Wording survey questions. Comment on each of the following as a potential sample survey question. Is the question clear? Is it slanted toward a desired response? (Survey questions on issues that one might regard as inflammatory are often prone to slanted wording.)

(a) Which of the following best represents your opinion on gun control?

   1. The government should take away our guns.

   2. We have the right to keep and bear arms.

(b) In light of skyrocketing gasoline prices, we should consider opening up a very small amount of Alaskan wilderness for oil exploration as a way of reducing our dependence on foreign oil. Do you agree or disagree?

(c) Do you think that the excessive restrictions placed on U.S. law enforcement agencies hampered their ability to detect the 9/11 terrorist plot before it occurred?

(d) Do you use drugs?

4.20 Bad survey questions. Write your own examples of bad sample survey questions.

(a) Write a biased question designed to get one answer rather than another.

(b) Write a question that is confusing, so that it is hard to answer.

4.21 Appraising a poll. In May 2018, an NBC News/SurveyMonkey poll asked about racism in American society and American politics. The question asked respondents to choose one of the following: "racism remains a major problem in our society, racism exists today but is not a major problem, racism once existed but no longer exists in our society, racism has never been a major problem in our society" The article noted 64% of Americans think racism is still a major problem in society today. News articles tend to be brief in describing sample surveys. NBC's description of this

poll explained that the poll was conducted in May 2018 and included 6,518 American adults from all adults (about 3 million!) who take SurveyMonkey polls each day. Note that this was not a probability sample. Polls of this size have a margin of error of plus or minus 1.5 percentage points.

Pages 75–76 list several "questions to ask" about an opinion poll. What answers does NBC News give to each of these questions?

4.22 Appraising a poll. In May–June 2018, an NBC News/GenForward poll asked the question "Do you personally know someone who has dealt with an opioid addiction?" The article noted 42% of millennials know someone who has dealt with opioid addiction. The description of this poll on the main article page explained that the poll was conducted in May 2018 and included 1,886 American adults between the ages of 18 and 34 who were recruited by NORC at the University of Chicago. Polls of this size and type have a margin of error of plus or minus 3.78 percentage points.

More specific information about the poll methodology was available by clicking a link. That information stated that the 1,886 respondents represented all 50 U.S. states and the District of Columbia. The respondents included 525 African Americans, 256 Asian Americas, 502 Latinx Americans, and 553 white Americans (another 50 young adults had other racial and ethnic backgrounds). Additionally, the methodology stated that the survey was conducted in either English or Spanish and through the web (93%) or telephone (7%). Also of note was that only 26% of surveys were completed.

Pages 75–76 list several "questions to ask" about an opinion poll. What answers does NBC/GenForward give to each of these questions?

4.23 Closed versus open questions. Two basic types of questions are closed questions and open questions. A closed question asks the subject for one or more of a fixed set of responses. An open question allows the subject to answer in his or her own words; the interviewer writes down the responses and classifies them later. An example of an open question is

What do you believe about the afterlife?

An example of a closed question is

What do you believe about the afterlife? Do you believe

a. *there is an afterlife and entrance depends only on your actions?*

b. *there is an afterlife and entrance depends only on your beliefs?*

c. *there is an afterlife and everyone lives there forever?*

d. *there is no afterlife?*

e. *I don't know.*

What are the advantages and disadvantages of open and closed questions?

4.24 Telling the truth? Many subjects don't give honest answers to questions about activities that are illegal or sensitive in some other way. One study divided a large group of white adults into thirds at random. All were asked if they had ever used cocaine. The first group was interviewed by telephone: 21% said "Yes." In the group visited at home by an interviewer, 25% said "Yes." The final group was interviewed at home but answered the question on an anonymous form that they sealed in an envelope. Of this group, 28% said they had used cocaine.

(a) Which result do you think is closest to the truth? Why?

(b) Give two other examples of behavior you think would be underreported in a telephone survey.

4.25 Did you vote? When the Current Population Survey asked the adults in its sample of 60,000 households if they voted in the 2016 presidential election, 56% said they had. The margin of error was less than 0.3%. In fact, only 55% of the adult population voted in that election. Why do you think the CPS result missed by 3 times the margin of error?

4.26 A party poll. At a party there are 20 students over age 21 and 40 students under age 21. You choose at random 2 of those over 21 and separately choose at random 4 of those under 21 to interview about attitudes toward alcohol. You have given every student at the party the same chance to be interviewed: what is that chance? Why is your sample not an SRS?

4.27 A stratified sample. A club has 30 student members and 10 faculty members. The students are

Aguirre	Cooper	Kemp	Peralta	Stankiewicz
Butterfield	Dobbs	Kessler	Risser	Steele
Caporuscio	Freeman	Koepnick	Rodriguez	Tong
Carlson	Girard	Macha	Ryndak	White
Chilson	Gonzales	Makis	Soria	Williams
Clement	Grebe	Palacios	Spiel	Zhang

The faculty members are

Atchade	Everson	Hansen	Nair	Romero
Craigmile	Fink	Murphy	Nguyen	Turkmen

The club can send 3 students and 2 faculty members to a convention. It decides to choose those who will go by random selection.

(a) Use the *Simple Random Sample* applet, other technology, or Table A to choose a stratified random sample of 3 students and 2 faculty members.

(b) What is the chance that the student named White is chosen? What is the chance that faculty member Romero is chosen?

4.28 A stratified sample. A state university has 4900 in-state students and 2100 out-of-state students. A financial aid officer wants to poll the opinions of a random sample of students. In order to give adequate attention to the opinions of out-of-state students, the financial aid officer decides to choose a stratified random sample of 200 in-state students and 200 out-of-state students. The officer has alphabetized lists of in-state and out-of-state students.

(a) Explain how you would assign labels and use random digits to choose the desired sample. Use the *Simple Random Sample* applet, other technology, or Table A at line 122 and give the first 5 in-state students and the first 5 out-of-state students in your sample.

(b) What is the chance that any one of the 4900 in-state students will be in your sample? What is the chance that any one of the 2100 out-of-state students will be in your sample?

4.29 Sampling by accountants. Accountants use stratified samples during audits to verify a company's records of such things as accounts receivable. The stratification is based on the dollar amount of the item and often includes 100% sampling of the largest items. One company reports 5000 accounts receivable. Of these, 100 are in amounts over $50,000; 500 are in amounts between $1000 and $50,000; and the remaining 4400 are in amounts under $1000. Using these groups as strata, you decide to verify all of the largest accounts and to sample 5% of the midsize accounts and 1% of the small accounts. How would you label the two strata from which you will sample? Use the *Simple Random Sample* applet, other technology, or Table A, starting at line 115, to select *only the first 5* accounts from each of these strata.

4.30 A sampling paradox? Example 8 compares two SRSs of a university's undergraduate and graduate students. The sample of undergraduates contains a smaller fraction of the population, 1 out of 90, versus 1 out of 15 for graduate students. Yet sampling 1 out of 90 undergraduates gives a smaller margin of error than sampling 1 out of 15 graduate students. Explain to someone who knows no Statistics why this happens.

4.31 Appraising a poll. Exercise 4.22 gives part of the description of a sample survey from the NBC News|GenForward. It appears that the sample was taken in several stages. Why can we say this? The first stage no doubt used a stratified sample, though the NBC survey does not say this. Explain why it would be bad practice to use an SRS from all possible telephone numbers and impossible to use an SRS of all possible Internet users rather than a stratified sample of telephone and Internet users.

4.32 Multistage sampling. An article in the journal *Science* looks at differences in attitudes toward genetically modified foods between Europe and the United States. This calls for sample surveys. The European survey chose a sample of 1000 adults in each of 17 European countries. Here's part of the description: "The Eurobarometer survey is a multistage, random-probability face-to-face sample survey."

(a) What does "multistage" mean?

(b) You can see that the first stage was stratified. What were the strata?

(c) What does "random-probability sample" mean?

4.33 Online courses in high schools? What do adults believe about requiring online courses in high schools? Are opinions different in urban, suburban, and rural areas? To find out, researchers wanted to ask adults this question:

It has become common for education courses after high school to be taken online. In your opinion, should public high schools in your community require every student to take at least one course online while in high school?

Because most people live in heavily populated urban and suburban areas, an SRS might contain few rural adults. Moreover, it is too expensive to choose people at random from a large region. We should start by choosing school districts rather than people. Describe a suitable sample design for this study, and explain the reasoning behind your choice of design.

4.34 Systematic random samples. The last stage of the Current Population Survey (Example 7) uses a **systematic random sample.** The following example will illustrate the idea of a systematic sample. Suppose that we must choose 4 rooms out of the 100 rooms in a dormitory. Because $100/4 = 25$, we can think of the list of 100 rooms as four lists of 25 rooms each. Choose 1 of the first 25 rooms at random, using Table A. The sample will contain this room and the rooms

25, 50, and 75 places down the list from it. If 13 is chosen, for example, then the systematic random sample consists of the rooms numbered 13, 38, 63, and 88. Use Table A to choose a systematic random sample of 5 rooms from a list of 200. Enter the table at line 120.

4.35 Systematic isn't simple. Exercise 4.34 describes a systematic random sample. Like an SRS, a systematic sample gives all individuals the same chance to be chosen. Explain why this is true, then explain carefully why a systematic sample is nonetheless *not* an SRS.

4.36 Planning a survey of students. The student government plans to ask a random sample of students their opinions about on-campus parking. The university provides a list of the 20,000 enrolled students to serve as a sampling frame.

(a) How would you choose an SRS of 200 students?

(b) How would you choose a systematic sample of 200 students? (See Exercise 4.34 to learn about systematic samples.)

(c) The list shows whether students live on campus (8000 students) or off campus (12,000 students). How would you choose a stratified sample of 50 on-campus students and 150 off-campus students?

4.37 Sampling students. You want to investigate the attitudes of students at your school toward the school's policy on extra fees for lab courses. You have a grant that will pay the costs of contacting about 500 students.

(a) Specify the exact population for your study. For example, will you include part-time students?

(b) Describe your sample design. For example, will you use a stratified sample with student majors as strata?

(c) Briefly discuss the practical difficulties that you anticipate. For example, how will you contact the students in your sample?

4.38 Mall interviews. Example 1 in Chapter 2 (page 22) describes mall interviewing. This is an example of a convenience sample. Why do mall interviews not produce probability samples?

4.39 Minorities and police. Here are three questions from an April 2018 Quinnipiac University Poll that deal with similar issues, along with the poll results:

Do you think the police in the United States are generally tougher on whites than on blacks, tougher on blacks than on whites, or do the police treat them both the same? Result: 1% tougher on whites, 48% tougher on blacks, 44% treat them the same.

When faced with a possible criminal situation, do you think police in the United States are more likely to shoot someone who is black, more likely to shoot someone who is white, or do you think police are equally likely to shoot someone of either race? website Result: 40% blacks, 2% whites, 53% either race

Is being the victim of police brutality something you personally worry about, or not? Result who worry about: 9% Republicans, 31% Democrats, 21% Independents; 12% whites, 64% blacks, 37% Hispanics

Using this example, discuss the difficulty of using responses to opinion polls to understand public opinion and come up with an overall conclusion.

Exploring the Web

Access these exercises on the text website: macmillanlearning.com/scc10e.

What's the Verdict?

The following "What's the verdict?" story explores unexpected results from a well-known mid-1990s survey and from a follow-up survey, as well as subsequent attempts by researchers to understand these results. The material in this chapter and Chapter 3 will help you appreciate issues discussed in these research papers.

In a well-known National Longitudinal Study of Adolescent Health in the mid-1990s, teenagers responded to survey questions about their health, including their sexual identity and behavior. When the results of this study were analyzed, researchers were surprised that 5 to 7% of teens were reporting that they identified as homosexual or bisexual, which was an increase from the 1% researchers had previously estimated.

Questions

WTV4.1. Researchers recognized that questions about sexuality are sensitive and thus worded their question in the study in terms of both-sex or same-sex romantic attraction. In general, what could be some challenges in asking teenagers about their sexual identity and behavior?

WTV4.2. How would you get a nationally representative sample of U.S. teenagers?

WTV4.3. How would you collect data about sexual behavior and identity to get the most accurate answers?

WTV4.4. Does your answer to the question above change if you want to follow up with these same teenagers as they become adults?

In follow-up surveys, researchers noticed that when these same teenagers were surveyed again as adults, over 70% of those teens (mostly males) who were identified as homosexual or bisexual replied that they were only heterosexual as adults. The researchers wondered why there would be such a large decrease. This would be the opposite of what other sexual development research had shown because people are more (not less) likely to be open about their non heterosexual identity as they get older.

A 2014 research paper (see Notes and Data Sources) discussed this issue. The authors of this paper proposed three explanations for the apparent inconsistency. First, perhaps gay adolescents went into the closet as young adults. Second, perhaps the teenagers were confused about the use of the phrase "romantic attraction" as a substitute for sexual orientation. Third, perhaps some of the teenagers "played a 'jokester' role by reporting same-sex attraction when none was present." The authors of the paper believed that the first explanation was very unlikely but did not dismiss the other two explanations as possible.

WTV4.5. What type of error is it when the survey question is worded in a confusing way?

The researchers did look at some of the other health questions that the same teenagers (and later adults) were asked in the surveys, and they found something else interesting: *"Most of the kids who first claimed to have artificial limbs [as teenagers] miraculously regrew arms and legs when researchers came back to interview them [as adults]."*

WTV4.6. Based on this additional piece of evidence, what is a possible explanation for the sudden decrease in the number of people who identify as homosexual or bisexual from teenagers to adulthood in this survey?

What's the verdict? Response errors can undermine the results of otherwise well-designed surveys.

Experiments, Good and Bad

5

Case Study: Online vs. Traditional Courses

In this chapter you will:

- Apply the language of experiments.
- Apply the logic of experiments.
- Determine the strengths and weaknesses of studies.

Colleges and universities offer many courses online. Are online courses better, or at least no worse, than traditional in-class courses? If so, replacing the traditional format of teaching college courses with an online format may save money for colleges and therefore provide justification for transitioning more courses to the on-line format.

A study at a Carnegie research extension university compared learning online with learning in a traditional course. Students enrolled in one of two sections of the same course. One section was taught online, and the other in a traditional classroom. Due to an error in the printing of the course schedule by the office of scheduling, neither section was defined as an online class. The students were all surprised in the online section when they found out it was an online course.

Both sections of the course were taught by the same instructor, had exactly the same reading assignments, and the same examinations. Powerpoint slides used in the traditional class, together with accompanying audio files of the lectures, were required materials for the online class. The researcher used the scores on the first hour exam in the class as a measure of a student's academic ability, amount of academic effort, and amount of time spent studying.

The score on a standardized final exam was used as a measure of student learning. Adjustments were made to account for differing student abilities, as measured by the first hour exam, as well as for the effects of sex. Is this a good study?

Talking about Experiments

Observational studies passively collect data. We observe, record, or measure, but we don't interfere. Experiments actively produce data. Experimenters intentionally intervene by imposing some treatment in order to see what happens. All experiments and many observational studies are interested in the effect that one variable has on another variable. Here is the vocabulary we use to distinguish the variable that acts from the variable that is acted upon.

You will often see explanatory variables called *independent variables* and response variables called *dependent variables*. The idea is that the response variables depend on the explanatory variables. We avoid using these older terms, partly because "independent" has other and very different meanings in statistics.

Example 1   *Learning on the Web*

An optimistic account of learning online reports a study at Nova Southeastern University, Fort Lauderdale, Florida. The authors of the study claim that students taking undergraduate courses online were "equal in learning" to students taking the same courses in class.

College students are the *subjects* in this study. The *explanatory variable* considered in the study is the setting for learning (in class or online). The *response variable* is a student's score on a test at the end of the course. Other variables were also measured in the study, including the score on a test on the course material before the courses started. Although this was not used as an explanatory variable in the study, prior knowledge of the course material might affect the response, and the authors wished to make sure this was not the case.

Example 2   *The effects of a sexual assault resistance program*

Young women attending universities are at risk of being sexually assaulted, primarily by male acquaintances. In an attempt to develop an effective strategy to reduce this risk, three universities in Canada investigated the effectiveness of a sexual assault resistance program. The program consists of four 3-hour units in which information is provided and skills are taught and practiced, with the goal of being able to assess risk from acquaintances, overcome emotional barriers in acknowledging danger, and engage in effective verbal and physical self-defense.

Frances Roberts/Alamy Stock Photo

First-year female students were randomly assigned to either the program or to attending a meeting providing access to brochures on sexual assault (as was common university practice). The result was that the sexual assault resistance program significantly reduced rapes as reported during one year of follow-up.

The Canadian study is an experiment in which the *subjects* are the 893 first-year female students. The experiment compares two *treatments*. The *explanatory variable* is the treatment a student received. Several *response variables* were measured. The primary one was completed rape as reported by participants after a one-year follow-up period.

How to Experiment Badly

Do students who take a course via the Web learn as well as those who take the same course in a traditional classroom? The best way to find out is to assign some students to the classroom and others to the Web. That's an experiment. The Nova Southeastern University study was not an experiment because it imposed no treatment on the student subjects. Students chose for themselves whether to enroll in a classroom or an online version of a course. The study simply measured their learning. It turns out that the students who chose the online course were very different from those students who chose the traditional course. For example, their average score on a test on the course material given before the courses started

Key Terms

A **response variable** is a variable that measures an outcome or result of a study.

An **explanatory variable** is a variable that we think explains or causes changes in the response variable.

The individuals studied in an experiment are often called **subjects.**

A **treatment** is any specific experimental condition applied to the subjects. If an experiment has several explanatory variables, a treatment is a combination of specific values of these variables.

Key Terms

A **lurking variable** is a variable that has an important effect on the relationship among the variables in a study but is not one of the explanatory variables studied.

Two variables are **confounded** when their effects on a response variable cannot be distinguished from each other. The confounded variables may be either explanatory variables or lurking variables.

was 40.70, against only 27.64 for the classroom students. It's hard to compare in-class versus online learning when the online students have a big head start. The effect of online versus in-class instruction is hopelessly mixed up with influences lurking in the background. Figure 5.1 (sometimes referred to as a causal diagram) shows the mixed-up influences in picture form.

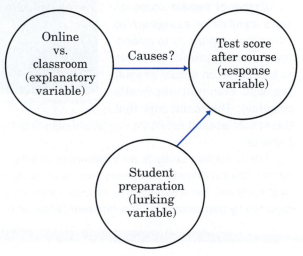

Figure 5.1 Confounding in the Nova Southeastern University study. The influence of course setting (the explanatory variable) cannot be distinguished from the influence of student preparation (a lurking variable).

Statistics in Your World

Causal inference Randomized comparative experiments are designed to allow us to draw cause-and-effect conclusions. However, many causal questions cannot be answered by randomized comparative experiments. Can one say anything about causality in these cases? Professor Judea Pearl at UCLA has a lot to say about the question of causality. He is a pioneer in the science of causal inference. His book, *The Book of Why: The New Science of Cause and Effect* is a very accessible discussion of the question of causality.

In the Nova Southeastern University study, student preparation (a lurking variable) is confounded with the explanatory variable. The study report claims that the two groups did equally well on the final test. We can't say how much of the online group's performance is due to their head start. That a group that started with a big advantage did no better than the more poorly prepared classroom students is not very impressive evidence of the wonders of web-based instruction. Here is another example, one in which a second experiment was proposed to untangle the confounding.

Both observational studies and one-track experiments often yield useless data because of confounding

Example 3 *Migraine*

Migraine is a prevalent disease characterized by headaches that are often severe and throbbing and accompanied by associated symptoms, such as nausea, vomiting, vertigo, and cognitive dysfunction. A drug, fremanezumab, may be an effective preventative treatment for migraine.

Experiments that study the effectiveness of medical treatments on actual patients are called **clinical trials.** A possible clinical trial would be to give migraine sufferers fremanezumab and see if the number of migraine days in a 12-week period is reduced. This would be a "one-track" design—that is, only a single treatment is applied:

Impose treatment → **Measure response**

Fremanezumab → **Reduced number of migraine days?**

If the patients do suffer fewer migraine days, we can't say that the fremanezumab caused the reduced symptoms. It might be just the **placebo effect.** A **placebo** is a dummy treatment with no active ingredients. Many patients respond favorably to *any* treatment, even a placebo. This response to a dummy treatment is the placebo effect. Perhaps the placebo effect is in our minds, based on trust in the doctor and expectations of a cure. Perhaps it is just a name for the fact that many patients improve for no visible reason. To determine if the results could be explained by the placebo effect, we would need to give a separate group of migraine sufferers a placebo and compare the response to the placebo to the response to fremanezumab. Unfortunately, the one-track design of the experiment meant that the placebo effect was confounded with any effect fremanezumab might have.

The researchers recognized this and used a better-designed experiment. Such an experiment might involve randomly dividing subjects with migraine into two groups. One group would be treated with fremanezumab as before. The other would receive a placebo. Subjects in both groups would not know which treatment they were receiving. Nor would the physicians recording the symptoms of the subjects know which treatment a subject received so that their diagnosis would not be influenced by such knowledge. An experiment in which neither subjects nor physicians recording the symptoms know which treatment was received is called **double-blind.**

with lurking variables. It is hard to avoid confounding when only observation is possible. Experiments offer better possibilities, as the migraine experiment shows. This experiment could be designed to include a group of subjects who receive only a placebo. This would allow us to see whether the treatment being tested does better than a placebo and so has more than the placebo effect going for it. Effective medical treatments pass the "placebo test" by outperforming a placebo.

Randomized Comparative Experiments

The first goal in designing an experiment is to ensure that it will show us the effect of the explanatory variables on the response variables. Confounding often prevents one-track experiments from doing this. The remedy is to *compare* two or more treatments. When confounding variables affect all subjects equally, any systematic differences in the responses of subjects receiving different treatments can be attributed to the treatments rather than to the confounding variables. This is the idea behind the use of a placebo. All subjects are exposed to the placebo effect because all receive some treatment. Here is an example of a new medical treatment that passes the placebo test in a direct comparison.

Example 4   *Sickle-cell anemia*

Sickle-cell anemia is an inherited disorder of the red blood cells that in the United States affects mostly blacks. It can cause severe pain and many complications. The National Institutes of Health carried out a clinical trial of the drug hydroxyurea for treatment of sickle-cell anemia. The subjects were 299 adult patients who had had at least three episodes of pain from sickle-cell anemia in the previous year. An episode of pain was defined to be a visit to a medical facility due to acute sickling-related pain that lasted more than four hours. The measurement of the length of the visit included all time spent after registration at the medical facility, including the time spent waiting to see a physician.

Simply giving hydroxyurea to all 299 subjects would confound the effect of the medication with the placebo effect and other lurking variables such as the effect of knowing that you are a subject in an experiment. Instead, approximately half of the subjects received hydroxyurea, and the other half received a placebo that looked and tasted the same. All subjects were treated exactly the same (same schedule of medical checkups, for example) except for the content of the medicine they took. Lurking variables, therefore, affected both groups equally and should not have caused any differences between their average responses.

The two groups of subjects must be similar in all respects before they start taking the medication. Just as in sampling, the best way to avoid bias in choosing which subjects get hydroxyurea is to allow impersonal chance to make the choice. A simple random sample of 152 of the subjects formed the hydroxyurea group; the remaining 147 subjects made up the placebo group. Figure 5.2 outlines the experimental design.

Victor Josan/Shutterstock

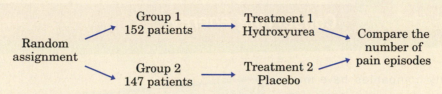

Figure 5.2 The design of a randomized comparative experiment to compare hydroxyurea with a placebo for treating sickle-cell anemia, Example 4.

The experiment was stopped ahead of schedule because the hydroxyurea group had many fewer pain episodes than the placebo group. This was compelling evidence that hydroxyurea is an effective treatment for sickle-cell anemia, good news for those who suffer from this serious illness.

Figure 5.2 illustrates the simplest **randomized comparative experiment,** one that compares just two treatments. The diagram outlines the essential information about the design: random assignment to groups, one group for each treatment, the number of subjects in each group (it is generally best to keep the groups similar in size), what treatment each group gets, and the response variable we compare. Random assignment of subjects to groups can use any of the techniques discussed in Chapter 2. For example, we could choose a simple random sample, labeling the 299 subjects 1 to 299, then using software to select the 152 subjects for Group 1. The remaining 147 subjects form Group 2. Lacking software, label the 299 subjects 001 to 299 and read three-digit groups from the table of random digits (Table A) until you have chosen the 152 subjects for Group 1. The remaining 147 subjects form Group 2.

The placebo group in Example 4 is called a **control group** because comparing the treatment and control groups allows us to control the effects of lurking variables. A control group need not receive a dummy treatment such as a placebo. In Example 2, the students who were randomly assigned to the session providing access to brochures on sexual assault (as was common university practice) were considered to be a control group. Clinical trials often compare a new treatment for a medical condition—not with a placebo, but with a treatment that is already on the market. Patients who are randomly assigned to the existing treatment form the control group. To compare more than two treatments, we can randomly assign the available experimental subjects to as many groups as there are treatments. Here is an example with three groups.

Example 5

Conserving energy

Many utility companies have introduced programs to encourage energy conservation among their customers. An electric company considers placing electronic meters in households to show what the cost would be if the electricity use at that moment continued for a month. Will meters reduce electricity use? Would cheaper methods work almost as well? The company decides to design an experiment.

One cheaper approach is to give customers an app and information about using the app to monitor their electricity use. The experiment compares these two approaches (meter,

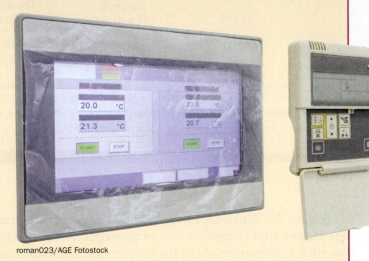

roman023/AGE Fotostock

app) and also a control. The control group of customers receives information about energy conservation but no help in monitoring electricity use. The response variable is total electricity used in a year. The company finds 60 single-family residences in the same city willing to participate, so it assigns 20 residences at random to each of the three treatments. Figure 5.3 outlines the design.

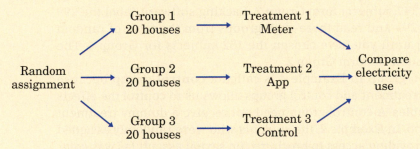

Figure 5.3 The design of a randomized comparative experiment to compare three programs to reduce electricity use by households, Example 5.

To carry out the random assignment, label the 60 households 01 to 60; then use software to select an SRS of 20 to receive the meters. From those not selected, use software to select the 20 to receive the app. The remaining 20 form the control group. Lacking software, label the 60 households 01 to 60. Enter Table A to select an SRS of 20 to receive the meters. Continue in Table A, selecting 20 more to receive the app. The remaining 20 form the control group.

Now it's your turn

5.1 Vitamin supplements. Do multivitamin supplements improve health? To answer this question researchers recruited 2000 adults. All were provided with supplies of capsules and were asked to take one capsule per day. One thousand of the adults received capsules that were multivitamin supplements and one thousand received capsules that were a placebo. The researchers used an extensive questionnaire to assess the health of all participants at the start of the study and after two years into the study. Outline the design of this study using a diagram like Figures 5.2 and 5.3.

 You now have enough information to read the "What's the verdict?" story on page 108, and to answer the first four questions.

WTV

The Logic of Experimental Design

The randomized comparative experiment is one of the most important ideas in statistics. It is designed to allow us to draw cause-and-effect conclusions. Be sure you understand the logic:

- Randomization produces groups of subjects that should be similar, on average, in all respects before we apply the treatments.

- Comparative design exposes all groups to similar conditions, other than the treatments they receive. This ensures that any additional lurking variables operate equally on all groups and, on average, groups differ only in the treatments they receive.

- Therefore, differences in the response variable must be due to the effects of the treatments.

We use chance to choose the groups in order to eliminate any systematic bias in assigning the subjects to groups. In the sickle-cell study, for example, a doctor might subconsciously assign the most seriously ill patients to the hydroxyurea group, hoping that the untested drug will help them. That would bias the experiment against hydroxyurea. Choosing an SRS of the subjects to be Group 1 gives everyone the same chance to be in either group. We expect the two groups to be similar in all respects—age, seriousness of illness, smoker or not, and so on. Chance tends to assign equal numbers of smokers to both groups, for example, even if we don't know which subjects are smokers.

What about the effects of lurking variables not addressed by randomization—for example, those that arise after subjects have been randomly assigned to groups? The placebo effect is such a lurking variable. Its effect occurs only after the treatments are administered to subjects. If the groups are treated at different times of the year, so that some groups are treated

Principles of experimental design

The basic principles of statistical design of experiments are:

1. **Control** the effects of lurking variables on the response by ensuring all subjects are affected similarly by these lurking variables. Then simply compare two or more treatments.
2. **Randomize**—use impersonal chance to assign subjects to treatments so treatment groups are similar, on average.
3. **Use enough subjects** in each group to reduce chance variation in the results.

during flu season and others not, higher exposure of some groups to the flu could be a lurking variable. In a comparative design, we try to ensure that these lurking variables operate similarly on all groups. All groups receive some treatment in order to ensure they are equally exposed to the placebo effect. All groups receive treatment at the same time, so all experience the same exposure to the flu.

It may not surprise you to learn that medical researchers adopted randomized comparative experiments only slowly—many doctors think they can tell "just by watching" whether a new therapy helps their patients. Not so. There are many examples of medical treatments that became popular on the basis of one-track experiments and were shown to be worth no more than a placebo when some skeptic tried a randomized comparative experiment. One search of the medical literature looked for therapies studied both by proper comparative trials and by trials with "historical controls." A study with historical controls compares the results of a new treatment, not with a control group, but with how well similar patients have done in the past. Of the 56 therapies studied, 44 came out winners with respect to historical controls. But only 10 passed the placebo test in proper randomized comparative experiments. Expert judgment is too optimistic even when aided by comparison with past patients. At present, U.S. law requires that new drugs be shown to be both safe and effective by randomized comparative trials. There is no such requirement for other medical treatments, such as surgery. A search of the Internet of "comparisons with historical controls" found recent studies for other medical treatments that have used historical controls.

There is one important caution about randomized experiments. Like random samples, they are subject to the laws of chance. Just as an SRS of voters might, by bad luck, choose people nearly all of whom have the same political party preference, a random assignment of subjects might, by bad luck, put nearly all the smokers in one group. We know that if we choose *large* random samples, it is very likely that the sample will match the population well. In the same way, if we use *many* experimental subjects, it is very likely that random assignment will produce groups that match closely. More subjects means that there is less chance variation among the treatment groups and less chance variation in the outcomes of the experiment. "Use enough subjects" joins "compare two or more treatments" and "randomize" as a basic principle of statistical design of experiments.

Statistical Significance

The presence of chance variation requires us to look more closely at the logic of randomized comparative experiments. We cannot say that *any* difference in the average number of pain episodes between the hydroxyurea group and the control group must be due to the effect of the drug. Even if both treatments are the same, there will always be some chance differences between the individuals in the control group and those in the treatment group. Randomization eliminates just the systematic differences between the groups.

The difference between the average number of pain episodes for subjects in the hydroxyurea group and the average for the control group was "highly statistically significant." That means that a difference of this size would almost never happen just by chance. We do indeed have strong evidence that hydroxyurea beats a placebo in helping sickle-cell disease sufferers. You will often see the phrase "statistically significant" in reports of investigations in many fields of study. It tells you that the investigators found good "statistical" evidence for the effect they were seeking.

Of course, the actual results of an experiment are more important than the seal of approval given by statistical significance. The treatment group in the sickle-cell experiment had an average of 2.5 pain episodes per year as opposed to 4.5 per year in the control group. That's a big enough difference to be important to people with the disease. A difference of 2.5 versus 2.8 would be much less interesting even if it were statistically significant.

How large an observed effect must be in order to be regarded as statistically significant depends on the number of subjects involved. A relatively small effect—one that might not be regarded as practically important—can be statistically significant if the size of the study is large. Thus, in the sickle-cell experiment, an average of 2.50 pain episodes per year versus 2.51 per year in the control group could be statistically significant if the number of subjects involved is sufficiently large. For a very large number of subjects, the average number of pain episodes per year should be almost the same if differences are due only to chance. It is also true that a very large effect may not be statistically significant. If the number of subjects in an experiment is small, it may be possible to observe large effects simply by chance. We will discuss these issues more fully in Parts III and IV.

Thus, in assessing statistical significance, it is helpful to know the magnitude of the observed effect and the number of subjects. Perhaps a better term than "statistically significant" might be "statistically dissimilar."

How to Live with Observational Studies

Do children who are bullied suffer depression as adults? Do doctors discriminate against women in treating heart disease? Are e-cigarettes safe? These are cause-and-effect questions, so we reach for our favorite tool, the randomized comparative experiment. Sorry. We refuse to require children to be bullied. We can't use random digits to assign heart disease patients to be men or women. We are reluctant to require people to use e-cigarettes because they may have harmful effects.

The best data we have about these and many other cause-and-effect questions come from observational studies. We know that observation is a weak second-best to experiment, but good observational studies are far from worthless and we will discuss this further in Chapter 15. What makes a good observational study?

First, good studies are **comparative** even when they are not experiments. We compare random samples of people who were bullied as children with those who were not bullied. We compare how doctors treat men and women patients. We might compare drivers talking on cell phones with the *same* drivers when they are not on the phone. We can often combine comparison with **matching** in creating a control group. To see the effects of taking a painkiller during pregnancy, we compare women who did so with women who did not. From a large pool of women who did not take the drug, we select individuals who match the drug group in age, education, number of children, and other lurking variables. We now have two groups that are similar in all these ways, so that these lurking variables should not affect our comparison of the groups. However, if other important lurking variables, not measurable or not thought of, are present, they will affect the comparison, and confounding will still be present.

Matching does not entirely eliminate confounding. People who were bullied as children may have characteristics that increase susceptibility to victimization as well as independently increasing the risk of depression. They are more likely to be female, have had concurrent emotional or mental health problems as a child, have parents who suffer from depression, or have experienced maltreatment at home as a child. Although matching can reduce some of these differences, direct comparison of rates of depression in young adults who were bullied as children and in young adults who were not bullied as children would still confound any effect of bullying with the effects of mental health issues in childhood, mental health issues of the parents, and maltreatment as a child. A good comparative study **measures and adjusts for confounding variables.** If we measure sex, the presence of mental health issues as a child, the presence of mental health issues in the parents, and aspects of the home environment, there are statistical techniques that reduce the effects of these variables on rates of depression so that (we hope) only the effect of bullying itself remains.

Example 6   *Bullying and depression*

A recent study in the United Kingdom examined data on 3898 participants in a large observational study for which they had information on both victimization by peers at age 13 and the presence of depression at age 18. The researchers also had information on many other variables, not just the explanatory variable (bullying at age 13) and the response variable (presence of depression at age 18). The research article said:

> Compared with children who were not victimized those who were frequently victimized by peers had over a twofold increase in the odds of depression . . . This association was slightly reduced when adjusting for confounders . . .

That "adjusting for confounders" means that the final results were adjusted for differences between the two groups. Adjustment reduced the association between bullying at age 13 and depression at age 18, but still left a nearly twofold increase in the odds of depression.

We note that the researchers do go on to mention that the use of observational data does not allow them to conclude the associations are causal.

Example 7 — Sex bias in treating heart disease?

Doctors are less likely to give aggressive treatment to women with symptoms of heart disease than to men with similar symptoms. Is this because doctors are sexist? Not necessarily. Women tend to develop heart problems much later than men so that female heart patients are older and often have other health problems. That might explain why doctors proceed more cautiously in treating them.

This is a case for a comparative study with statistical adjustments for the effects of confounding variables. There have been several such studies, and they produce conflicting results. Some show, in the words of one doctor, "When men and women are otherwise the same and the only difference is gender, you find that treatments are very similar." Other studies find that women are undertreated even after adjusting for differences between the female and male subjects.

As Example 7 suggests, statistical adjustment is complicated. Randomization creates groups that are similar in *all* variables known and unknown. Matching and adjustment, on the other hand, can't work with variables the researchers didn't think to measure. Even if you believe that the researchers thought of everything, you should be a bit skeptical about statistical adjustment. There's lots of room for cheating in deciding which variables to adjust for. And the "adjusted" conclusion is really something like this:

> If female heart disease patients were younger and healthier than they really are, and if male patients were older and less healthy than they really are, then the two groups would get the same medical care.

This may be the best we can get, and we should thank statistics for making such wisdom possible. But we end up longing for the clarity of a good experiment.

Watch the Cheerios commercial mentioned in the "What's the verdict?" story on page 108, and answer the remaining questions.

Chapter 5: Statistics in Summary

- Statistical studies often try to show that changing one variable (the **explanatory variable**) causes changes in another variable (the **response variable**).

- In an **experiment,** we actually set the explanatory variables ourselves rather than just observe them.

- Observational studies and one-track experiments that simply apply a single treatment often fail to produce useful data because **confounding** with **lurking variables** makes it impossible to say what the effect of the treatment was.

- In a **randomized comparative experiment** we compare two or more treatments, use chance to decide which subjects get each treatment, and use enough subjects so that the effects of chance are small.

- Comparing two or more treatments **controls** lurking variables affecting all subjects, such as the **placebo effect,** because they act on all the treatment groups.

- Differences among the effects of the treatments so large that they would rarely happen just by chance are called **statistically significant.**

- Observational studies of cause-and-effect questions are more impressive if they **compare matched groups** and measure as many lurking variables as possible to allow **statistical adjustment.**

This chapter summary will help you evaluate the Case Study.

Link It

In Chapter 1 we saw that experiments are best suited for drawing conclusions about whether a treatment causes a change in a response. In this chapter, we learned that only well-designed experiments, in particular randomized comparative experiments, provide a sound basis for such conclusions. Statistically significant differences among the effects of treatments are the best available evidence that changing the explanatory variable really *causes* changes in the response.

When it is not possible to do an experiment, observational studies that measure as many lurking variables as possible and make statistical adjustments for their effects are sometimes used to answer cause-and-effect questions. However, they remain a weak second-best to well-designed experiments.

Case Study Evaluated

Use what you have learned in this chapter to evaluate the Case Study that opened the chapter. Start by reviewing the Chapter Summary. Then answer each of the following questions in complete sentences. Be sure to communicate clearly enough for any of your classmates to understand what you are saying.

First, here are the results of the study. Controlling for differences in student's abilities and sex, there was no statistically significant difference between exam scores in the two sections.

1. Is this study an experiment or an observational study?

2. What confounding variables did the researcher adjust for?

3. Explain what the phrase "no statistically significant difference" means.

4. The researcher stated that the error in scheduling resulted in a research design approaching that of a randomized experiment. What do you think the researcher meant by this statement? Do you agree? Explain your answer.

In this chapter you:

- Applied the language of experiments, by the use of terms such as explanatory and response variable, subjects, and treatments.

- Applied the logic of experiments through the use of randomized comparative studies and statistical significance to design and evaluate experiments.

- Determined the strengths and weaknesses of studies by assessing whether a study has controlled for lurking variables.

macmillan learning **Online Resources**

- The Snapshots video, *Types of Studies*, and the StatClips video, *Types of Studies*, both review the differences between experiments and observational studies.

- The Snapshots video, *Introduction to Statistics*, describes real-world situations for which knowledge of statistical ideas are important.

- The StatBoards video, *Factors and Treatments*, identifies subjects, factors, treatments, and response variables in additional experiments.

- The StatBoards video, *Outlining an Experiment*, provides additional examples of outlining an experiment using figures similar to those given in this chapter.

Check the Basics

For Exercise 5.1, see page 95.

5.2 Explanatory and response variables. Does church attendance lengthen people's lives? One study of the effect of attendance at religious services gathered data from 2001 obituaries. The researchers measured whether the obituary mentioned religious activities and length of life. Which of the following is true?

(a) In this study, length of life is the explanatory variable and mention of religious activities is the response variable.

(b) In this study, mention of religious activities is the explanatory variable and length of life is the response variable.

(c) In this study, the 2001 obituaries are the explanatory variable and the information in the obituary is the response variable.

(d) In this study, there are no explanatory and response variables because these data come from a survey of obituaries.

5.3 Observational study or experiment? The study described in Exercise 5.2 is

(a) a randomized comparative experiment.

(b) an experiment, but not a randomized experiment.

(c) an observational study.

(d) neither an experiment nor an observational study but, instead, a sample survey.

5.4 Lurking variables. People who are active in religious activities are less likely to smoke or drink excessively than people who are not active in religious activities. In the study described in Exercise 5.2,

(a) smoking is a lurking variable, but excessive drinking is not.

(b) excessive drinking is a lurking variable, but smoking is not.

(c) smoking and excessive drinking are both lurking variables.

(d) neither smoking nor excessive drinking are lurking variables.

5.5 Statistical significance. In the study described in Exercise 5.2, researchers found that there was a statistically significant difference in longevity between those whose obituary mentioned religious activities and those whose obituary did not. Those whose obituary mentioned religious activities lived more than 5 years longer. Statistical significance here means

(a) the size of the observed difference in longevity is not likely to be due to chance.

(b) the size of the observed difference in longevity is likely to be due to chance.

(c) the size of the observed difference in longevity has a 5 times greater chance of occurring.

(d) the size of the observed difference in longevity has a one-fifth chance of occurring.

5.6 Randomized comparative experiment? For which of the following studies would it be possible to conduct a randomized comparative experiment?

(a) A study to determine if the month you were born in affects how long you will live.

(b) A study to determine if taking Tylenol dulls your emotions.

(c) A study to determine if a person's sex affects their salary.

(d) A study to determine if the wealth of parents affects the wealth of their children.

Chapter 5 Exercises

5.7 Digital media and ADHD. Researchers identified 2587 10th grade students in the Los Angeles area who did not have significant symptoms of ADHD (Attention-Deficit/Hyperactivity Disorder). These students were followed for approximately two years. The frequency with which each student used digital media over the two-year period was recorded. At the end of the two-year period, students were again tested for symptoms of ADHD. Researchers found that the higher the frequency of digital media use, the more likely the student was to have developed symptoms of ADHD at the end of the study.

(a) What are the explanatory and response variables?

(b) Explain carefully why this study is not an experiment.

(c) Explain why confounding prevents us from concluding that the more one uses digital media, the more likely one is to develop ADHD.

5.8 Birth month and health. A *Columbus Dispatch* article reported that researchers at the Columbia University Department of Medicine examined records for an incredible 1.75 million patients born between 1900 and 2000 who had been treated at Columbia University Medical Center. Using statistical analysis, the researchers found that for cardiovascular disease, those born in the fall (September through December) were more protected, while those born in winter and spring (January to June) had higher risk. And because so many lives are cut short due to cardiovascular diseases, being born in the autumn was actually associated with living longer than being born in the spring. Is this conclusion the result of an experiment? Why or why not? What are the explanatory and response variables?

5.9 Unhappy marriage, unhappy gut. An article in *Newsweek* reported that, to investigate how an unhappy marriage can affect an individual's health, scientists recruited 43 healthy couples between 24 and 61 years old who had been married for at least three years. The researchers asked couples to discuss touchy topics likely to spark disagreement, such as money or in-laws, and taped the conversations. They used this footage to analyze verbal and nonverbal modes of conflict, including eye rolling. The team also took blood samples from the couples before and after arguing, and found those who were most hostile toward their spouses had higher levels of LPS-binding protein, a biomarker for a leaky gut. Scientists found the highest levels of LPS-binding protein in participants who had the nastiest fights and a history of mood disorders such as depression. The biomarker was also linked to inflammation in the body. Couples choose to argue and engage in hostile behavior when discussing touchy subjects. And anger and unhappiness that can lead to fighting may be symptoms of a physiological or mental health problem. Explain why these facts make any conclusion about cause and effect untrustworthy. Use the language of lurking variables and confounding in your explanation, and draw a picture like Figure 5.1 to illustrate your explanation.

5.10 Is obesity contagious? A study closely followed a large social network of 12,067 people for 32 years, from 1971 until 2003. The researchers found that when a person gains weight, close friends tend to gain weight, too. The researchers reported that obesity can spread from person to person, much like a virus.

Explain why, when a person gains weight, close friends also tend to gain weight does not necessarily mean that weight gain in a person causes weight gains in close friends. In particular, identify some lurking variables whose effect on weight gain may be confounded with the effect of weight gains in close friends. Draw a picture like Figure 5.1 to illustrate your explanation.

5.11 Aspirin and heart attacks. Can aspirin help prevent heart attacks? The Physicians' Health

Study, a large medical experiment involving 22,000 male physicians, attempted to answer this question. One group of about 11,000 physicians took an aspirin every second day, while the rest took a placebo. After several years, the study found that subjects in the aspirin group had significantly fewer heart attacks than subjects in the placebo group.

(a) Identify the experimental subjects, the explanatory variable and the values it can take, and the response variable.

(b) Use a diagram to outline the design of the Physicians' Health Study. (When you outline the design of an experiment, be sure to indicate the size of the treatment groups and the response variable. The diagrams in Figures 5.2 and 5.3 are models.)

(c) What do you think the term "significantly" means in "significantly fewer heart attacks"?

5.12 The pen is mightier than the keyboard. Is longhand note-taking more effective for learning than taking notes on a laptop? Researchers at two universities studied this issue. In one of the studies, 65 students listened to five talks. Students were randomly assigned either a laptop or a notebook for purposes of taking notes. Assume that 33 students were assigned to use laptops and 32 longhand. Whether taking notes on a laptop or by hand in a notebook, students were instructed to use their normal note-taking strategy. Thirty minutes after the lectures, participants were tested with conceptual application questions based on the lectures. Those taking notes by hand performed better than those taking notes on a laptop. Why is instructing students to use their normal note-taking strategy a problem if the goal is to determine the effect on learning of note-taking on a laptop as compared to note-taking by hand?

5.13 Neighborhood's effect on grades. To study the effect of neighborhood on academic performance, 1000 families were given federal housing vouchers to move out of their low-income neighborhoods. No improvement in the academic performance of the children in the families was found one year after the move.

Explain clearly why the lack of improvement in academic performance after one year does not necessarily mean that neighborhood does not affect academic performance. In particular, identify some lurking variables whose effect on academic performance may be confounded with the effect of neighborhood. Use a picture like Figure 5.1 to illustrate your explanation.

5.14 The pen is mightier than the keyboard, continued.

(a) Outline the design of Exercise 5.12 for the experiment to compare the two treatments (laptop note-taking and longhand note-taking) that students received for taking notes. When you outline the design of an experiment, be sure to indicate the size of the treatment groups and the response variable. The diagrams in Figures 5.2 and 5.3 are models.

(b) If you have access to statistical software, use it to carry out the randomization required by your design. Otherwise, use Table A, beginning at line 119, to do the randomization your design requires.

5.15 Learning on the Web. The discussion following Example 1 notes that the Nova Southeastern University study does not tell us much about Web versus classroom learning because the students who chose the Web version were much better prepared. Describe the design of an experiment to get better information.

5.16 Do antioxidants prevent cancer? People who eat lots of fruits and vegetables have lower rates of colon cancer than those who eat little of these foods. Fruits and vegetables are rich in "antioxidants" such as vitamins A, C, and E. Will taking antioxidants help prevent colon cancer? A clinical trial studied this question with 864 people who were at risk for colon cancer. The subjects were divided into four groups: daily beta-carotene, daily vitamins C and E, all three vitamins every day, and daily placebo. After four years, the researchers were surprised to find no significant difference in colon cancer among the groups.

(a) What are the explanatory and response variables in this experiment?

(b) Outline the design of the experiment. (The diagrams in Figures 5.2 and 5.3 are models.)

(c) Assign labels to the 864 subjects. If you have access to statistical software, use it to choose the *first five* subjects for the beta-carotene group. Otherwise, use Table A, starting at line 118, to choose the *first five* subjects for the beta-carotene group.

(d) What does "no significant difference" mean in describing the outcome of the study?

(e) Suggest some lurking variables that could explain why people who eat lots of fruits and vegetables have lower rates of colon cancer. The results of the experiment suggest that these variables, rather than the antioxidants, may be responsible for the observed benefits of fruits and vegetables.

5.17 Conserving energy. Example 5 describes an experiment to learn whether providing households with electronic meters or with an app will reduce their electricity consumption. An executive of the electric company objects to including a control group. He says, "It would be cheaper to just compare electricity use last year [before the meter or app was provided] with consumption in the same period this year. If households use less electricity this year, the meter or app must be working." Explain clearly why this design is inferior to that in Example 5.

5.18 The safest level of drinking is none. The news site, *Vox*, reported on a study in the journal *Lancet*. Researchers looked at 700 studies from around the world, involving millions of people, and concluded that "the safest level of drinking is none." The study found that the more people drank across the globe, the greater their risk of cancer rose. In their paper, the researchers stated, "Alcohol use is a leading risk factor for global disease burden and causes substantial health loss," and that "the level of consumption that minimizes health loss is zero."

However, the article in *Vox* goes on to say that "the data in the paper do not support a zero drinks recommendation." Why do you think *Vox* makes this statement?

5.19 Sounds big. Does a lower pitch of a voice in an ad lead consumers to envision a bigger product? To test this, researchers had students listen to a radio advertisement for the new Southwest Turkey Club Sandwich at a fictitious sandwich chain, Cosmo. Half the students were randomly assigned to hear the ad spoken at a high pitch and the other half at a low pitch. In all other respects, the ads were identical and no clues were given as to the size of the sandwich. After hearing the ad, students were asked to rate the perceived size of the sandwich on a 7 point scale, ranging from –3 (much smaller than average) to +3 (much larger than average).

(a) What is the explanatory variable?

(b) What is the response variable, and what values does it take?

(c) Could the researchers have used a placebo in this experiment? Explain.

5.20 Reducing health care spending. Will people spend less on health care if their health insurance requires them to pay some part of the cost themselves? An experiment on this issue asked if the percentage of medical costs that is paid by health insurance has an effect both on the amount of medical care that people use and on their health. The treatments were four insurance plans. Each plan paid all medical costs above a ceiling. Below the ceiling, the plans paid 100%, 75%, 50%, or 0% of costs incurred.

(a) Outline the design of a randomized comparative experiment suitable for this study.

(b) Briefly describe the practical and ethical difficulties that might arise in such an experiment.

5.21 Sounds big. Consider again the voice pitch experiment of Exercise 5.19.

(a) Use a diagram to describe a randomized comparative experimental design for this experiment.

(b) Assume there were 20 subjects used in the experiment. Use software or Table A, starting at line 120, to do the randomization required by your design.

5.22 Treating drunk drivers. Once a person has been convicted of drunk driving, one purpose of court-mandated treatment or punishment is to prevent future offenses of the same kind. Suggest three different treatments that a court might require. Then outline the design of an experiment to compare their effectiveness. Be sure to specify the response variables you will measure.

5.23 Statistical significance. Exercise 5.19 describes a randomized comparative experiment to determine the effect of the pitch of a voice in a radio advertisement for a new sandwich on the perceived size of the sandwich. The researchers concluded that there was a statistically significant effect of pitch on perceived size, with those who heard the lower pitched voice perceiving the sandwich to be larger than those who heard the higher pitched voice. Explain what "statistically significant" means in the context of this experiment, as if you were speaking to a person who knows no statistics.

5.24 Statistical significance. A study, mandated by Congress when it passed No Child Left Behind in 2002, evaluated 15 reading and math software products used by 9424 students in 132 schools across the country during the 2004–2005 school year. It is the largest study that has compared students who received the technology with those who did not, as measured by their scores on standardized tests. There were no statistically significant differences between students who used software and those who did not. Explain the meaning of "no statistically significant differences" in plain language.

5.25 Let them eat chocolate. There is some evidence that cocoa has beneficial effects on heart health. To study this, researchers decide to give subjects either a cocoa pill or a placebo daily for a two-year period. Measurements of the subjects' heart health before and after the two-year period are to be compared. You have 50 people who are willing to serve as subjects.

(a) Outline an appropriate design for the experiment.

(b) The names of the subjects appear below. If you have access to statistical software, use it to carry out the randomization required by your design. Otherwise, use Table A, beginning at line 131, to do the randomization required by your design. List the subjects you will assign to the group who will do 30 minutes of daily exercise.

Aaron	Gehrig	Koufax	Paige	Terry
Alexander	Gibson	Lajoie	Palmer	Tinker
Banks	Greenberg	Lemon	Robinson	Traynor
Berra	Herman	Lombardi	Ruffing	Vance
Campanella	Hornsby	Mantle	Ruth	Wagner
Cobb	Hubbell	Mays	Seaver	Waner
Dean	Jackson	Mize	Sisler	Williams
Duffy	Johnson	Musial	Snider	Wynn
Feller	Kaline	Newhouser	Spahn	Yastrzemski
Foxx	Kiner	Ott	Speaker	Young

5.26 Treating prostate disease. A large study used records from Canada's national health care system to compare the effectiveness of two ways to treat prostate disease. The two treatments are traditional surgery and a new method that does not require surgery. The records described many patients whose doctors had chosen one or the other method. The study found that patients treated by the new method were significantly more likely to die within eight years.

(a) Further study of the data showed that this conclusion was wrong. The extra deaths among patients treated with the new method could be explained by lurking variables. What lurking variables might be confounded with a doctor's choice of surgical or nonsurgical treatment? For example, why might a doctor avoid assigning a patient to surgery?

(b) You have 300 prostate patients who are willing to serve as subjects in an experiment to compare the two methods. Use a diagram to outline the design of a randomized comparative experiment.

5.27 Prayer and meditation. You read in a magazine that "nonphysical treatments such as meditation and prayer have been shown to be effective in controlled scientific studies for such ailments as high blood pressure, insomnia, ulcers, and asthma." Explain in simple language what the article means by "controlled scientific studies" and why such studies might show that meditation and prayer are effective treatments for some medical problems.

5.28 Exercise and bone loss. Does regular exercise reduce bone loss in postmenopausal women? Here are two ways to study this question. Which design will produce more trustworthy data? Explain why.

1. A researcher finds 1000 postmenopausal women who exercise regularly. She matches each with a similar postmenopausal woman who does not exercise regularly, and she follows both groups for five years.

2. Another researcher finds 2000 postmenopausal women who are willing to participate in a study. She assigns 1000 of the women to a regular program of supervised exercise. The other 1000 continue their usual habits. The researcher follows both groups for five years.

5.29 Traumatic brain injuries and suicide. Traumatic brain injuries (TBI) can have serious long-term consequences, including psychiatric disorders. To determine if there is a relation between TBI and the risk of suicide, researchers examined the medical records of 7,418,391 individuals living in Denmark from 1980 to 2014. The researchers found that suicide rates were statistically significantly higher in those individuals who had medical contact for TBI compared to those without TBI. However, the medical records did not contain information about TBI suffered prior to 1977, nor did the records indicate what treatment patients with TBI received. The sample size in this study is very large, so is this good evidence that people with TBI are at greater risk for suicide than those without TBI? Explain your answer.

5.30 Randomization at work. To demonstrate how randomization reduces confounding, consider the following situation. A nutrition experimenter intends to compare the weight gain of prematurely born infants fed Diet A with those fed Diet B. To do this, she will feed each diet to 10 prematurely born infants whose parents have enrolled them in the study. She has available 10 baby girls and 10 baby boys. The researcher is concerned that baby boys may respond more favorably to the diets, so if all the baby boys were fed diet A, the experiment would be biased in favor of Diet A.

(a) Label the infants 00, 01, . . . , 19. Use Table A to assign 10 infants to Diet A. Or, if you have access to statistical software, use it to assign 10 infants to Diet A. Do this four times, using different parts of the table (or different runs of your software), and write down the four groups assigned to Diet A.

(b) The infants labeled 10, 11, 12, 13, 14, 15, 16, 17, 18, and 19 are the 10 baby boys. How many of these infants were in each of the four Diet A groups that you generated? What was the average number of baby boys assigned to Diet A? What does this suggest about the effect of randomization on reducing confounding?

Exploring the Web

Access these exercises on the text website: macmillanlearning.com/scc10e.

What's the Verdict?

The *American Journal of Clinical Nutrition* published a research study to investigate the effect of a diet rich in whole grains on metabolism. Fifty adults with metabolic syndrome were randomly divided into two groups. Both groups had a diet of reduced calories, but for one of the groups, all of their grains were whole grains (brown rice, whole wheat bread, etc.) and the other group got all their grains from refined grains (white bread, white rice, etc.). Both groups lost weight at the end of 12 weeks. On average, those in the refined-grain group lost 11 pounds while those in the whole-grain group lost 8 pounds but lost more body fat from their abdomens and had other health benefits than those in the refined grain group.

Questions

WTV5.1. Is this an experiment or observational study?

WTV5.2. What are the treatments?

WTV5.3. What are the response variables?

WTV5.4. Draw a diagram like Figures 5.2 and 5.3 to illustrate the design of this study.

Watch this commercial for Cheerios and pay close attention to the wording that they use with their claims: **https://www.youtube.com/watch?v=_49QCb2-258**

WTV5.5. What is the exact wording of the claim in the Cheerios commercial?

(a) People who eat Cheerios tend to weigh less.

(b) People who eat whole grains tend to weigh less.

(c) People who choose whole grains tend to weigh less.

Let's investigate what this claim might mean. Could this be a true statement? What is the evidence? What possible lurking variables might be involved?

WTV5.6. What could "choose" mean? Indicate all that apply.

(a) Picking whole grains on a survey about which foods are healthiest.

(b) Picking whole grains on a survey about which foods they are most likely to eat.

(c) Picking whole grains in a focus group about healthy foods.

(d) Reporting what foods are in their pantry at home.

(e) Purchasing whole-grain foods at the grocery store.

WTV5.7. What could "tend to weigh less" mean? Indicate all that apply.

(a) The fiber in whole grains build muscle.

(b) The fiber in whole grains help you to lose weight.

(c) Whole grains stunt your growth.

(d) Whole grains are more likely to be selected by moms, and women, who on average, weigh less than men.

(e) Eating whole grains are the only thing that could make you lose weight.

(f) People who chose whole grains are probably also doing other healthy things like eating more fruits and vegetables and getting regular exercise.

WTV5.8. What are some possible lurking variables? (Lurking variables that could be involved in the relationship between whole grains and weighing less that weren't listed in the original claim.) Indicate all that apply.

(a) Other things people are eating besides whole grains.

(b) Exercise.

(c) Genetics.

(d) How much somebody likes Cheerios.

(e) Whether somebody eats their Cheerios at breakfast or later in the day.

WTV5.9. Do you think that the study by the *American Journal of Clinical Nutrition* backs up the claim made in the Cheerios ad? Why or why not?

(a) Yes, because the whole-grain group lost more weight than the refined-grain group.

(b) No, because the whole-grain group lost less weight than the refined-grain group.

(c) No, because this study didn't use Cheerios.

(d) Yes, because abdominal fat is the most important kind to reduce.

Experiments in the Real World

Case Study: Caffeine Dependence

In this chapter you will:

- Identify problems that can arise in experiments in the real world.

- Evaluate the strengths and weaknesses of experimental methods that have been developed to address some of these problems.

Is caffeine dependence real? Researchers at the Johns Hopkins University School of Medicine wanted to determine if some individuals develop a serious addiction called caffeine dependence syndrome. Eleven volunteers were recruited who were diagnosed as caffeine dependent. For a two-day period these volunteers were given a capsule that either contained their daily amount of caffeine or a fake (nonactive) substance. Over another two-day period, at least one week after the first, the contents of the capsules received were switched. Whether the subjects first received the capsule containing caffeine or the capsule with the fake substance was determined by randomization. The subjects' diets were restricted during the study periods. All products with caffeine were prohibited, but to divert the subjects' attention from caffeine, products containing ingredients such as artificial sweeteners were also prohibited. Questionnaires assessing depression, mood, and the presence of certain physical symptoms were administered at the end of each two-day period. The subjects also completed a tapping task in which they were instructed to press a button 200 times as fast as they could. Finally, subjects were interviewed by a researcher, who did not know what was in the capsules the subjects had taken, to find other evidence of functional impairment. The EESEE story "Is Caffeine Dependence Real? which can be found on the text website," contains more information about this study. Is this a good study?

Equal Treatment for All

Probability samples are a big idea, but sampling in practice has difficulties that just using random samples doesn't solve. Randomized comparative experiments are also a big idea, but they don't solve all the difficulties of experimenting. A sampler must know exactly what information she wants and must compose questions that extract that information from her sample. An experimenter must know exactly what treatments and responses he wants information about, and he must construct the apparatus needed to apply the treatments and measure the responses. This is what psychologists, medical researchers or engineers mean when they talk about "designing an experiment." We are concerned with the *statistical* side of designing experiments, ideas that apply to experiments in psychology, medicine, engineering, and other areas as well. Even at this general level, you should understand the practical problems that can prevent an experiment from producing useful data.

The logic of a randomized comparative experiment assumes that all the subjects are treated alike except for the treatments that the experiment is designed to compare. Any other unequal treatment can cause bias. Treating subjects exactly alike is hard to do.

Example 1   *Rats and rabbits*

Rats and rabbits that are specially bred to be uniform in their inherited characteristics are the subjects in many experiments. However, animals, like people, can be quite sensitive to how they are treated. Here are three amusing examples of how unequal treatment can create bias.

Does a new breakfast cereal provide good nutrition? To find out, compare the weight gains of young rats fed the new product and rats fed a standard diet. The rats are randomly assigned to diets and are housed in large racks of cages. It turns out that rats in upper cages grow a bit faster than rats in bottom cages. If the experimenters put rats fed the new product at the top and those fed the standard diet below, the experiment is biased in favor of the new product. Solution: randomly assign the rats to cages.

Another study looked at the effects of human affection on the cholesterol level of rabbits. All the rabbit subjects ate the same diet. Some (chosen at random) were regularly removed from their cages to have their furry heads scratched by friendly people. The rabbits who received affection had lower cholesterol. So affection for some but not other rabbits could bias an experiment in which the rabbits' cholesterol level is a response variable.

Finally, a paper in *Nature* reported that exposure of rats to male but not female experimenters produces pain inhibition. This effect could even be replicated with T-shirts worn by men and bedding material from other male mammals. Thus, the sex of an experimenter can affect apparent baseline responses in behavioral testing of rats.

Double-Blind Experiments

Placebos "work." That bare fact means that medical studies must take special care to show that a new treatment is not just a placebo. Part of equal treatment for all is to be sure that the placebo effect operates on all subjects.

Example 2

The powerful placebo

Want to help balding men keep their hair? Give them a placebo—one study found that 42% of balding men maintained or increased the amount of hair on their heads when they took a placebo. Another study told 13 people who were very sensitive to poison ivy that the stuff being rubbed on one arm was poison ivy. It was a placebo, but all 13 broke out in a rash. The stuff rubbed on the other arm really was poison ivy, but the subjects were told it was harmless—and only 2 of the 13 developed a rash.

Even more striking is the following example. Researchers randomly assigned patients with chronic lower back pain to one of two groups. One continued with treatment as usual: the drugs they were currently taking to prevent pain. The other group received their usual pain drugs as well as placebo pills in a typical prescription medicine bottle. They were told that a placebo pill contained no medication, and the bottle's label read "Placebo pills. Take 2 pills twice a day." After three weeks, both groups filled out a second questionnaire about their pain. On average, the pain medication group experienced a 9% reduction in usual pain, a 16% reduction in maximum pain, and no reduction in disability. By contrast, the placebo group averaged a 30% reduction in both usual and maximum pain, while reporting nearly the same reduction, 29%, in disability.

When the ailment is vague and psychological, like depression, some experts think that about three-quarters of the effect of the most widely used drugs is just the placebo effect. Others disagree (see Exercise 6.31). The strength of the placebo effect in medical treatments is hard to pin down because it depends on the exact environment. How enthusiastic the doctor is seems to matter a lot. But "placebos work" is a good place to start when you think about planning medical experiments.

The strength of the placebo effect is a strong argument for randomized comparative experiments. In the baldness study, 42% of the placebo group kept or increased their hair, but 86% of the men getting a new drug to fight baldness did so. The drug beats the placebo, so it has something besides the placebo effect going for it. Of course, the placebo effect is still part of the reason this and other treatments work.

Because the placebo effect is so strong, it would be foolish to tell subjects in a medical experiment whether they are receiving a new drug or a placebo. Knowing that they are getting "just a placebo" might weaken the placebo effect and bias the experiment in favor of the other treatments. It is also foolish to tell doctors and other medical personnel what treatment each subject is receiving. If they know that a subject is getting "just a placebo," they may expect less than if they know the subject is receiving a promising experimental drug. Doctors' expectations change how they interact with patients and even the way they diagnose a patient's condition. Whenever possible, experiments with human subjects should be *double-blind*.

Until the study ends and the results are in, only the study's statistician knows for sure. Reports in medical journals regularly begin with words like these, from a study of a flu vaccine given as a nose spray: "This study was a randomized, double-blind, placebo-controlled trial. Participants were enrolled from 13 sites across the continental United States between mid-September and mid-November 1997." Doctors are expected to know what "randomized," "double-blind," and "placebo-controlled" mean. Now you also know.

> **Key Terms**
>
> In a **double-blind experiment,** neither the subjects nor the people who work with them know which treatment each subject is receiving.

Refusals, Nonadherers, and Dropouts

Sample surveys suffer from nonresponse due to failure to contact some people selected for the sample and the refusal of others to participate. Experiments with human subjects suffer from similar problems.

Example 3   *Minorities in clinical trials*

Refusal to participate is a serious problem for medical experiments on treatments for major diseases such as cancer. As in the case of samples, bias can result if those who refuse are systematically different from those who cooperate.

Clinical trials are medical experiments involving human subjects. Minorities, women, people living in poverty, and the elderly have long been underrepresented in clinical trials. In many cases, they weren't asked. The law now requires representation of women and minorities, and data show that most clinical trials now have fair representation. But refusals remain a problem. Minorities, especially blacks, are more likely to refuse to participate. The government's Office of Minority Health says, "Though recent studies have shown that African Americans have increasingly positive attitudes toward cancer medical research, several studies corroborate that they are still cynical about clinical trials. A major impediment for lack of participation is a lack of trust in the medical establishment." Some remedies for lack of trust are complete and clear information about the experiment, insurance coverage for experimental treatments, participation of black researchers, and cooperation with doctors and health organizations in black communities.

Subjects who participate but don't follow the experimental treatment, called **nonadherers,** can also cause bias. AIDS patients who participate in trials of a new drug sometimes take other treatments on their own, for example. In addition, some AIDS subjects have their medication tested and drop out or add other medications if they were not assigned to the new drug. This may bias the trial against the new drug.

Experiments that continue over an extended period of time also suffer **dropouts,** subjects who begin the experiment but do not complete it. If the reasons for dropping out are unrelated to the experimental treatments, no harm is done other than reducing the number of subjects. If subjects drop out because of their reaction to one of the treatments, bias can result.

Example 4

Dropouts in a medical study

Orlistat is a drug that may help reduce obesity by preventing absorption of fat from the foods we eat. As usual, the drug was compared with a placebo in a double-blind randomized trial. Here's what happened.

The subjects were 1187 obese subjects. They were given a placebo for four weeks, and the subjects who wouldn't take a pill regularly were dropped. This addressed the problem of nonadherers. There were 892 subjects left. These subjects were randomly assigned to orlistat or a placebo, along with a weight-loss diet. After a year devoted to losing weight, 576 subjects were still participating. On average, the Orlistat group lost 3.15 kilograms (about 7 pounds) more than the placebo group. The study continued for another year, now emphasizing maintaining the weight loss from the first year. At the end of the second year, 403 subjects were left. That's only 45% of the 892 who were randomized. Orlistat again beat the placebo, reducing the weight regained by an average of 2.25 kilograms (about 5 pounds).

Can we trust the results when so many subjects dropped out? The overall dropout rates were similar in the two groups: 54% of the subjects taking Orlistat and 57% of those in the placebo group dropped out. Were dropouts related to the treatments? Placebo subjects in weight-loss experiments often drop out because they aren't losing weight. This would bias the study against Orlistat because the subjects in the placebo group at the end may be those who could lose weight just by following a diet. The researchers looked carefully at the data available for subjects who dropped out. Dropouts from both groups had lost less weight than those who stayed, but careful statistical study suggested that there was little bias. Perhaps so, but the results aren't as clean as our first look at experiments promised.

Can We Generalize?

A well-designed experiment tells us that changes in the explanatory variable cause changes in the response variable. More exactly, it tells us that this happened for specific subjects in the specific environment of this specific experiment. No doubt we had grander things in mind. We want to proclaim that our new method of teaching math does better for high school students in general or that our new drug beats a placebo for some broad class of patients. Can we generalize our conclusions from our little group of subjects to a wider population?

The first step is to be sure that our findings are *statistically significant,* that they are too strong to often occur just by chance. That's important, but it's a technical detail that the study's statistician can reassure us about. The serious threat is that the treatments, the subjects, or the environment of our experiment may not be realistic. Let's look at some examples.

Example 5
Pitch and perception

Researchers had undergraduate students, participating in exchange for course credit, listen to the audio portion of an advertisement for a laptop computer with the voice-over removed. An obscure instrumental rock song was played in the background. Half the students were randomly assigned to hear the ad with the song at a high pitch and the other half at a lower pitch. After hearing the ad, students were asked to rate the advertised laptop on how lightweight they thought it was.

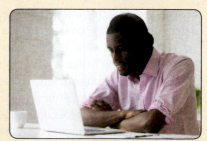
fizkes/Getty Images

Listening to an audio portion of an ad with voice-over removed in a laboratory is a long way from determining the effect of pitch on perceived size of products in actual ads. Does the behavior of the students in the lab tell us much about the behavior of all adults?

In Example 5, the subjects (students who know they are subjects in an experiment), the treatment (the pitch of an obscure rock song), and the environment (a psychology lab) are all unrealistic if the researchers' goal is to reach conclusions about the effects of pitch on perceived product size in advertisements. Psychologists do their best to devise realistic experiments for studying human behavior, but lack of realism limits the ability to generalize beyond the environment and subjects in their study, and hence the usefulness of some experiments in this area.

Example 6

The effects of day care

Should the government provide day care for low-income, preschool children? If day care helps these children stay in school and hold good jobs later in life, the government would save money by paying less welfare and collecting more taxes. Even those who are concerned only about the cost to the government might support day care programs. The Carolina Abecedarian Project (the name suggests learning the ABCs) has followed a group of children since 1972.

The Abecedarian Project is an experiment involving 111 people who in 1972 were healthy but low-income black infants in Chapel Hill, North Carolina. All the infants received nutritional supplements and help from social workers. Approximately half, chosen at random, were also placed in an intensive preschool program. The experiment compares these two treatments. Many response variables were recorded over more than 30 years, including academic test scores, college attendance, and employment.

This long and expensive experiment does show that intensive day care has substantial benefits in later life. The day care in the study was intensive indeed—lots of highly qualified staff, lots of parent participation, and detailed activities starting at a very young age, all costing about $11,000 per year for each child. It's unlikely that society will decide to offer such care to all low-income children, so the level of care in this experiment is somewhat unrealistic. The unanswered question is a big one: how good must day care be to really help children succeed in life?

Example 7

Are subjects treated too well?

Surely medical experiments are realistic? After all, the subjects are real patients in real hospitals really being treated for real illnesses.

Even here, there are some questions. Patients participating in medical trials get better medical care than most other patients, even if they are in the placebo group. Their doctors are specialists doing research on their specific ailment. They are watched more carefully than other patients. They are more likely to take their pills regularly because they are constantly reminded to do so. Providing "equal treatment for all" except for the experimental and control therapies translates into "provide the best possible medical care for all." The result: ordinary patients may not do as well as the clinical trial subjects when the new therapy comes into general use. It's likely that a therapy that beats a placebo in a clinical trial will beat it in ordinary medical care, but "cure rates" or other measures of success estimated from the trial may be optimistic.

When experiments are not fully realistic, statistical analysis of the experimental data cannot tell us how far the results will generalize. Experimenters generalizing from students in a lab to workers in the real world must argue based on their understanding of how people function, not based just on the data. It is even harder to generalize from rats in a lab to people in the real world. This is one reason why a single experiment is rarely completely convincing, despite the compelling logic of experimental design. The true scope of a new finding must usually be explored by a number of experiments in various settings.

A convincing case that an experiment is sufficiently realistic to produce useful information is based not on statistics but on the experimenter's knowledge of the subject matter of the experiment. The attention to detail required to avoid hidden bias also rests on subject-matter knowledge. Good experiments combine statistical principles with understanding of a specific field of study.

Statistics in Your World

Meta-analysis A single study of an important issue is rarely decisive. We often find several studies in different settings, with different designs, and of different quality. Can we combine their results to get an overall conclusion? That is the idea of "meta-analysis." Of course, differences among the studies prevent us from just lumping them together. Statisticians have more sophisticated ways of combining the results. Meta-analysis has been applied to issues ranging from the effect of secondhand smoke to whether coaching improves SAT scores.

Key Terms

In a **completely randomized** experimental design, all the experimental subjects are allocated at random among all the treatments.

Experimental Design in the Real World

The experimental designs we have met all have the same pattern: divide the subjects at random into as many groups as there are treatments, then apply each treatment to one of the groups. These are *completely randomized* designs.

What is more, our examples to this point have had only a single explanatory variable (for example, drug versus placebo, classroom versus Web instruction). A completely randomized design can have any number of explanatory variables. Here is an example with two.

Example 8

Can low-fat food labels lead to obesity?

What are the effects of low-fat food labels on food consumption? Do people eat more of a snack food when the food is labeled as low-fat? The answer may depend both on whether the snack food is labeled low-fat and whether the label includes serving-size information. An experiment investigated this question using university staff, graduate students, and undergraduate students at a large university as subjects. Over 10 late-afternoon sessions, all subjects viewed episodes of a 60-minute, made-for-television program in a theater on campus and were asked to rate the episodes. They were also told that because it was late in the afternoon, they would be given a cold 24-ounce bottle of water and a bag of granola from a respected campus restaurant called the Spice Box. They were told to enjoy as much or as little of it as they wanted. Each participant received 640 calories (160 grams) of granola in ziplock bags that were labeled with an attractive 3.25 × 4 inch color label. Depending on the condition randomly assigned to the subjects, the bags were labeled

Science History Images/Alamy Stock Photo

either "Regular Rocky Mountain Granola" or "Low-Fat Rocky Mountain Granola." Below this, the label indicated "Contains 1 Serving" or "Contains 2 Servings," or it provided no serving-size information. As participants left the theater, they were asked how many serving sizes they believed their package contained. Out of sight of the participants, the researchers also weighed each granola bag. Participants' statements about serving size and the actual weights of the granola bags are the response variables.

		Variable B Serving size		
		No information	1 serving	2 servings
Variable A Fat content	Regular	Treatment 1	Treatment 2	Treatment 3
	Low-fat	Treatment 4	Treatment 5	Treatment 6

Figure 6.1 The treatments in the experiment of Example 8. Combinations of two explanatory variables form six treatments.

This experiment has two explanatory variables: fat content, with 2 levels, and serving size, with 3 levels. The 6 combinations of 1 level of each variable form 6 treatments. Figure 6.1 shows the layout of the treatments.

Now it's your turn

6.1 Tweeting and TV. A large broadcast company wishes to study the effect of tweeting on TV viewing. Followers of the company are to be sent no tweet, a single tweet, or a tweet with a retweet an hour later, about a TV show. In addition, two shows are used, one a weekday show and the other a weekend show. Followers of the company are randomly assigned to one of the tweeting–TV show combinations. Subjects are to be surveyed the following week to see whether they watched the shows.

What are the explanatory variables and the response variable for this experiment?

Make a diagram like Figure 6.1 to describe the treatments. How many treatments are there?

 This might be a good place to read the "What's the verdict?" story on page 133, and to answer the questions. These questions involve material from all the preceding chapters to assess an experiment from the television series *Mythbusters*.

Experimenters often want to study the combined effects of several variables simultaneously. The interaction of several factors can produce effects that could not be predicted from looking at the effect of each factor alone. Perhaps longer commercials increase interest in a product, and more commercials also increase interest, but if we both make a commercial longer and show it more often, viewers get annoyed and their interest in the product drops. An experiment similar to that in Example 8 will help us find out.

Matched Pairs and Block Designs

Completely randomized designs are the simplest statistical designs for experiments. They illustrate clearly the principles of control and randomization. However, completely randomized designs are often inferior to more elaborate statistical designs. In particular, matching the subjects in various ways can produce more precise results than simple randomization.

One common design that combines matching with randomization is the **matched pairs design.** A matched pairs design compares just two treatments. Choose pairs of subjects that are as closely matched as possible. Assign one of the treatments to each subject in a pair by tossing a coin or reading odd and even digits from Table A. Sometimes each "pair" in a matched pairs design consists of just one subject who gets both treatments together (for example, each on a different arm or leg) or one after the other. Each subject serves as his or her own control. The *order* of the treatments can influence the subject's response, so we randomize the order for each subject, again by a coin toss.

Example 9
Testing insect repellants

Consumers Reports describes a method for comparing the effectiveness of two insect repellants. The active ingredient in one is 15% Deet. The active ingredient in the other is oil of lemon eucalyptus. Repellants are tested on several volunteers. For each volunteer, the left arm is sprayed with one of the repellants and the right arm with the other. This is a matched pairs design in which each subject compares two insect repellants. To guard against the possibility that responses may depend on which arm is sprayed, which arm receives which repellant is determined randomly. Beginning 30 minutes after applying the repellants, once every hour volunteers put each arm in separate 8 cubic foot cages containing 200 disease-free female mosquitoes in need of a blood meal to lay their eggs. Volunteers leave their arms in the cages for 5 minutes. The repellant is considered to have failed if a volunteer is bitten two or more times in a 5-minute session. The response is the number of one-hour sessions until a repellant fails.

dokurose/Shutterstock

Matched pairs designs use the principles of comparison of treatments and randomization. However, the randomization is not complete—we do not randomly assign all the subjects at once to the two treatments. Instead, we randomize only within each matched pair. This allows matching to reduce the effect of variation among the subjects. Matched pairs are an example of *block designs*.

A block design combines the idea of creating equivalent treatment groups by matching with the principle of forming treatment groups at random. Blocks are another form of *control*. They control the effects of some outside variables by bringing those variables into the experiment to form the blocks. Here are some typical examples of block designs.

Key Terms

A **block** is a group of experimental subjects that are known before the experiment to be similar in some way that is expected to affect the response to the treatments. In a **block design,** the random assignment of subjects to treatments is carried out separately within each block.

Statistics in Your World

Hawthorne effect The Hawthorne effect is a term referring to the tendency of some people to work harder and perform better when they are participants in an experiment. Individuals may change their behavior due to the attention they are receiving from researchers rather than because of any manipulation of independent variables.

The effect was first described in the 1950s by researcher Henry A. Landsberger during his analysis of experiments conducted during the 1920s and 1930s at the Hawthorne Works electric company.

The electric company had commissioned research to determine if there was a relationship between productivity and work environment.

The focus of the original studies was to determine if increasing or decreasing the amount of light workers received would have an effect on worker productivity. Employee productivity seemed to increase due to the changes, but then decreased after the experiment was over. Researchers suggested that productivity increased due to attention from the research team and not because of changes to the experimental variables. Landsberger defined the Hawthorne effect as a short-term improvement in performance caused by observing workers.

Later research into the Hawthorne effect has suggested that the original results may have been overstated. In 2009, researchers at the University of Chicago reanalyzed the original data and found that other factors also played a role in productivity and that the effect originally described was weak at best.

Example 10   *Men, women, and advertising*

Women and men respond differently to advertising. An experiment to compare the effectiveness of three television commercials for the same product will want to look separately at the reactions of men and women, as well as assess the overall response to the ads.

A completely randomized design considers all subjects, both men and women, as a single pool. The randomization assigns subjects to three treatment groups without regard to their sex. This ignores the differences between men and women. A better design considers women and men separately. Randomly assign the women to three groups, one to view each commercial. Then separately assign the men at random to three groups. Figure 6.2 outlines this improved design.

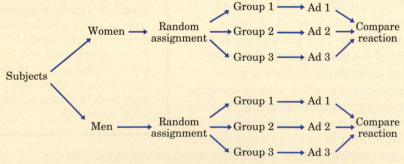

Figure 6.2 A block design to compare the effectiveness of three TV advertisements, Example 10. Female and male subjects form two blocks.

Example 11

Comparing welfare systems

A social policy experiment will assess the effect on family income of several proposed new welfare systems and compare them with the present welfare system. Because the future income of a family is strongly related to its present income, the families who agree to participate are divided into blocks of similar income levels. The families in each block are then allocated at random among the welfare systems.

A block is a group of subjects formed before an experiment starts. We reserve the word "treatment" for a condition that we impose on the subjects. We don't speak of 6 treatments in Example 10 even though we can compare the responses of 6 groups of subjects formed by the 2 blocks (men, women) and the 3 commercials. Block designs are similar to stratified samples, which we discussed in Chapter 4. Blocks and strata both group similar individuals together. We use two different names only because the idea developed separately for sampling and experiments. The advantages of block designs are the same as the advantages of stratified samples. Blocks allow us to draw separate conclusions about each block—for example, about men and women in the advertising study in Example 10. Blocking also allows more precise overall conclusions because the systematic differences between men and women can be removed when we study the overall effects of the three commercials. The idea of blocking is an important additional principle of statistical design of experiments. A wise experimenter will form blocks based on the most important unavoidable sources of variability among the experimental subjects. Randomization will then average out the effects of the remaining variation and allow an unbiased comparison of the treatments.

Now it's your turn

6.2 Car repairs. Does your sex affect the price you are quoted for car repairs? A car with a specific problem is to be taken to repair shops in a large city for a quote on how much it will cost to have the problem fixed. For some repair shops, the person bringing in the car is a woman and for others a man. The researchers thought that there might be a difference between repair shops that are part of a dealership and those that are not. So the researchers decided to block on whether or not the repair shop was part of a dealership. Use a diagram to outline a block design for this experiment. Use Figure 6.2 as a model.

Like the design of samples, the design of complex experiments is a job for experts. Now that we have seen a bit of what is involved, for the remainder of the text we will usually assume that most experiments were completely randomized.

Is It or Isn't It a Placebo?

Natural supplements are big business: creatine and amino acid supplements to enhance athletic performance; green tea extract to boost the immune system; yohimbe bark to help your sex life; grapefruit extract and apple cider vinegar to support weight loss; white kidney bean extract to block carbs. Store shelves and websites are filled with exotic substances claiming to improve your health.

Thamkc/Deposit Photos

A therapy that has not been compared with a placebo in a randomized experiment may itself be just a placebo. In the United States, the law requires that new prescription drugs and new medical devices show their safety and effectiveness in randomized trials.

What about those "natural remedies"? The law allows packagers of herbs, vitamins, and dietary supplements to claim without any evidence that they are safe and will help "natural conditions." They can't claim to treat "diseases." Of course, the boundary between natural conditions and diseases is vague. Without any evidence whatsoever, we can claim that Dr. Moore's Old Indiana Extract promotes healthy hearts. But without clinical trials and an okay by the Food and Drug Administration (FDA), we can't claim that it reduces the risk of heart disease. No doubt lots of folks will think that "promotes healthy hearts" means the same thing as "reduces the risk of heart disease" when they see our advertisements. We also don't have to worry about what dose of Old Indiana Extract our pills contain or about what dose might actually be toxic.

Should the FDA require natural remedies to meet the same standards as prescription drugs? What does your statistical training tell you about claims not backed up by well-designed experiments? What about the fact that sometimes these natural remedies have real effects? Should that be sufficient for requiring FDA approval on natural remedies?

Chapter 6: Statistics in Summary

- Because the **placebo effect** is strong, **clinical trials** and other experiments with human subjects should be **double-blind** whenever this is possible.

- The double-blind method helps achieve a basic requirement of comparative experiments: **equal treatment for all subjects** except for the actual treatments the experiment is comparing.

- The most common weakness in experiments is that we can't **generalize** the conclusions widely. Some experiments apply unrealistic treatments, some use subjects from some special group such as college students, and all are performed at some specific place and time. We want to see similar experiments at other places and times confirm important findings.

- Many experiments use designs that are more complex than the basic **completely randomized design,** which divides all the subjects among all the treatments in one randomization. **Matched pairs designs** compare two treatments by giving one to each of a pair of similar subjects or by giving both to the same subject in random order. **Block designs** form blocks of similar subjects and assign treatments at random separately in each block.

- The big ideas of **randomization, control,** and **adequate numbers of subjects** remain the keys to convincing experiments.

This chapter summary will help you evaluate the Case Study.

Link It

In Chapter 5 we learned that well-designed randomized comparative experiments provide a sound basis for determining if a treatment causes changes in a response. In the real world, simple randomized comparative experiments don't solve all the difficulties of experimenting. The placebo effect and researchers' expectations can introduce biases that undermine our conclusions. Just as samples suffer from nonresponse, experiments suffer from uncooperative subjects. Some subjects refuse to participate; others drop out before the experiment is complete; others don't follow instructions, as when some subjects in a drug trial don't take their pills. More complex designs and techniques, some of which are discussed in this chapter, are used to overcome real-world difficulties. We must pay careful attention to every aspect of an experiment to ensure that the conclusions we make are valid. And when reading about the results of experiments, you should use the ideas provided in this chapter to assess the quality of the conclusions.

Case Study Evaluated

Use what you have learned in this chapter to evaluate the Case Study that opened the chapter. Start by reviewing the Chapter Summary. You can also read the EESEE story "Is Caffeine Dependence Real?" for additional information. Then answer each of the following questions in complete sentences. Be sure to communicate clearly enough for any of your classmates to understand what you are saying.

First, here are the results of the study. The number of subjects who showed withdrawal symptoms during the period in which they took capsules that did not contain caffeine and the magnitude of their symptoms were considered statistically significant.

1. Explain what the phrase "statistically significant" means.

2. Explain why the researchers gave subjects capsules with a fake substance rather than just having them take nothing during one of the periods.

3. What advantage is gained by having subjects take both a capsule with caffeine and a capsule with a fake substance rather than having some of the subjects just take a capsule with caffeine and the remaining subjects just take a capsule with a fake substance?

In this chapter you:

- Identified problems such as the placebo effect, unequal treatment of subjects, small sample sizes, and the inability to generalize results that can arise in the real world.

- Evaluated the strengths and weaknesses of experimental methods such as completely randomized designs, double-blind experiments, matched pairs designs, and block designs that have been developed to address these problems.

macmillan learning Online Resources

■ The Snapshots Video, *Introduction to Statistics,* describes real world situations for which knowledge of statistical ideas are important.

Check the Basics

For Exercise 6.1, see page 120; for Exercis e 6.2, see page 123.

6.3 **Prayer and healing.** To study the effect of prayer on healing, patients with health problems are randomly divided into two groups. In one group, intercessors pray for the health of the patients. In the other group, patients are not prayed for. Patients do not know that they are being prayed for, and the persons who are praying do not come in contact with the patients for whom they pray. Medical outcomes in the two groups of patients are compared. Finally, the medical treatment team is also blind to the prayer group status of individual patients. This experiment is an example of

(a) a nonrandomized clinical trial.

(b) a double-blind experiment.

(c) a matched pairs design.

(d) a block design.

6.4 **Recycling.** Researchers recruited 60 undergraduate students, in exchange for course credit, for a study on the effect of recycling on how much wrapping paper subjects used to wrap a gift. The subjects were randomly assigned to one of two rooms. In one room there was a large recycling bin and in the other a large trash bin. Subjects were asked to wrap a gift. Unknown to the students, the researchers were interested in how much paper the students used. The researchers found that students in the room with the recycling bin used (statistically) significantly more paper than those in the room with a trash bin. The researchers had hypothesized that people in general would rather recycle than trash, and hence will use less disposable resources when recycling is not available. In this experiment, the amount of wrapping paper used is

(a) the response.

(b) the blocking variable.

(c) the lurking variable.

(d) the explanatory variable.

6.5 **Recycling.** Which of the following is an important weakness of the experiment described in Exercise 6.4?

(a) This was not a matched pairs design.

(b) Because undergraduate students were used as subjects, the results may not generalize to all adults and all situations involving disposable items.

(c) This was not a double-blind experiment.

(d) This experiment did not use a placebo.

6.6 Reducing smoking. The Community Intervention Trial for Smoking Cessation asked whether a community-wide advertising campaign would reduce smoking. The researchers located 11 pairs of communities, with each pair similar in location, size, economic status, and so on. One community in each pair was chosen at random to participate in the advertising campaign and the other was not. This is

(a) an observational study.

(b) a matched pairs experiment.

(c) a completely randomized experiment.

(d) a randomized block design.

6.7 Multivitamin supplements. To explore the effects of multivitamin supplements on health, you recruit 100 volunteers. Half are to take a multivitamin supplement daily. The other half are to take a placebo daily. Multivitamin supplements may have different effects on men and women because their nutritional needs may differ. Forty of the volunteers are women and 60 are men, so you separately randomly assign half the women to the multivitamin group and half the men to the multivitamin group. The remaining volunteers are assigned to the placebo group. This is an example of

(a) a completely randomized design.

(b) a matched pairs design.

(c) a block design.

(d) an observational study.

Chapter 6 Exercises

6.8 Magic mushrooms. A *Washington Post* article reported that psilocybin, the active ingredient of "magic mushrooms," promoted a mystical experience in two-thirds of people who took it for the first time, according to a study published in the online journal *Psychopharmacology*. The authors of the article stated that their "double-blind study evaluated the acute and longer-term psychological effects of a high dose of psilocybin relative to a comparison compound administered under comfortable, supportive conditions." Explain to someone who knows no statistics what the term "double-blind" means here.

6.9 Do antidepressants help? A researcher studied the effect of an antidepressant on depression. He randomly assigned subjects with moderate levels of depression to two groups. One group received the antidepressant and the other a placebo. Subjects were blinded with respect to the treatment they received. After four weeks, the researcher interviewed all subjects and rated the change in their symptoms based on the comments of subjects during the interview. Critics said that the results were suspect because the ratings were not blind. Explain what this means and how lack of blindness could bias the reported results.

6.10 Eggs and cholesterol. An article in a medical journal reports on an experiment to see the effect of eating three whole eggs per day on cholesterol levels as compared to eating the equivalent amount of a yolk-free egg substitute. The article describes the experiment as a randomized, single-blinded experiment of 37 subjects with metabolic syndrome. What do you think "single-blinded" means here? Why isn't a double-blind experiment possible?

6.11 Bright bike lights. Will requiring bicyclists to use bright, high-intensity xenon lights mounted on the front and rear of the bike reduce accidents with cars by making bikes more visible?

(a) Briefly discuss the design of an experiment to help answer this question. In particular, what response variables will you examine?

(b) Suppose your experiment demonstrates that using high-intensity xenon lights reduces the number of accidents. What concerns might you have about whether your experimental results will reduce the number of accidents with cars if all bicyclists are required to use such lights? (*Hint:* To help you answer this question, consider the following example. A 1980 report by the Highway Traffic Safety Administration found that adding a center brake light to cars reduced rear-end collisions by as much as 50%. These findings were the result of a randomized comparative experiment. As a result, center brake lights have been required on all cars sold since 1986. Ten years later, the Insurance Institute found only a 5% reduction in rear-end collisions. Apparently, when the study was originally carried out, center brake lights were unusual and caught the eye of following drivers. By 1996, center brake lights were common and no longer captured attention.)

6.12 A high-fat diet prevents obesity? A *Science News* article reported that according to a study conducted by researchers at Hebrew University of Jerusalem, a high-fat diet could reset the metabolism and prevent obesity. In the study, researchers fed a group of mice a high-fat diet on a fixed schedule (eating at the same time and for the same length of time every day) for 18 weeks. They compared these mice to three other experimental groups: one that ate a low-fat diet on a fixed schedule, one that ate an unscheduled low-fat diet (in the quantity and frequency of its choosing), and one that ate an unscheduled high-fat diet. All four groups of mice gained weight throughout the experiment. However, the mice on the scheduled high-fat diet had a lower final average body weight than the mice that ate an unscheduled low-fat diet, even though both groups consumed the same amount of calories. In addition, the mice on the scheduled high-fat diet exhibited a unique metabolic state in which the fats they ingested were not stored, but rather utilized for energy at times when no food was available, such as between meals. Does this experiment provide good evidence that a scheduled high-fat diet is beneficial for humans? Briefly discuss the questions that arise in using this experiment to decide the benefits of a scheduled high-fat diet for humans.

6.13 Blood-chilling and strokes. A *Science News* article reported a study of the effect of cooling the blood of stroke patients on the extent of recovery ninety days after the stroke. Researchers randomly assigned 58 severe-stroke patients to receive either tPA (the standard treatment for stroke) or tPA plus blood-chilling. Regulators overseeing the study required a one-hour delay from the point at which tPA was given before cooling could be started. The researchers found no significant difference in the effects of the two treatments on recovery. Researchers also noted that the recovery rate for both groups was worse than the average seen in stroke patients nationwide, but were not concerned. Why were they unconcerned?

6.14 Beating sunburn with broccoli. Some recent studies suggest that compounds in broccoli may be helpful in combating the effects of overexposure to ultraviolet radiation. Based on these studies we hope to show that a cream consisting of a broccoli extract reduces sunburn pain. To do this, we recruit 60 patients suffering from severe sunburn and needing pain relief. We will apply the cream to the sunburn of each patient and ask them an hour later, "About what percent of pain relief did you experience?"

(a) Why should we not simply apply the cream to all 60 patients and record the responses?

(b) Outline the design of an experiment to compare the cream's effectiveness with that of an over-the-counter product for sunburn relief and that of a placebo.

(c) Should patients be told which remedy they are receiving? How might this knowledge affect their reactions?

(d) If patients are not told which treatment they are receiving, but the researchers assessing the effect of the treatment know, the experiment is single-blind. Should this experiment be double-blind? Explain.

6.15 Testing a natural remedy. The statistical controversy presented in this chapter discusses issues surrounding the efficacy of natural remedies. The National Institutes of Health has now begun sponsoring proper clinical trials of some natural remedies. The following is an example of one such study. At Duke University, 330 patients with mild depression were enrolled in a trial to compare St. John's Wort with a placebo and with Zoloft, a common prescription drug for depression. One method for measuring the severity of depression is the Beck Depression Inventory that rates the severity of depression on a 0 to 3 scale.

(a) What would you use as the response variable to measure *change* in depression after treatment?

(b) Outline the design of a completely randomized clinical trial for this study.

(c) What other precautions would you take in this trial?

6.16 The placebo effect. A survey of physicians found that some doctors give a placebo to a patient who complains of pain for which the physician can find no cause. If the patient's pain improves, these doctors conclude that it had no physical basis. The medical school researchers who conducted the survey claimed that these doctors do not understand the placebo effect. Why?

6.17 The best painkiller for children. A *Washington Post* article reported a study comparing the effectiveness of three common painkillers for children. Three hundred children, aged 6 to 17, were randomly assigned to three groups. Group A received a standard dose of ibuprofen. Group B received a standard dose of acetaminophen. Group C received a standard dose of codeine. The youngsters rated their pain on a 100-point scale before and after taking the medicine.

(a) Outline the design of this experiment. You do not need to do the randomization that your design requires.

(b) You read that "the children and physicians were blinded" during the study. What does this mean?

(c) You also read that there was a significantly greater decrease in pain ratings for Group A than for Groups B and C, but there was no significant difference in the decrease of pain ratings for Groups B and C. What does this mean? What does this finding lead you to conclude about the use of ibuprofen as a painkiller?

6.18 Flu shots. A *New York Times* article reported a study that investigated whether giving flu shots to schoolchildren protects a whole community from the disease. Researchers in Canada recruited 49 remote Hutterite farming colonies in western Canada for the study. In 25 of the colonies, all children aged 3 to 15 received flu shots in late 2008; in the 24 other colonies, they received a placebo. Which colonies received flu shots and which received the placebo was determined by randomization, and the colonies did not know whether they received the flu shots or the placebo. The researchers recorded the percentage of all children and adults in each colony who had laboratory-confirmed flu over the ensuing winter and spring.

(a) Outline the design of this experiment. You do not need to do the randomization that your design requires.

(b) The placebo was actually the hepatitis A vaccine and that "hepatitis was not studied, but to keep the investigators from knowing which colonies received flu vaccine, they had to offer placebo shots, and hepatitis shots do some good while sterile water injections do not." In addition, the article mentions that the colonies were studied "without the investigators being subconsciously biased by knowing which received the placebo." Why was it important that investigators not be subconsciously biased by knowing which received the placebo?

(c) By June 2009, more than 10% of all the adults and children in colonies that received the placebo had laboratory-confirmed seasonal flu. Less than 5% of those in the colonies that received flu shots had. This difference was statistically significant. Explain to someone who knows no statistics what "statistically significant" means in this context.

6.19 Ibuprofen and atherosclerosis. The theory of atherosclerosis (hardening and narrowing of the arteries) emphasizes the role of inflammation in the vascular walls. Since ibuprofen is known to possess a wide range of anti-inflammatory actions, it was hypothesized that it might help in the prevention of atherosclerotic lesion development. Both a low cholesterol and a high cholesterol diet were used, as the extent of atherosclerosis is also affected by diet. Thirty-two New Zealand rabbits served as subjects in the experiment and, after three months, the percentage of the surface covered by atherosclerotic plaques in a region of the aorta was evaluated. Although ibuprofen did suppress the expression of a gene thought to be related to atherosclerosis, it was not shown to have an effect on the extent of fat-induced atherosclerotic lesions.

(a) What are the individuals and the response variable in this experiment?

(b) How many explanatory variables are there? How many treatments? Use a diagram like Figure 6.1 to describe the treatments.

(c) Use a diagram to describe a completely randomized design for this experiment. (Don't actually do the randomization.)

6.20 Price change and fairness. A marketing researcher wishes to study what factors affect the perceived fairness of a change in the price of an item from its advertised price. In particular, does the type of change in price (an increase or decrease) and the source of the information about the change (from a store clerk or from the price tag on the item) affect the perceived fairness? In an experiment, 20 subjects interested in purchasing a new rug are recruited. They are told that the price of a rug in a certain store was advertised at $500. Subjects are sent, one at a time, to the store, where they learn that the price has changed. Five subjects are told by a store clerk that the price has *increased* to $550. Five subjects learn that the price has *increased* to $550 from the price tag on the rug. Five subjects are told by a store clerk that the price has *decreased* to $450. Five subjects learn that the price has *decreased* to $450 from the price tag on the rug. After learning about the change in price, each subject is asked to rate the fairness of the change on a 10-point scale with 1 = "very unfair" to 10 = "very fair."

(a) What are the explanatory variables and the response variables for this experiment?

(b) Make a diagram like Figure 6.1 to describe the treatments. How many treatments are there?

(c) In the experiment, the first five subjects learn from a store clerk that the price has increased to $550, the next five learn that the price has increased to $550 from the price tag on the rug, and so on. Would it be better to determine the order in which subjects are sent to the store and which scenario they will encounter (type of change and source of information about the change) randomly? Explain your answer.

6.21 Liquid water enhancers. Bottled water, flavored and plain, is expected to become the largest segment of the liquid refreshment market by the end of the decade, surpassing traditional carbonated soft drinks. Kraft's MiO, a liquid water enhancer, comes in a variety of flavors and a few drops added to water gives a zero calorie flavored water drink. You wonder if those who drink flavored water like the taste of MiO as well as they like the taste of a competing flavored water product that comes ready to drink. Describe a matched pairs design to answer this question. Be sure to include any blinding of your subjects. What is your response variable going to be?

6.22 Athletes take oxygen. We often see players on the sidelines of a football game inhaling oxygen. Their coaches think this will speed their recovery. We might measure recovery from intense exertion as follows: Have a football player run 100 yards three times in quick succession. Then allow three minutes to rest before running 100 yards again. Time the final run. Describe the design of two experiments to investigate the effect of inhaling oxygen during the rest period. One of the experiments is to be a completely randomized design and the other a matched pairs design in which each student serves as his or her own control. Suppose you have 20 football players available as subjects. For both experiments, carry out the randomization of the 20 football players to treatments as required by the design.

6.23 Handwriting versus keyboard writing. Do people who write by hand have a better memory of what they write than those who write using a keyboard? To test this, researchers had 36 participants in a study write down a long list of words read out loud to them. They were then asked to put their list aside and try to recall as many of the words as possible. Two methods of writing down words were used. One was using a blue-ink regular ball pen and a notepad. The other was using a laptop equipped with a full-size keyboard. The number of words correctly recalled was the response.

(a) Outline a completely randomized design to learn the effect of method of writing words on number of words correctly recalled.

(b) Describe in detail the design of a matched pairs experiment, using the same 36 subjects, in which each subject serves as his or her own control.

6.24 Technology for teaching statistics. The Brigham Young University statistics department performed randomized comparative experiments to compare teaching methods. One study compared two levels of technology for large lectures: standard (overhead projectors and chalk) and multimedia. The "individuals" in the study were the eight lectures in a basic statistics course. There were four instructors, each of whom taught two lectures. Because lecturers differed, a block design was used with the lecturers forming four blocks and each lecturer using the standard technology in one lecture and the multimedia technology in the other. The average of the final exam scores of the students in the lecture was used as the response for each lecture. Suppose the lectures and lecturers were as follows.

Lecture	Lecturer	Lecture	Lecturer
1	Barney	5	Heaton
2	Page	6	Page
3	Heaton	7	Barney
4	Richardson	8	Richardson

Use a diagram to outline a block design for this experiment. Figure 6.2 is a model.

6.25 Comparing weight-loss treatments. Twenty overweight males have agreed to participate in a study of the effectiveness of supplements that claim to produce weight loss. Four treatments are used: three supplements and a placebo. The three supplements are green coffee bean extract, raspberry ketones, and glucomannan. The researcher first calculates how overweight each subject is by comparing the subject's actual weight with his "ideal" weight. The subjects and their excess weights in pounds are

Andrews	21	Johnson	25	Peters	34	Waddle	42
Brooks	34	Karras	33	Pryor	28	Warmack	33
Brown	30	Kelce	28	Schwenke	30	Weathersby	35
Cannon	25	Mailata	32	Thuney	30	Wisniewski	29
John	24	Mason	39	Vaitai	27	Wynn	35

The response variable is the weight lost after 8 weeks of treatment. Because a subject's excess weight will influence the response, a block design is appropriate.

(a) Arrange the subjects in order of increasing excess weight. Form 5 blocks of 4 subjects each by grouping the 4 least overweight, then the next 4, and so on.

(b) Use Table A (or statistical software) to randomly assign the 4 subjects in each block to the 4 weight-loss treatments. Be sure to explain exactly how you used the table.

6.26 Algal blooms. Algal blooms have become a recurring problem on many American lakes. Among other things, they can cause damage to a person's liver, kidneys, and nervous system. Phosphorus runoff from farms is one factor that contributes to algal blooms. Will inserting fertilizer into fields rather than spreading it across the surface help reduce runoff? To study this, researchers compare the effects of these two methods of fertilizing fields on the amount of phosphorus in runoff. Specific features of a field, such as slope of the ground and nature of the soil, can affect runoff, so the researchers divide four fields into two plots of equal size in such a way that the runoff from each plot can be measured separately. They use a block design with the four fields as blocks.

(a) Draw a sketch of the four fields, displaying each as a rectangle. Divide each field (rectangle) in half, each half representing one of the two plots. Label the two plots for each field as Block 1 and Block 2.

(b) Do the randomization required by the block design. That is, randomly assign the two treatments to the two plots in each block. Mark on your sketch which treatment is used in each plot.

6.27 Better sleep? Is the number of times you awaken during the night affected by whether you have a glass of wine before bed and whether you have a snack before you go to bed? Describe briefly the design of an experiment with two explanatory variables, whether or not you have a glass of wine and whether or not you have a snack before going to bed, to investigate this question. Be sure to specify what the response variable will be. Also tell how you will handle lurking variables such as amount of sleep the previous night.

6.28 Five Guys versus In-N-Out Burgers. Five Guys and In-N-Out Burgers are often rated among the top hamburger chains in the country. Do consumers prefer the taste of a hamburger from Five Guys or from In-N-Out burgers in a blind test in which neither chain is identified? Describe briefly the design of a matched pairs experiment to investigate this question.

6.29 What do you want to know? The previous two exercises illustrate the use of statistically designed experiments to answer questions that arise in everyday life, such as how to improve sleep or what hamburger tastes best. Select a question of interest to you that an experiment might answer and briefly discuss the design of an appropriate experiment.

6.30 Political polarization and social media. On September 1, 2018, *The Columbus Dispatch* reported on a study about political polarization and social media. In this study, 901 Democrats and 751 Republicans were recruited. The Democrats were randomly divided into two groups. All were asked to follow an automated Twitter account (Twitter bot) each day for one month. One group received tweets with a liberal point of view, and the other tweets with a conservative point of view. Likewise, Republicans were randomly divided into two groups and received the same two treatments (liberal or conservative tweets). All subjects were given a test, both before and after the experiment, that scored them on a liberal/conservative scale. Changes in scores were the response variable.

(a) Is the political affiliation of the subjects (Democrat or Republican) a treatment variable or a block? Why?

(b) Is the type of tweet (liberal or conservative) a treatment variable or a block? Why?

Exploring the Web

Access these exercises on the text website: macmillanlearning.com/scc10e.

What's the Verdict?

In the Discovery Channel's popular series *Mythbusters,* the hosts investigate common beliefs to determine whether there is evidence to verify them as true. In this segment, the hosts are investigating whether yawning is contagious.

https://www.discovery.com/tv-shows/mythbusters/videos/is-yawning-contagious

Questions

WTV6.1. *Mythbusters* recruited 50 people by posting an ad. What type of sampling did they use?

WTV6.2. If they also recruited by stopping people walking by the trailer at the market where they performed the research, what type of sampling would this be?

WTV6.3. The hosts randomly assigned the individuals to one of three rooms: two with a "yawn seed" (a person who yawns at them as they enter the room) planted and one without. Is this anecdotal evidence, an observational study, or an experiment? Why?

WTV6.4. Did the hosts follow the three principles of good experimental design?

WTV6.5. The hosts then recorded whether each person yawned or did not yawn. Is this a categorical or quantitative variable?

Data Ethics

Case Study: Marijuana and Driving Performance

In this chapter you will:

- Study data ethics in general.
- Learn why institutional review boards are important in academic research.
- Understand informed consent (and who can give informed consent).
- Learn about confidentiality and anonymity in research.

What is the relationship between the dose of marijuana and driving performance? How long is driving impaired after using marijuana? How can marijuana use be tested by law enforcement in the field? Researchers at the Center for Medicinal Cannabis Research at the University of California, San Diego, recruited participants between January 2017 and June 2019 to determine the answers to these questions. Participants had to be licensed drivers aged 21–55 who have 20/40 or better vision, with or without correction. Participants could not join the study if an examining physician excluded them, if they were unwilling to abstain from marijuana use for two days prior to the screening and the experimental visits, if they were pregnant, if they tested positive for a selection of drugs, and/or if they were unwilling to refrain from driving or operating heavy machinery for four hours after taking the study medication. The study medication was randomized to study participants in the following doses: placebo (0% THC), 5.9% THC, 13.4% THC.

Many adults would like to be participants in such an experiment. But is it ethical to have subjects consume either the 5.9% or 13.4% THC so that their judgment is impaired?

First Principles

The production and use of data, like all human endeavors, raise ethical questions. We won't discuss the telemarketer who begins a telephone sales pitch with "I'm conducting a survey," when the goal is to sell you something rather than collect useful information. Such deception is clearly unethical. It enrages legitimate survey organizations, which find the public less willing to talk with them. Neither will we discuss those few researchers who, in the pursuit of professional advancement, publish fake data. There is no ethical question here—faking data to advance your career is just wrong. It will end your career when uncovered. But just how honest must researchers be about real, unfaked data? Here is an example that suggests the answer is "More honest than they often are."

Example 1 *Missing details*

Papers reporting scientific research are supposed to be short, with no extra baggage. Brevity can allow the researchers to avoid complete honesty about their data. Did they choose their subjects in a biased way? Did they report data on only some of their subjects? Did they try several statistical analyses and report only the ones that supported what the researchers hoped to find? Statistician John Bailar screened more than 4000 medical papers in more than a decade as consultant to the *New England Journal of Medicine*. He wrote, "When it came to the statistical review, it was often clear that critical information was lacking, and the gaps nearly always had the practical effect of making the authors' conclusions look stronger than they should have." The situation is no doubt worse in fields that screen published work less carefully.

The most complex issues of data ethics arise when we collect data from people (but research with animals also raises ethical issues). The ethical difficulties are more severe for experiments that impose some treatment on people than for sample surveys that simply gather information. Trials of new medical treatments, for example, can do harm as well as good to their subjects. Here are some basic standards of data ethics that must be obeyed by any study that gathers data from human subjects, whether sample survey or experiment.

Basic data ethics

The organization that carries out the study must have an **institutional review board** that reviews all planned studies in advance in order to protect the subjects from possible harm.

All individuals who are subjects in a study must give their **informed consent** before data are collected.

All individual data must be kept **confidential.** Only statistical summaries for groups of subjects may be made public.

If subjects are minor children, then their *assent* (agreement to participate in the research) is needed in addition to permission from their parent(s) or guardian(s). The parental/guardian permission must meet the standards of informed consent.

Many journals have a formal requirement of explicitly addressing human subjects issues if the study is classified as human subjects research. For example, here is a statement from the instructions for authors for *JAMA* (*The Journal of the American Medical Association*):

> For all manuscripts reporting data from studies involving human participants or animals, formal review and approval, or formal review and waiver, by an appropriate institutional review board or ethics committee is required and should be described in the Methods section.

For situations where a formal ethics review committee does not exist, the principles of the Declaration of Helsinki should be followed. Researchers working with human subjects should state in the Methods section how informed consent was obtained from the study participants (e.g., oral or written) and whether a stipend was given to study participants. Journal editors may request that manuscript authors provide documentation of the formal review and recommendation from the institutional review board or ethics committee responsible for oversight of the study. Also, the law requires that studies funded by the federal government obey these principles. But neither the law nor the consensus of experts is completely clear about the details of the application of these principles.

Institutional Review Boards

The purpose of an institutional review board (often abbreviated IRB) is not to decide whether a proposed study will produce valuable information or whether it is statistically sound. The board's purpose is, in the words of one university's board, "to protect the rights and welfare of human subjects (including patients) recruited to participate in research activities." The board reviews the plan of the study and can require changes. It reviews the consent form to ensure that subjects are informed about the nature of the study and about any potential risks. Once research begins, the board may monitor progress of the study.

The most pressing issue concerning institutional review boards is whether their workload has become so large that the IRB itself is no longer effective at protecting research subjects. When the government temporarily stopped human-subject research at Duke University Medical Center in 1999 due to inadequate protection of subjects, more than 2000 studies at Duke were in progress. That's a lot of review work. There are shorter review procedures for projects that involve only minimal risks to subjects, such as most sample surveys or research involving commonly accepted educational practices. When a board is overloaded, there is a temptation to put more proposals in the minimal-risk category to speed the work.

Informed Consent

Both words in the phrase "informed consent" are important, and both can be controversial. Subjects must be *informed* in advance about the nature of a study and any risk of harm it may bring. In the case of a sample survey, physical harm is not possible. The subjects should be told what kinds of questions the survey will ask and about how much of their time it will take. Experimenters must tell subjects the nature and purpose of the study and outline possible risks. Subjects must then *consent*, usually in writing.

Example 2

Who can consent?

Are there some subjects who can't give informed consent? It was once common, for example, to test new vaccines on prison inmates who gave their consent in return for good-behavior credit. Now we worry that prisoners are not really free to refuse, and the law forbids medical experiments in prisons.

Children can't give fully informed consent, so the usual procedure is to ask them for assent and their parent(s) or guardian(s) for permission that meets the standards of informed consent. A study of new ways to teach reading is about to start at a local elementary school, so the study team sends consent forms home to parents. Many parents don't return the forms. Can their children take part in the study because the parents did not say "No," or should we allow only children whose parents returned the form and said "Yes"? Should the children whose parents did not return the form be given the placebo treatment (the standard way to teach reading) and the children whose parents returned the form and said "Yes" be taught with the new way to teach reading?

What about research into new medical treatments for people with mental health disorders? What about research into new teaching techniques for people with an intellectual disability? What about studies of new ways to help emergency room patients who may be unconscious or have suffered a stroke? In most cases, like an emergency room, there is no time even to get the consent of the family. Does the principle of informed consent prohibit realistic trials of new treatments for unconscious patients?

These are questions without clear answers. Reasonable people differ strongly on all of them. There is nothing simple about informed consent.

Now it's your turn

7.2 Informed consent? A 72-year-old man with multiple sclerosis is hospitalized. His doctor feels he may need to be placed on a feeding tube soon to ensure adequate nourishment. He asks the patient about this in the morning and the patient agrees. However, in the evening (before the tube has been placed), the patient becomes disoriented and seems confused about his decision to have the feeding tube placed. He tells the doctor he doesn't want it in. The doctor revisits the question in the morning, when the patient is again lucid. Unable to recall his state of mind from the previous evening, the patient again agrees to the procedure. Do you believe the patient has given informed consent to the procedure? Explain your reasoning.

The difficulties of informed consent do not vanish even for capable subjects. Some researchers, especially in medical trials, regard consent as a barrier to getting patients to participate in research. The researchers may not explain all possible risks; they may not point out that there are other therapies or treatments that might be better than those being studied; they may be too optimistic when talking with patients even when the consent form has all the right details. On the other hand, mentioning every possible risk leads to very long consent forms that really are barriers. Long consent forms are like the very detailed documents we are asked to read when we upgrade to a new operating system on our smartphones. One lawyer said that consent forms "are like rental car contracts." Some subjects don't read forms that run five or six printed pages. Some subjects may not be able to read and must have the consent forms read to them. However, these subjects might not want to tell the researchers that they are illiterate. Other subjects are frightened by the large number of possible (but unlikely) disasters that might happen and so refuse to participate. Of course, unlikely disasters sometimes happen. When they do, lawsuits follow, and the consent forms become yet longer and more detailed.

Confidentiality

Ethical problems do not disappear once a study has been cleared by the institutional review board and once the researchers have obtained consent from their subjects

Statistics in Your World

Statisticians, honest and dishonest Developed nations rely on government statisticians to produce honest data. We trust the monthly unemployment rate, for example, to guide both public and private decisions. Honesty can't be taken for granted, however. In 1998, the Russian government arrested top statisticians in the State Committee for Statistics, who were accused of taking bribes to fudge data to help companies avoid taxes. In 2017, Andreas Georgiou, the former top statistician of Greece, was convicted and given a suspended sentence for a "violation of duty." Did Georgiou really violate his duty? According to *The Independent*, "Georgiou's real 'crime' was, finally, starting to tell the truth about the extent of Greece's public sector borrowing in 2010."

and have actually collected data from the subjects. It is important to protect the subjects' privacy by keeping all data about individuals confidential. In other words, while the researcher knows the identity of the research subjects, the researcher keeps all identifying information private to protect any subject's identity from being discovered by others. For example, the report of an opinion poll may say what percentage of the 1500 respondents felt that legal immigration should be reduced. It may not report what *you* said about this or any other issue.

Confidentiality is not the same as **anonymity.** Anonymity means that subjects are anonymous—the subjects' names are not known to anyone involved in the study, not even to the director of the study. In other words, there are no identifying values in the study that would allow the researcher to identify a specific subject. It is not possible to determine which subject produced which data. Anonymity is rare in statistical studies. Even where anonymity is possible (mainly in surveys conducted by mail, email, or the Internet), it prevents any follow-up to improve nonresponse or inform subjects of results.

Any breach of confidentiality is a serious violation of data ethics. The best practice is to separate the identity of the subjects from the rest of the data at once. Typically, the researchers de-identify the data by separating the personally identifiable data from the data of interest to the study. The de-identified data can then be analyzed. For example, sample surveys use the identification only to check on who did or did not respond. But, in this era of advanced technology, it is no longer enough to be sure that each individual set of data protects people's privacy. The U.S. government, for example, maintains a vast amount of information about citizens in many separate databases—census responses, tax returns, Social Security information, data from surveys such as the Current Population Survey, and so on. Many of these databases can be searched by computers for statistical studies. A clever computer search of several databases might be able, by combining information, to identify you and learn a great deal about you even if your name and other identification have been removed from the data available for search. A colleague from Germany once remarked that "female full professor of statistics with a PhD from the United States" was enough to identify her among all the 83 million residents of Germany. Privacy and confidentiality of data are hot issues among statisticians in the computer age. Computer hacking and thefts of laptops containing data add to the difficulties. Is it even possible to guarantee confidentiality of data stored in databases that can be hacked or stolen? Figure 7.1 displays the Internet privacy policy that appears on the U.S. Social Security website.

Statistics in Your World

Who owns published data?
A researcher gathers data and publishes it. Who owns the data? The U.S. Supreme Court has ruled that "data" are facts and cannot be copyrighted. However, compilations of facts are generally copyrightable. So the answer to who owns data is not always clear. No permission is required for the use of published data or the creative use of a subset of data. Data from a table used to make a graphical presentation or data read from a graph can be used freely without permission. But beyond these guidelines, there is tremendous variation in determining whether permission is needed.

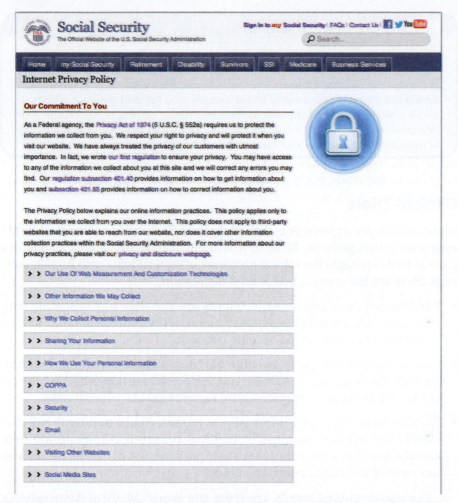

Figure 7.1 The privacy policy of the U.S. government's Social Security Administration website. (*Source*: Social Security Administration.)

Example 3

Use of government databases

Citizens are required to give information to the government. Think of tax returns and Social Security contributions, for example, in the United States. The government needs these data for administrative purposes—to see if we paid the right amount of tax and how large a Social Security benefit we are owed when we retire. Some people feel that individuals should be able to forbid any other use of their data, even with all identification removed. This would prevent using government records to study, say, the ages, incomes, and household sizes of Social Security recipients. Such a study could well be vital to debates on reforming Social Security.

Now it's your turn

7.3 Anonymous or confidential? Consider two at-home tests for human immu-nodeficiency virus (HIV). The first is a swab of your gums that gives you results in 20 minutes. Another at-home HIV test has you prick your finger to get a small amount of blood. You mail your blood sample to a lab and get your results in about a week. With both of these at-home HIV tests, nothing is reported to your insurance or placed in your medical records. Does this practice offer anonymity or confidentiality? Explain your reasoning.

Clinical Trials

Clinical trials are experiments that study the effectiveness of medical treat-ments on actual patients. Medical treatments can harm as well as heal, so clinical trials spotlight the ethical problems of experiments with human sub-jects. Here are the starting points for a discussion:

- Randomized comparative experiments are the only way to see the true effects of new treatments. Without them, risky treatments that are no better than placebos could become common.

- Clinical trials produce great benefits, but most of these benefits go to future patients. The trials also pose risks, and these risks are borne by the subjects of the trial. So we must balance future benefits against present risks.

- Both medical ethics and international human rights standards say that "While the primary purpose of medical research is to generate new knowledge, this goal can never take precedence over the rights and interests of individual research subjects."

The above quoted words are from the World Medical Association's Declaration of Helsinki (first adopted in 1964, most recently updated in 2013), which is the most respected international standard. The most outrageous examples of unethical experiments are those that ignore the interests of the subjects.

Example 4   *The Tuskegee syphilis study*

In the 1930s, syphilis was common among black men in the rural South of the United States, a group that had almost no access to medical care. The Public Health Service recruited 399 poor black sharecroppers with syphilis and 201 others without the disease in order to observe how syphilis progressed when no treatment was given. Beginning in 1943, penicillin became available to treat syphilis. However, the study subjects were not treated, even after penicillin became a standard treatment for syphilis. In fact, the Public Health Service tried to prevent any treatment until word leaked out and forced an end to the study in 1972.

The Tuskegee study is an extreme example of investigators following their own interests and ignoring the well-being of their subjects. A 1996 review said, "It has come to symbolize racism in medicine, ethical misconduct in human research, paternalism by physicians, and government abuse of vulnerable people." In 1997, President Clinton formally apologized to the surviving participants in a White House ceremony.

The Tuskegee study helps explain to many of us the lack of trust that lies behind the reluctance of many African Americans to take part in clinical trials: "From a historical perspective, the Tuskegee syphilis study is widely recognized as a reason for mistrust because of the extent and duration of deception and mistreatment and the study's impact on human subject review and approval." Unfortunately, African Americans have not only been impacted by the Tuskegee study. There have been enough studies that have taken advantage of African Americans for Harriet S. Washington to write the 528-page book *Medical Apartheid: The Dark History of Medical Experimentation on Black Americans from Colonial Times to the Present.* Washington's documentation of the mistreatment that African Americans have experienced in the name of research explains even more why many African Americans are reluctant to participate in clinical trials.

Because "the interests of the subject must always prevail," medical treatments can be tested in clinical trials only when there is reason to hope that they will help the patients who are subjects in the trials. Future benefits alone aren't enough to justify any experiment with human subjects. Of course, if there is already strong evidence that a treatment works and is safe, it is unethical *not* to give it. Dr. Charles Hennekens of the Harvard Medical School, who directed the large clinical trial that showed that aspirin reduces the risk of heart attacks in men, discussed the issue of when to do or not do a randomized trial. Here are his words:

> On the one hand, there must be sufficient belief in the agent's potential to justify exposing half the subjects to it. On the other hand, there must be sufficient doubt about its efficacy to justify withholding it from the other half of subjects who might be assigned to placebos.

Why is it ethical to give a control group of patients a placebo? Well, we know that placebos often work. Patients on placebos often show real improvement. What is more, placebos have no harmful side effects. So in the state of balanced doubt described by Dr. Hennekens, the placebo group may be getting a better treatment than the drug group. If we *knew* which treatment was better, we would give it to everyone. When we don't know, it is ethical to try both and compare them. Here are some harder questions about placebos, with arguments on both sides.

Example 5

Placebo controls?

You are testing a new drug. Is it ethical to give a placebo to a control group if an effective drug already exists?

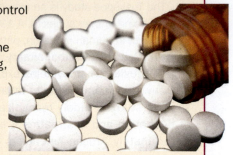

STILLFX/Shutterstock

Yes: The placebo gives a true baseline for the effectiveness of the new drug. There are three groups: new drug, best existing drug, and placebo. Every clinical trial is a bit different, and not even genuinely effective treatments work in every setting. The placebo control helps us see if the study is flawed so that even the best existing drug does not beat the placebo. Sometimes the placebo wins, so the doubt about the efficacy of the new and the existing drugs is justified. Placebo controls are ethical except for life-threatening conditions.

No: It isn't ethical to deliberately give patients an inferior treatment. We don't know whether the new drug is better than the existing drug, so it is ethical to give both in order to find out. If past trials showed that the existing drug is better than a placebo, it is no longer right to give patients a placebo. After all, the existing drug includes the placebo effect. A placebo group is ethical only if the existing drug is an older one that did not undergo proper clinical trials, or doesn't work well, or is dangerous.

Example 6

Sham surgery

"Randomized, double-blind, placebo-controlled trials are the gold standard for evaluating new interventions and are routinely used to assess new medical therapies." So says an article in the *New England Journal of Medicine* that discusses the treatment of Parkinson's disease. The article isn't about the new treatment, which offers hope of reducing the tremors and lack of control brought on by the disease, but about the ethics of studying the treatment.

The law requires well-designed experiments to show that new drugs work and are safe. Not so with surgery—only about 7% of studies of surgery use randomized comparisons. Surgeons think their operations succeed, but innovators always think their innovations work. Even if the patients are helped, the placebo effect may deserve most of the credit. So we don't really know whether many common surgeries are worth the risk they carry. To find out, do a proper experiment. That includes a "sham surgery" to serve as a placebo. In the case of Parkinson's disease, the promising treatment involves surgery to implant new cells. The placebo subjects get the same surgery, but the cells are not implanted.

Placebos work. Patients on placebos often show improvement, and their inclusion produces a better experiment. As more doctors recognize this fact, more begin to ask, "If we accept a placebo in drug trials, why don't we accept it in surgery trials?" This is a very controversial question. Here are two arguments about whether placebos should be used in surgery trials.

Yes: Most surgeries have not been tested in comparative experiments, and some are no doubt just placebos. Unlike placebo pills, these surgeries carry risks. Comparing real surgeries to placebo surgeries can eliminate thousands of unnecessary operations and save many lives. The placebo surgery can be made quite safe. For example, placebo patients can be given a safe drug that removes their memory of the operation rather than a more risky anesthetic required for the more serious real surgery. Subjects are told that they are in a placebo-controlled trial and they agree to take part. Placebo-controlled trials of surgery are ethical (except for life-threatening conditions) if the risk to the placebo group is small and there is informed consent.

No: Placebo surgery, unlike placebo drugs, always carries some risk, such as postoperative infection. Remember that "the interests of the subject must always prevail." Even great future benefits can't justify risks to subjects today unless those subjects receive some benefit. We might give a patient a placebo drug as a medical therapy because placebos work and are not risky. No doctor would do a sham surgery as ordinary therapy because there is some risk. If we would not use it in medical practice, it isn't ethical to use it in a clinical trial.

STATISTICAL CONTROVERSIES

Hope for Sale?

We have pointed to the ethical problems of experiments with human subjects, clinical trials in particular. *Not* doing proper experiments can also pose problems. Here is an example. Women with advanced breast cancer will eventually die. A promising but untried treatment appears.

The promising treatment is bone marrow transplant (BMT for short). The idea of BMT is to harvest a patient's bone marrow cells, blast the cancer with very high doses of drugs, then return the harvested cells to keep the drugs from killing the patient. BMT has become popular, but it is painful, expensive, and dangerous.

New anticancer drugs are first available through clinical trials, but there is no constraint on therapies such as BMT. When small, uncontrolled trials seemed to show success, BMT became widely available. The economics of medicine had a lot to do with this. The early leaders in offering BMT were for-profit hospitals that advertise heavily to attract patients. Others soon jumped in. The *New York Times* reported: "Every entity offering the experimental procedure tried a different sales pitch. Some promoted the prestige of their institutions, others the convenience of their locations, others their caring attitudes and patient support." The profits for hospitals and doctors are high.

Should we have waited for controlled clinical trials to show that the treatment works, or was it right to make it available immediately? What do you think? What are some of the issues one should consider?

Now it's your turn

7.4 Ethics and scientific research. The authors of a paper on clinical research and ethics stated the following:

> For a clinical research protocol to be ethical, [it] must have a clear scientific objective; be designed using accepted principles, methods, and reliable practices; have sufficient power to definitively test the objective; and offer a plausible data analysis plan. [And], it must be possible to execute the proposed study.

Do you think this rules out observational studies as "ethical"? Explain your reasoning.

Behavioral and Social Science Experiments

When we move from medicine to the behavioral and social sciences, the direct risks to experimental subjects are less acute, but so are the possible benefits to the subjects. Consider, for example, the experiments conducted by psychologists in their study of human behavior.

Example 7

Keep out of my space

Psychologists observe that people have a "personal space" and get annoyed if others come too close to them. We don't like strangers to sit at our table in a coffee shop if other tables are available, and we see people move apart in elevators if there is room to do so. Americans tend to require more personal space than people in most other cultures. Can violations of personal space have physical, as well as emotional, effects?

Olena Yakobchuk/Shutterstock

In the 1970s, investigators set up shop in a men's public restroom. They blocked off urinals to force men walking in to use either a urinal next to an experimenter (treatment group) or a urinal separated from the experimenter (control group). Another experimenter, using a periscope from a toilet stall, measured how long the subject took to start urinating and how long he kept at it.

This personal space experiment illustrates the difficulties facing those who plan and review behavioral studies.

- There is no risk of harm to the subjects, although they would certainly object to being watched through a periscope. What should we protect subjects from when physical harm is unlikely? Possible emotional harm? Undignified situations? Invasion of privacy?

- What about informed consent? The subjects in Example 7 did not even know they were participating in an experiment. Many behavioral experiments rely on hiding the true purpose of the study. The subjects would change their behavior if told in advance what the investigators were looking for. Subjects are asked to consent on the basis of vague information. They receive full information only after the experiment.

The "Ethical Principles" of the American Psychological Association require consent unless a study merely observes behavior in a public place. They allow deception only when it is necessary to the study, does not hide information that might influence a subject's willingness to participate, and is explained to subjects as soon as possible. The personal space study of Example 7 (from the 1970s) *does not* meet current ethical standards.

We see that the basic requirement for informed consent is understood differently in medicine and psychology. Here is an example of another setting with yet another interpretation of what is ethical. The subjects get no information and give no consent. They don't even know that an experiment may be sending them to jail for the night.

Example 8

Domestic violence

How should police respond to domestic violence calls? In the past, the usual practice was to remove the offender and order him to stay out of the household overnight. Police were reluctant to make arrests because the victims rarely pressed charges. Women's groups argued that arresting offenders would help prevent future violence even if no charges were filed. Is there evidence that arrest will reduce future offenses? That's a question that experiments have tried to answer.

A typical domestic violence experiment compares two treatments: arrest the suspect and hold him overnight or warn the suspect and release him. When police officers reach the scene of a domestic violence call, they calm the participants and investigate. Weapons or death threats require an arrest. If the facts permit an arrest but do not require it, an officer radios headquarters for instructions. The person on duty opens the next envelope in a file prepared in advance by a statistician. The envelopes contain the treatments in random order. The police either arrest the suspect or warn and release him, depending on the contents of the envelope. The researchers then monitor police records and visit the victim to see if the domestic violence reoccurs.

The first such experiment appeared to show that arresting domestic violence suspects does reduce their future violent behavior. As a result of this evidence, arrest has become the common police response to domestic violence.

The domestic violence experiments shed light on an important issue of public policy. Because there is no informed consent, the ethical rules that govern clinical trials and most social science studies would forbid these experiments. They were cleared by review boards because, in the words of one domestic violence researcher, "These people became subjects by committing acts that allow the police to arrest them. You don't need consent to arrest someone."

This might be a good place to read the "What's the verdict?" story on page 158, and to answer the questions. These questions ask you to consider ethical issues surrounding an experiment conducted by Facebook in 2012 involving its users.

Chapter 7: Statistics in Summary

■ Data ethics begin with some principles that go beyond just being honest. Studies with human subjects must be screened in advance by an **institutional review board.**

■ All subjects must give their **informed consent** before taking part. If the subjects are minor children, the children must *assent* before taking part in a research study, but their parent(s) or guardian(s) need to give permission that meets the standards of informed consent.

■ All information about individual subjects must be kept **confidential.**

This chapter summary will help you evaluate the Case Study.

Link It

The production and use of data to make decisions, like all human endeavors, raise ethical questions. In real-world applications of statistics, these must be addressed as part of the process of reasoning from data to a conclusion. The principles discussed in this chapter are a good start in addressing these questions, but many ethical debates remain, especially in the area of experiments with humans. Many of the debates concern the right balance between the welfare of the subjects and the future benefits of the experiment. Remember that randomized comparative experiments can answer questions that can't be answered without them. Also remember that "the interests of the subject must always prevail over the interests of science and society."

Case Study Evaluated

Use what you have learned in this chapter to evaluate the Case Study that opened the chapter. Start by reviewing the Chapter Summary. Then communicate clearly enough for any of your classmates to understand your reasoning. In particular, consider the following:

1. What does it mean to you for research to be ethical? How might that impact your answers to the following questions?

2. Based on the principles discussed in this chapter, would you consider the experiment to be ethical?

3. The Center for Medicinal Cannabis Research (CMCR) at the University of California, San Diego will have just finished recruiting study participants when this textbook is published. Consider this research study and a research study on the health benefits of dark chocolate. For which study would confidentiality be more important?

4. Federal regulations say that "minimal risk" means that the risks are no greater than "those ordinarily encountered in daily life or during the performance of routine physical or psychological examinations or tests." Do you think this study qualifies as "minimal risk"?

In this chapter you:

- Studied data ethics in general.
- Learned why institutional review boards are important in academic research.
- Understood informed consent (and who can give informed consent).
- Learned about confidentiality and anonymity in research.

macmillan learning Online Resources

- The StatBoards Video, *Informed Consent and Psychological Experimentation*, discusses a real example involving issues of informed consent.

Check the Basics

Most of the exercises in this chapter pose issues for discussion. There are no right or wrong answers, but there are more and less thoughtful answers.

For Exercise 7.1, see page 138; for Exercise 7.2, see page 139; for Exercise 7.3, see page 142; for Exercise 7.4, see page 146.

7.5 Institutional review board. The purpose of an institutional review board is

(a) to decide whether a proposed study will produce valuable information.

(b) to protect the rights of human subjects (including patients) recruited to participate in research activities.

(c) to decide whether a proposed study is statistically sound.

(d) All of the above.

7.6 Informed consent. Informed consent should include

(a) consent by the subject, usually in writing.

(b) information, in advance, about the nature of a study.

(c) information, in advance, about possible risks.

(d) All of the above.

7.7 Confidentiality? If, in a study, it is not possible to determine which subjects produced which data, we would say

(a) the subjects are anonymous.

(b) the study is confidential, but subjects are not anonymous.

(c) the study is double blind.

(d) the study is blind, but not double blind.

7.8 Clinical trials. A clinical trial is

(a) an observational study held in a controlled, clinical environment.

(b) an experiment to study the effectiveness of medical treatments on actual patients.

(c) any study performed in a medical clinic.

(d) the review, by a court, of ethical violations in medical studies.

7.9 Ethics. Which of the following would be considered unethical in an experiment?

(a) Failure to obtain informed consent from subjects.

(b) Promising confidentiality to subjects but failing to protect it.

(c) Placing the interests of science over the interests of patients.

(d) All of the above.

Chapter 7 Exercises

Most of the exercises in this chapter pose issues for discussion. There are no right or wrong answers, but there are more and less thoughtful answers.

7.10 Minimal risk? You are a member of your college's institutional review board. You must decide whether several research proposals qualify for less rigorous review because they involve only minimal risk to subjects. Federal regulations say that "minimal risk" means the risks are no greater than "those ordinarily encountered in daily life or during the performance of routine physical or psychological examinations or tests." That's vague. Which of these do you think qualifies as "minimal risk"? Explain your reasoning.

(a) Take hair and nail clippings in a non-disfiguring manner.

(b) Draw a drop of blood by pricking a finger in order to measure blood sugar.

(c) Draw blood from the arm for a full set of blood tests.

(d) Insert a tube that remains in the arm so that blood can be drawn regularly.

(e) Take extra specimens from a subject who is undergoing an invasive clinical procedure such as a bronchoscopy (a procedure in which a physician views the inside of the airways for diagnostic and therapeutic purposes using an instrument that is inserted into the airways, usually through the nose or mouth).

7.11 Who serves on the review board? Government regulations require that institutional review boards consist of at least five people, including at least one scientist, one nonscientist, and one person from outside the institution. Most boards are larger, but many contain just one outsider.

(a) Why should review boards contain at least one person who is not a scientist?

(b) Why should review boards contain at least one outsider?

(c) Do you think that one outside member is enough? How would you choose that member? (For example, would you prefer a medical doctor? A member of the clergy? An activist for patients' rights?)

7.12 Institutional review boards. If your college or university has an institutional review board that screens all studies that use human subjects, get a copy of the document that describes this board (you can probably find it online). At larger institutions you may find multiple institutional review boards—for example, separate boards for medical studies and for studies in the social sciences. If that is the case, choose the institutional review board that is of most interest to you. *Note: If you are looking for institutional review board information after January 21, 2019, you might find only one IRB, even at a large institution. This is a change that was instituted by the U.S. Department of Health and Human Services as part of the Federal Policy for the Protection of Human Subjects (also known as the "Common Rule").*

(a) How many different institutional review boards does your institution have?

(b) According to the document you selected, what are the duties of the board?

(c) How are members of the board chosen? How many members are not scientists? How many members are not employees of the institution? Do these members have some special expertise, or are they simply members of the "general public"?

7.13 Informed consent. A researcher suspects that people who are abused as children tend to be more prone to severe depression as young adults. She prepares a questionnaire that measures depression and that also asks many personal questions about childhood experiences. Write a description of the purpose of this research to be read by subjects in order to obtain their informed consent. You must balance the conflicting goals of not deceiving the subjects as to what the questionnaire will tell about them and of not biasing the sample by scaring off people with painful childhood experiences.

7.14 Is consent needed? In which of the circumstances below would you allow collecting personal information without the subjects' consent? Why?

(a) A government agency takes a random sample of income tax returns to obtain information on the marital status and average income of people who identify themselves as belonging to an ultraconservative political group. Only the marital status and income are recorded from the returns, not the names.

(b) A social psychologist attends public meetings of an ultraconservative political group to study the behavior patterns of members.

(c) A social psychologist pretends to be converted to membership in an ultraconservative political group and attends private meetings to study the behavior patterns of members.

7.15 Coercion? The U.S. Department of Health and Human Services regulations for informed consent state that "An investigator shall seek such consent only under circumstances that provide the prospective subject or the representative sufficient opportunity to consider whether or not to participate and that minimize the possibility of coercion or undue influence." Coercion occurs when an overt or implicit threat of harm is intentionally presented by one person to another in order to obtain compliance. Which of the following circumstances do you believe constitutes coercion? Explain your reasoning.

(a) An investigator tells a prospective subject that the subject will lose access to needed health services if they do not participate in the research.

(b) An employer asks employees to participate in a research study. Although the employer has assured employees that participation is voluntary, several employees are concerned that a decision to not participate could affect performance evaluations or job advancement.

7.16 Undue influence? Undue influence in obtaining informed consent often occurs through an offer of an excessive or inappropriate reward or other overture in order to obtain compliance. Which of the following circumstances do you believe constitute undue influence? Explain why you answered the way that you did using complete sentences.

(a) The patients of a physician are asked to participate in a study in which the physician is also the investigator.

(b) A professor asks a student to participate in a research study. He tells the student that everyone else in the class has agreed to participate.

(c) Research subjects are paid in exchange for their participation.

(d) A health insurance representative asks insured persons to take a battery of health and blood tests that will help inform future health insurance pricing. The health insurance representative says that people who agree to the tests will get a 10% reduction in their insurance rates for the coming year.

7.17 Students as subjects. Students taking Psychology 001 are required to serve as experimental subjects. Students in Psychology 002 are not required to serve, but they are given extra credit if they do so. Students in Psychology 003 are required either to sign up as subjects or to write a term paper. Serving as an experimental subject may be educational, but current ethical standards frown on using "dependent subjects" such as prisoners or charity medical patients. Students are certainly somewhat dependent on their teachers.

(a) Do you object to any of these course policies? If so, which one(s)? Explain your reasoning.

(b) The University of Virginia's Institutional Review Board for Health Sciences Research in its information about subject recruitment states "Investigators proposing to recruit their students, employees or patients as research subjects should justify in the

protocol the necessity for the inclusion of the dependent subject. In addition, the IRB will closely scrutinize the precautions in place to prevent the appearance of coercion in the recruitment of subjects." As a student, explain why you think a policy about recruitment of students as subjects is important.

7.18 How common is HIV infection? Researchers from Yale University, working with medical teams in Tanzania, wanted to know how common infection with the AIDS virus is among pregnant women in that African country. To do this, they planned to test blood samples drawn from pregnant women.

Yale's institutional review board insisted that the researchers get the informed consent of each woman and tell her the results of the test. This is the usual procedure in developed nations. The Tanzanian government did not want to tell the women why blood was drawn or tell them the test results. The government feared panic if many people turned out to have an incurable disease for which the country's medical system could not provide care. The study was canceled. Do you think that Yale was right to apply its usual standards for protecting subjects? Explain your answer.

7.19 Anonymous or confidential? One of the most important nongovernment surveys in the United States is the General Social Survey (see Example 7 in Chapter 1). The GSS regularly monitors public opinion on a wide variety of political and social issues. Interviews are conducted in person in the subject's home. Are a subject's responses to GSS questions anonymous, confidential, or both? Explain your answer. You may wish to visit the GSS website at http://gss.norc.org/.

7.20 Anonymous or confidential? The website for STDHELP.org contains the following information about one method offered for HIV testing: "The clinic will require you to provide some information that allows them to deliver your results. Typically a random numeric code is used for identification, and your name or social security number are never used in the process. There are no written results that are documented . . ."

Does this practice offer anonymity or just confidentiality? Explain your reasoning.

7.21 Anonymous or confidential? A website is looking for volunteers for a research study involving methicillin-resistant *Staphylococcus aureus* (MRSA), a bacterial infection that is highly resistant to some antibiotics. The website contains the following information about the study: "The Alliance for the Prudent Use of Antibiotics is looking for individuals who have or have had MRSA to fill out an anonymous survey and provide suggestions on how to improve treatment. The survey will help us to find out more about the concerns of people affected by MRSA . . ." Following the announcement is a web link that takes you to the questionnaire. Does this study really provide anonymity or just confidentiality? Explain your reasoning.

7.22 Sunshine laws. All U.S. states have open records laws, sometimes known as "Sunshine Laws," that give citizens access to government meetings and records. This includes, for example, reports of crimes and recordings of 911 calls. Crime reports will include the names of anyone accused of the crime. Suppose a 10-year-old juvenile is accused of committing a crime. A reporter from the local newspaper asks for a copy of the crime report. The sheriff refuses to provide the report because the accused is a juvenile and the sheriff believes the name of the accused should be confidential. Is this an issue of confidentiality? Explain your reasoning.

7.23 https. Generally, secure websites use encryption and authentication standards to protect the confidentiality of web transactions. The most commonly used protocol for web security has been TLS, or Transport Layer Security. This technology is still commonly referred to as SSL (Secure Sockets Layer). According to Symantec Corporation, "HTTPS (Hyper Text Transfer Protocol Secure) appears in the URL when a website is secured by an SSL certificate."

(a) Do you believe that https websites provide true confidentially? Do you think it is possible to guarantee the confidentiality of data on any website? Explain your reasoning.

(b) As of July 2018, Google's Chrome browser began showing "http" websites as "Not secure." ("https" websites do not carry this warning.) In what ways do you think this move by Google is ethical? Unethical? Explain your reasoning.

7.24 Not really anonymous. Some common practices may appear to offer anonymity while actually delivering only confidentiality. Market researchers often use mail, email, or Internet surveys that do not explicitly ask the respondent's identity but contain hidden codes on the questionnaire that identify the respondent. A false claim of anonymity is clearly unethical. If only confidentiality is promised, is it also unethical to say nothing about the identifying code, perhaps causing respondents to believe their replies are anonymous? Explain your reasoning.

7.25 Human biological materials. Long ago, doctors drew a blood specimen from you as part of treating minor anemia. Unknown to you, the sample was stored. Now researchers plan to use stored samples from you and many other people to look for genetic factors that may influence anemia. It is no longer possible to ask your consent because you are no longer alive. Modern technology can read your entire genetic makeup from the blood sample.

(a) Do you think it violates the principle of informed consent to use your blood sample if your name is on it but you were not told that it might be saved and studied later? Explain your reasoning.

(b) Suppose that your name is not attached to the sample. The blood sample is known only to come from (say) "a 20-year-old white female being treated for anemia." Is it now okay to use the sample for research? Explain your reasoning.

(c) Perhaps we should use biological materials such as blood samples only from patients who have agreed to allow the material to be stored for later use in research. It isn't possible to say in advance what kind of research, so this falls short of the usual standard for informed consent. Is this practice nonetheless acceptable, given complete confidentiality and the fact that using the sample can't physically harm the patient? Explain your reasoning.

7.26 Equal treatment. Researchers on depression proposed to investigate the effect of supplemental therapy and counseling on the quality of life of adults with depression. Eligible patients on the rolls of a large medical clinic were to be randomly assigned to treatment and control groups. The treatment group would be offered dental care, vision testing, transportation, and other services not available without charge to the control group. The review board felt that providing these services to some but not other persons in the same institution raised ethical questions. Do you agree? Explain your answer.

7.27 Sham surgery? Clinical trials like the Parkinson's disease study mentioned in Example 6 are becoming more common. One medical researcher says, "This is just the beginning. Tomorrow, if you have a new procedure, you will have to do a double-blind placebo trial." Example 6 outlines the arguments for and against testing surgery just as drugs are tested. When would you allow sham surgery in a clinical trial of a new surgery? Explain your reasoning.

7.28 The Willowbrook hepatitis studies. In the 1960s, children entering the Willowbrook State School, an institution for the mentally retarded, were deliberately infected with hepatitis. The researchers argued that almost all children in the institution quickly became infected anyway. The studies showed for the first time that two strains of hepatitis existed. This finding contributed to the development of effective vaccines. Despite these valuable results, the Willowbrook studies are now considered an example of unethical research. Explain why, according to current ethical standards, useful results are not enough to allow a study.

7.29 AIDS clinical trials. Now that effective treatments for AIDS are available, is it ethical to test treatments that may be less effective? Combinations of several powerful drugs reduce the level of HIV in the blood and at least delay illness and death from complications due to AIDS. But effectiveness depends on how damaged the patient's immune system is and what drugs the patient has previously taken. There are strong side effects, and patients must be able to take more than a dozen pills at one time every day. Because AIDS is often fatal and the combination therapy works, we might argue that it isn't ethical to test any new treatment for AIDS that might possibly be less effective. But that might prevent discovery of better treatments. This is a strong example of the conflict between doing the best we know for patients now and finding better treatments for other patients in the future. How can we ethically test new drugs for AIDS?

7.30 AIDS trials in Africa. Effective drugs for treating AIDS are very expensive, so most African nations cannot afford to give them to large numbers of people. Yet AIDS is more common in parts of Africa than anywhere else. A few clinical trials are looking at ways to prevent pregnant mothers infected with HIV from passing the infection to their unborn children, a major source of HIV infections in Africa. Some people say these trials are unethical because they do not give effective AIDS drugs to their subjects, as would be required in rich nations. Others reply that the trials are looking for treatments that can work in the real world in Africa, and that they promise benefits at least to the children of their subjects. What do you think?

7.31 AIDS trials in Africa. One of the most important goals of AIDS research is to find a vaccine that will protect against HIV. Because AIDS is so common in parts of Africa, that is the easiest place to test a vaccine. It is likely, however, that a vaccine would be so expensive that it could not (at least at first) be widely used in Africa. Is it ethical to test in Africa if the benefits go mainly to rich countries? The treatment group of subjects would get the vaccine, and the placebo group would later be given the vaccine if it proved effective. So the actual subjects would benefit—it is the future benefits that would go elsewhere. What do you think? Explain your answer.

7.32 Henrietta Lacks. You may have heard of the book *The Immortal Life of Henrietta Lacks* by Rebecca Skloot (2010) or the Oprah Winfrey movie based on the book. Henrietta Lacks came from an impoverished tobacco farming family in Virginia. In 1951, when Henrietta was 31, she died at the Johns Hopkins Hospital from complications due to cervical cancer. While Henrietta died in 1951, some of her cells live on to this day. The cells were originally taken without Henrietta Lacks's permission. These same cells "became the immortal He-La cell line used for extensive bio-medical research and then commodified in a multi-million dollar industry." The He-La cells have been used to develop the polio vaccine and flu treatments and in HIV/AIDS, leukemia, tuberculosis, and Parkinson's disease research, just to name a few applications.

(a) The research from He-La cells has saved hundreds of thousands, if not millions, of people. Does the benefit society received from the "immortal" cells of Henrietta Lacks outweigh the ethics of anyone in the Lacks family, including Henrietta herself, never being asked for permission to use the cells? Explain your reasoning.

(b) Suppose that you found out that you had "immortal" cells that would allow researchers to develop vaccines or treatments for HIV, tuberculosis, Alzheimer's disease, etc.? What information would you want from the doctor before the doctor shared the discovery of your immortal cells with others? What payment do you think you should get from the biomedical research done and pharmaceutical industry?

7.33 The Stanford Prison Experiment. You may have seen or heard about the film *The Stanford Prison Experiment* (2015), which centered around Philip Zimbardo's 1971 research on psychologically healthy individuals placed into a stressful situation, to answer the question about how context impacts human behavior. Zimbardo wondered, "what happens when you put good people in an evil place? Does humanity win over evil, or does evil triumph?" The study was planned for two weeks, but ended shortly before one week had passed because of experimenter concerns about signs of sadism in the guards and signs of extreme depression and stress in the prisoners. You can find more information about the actual experiment at http://www.prisonexp.org/.

(a) The consent form subjects signed can be found at http://pdf.prisonexp.org/consentpdf. Does this meet the standards of informed consent that you have studied in this chapter? Explain your reasoning.

(b) According to the Stanford Prison Experiment website, "despite suffering extreme emotional stress during the experiment, all participants appear to have regained their baseline emotional states after the study. Extensive follow-up testing revealed no lasting trauma to participants." Does this mitigate what the subjects experienced in the experiment? Explain your reasoning.

7.34 Opinion polls. The congressional campaigns are in full swing, and the candidates have hired polling organizations to take regular polls to find out what the voters think about the issues. What information should the pollsters be required to give out?

(a) What does the standard of informed consent, as discussed in this chapter, require the pollsters to tell potential respondents?

(b) The standards accepted by polling organizations also require giving respondents the name and address of the organization that carries out the poll. Why do you think this is required?

(c) The polling organization usually has a professional name such as "Samples Incorporated," so respondents don't know that the poll is being paid for by a political party or candidate. Would revealing the sponsor to respondents bias the poll? Should the sponsor always be announced whenever poll results are made public?

7.35 A right to know? Some people think that the law should require that all political poll results be made public. Otherwise, the possessors of poll results can use the information to their own advantage. They can act on the information, release only selected parts of it, or time the release for best effect. A candidate's organization replies that they are paying for the poll in order to gain information for their own use, not to amuse the public. Do you favor requiring complete disclosure of political poll results? What about other private surveys, such as market research surveys of consumer tastes?

7.36 Telling the government. The 2010 census was a short-form-only census. The decennial long form was eliminated. The American Community Survey (ACS) replaced the long form in 2010 and will collect long-form-type information throughout the decade rather than only once every 10 years. The 2010 ACS asked detailed questions, for example:

Does this house, apartment, or mobile home have a) hot and cold piped water?; b) a flush toilet?; c) a bathtub or shower?; d) a sink or faucet?; e) a stove or range?; f) a refrigerator?; and g) telephone service from which you can both make and receive calls? Include cell phones.

The form also asked your individual income in dollars, broken down by source, and whether any "physical, mental, or emotional condition" causes you difficulty in "concentrating, remembering, or making decisions."

Give brief arguments for and against the use of the ACS form: the government has legitimate uses for such information, but the questions seem to invade people's privacy.

7.37 Charging for data? Data produced by the government are often available free or at low cost to private users. For example, satellite weather data produced by the U.S. National

Weather Service are available free to TV stations for their weather reports and to anyone on the Web.

Opinion 1: Government data should be available to everyone at minimal cost. European governments, on the other hand, charge TV stations for weather data.

Opinion 2: The satellites are expensive, and the TV stations are making a profit from their weather services, so they should share the cost. Which opinion do you support, and why?

7.38 Surveys of youth. The Centers for Disease Control and Prevention, in a survey of teenagers, asked the subjects if they had ever had sexual intercourse. Males who said "Yes" were then asked, "That very first time that you had sexual intercourse with a female, how old were you?" and "Please tell me the name or initials of your first sexual partner so that I can refer to her during the interview." Should consent of parents be required to ask minors about sex, drugs, and other such issues, or is consent of the minors themselves enough? Give reasons for your opinion.

7.39 Deceiving subjects. Students sign up to be subjects in a psychology experiment. When they arrive, they are placed in a room and assigned a task. During the task, the subject hears a loud thud from an adjacent room and then a piercing cry for help. Some subjects are placed in a room by themselves. Others are placed in a room with "confederates" (a research methods term for accomplices) who have been instructed by the researcher to look up upon hearing the cry, then return to their task. The treatments being compared are whether the subject is alone in the room or in the room with confederates. Will the subject ignore the cry for help?

The students had agreed to take part in an unspecified study, and the true nature of the experiment is explained to them afterward. Do you think this study is ethically okay?

7.40 Tempting subjects. A psychologist conducts the following experiment: he measures the attitude of subjects toward cheating, then has them take a mathematics skills exam in which the subjects are tempted to cheat. Subjects are told that high scores will receive a $100.00 gift certificate and that the purpose of the experiment is to see if rewards affect performance. The exam is computer-based and multiple choice. Subjects are left alone in a room with a computer on which the exam is available and are told that they are to click on the answer they believe is correct. However, when subjects click on an answer a small pop-up window appears with the correct answer indicated. When the pop-up window is closed, it is possible to change the answer selected. The computer records—unknown to the subjects—whether or not they change their answers after closing the pop-up window. After completing the exam, attitude toward cheating is retested.

Subjects who cheat tend to change their attitudes to find cheating more acceptable. Those who resist the temptation to cheat tend to condemn cheating more strongly on the second test of attitude. These results confirm the psychologist's theory.

This experiment tempts subjects to cheat. The subjects are led to believe that they can cheat secretly when in fact they are observed. Is this experiment ethically objectionable? Explain your position.

Exploring the Web

Access these exercises on the text website: macmillanlearning.com/scc10e.

What's the Verdict?

The following "What's the verdict?" story explores issues surrounding an experiment conducted by Facebook in 2012. You will use what you have learned in this chapter to explore ethical issues associated with this experiment,

In January 2012, Facebook performed an experiment on over 689,000 users without informing them even after the experiment was over. Facebook adjusted people's newsfeeds so that half of these individuals only saw happy posts from their friends and the other half only saw sad posts from their friends. Facebook then determined the mood of the user by judging the quality of their own posts. Why would Facebook want to learn how to manipulate emotions? It is well known that sad people tend to shop more, and Facebook sells advertisement space on their site.

Did Facebook behave ethically? That is an interesting question in today's social media world.

Questions

WTV7.1. Review board approval: Any organization that receives federal funding must receive review board approval for research with humans. Facebook does not receive federal funding. Facebook partnered with Cornell University to write the article and analyze the data after the experiment was already performed. (**www.pnas.org/content/111/24/8788.full**) The researcher at Cornell consulted his institutional review board to get approval for his part of this work, but since his involvement started after the experiment was already completed, his review board said that he did not need approval from them. How do you feel about this experiment happening without a review board?

WTV7.2. Confidentiality: Facebook did take an unusual step for a business by publishing their results from this experiment in the *Proceedings of the National Academy of Sciences*, a prestigious journal. Facebook knew who all of the individuals were and what they had posted, but Facebook did not publish any individual information in the article. Did they use confidentiality?

WTV7.3. Informed consent: Facebook claims that their data privacy policy covered this experiment because it included this line: "For example, in addition to helping people see and find things that you do and share, we may use the information we receive about you . . . for internal operations, including troubleshooting, data analysis, testing, research and service improvement." Do you agree that this policy does enough to count as informed consent? Discuss your reasoning.

WTV7.4. Sometimes exceptions can be made to the informed consent process. Examples include education research studies with normal classroom activities posing no unusual risks (like trying a lecture vs. an active learning activity to teach a new concept) or behavioral studies in a public place. These ethical guidelines were written in the middle of the twentieth century, well before the Internet and social media existed. Do you believe that Facebook and other social media sites count as a "public place"? If so, does that change your answer to whether informed consent was necessary for this experiment?

WTV7.5. Social media can be a powerful research tool with large quantities of data available. However, we have not developed many rules for how this data can be used or analyzed yet. Read about the European Union's General Data Protection Rules (**https://ec.europa.eu /commission/priorities/justice-and-fundamental- rights/data-protection/2018-reform-eu-data -protection-rules_en**) to see if you agree with them. Can you think of any other rules you would like to have for researchers working with social media data?

What's the verdict? Social media research may or may not fit into our basic ethical rules developed in the twentieth century, and as a society we will have to decide if new rules should be developed to protect us while still allowing us to learn from the rich data available.

Measuring

In this chapter you will:

- Learn the basics of measurement.
- See how to question whether measurements are valid or accurate.
- Consider how we can improve reliability and reduce bias.

Are people with larger brains more intelligent? People have investigated this question throughout history. To answer it, we must **measure** "intelligence." This requires us to reduce a complex construct to a number that can go up or down. The first step is to specify what we mean by intelligence. Does a vast knowledge of many subjects constitute intelligence? How about the ability to solve difficult puzzles or do complicated mathematical calculations? Or is it some combination of these?

Once we decide what intelligence is, we must actually produce the numbers. Should we use the score on a written test? Maybe we could collect data online from people who click on a link that promises them an answer to "Which celebrities have your same IQ?" Perhaps a formula that also includes grades in school would be better. Not only is it hard to say exactly what "intelligence" is, but it's hard to attach a number to measure whatever we say it is. And in the end, can we even trust the number we produce? The material in this chapter will provide you with the tools to address the Case Study and will help you understand the process of measurement and determine whether you can trust the resulting numbers.

Measurement Basics

Statistics deals with data, and the data may or may not be numbers. For example, planning the production of data through a sample or an experiment does not by itself produce numbers. Once we have our sample respondents or our experimental subjects, we must still *measure* whatever characteristics interest us. First, think broadly: Are we trying to measure the right things? Are we overlooking some outcomes that are important even though they may be hard to measure?

Example 1 — *Saving lives and money*

A recent study of patients with cardiovascular disease (CVD) investigated the impact of a program on survival rates and treatment costs. As reported, the Coaching On Achieving Cardiovascular Health (COACH) Program improved both biomedical and lifestyle CVD risk factors. The study focused on measuring the long-term impact of the COACH Program on overall survival and hospital costs in patients with CVD. The researchers needed to determine how to measure both survival and hospital costs for the patients with CVD who participated in the study.

Once we have decided what properties we want to measure, we can think about how to do the measurements.

Measurement basics

We **measure** a property of a person or thing when we assign a value to represent the property.

We often use an **instrument** to make a measurement. We may have a choice of the **units** we use to record the measurements.

The result of measurement is a numerical **variable** that takes different values for people or things that differ in whatever we are measuring.

Example 2 — *Length, college readiness, highway safety*

To measure the length of a room, you can use a digital laser measuring device as the *instrument*. You can choose either feet or meters as the *unit of measurement*. If you choose meters, your *variable* is the length of the room in meters.

To measure a student's readiness for college, you might ask the student to take the SAT exam. The exam is the *instrument*. The *variable* is the student's score in points, somewhere between 400 and 1600 if you combine the Evidence-Based Reading and Writing and the Mathematics sections of the SAT. "Points" are the *units of measurement*, but these are determined by a complicated scoring system described at the SAT website (www.collegeboard.com).

How can you measure the safety of traveling on the highway? You might decide to use the number of people who die in motor vehicle accidents in a year as a *variable* to measure highway safety. The National Highway Traffic Safety Administration's (NHTSA) Fatality Analysis Reporting System (FARS) collects data on all fatal traffic crashes. The *unit of measurement* is the number of people who died, and the FARS serves as our measuring *instrument*.

Here are some questions you should ask about the variables in any statistical study:

1. Exactly how is the variable defined?

2. Is the variable an accurate way to describe the property it claims to measure?

3. How dependable are the measurements?

We don't often design our own measuring devices—we use the results of the SAT or the FARS, for example—so we won't go deeply into that aspect of measurement. Any consumer of numbers, however, should know a bit about how those numbers are produced.

Know Your Variables

Measurement is the process of turning concepts like length or employment status into precisely defined variables. Using a digital laser measuring device to turn the idea of "length" into a number is straightforward because we know exactly what we mean by length. Measuring college readiness is controversial because it isn't clear exactly what makes a student ready for college. Using SAT scores at least says exactly how we will get numbers, but does it take into account all aspects that make a person "ready" for college? Measuring free time requires that we first say what activities count as "free" time. Even counting highway deaths requires us to say exactly what counts as a highway death: Pedestrians hit by cars? People in cars hit by a train at a crossing? People who die from injuries 6 months after an accident? We can simply accept the NHSTA's counts, but someone had to answer those and other questions in order to know what to count. For example, to

Statistics in Your World

What are your units? Not paying attention to units of measurement can get you into trouble. This example may seem old to you, but it is a classic example of issues with measurement. In 1999, the *Mars Climate Orbiter* burned up in the Martian atmosphere. It was supposed to be 93 miles (150 kilometers) above the planet's surface but was, in fact, only 35 miles (57 kilometers) above the surface. It seems that Lockheed Martin, which built the *Orbiter*, specified important measurements in English units (pounds, miles). The National Aeronautics and Space Administration team, who flew the *Orbiter*, thought the numbers were in metric system units (kilograms, kilometers). There went $125 million, which is about $189 million in 2018.

be included in the FARS, "a crash must involve a motor vehicle traveling on a trafficway customarily open to the public and must result in the death of at least one person (occupant of a vehicle or a non-motorist) within 30 days of the crash." The details of when a death is counted as a highway death are necessary because they can make a difference in the data.

Example 3   *Measuring unemployment*

Each month the Bureau of Labor Statistics (BLS) announces the *unemployment rate* for the previous month. People who are not available for work (retired people, for example, or students who do not work while in school) should not be counted as unemployed just because they don't have a job. To be unemployed, a person must first be in the labor force. That is, the person must be available for work and looking for work. The unemployment rate is

$$\text{unemployment rate} = \frac{\text{number of people unemployed}}{\text{number of people in the labor force}}$$

To complete the exact definition of the unemployment rate, the BLS has very detailed descriptions of what it means to be "in the labor force" and what it means to be "employed." For example, if you are on strike but expect to return to the same job, you count as employed. If you are not working and did not look for work in the last four weeks, you are not in the labor force. So people who say they want to work but are too discouraged to keep looking for a job don't count as unemployed. According to the BLS, "passive methods of job search do not have the potential to connect job seekers with potential employers and therefore do not qualify as active job search methods. Examples of passive methods include attending a job training program or course, or merely reading about [posted] job openings." The details matter. The official unemployment rate would be different if the BLS used a different definition of unemployment.

The BLS estimates the unemployment rate based on interviews with the sample in the monthly Current Population Survey. The interviewer can't simply ask, "Are you in the labor force?" and "Are you employed?" Many questions are needed to classify a person as employed, unemployed, or not in the labor force. Changing the questions can change the unemployment rate. For example, at the beginning of 1994, after several years of planning, the BLS introduced computer-assisted interviewing and improved its questions. Figure 8.1 is a graph of the unemployment rate that appeared on the front page of the BLS monthly news release on the employment situation. There is a gap in the graph before January 1994 because of this change in the interviewing process. The unemployment rate would have been 6.3% under the old system, but jumped to 6.7% under the new system. That's a big enough difference to make politicians unhappy.

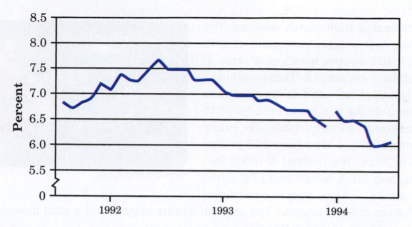

Figure 8.1 The unemployment rate from August 1991 to July 1994. The gap shows the effect of a change in how the government measures unemployment.

Measurements, Valid and Invalid

No one would object to using a digital laser measuring device reading in meters to measure the length of a room. Many people object to using SAT scores to measure readiness for college. Let's shortcut that debate: just measure the height in meters of all applicants and accept the tallest. Bad idea, you say. Why? Because height has nothing to do with being prepared for college. In more formal language, height is not a *valid* measure of a student's academic background.

> **Key Terms**
>
> A variable is a **valid** measure of a property if it is relevant or appropriate as a representation of that property.

It is valid to measure length with a digital laser measuring device. It isn't valid to measure a student's readiness for college by recording her height. The BLS unemployment rate is a valid measure, even though changes in the official definitions would give a somewhat different measure. Let's think about measures, both valid and invalid, in some other settings.

Example 4
Measuring highway safety

Roads got better. Speed limits increased. Big SUVs and crossovers have replaced some cars, while smaller cars, hybrids, and electric vehicles have replaced others. Enforcement campaigns reduced impaired and distracted driving. How did highway safety change between 2007 and 2016 in this changing environment?

We could just count deaths from motor vehicles. The Fatality Analysis Reporting System says there were 41,259 deaths in 2007 and 37,461 deaths nine years later in 2016. The number of deaths decreased. These numbers alone show progress. However, we need to keep in mind other things that happened during this same time frame to determine how much progress has been made. For example, the number of licensed drivers rose from 206 million in 2007 to 222 million in 2016. The number of miles that people drove increased from 3031 billion to 3174 billion during this same time period. If more people drive fewer miles, should we expect more or fewer deaths? The count of deaths alone is not a valid measure of highway safety. So what should we use instead?

Photo Spirit/Shutterstock

Rather than a *count,* we should use a *rate*. The number of deaths per mile driven takes into account the fact that more people drive more miles than in the past. In 2016, it is estimated that vehicles drove 3,174,000,000,000 miles in the United States. Because this number is so large, it is usual to measure safety by deaths per 100 million miles driven rather than deaths per mile. For 2016, this death rate is

$$\frac{\text{motor vehicle deaths}}{\text{100s of millions of miles driven}} = \frac{37,461}{31,740}$$

$$= 1.2$$

The death rate fell from 1.4 deaths per 100 million miles in 2007 to 1.2 in 2016. That's a decrease—there were 15% fewer deaths per mile driven in 2016 than in 2007. Driving became safer during this time period even though there were more drivers on the roads.

Now it's your turn

8.1 Driver fatigue. A researcher studied the number of traffic accidents that were attributed to driver fatigue at different times of the day. He noticed that the number of accidents was higher in the late afternoon (between 5 and 6 P.M.) than in the early afternoon (between 1 and 2 P.M.). He concluded that driver fatigue plays a more prominent role in traffic accidents in the late afternoon than in the early afternoon. Do you think this conclusion is justified? Explain your reasoning.

Key Terms

Often a **rate** (a fraction, proportion, or percentage) at which something occurs is a more valid measure than a simple **count** of occurrences.

Using height to measure readiness for college and using counts when rates are needed are examples of clearly invalid measures. The tougher questions concern measures that are neither clearly invalid nor obviously valid.

Example 5 — *Assessing the validity of achievement tests*

When you take a chemistry exam, you hope that it will ask you about the main points of material listed in the course syllabus. If it does, the exam is a valid measure of how much you know about the course material. The College Board, which administers the SAT, also offers Advanced Placement (AP) exams in a variety of disciplines. These AP exams are not very controversial. Experts can judge validity by comparing the test questions with the syllabus of material the questions are supposed to cover.

Example 6 — *Assessing the validity of IQ tests*

Psychologists would like to measure aspects of the human personality that can't be observed directly, such as "intelligence" or "authoritarian personality." Does an IQ test measure intelligence? Some psychologists say "Yes" rather loudly. There is such a thing as general intelligence, they argue, and the various standard IQ tests do measure it, though not perfectly. For example, an adult's general intelligence can be measured with the Wechsler Adult Intelligence Scale. Other experts say "No" equally loudly. Famously, in the 1980s Howard Gardner proposed that there is no single intelligence, just a variety of mental abilities (for example, logical, linguistic, spatial, musical, kinesthetic, interpersonal, and intrapersonal) that no one instrument can measure. And still other psychologists criticize Gardner's approach, especially since Gardner has never developed an assessment for his multiple intelligences because he believes such experimental evidence to be inappropriate.

The disagreement over the validity of IQ tests is rooted in disagreement over the nature of intelligence. If we can't agree on exactly what intelligence is, we can't agree on how to measure it.

Example 7 — *Assessing the validity of world happiness*

Each year, the United Nations Sustainable Development Solutions Network (UNSDSN) publishes the "World Happiness Report." According to the 2018 report, Finland is the happiest country, measured by Gallup World Polls from 2015–2017. (Finland's immigrants are also the happiest of immigrant populations.) How was happiness measured? The UNSDSN bases the happiness rankings on one question that asks people from 156 countries to "rate their lives on a scale of zero to 10 – with zero being the worst possible life and 10 being the best possible life. . . . While the metric is known as 'life satisfaction,' there is some disagreement as to whether it should actually be labeled as 'happiness.'" Many people might disagree with this way to measure happiness.

Statistics is little help in these examples. The examples start with an idea like "knowledge of chemistry" or "intelligence" or "happiness." If the idea is vague, validity becomes a matter of opinion. However, statistics can help a lot if we refine the idea of validity a bit.

Example 8 — Assessing the validity of the SAT, again

The SAT has been criticized for having both cultural and gender bias. Writing for the American Enterprise Institute, Mark J. Perry said "my main interest in the annual SAT test results is the ongoing gender gap in favor of boys for the SAT math test." According to the College Board, in 2016, the gender gap for college-bound seniors was larger on the math part of the SAT, where women averaged 494 and men averaged 524. If we examine the average math scores by ethnicity and gender, we find that Asian or Asian American men scored the highest on the math part of the SAT, with an average of 614, while black or African American women scored the lowest on the math part of the SAT, with an average of 422.

The College Board, which administers the SAT, replies that there are many reasons some groups have lower average scores than others. For example, more women than men from families with low incomes and little education sign up for the SAT. Students whose parents have low incomes and little education have, on the average, fewer advantages at home and in school than more affluent students. They have lower SAT scores because their backgrounds have not prepared them as well for college. The mere fact of lower scores doesn't imply that the test is not valid.

Is the SAT a valid measure of readiness for college? "Readiness for college academic work" is a vague concept that probably combines innate intelligence (whatever we decide that is), learned knowledge, study and test-taking skills, and motivation to work at academic subjects. Opinions will always differ about whether SAT scores (or any other measure) accurately reflect this vague concept.

Instead, we ask a simpler and more easily answered question: do SAT scores help predict students' success in college? Success in college is a clear concept, measured by whether students graduate and by their college grades. Students with high SAT scores are more likely to graduate and earn (on the average) higher grades than students with low SAT scores. We say that SAT scores have *predictive validity* as measures of readiness for college. This is the only kind of validity that data can assess directly.

> ## Key Terms
>
> A measurement of a property has **predictive validity** if it can be used to predict success on tasks that are related to the property measured.

Predictive validity is the clearest and most useful form of validity from the statistical viewpoint. "Do SAT scores help predict college grades?" is a much clearer question than "Do IQ test scores measure intelligence?" However, predictive validity is not a yes-or-no idea. We must ask *how accurately* SAT scores predict college grades. Moreover, we must ask *for what groups* the SAT has predictive validity. It is possible, for example, that the SAT predicts college grades well for men but not for women. There are statistical ways to describe "how accurately." The Statistical Controversies feature in this chapter asks you to think about these issues.

STATISTICAL CONTROVERSIES

Six-Year Graduation Rates, High School GPA, and Standardized Tests

Colleges use a variety of measures to make admissions decisions. More and more colleges have gone to "test-optional" for applicants to their institutions. According to a 2018 study by the National Association for College Admission Counseling, test-optional policies (TOPs) may be "a tool [admissions offices] employed in the hope of increasing applications from a more diverse range of students." Students who were identified as *any* underrepresented minority (URM) group were identified as students who would

VIPDesignUSA/Depositphotos

expand diversity in college enrollments. These URMs include first-generation college students, students from lower socioeconomic statuses, and students from racial and ethnic groups that have traditionally been underrepresented in college populations.

The researchers found that "a well-executed test-optional admission policy can lead to an increase in overall applications as well as an increase in the representation of URM students (both numeric and proportionate) in the applicant pool and the freshman class." Success in college may be measured by six-year graduation rates. The following table shows the six-year college graduation rate for the relationship between high school GPA and SAT or ACT scores. Note that ACT scores have been converted into SAT score equivalents.

HS GPA	SAT or ACT Score				
	<800	800–890	900–990	1000–1090	>1100
<2.67	26%	32%	34%	35%	35%
2.67–3.00	29%	40%	40%	43%	39%
3.01–3.32	39%	49%	52%	54%	51%
3.33–3.66	45%	54%	58%	59%	57%
3.67–4.00	47%	62%	62%	66%	72%

How well do you think high school GPA alone predicts a college's six-year graduation rate? How well do you think SAT or ACT scores alone predict a college's six-year graduation rate? Do you think one measurement performs better than the other or would you suggest that they be used together? If a college becomes test optional, how do graduation rates change? Do you think the expanded diversity that comes with TOPs is worth the change in six-year graduation rates?

Measurements, Accurate and Inaccurate

Using a bathroom scale to measure your weight is valid. If your scale is like many commonly used ones, however, the measurement may not be very accurate. It measures weight, but it may not give the true weight. This can be true whether your scale is digital or analog. Let's say that originally your scale always read 2 pounds too high, so

measured weight = true weight + 2 pounds

If that is the whole story, the scale will always give the same reading for the same true weight. Most scales vary a bit—they don't always give the same reading when you step off and step right back on. Your scale now is somewhat old and is not sitting evenly on the floor. It still always reads 2 pounds too high because its aim is off, but now it is also erratic so readings deviate from 2 pounds. This morning, it reads 1 pound too low for that reason. So the reading is

measured weight = true weight + 2 pounds − 1 pound

When you step off and step right back on, the scale sticks in a different spot that makes it read 0.4 pound too high. The reading you get is now

measured weight = true weight + 2 pounds + 0.4 pound

You don't like the fact that this second reading is higher than the first, so you again step off and step right back on. The scale gives yet another reading:

$$\text{measured weight} = \text{true weight} + 2 \text{ pounds} - 1.2 \text{ pounds}$$

If you have nothing better to do than keep stepping on and off the scale, you will keep getting different readings. They center on a reading 2 pounds too high, but they vary about that center.

Your scale has two kinds of errors. If it wasn't uneven on the floor, the scale would always read 2 pounds high. That is true every time anyone steps on the scale. This systematic error that occurs every time we make a measurement is called *bias*. Your scale also reads inaccurately because it is uneven on the floor—but how much this changes the reading differs every time someone steps on the scale. Sometimes the unevenness pushes the scale reading up; sometimes it pulls it down. The result is that the scale weighs 2 pounds too high on the average, but its reading varies when we weigh the same thing repeatedly. We can't predict the error due to "stickiness," so we call it *random error*.

Errors in measurement

We can think about errors in measurement this way:

$$\text{measured value} = \text{true value} + \text{bias} + \text{random error}$$

A measurement process has **bias** if it systematically tends to overstate or understate the true value of the property it measures.

A measurement process has **random error** if repeated measurements on the same individual give different results. If the random error is small, we say the measurement is **reliable.**

To determine if the random error is small, we can use a quantity called the **variance.** The variance of n repeated measurements on the same individual is computed as follows:

1. Find the arithmetic average of these n measurements.
2. Compute the difference between each observation and the arithmetic average and square each of these differences.
3. Average the squared differences by dividing their sum by $n - 1$. This average squared difference is the variance.

A reliable measurement process will have a small variance.

For the three measurements on our uneven scale, suppose that our true weight is 140 pounds. Then the three measurements are

$$140 + 2 - 1 = 141 \text{ pounds}$$
$$140 + 2 + 0.4 = 142.4 \text{ pounds}$$
$$140 + 2 - 1.2 = 140.8 \text{ pounds}$$

The average of these three measurements is

$$(141 + 142.4 + 140.8)/3 = 424.2/3 = 141.4 \text{ pounds}$$

The differences between each measurement and the average are

$$141 - 141.4 = -0.4$$
$$142.4 - 141.4 = 1$$
$$140.8 - 141.4 = -0.6$$

The sum of the squares of these differences is

$$(-0.4)^2 + (1)^2 + (-0.6)^2 = 0.16 + 1 + 0.36 = 1.52$$

and so the variance of these random errors is

$$1.52/(3-1) = 0.76$$

A reliable measurement process will have a small variance. In this example, the variance of 0.76 is quite small relative to the actual weights, so it appears that the measurement process using this scale is reliable.

A scale that always reads the same when it weighs the same item is perfectly reliable even if it is biased. For such a scale, the variance of the measurements will be 0.

Reliability says only that the result is dependable. Bias means that in repeated measurements the *tendency* is to systematically either overstate or understate the true value. It does not necessarily mean that every measurement overstates or understates the true value. Bias and lack of reliability are different kinds of error. And don't confuse reliability with validity just because both sound like good qualities. Using a scale to measure weight is valid even if the scale is not reliable.

Here are two examples of measurements that are reliable but not valid.

Example 9

Do big skulls house smart brains?

In the mid-nineteenth century, it was thought that measuring the volume of a human skull would measure the intelligence of the skull's owner. It was difficult to measure a skull's volume reliably, even after it was no longer attached to its owner. In the 1870s, Paul Broca, a professor of surgery, showed that filling a skull with small lead shot, then pouring out the shot and weighing it, gave quite reliable measurements of the skull's volume. These accurate measurements do not, however, give a valid measure of intelligence. Skull volume turned out to have no relation to intelligence or achievement.

Example 10   *The errant clock*

Suppose that you set your alarm clock for 7:00 A.M., but it goes off every morning at 6:45 A.M.

The alarm clock is highly reliable but is not valid.

Alchie/Shutterstock

Now it's your turn

8.2 Android versus iOS. You and your friends have probably debated who has the better cell phone operating system. You know that you are correct, but your friends insist that they are. According to Consumer Report's Cell Phones & Services 2018 report, the Samsung Galaxy Note 9 ranked first and the Apple iPhone XS Max came in second. The overall scores for the phones were 83 for the Samsung (running Android) and 82 for the iPhone (running iOS). Consumer Reports gave the edge to Samsung for handset capabilities but gave the iPhone the edge for the rear camera and the navigation system. Does this information settle the debate between you and your friends? Do you think these ratings are biased, unreliable, or both? Explain your answer.

Improving Reliability, Reducing Bias

What time is it? Much modern technology, such as the Global Positioning System, which uses satellite signals to tell you where you are, requires very exact measurements of time. In 1967, the International Committee for Weights and Measures defined the second to be the time required for 9,192,631,770 vibrations of a cesium atom. The cesium atom is not impacted by changes in temperature, humidity, and air pressure like physical clocks are. To this day, the definition of the second has remained unchanged. The National Institute of Standards and Technology (NIST) has the world's most accurate atomic clock and broadcasts the results (with some loss in transmission) by radio, telephone, and Internet.

Example 11 — *Really accurate time*

NIST's atomic clock (Figure 8.2) is very accurate but not perfectly accurate. The world standard is Coordinated Universal Time, compiled by the International Bureau of Weights and Measures (BIPM) in Sèvres, France. BIPM doesn't have a better clock than NIST. It calculates the time by averaging the results of more than 200 atomic clocks around the world. NIST tells us (after the fact) by how much it misses the correct time. Here are the last 12 errors, in seconds:

0.0000000075	0.0000000012
0.0000000069	−0.0000000020
0.0000000067	−0.0000000045
0.0000000063	−0.0000000046
0.0000000041	−0.0000000042
0.0000000032	−0.0000000036

Figure 8.2 This atomic clock at the National Institute of Standards and Technology is accurate to 1 second in 6 million years.

In the long run, NIST's measurements of time are not biased. The NIST second is sometimes shorter than the BIPM second and sometimes longer, not always off in the same direction. NIST's measurements are very reliable, but the numbers above do show some variation. There is no such thing as a perfectly reliable measurement. The average (mean) of several measurements is more reliable than a single measurement. That's one reason BIPM combines the time measurements of many atomic clocks.

Use averages to improve reliability

No measuring process is perfectly reliable. The **average** of several repeated measurements of the same individual is more reliable (less variable) than a single measurement.

Scientists everywhere repeat their measurements and use the average to get more reliable results. Even students in a chemistry lab often do this. Just as larger samples reduce variation in a sample statistic, averaging over more measurements reduces variation in the final result.

Unfortunately, there is no similarly straightforward way to reduce the bias of measurements. Bias depends on how good the measuring instrument is. To reduce the bias, you need a better instrument. The atomic clock at NIST is accurate to 1 second in 6 million years but is a bit large to put beside your bed. (Fortunately the clock on your cell phone is pretty accurate!)

Example 12

Measuring unemployment again

Measuring unemployment is also "measurement." The concepts of bias and reliability apply here just as they do to measuring length or time.

The Bureau of Labor Statistics checks the *reliability* of its measurements of unemployment by having supervisors reinterview about 5% of the sample. This is repeated measurement on the same individual, just as a student in a chemistry lab measures a weight several times.

The BLS attacks *bias* by improving its instrument over time. That's what happened in 1994, when the Current Population Survey was given its biggest overhaul in more than 50 years. The old system for measuring unemployment, for example, underestimated unemployment among women because the detailed procedures had not kept up with changing patterns of women's work. The new measurement system corrected that bias—and raised the reported rate of unemployment.

Measuring Psychological and Social Factors

Statisticians think about measurement much the same way as they think about sampling. In both settings, the big idea is to ask, "What would happen if we did this many times?" In sampling, we want to estimate a population parameter, and we worry that our estimate may be biased or vary too much from sample to sample. Now we want to measure the true value of some property, and we worry that our measurement may be biased or vary too much when we repeat the measurement on the same individual. Bias is systematic error that happens every time; high variability (low reliability) means that our result can't be trusted because it isn't repeatable.

Thinking of measurement this way is pretty straightforward when you are measuring your weight. To start with, you have a clear idea of what your "true weight" is. You know that there are really good scales around: start at the doctor's office, go to the physics lab, end up at NIST. You can measure your weight as accurately as you wish. This makes it easy to see that your bathroom scale always reads 2 pounds too high. Reliability is also easy to describe—step on and off the scale many times and see how much its readings vary.

Asking "What would happen if we did this many times?" is a lot harder to put into practice when we want to measure "intelligence" or "readiness for college" or "happiness." Any social construct or ideological belief is difficult to measure. Consider political psychologists who tried to determine the relationship between two psychological and social factor scales and who likely voters would vote for in the 2016 U.S. election.

Example 13 — *Psychology and the 2016 election*

Polls conducted in 2016 found that Americans who scored higher on authoritarianism were more likely to vote for Donald Trump than for Hillary Clinton. Researchers in Canada and the United Kingdom wondered if right-wing authoritarianism (RWA) and Social Dominance Orientation (SDO) might indicate how Americans intended to vote in the 2016 presidential election. The researchers also posited that cognitive ability might underly the ideologies of RWA and SDO.

One goal of the research was to determine if people who were more authoritarian in nature (those higher in RWA and SDO) were more likely to support Trump in the 2016 presidential election, while those lower in RWA and SDO might support Clinton in the 2016 presidential election. The researchers pointed out that political party affiliation was controlled for in all of its analyses.

Here is one example from each of the RWA and SDO scales:

- On a 12-item version of the RWA scale on a scale from 1 (strongly disagree) to 7 (strongly agree), one statement was "Our country will be destroyed someday if we do not smash the perversions eating away at our moral and traditional beliefs." Items were averaged, and higher scores denoted greater endorsement of right-wing authoritarianism.

- On a 16-item version of the SDO scale from 1 (do not at all agree) to 7 (strongly agree), one statement was "Inferior groups should stay in their place." As with the RWA, items were averaged, and higher scores denoted greater endorsement of social dominance orientation.

The researchers from this study found that those higher in the two authoritarian measures were more likely to support Donald Trump in the 2016 presidential election. The idea of authoritarianism and social dominance are prominent in Psychology, especially in studies of prejudice and right-wing extremist movements.

Here are some questions we might ask about using either the RWA or the SDO to measure "authoritarian personality." The same questions come to mind when we think about IQ tests or the SAT exam.

1. Just what is "right-wing authoritarianism" or "social dominance orientation"? We understand these much less well than we understand our weight. The answer in practice seems to be "whatever the RWA and SDO measure." Any claim for validity must rest on what kinds of behavior high RWA and SDO scores go along with. That is, we fall back on predictive validity.

2. As the questions in Example 13 suggest, people who hold strong moral and traditional beliefs and who think that some groups are inferior are likely to get higher scores on both the RWA and the SDO than similar people who don't hold those beliefs. Do the instruments reflect the beliefs of those who developed them? That is, would people with different beliefs come up with a quite different instrument?

3. You think you know what your true weight is. What is the true value of your RWA or SDO score? The measuring devices at NIST can help us find a true weight but not a true RWA or SDO score. If we suspect that the RWA instrument is biased as a measure of "right-wing authoritarianism" because it penalizes those with strong traditional beliefs, how can we check that?

4. You can weigh yourself many times to learn the reliability of your bathroom scale. If you take the RWA or SDO tests many times, you remember what answers you gave the first time. That is, repeats of the same psychological measurement are not really repeats. So reliability is hard to check in practice. Psychologists sometimes develop several forms of the same instrument in order to repeat their measurements. But how do we know these forms are really equivalent?

The point is not that psychologists lack answers to these questions. The first two are controversial because not all psychologists think about human personality in the same way. The second two questions have at least partial answers but not simple answers. The point is that "measurement," which seems so straightforward when we measure weight, is complicated indeed when we try to measure human personality constructs.

There is a larger lesson here. Be wary of statistical "facts" about complicated constructs like personality, intelligence, and even readiness for college. The numbers look solid, as numbers always do. But data are a human product and reflect human desires, prejudices, and weaknesses. This does not mean that there is not validity and reliability in these measures, but if we don't understand and agree on what we are measuring, the numbers may produce more disagreement than enlightenment.

Chapter 8: Statistics in Summary

- To **measure** something means to assign a number to some property of an individual.

- When we measure many individuals, we have values of a **variable** that describe them.

- Variables are recorded in **units.**

- When you work with data or read about a statistical study, ask if the variables are **valid** as numerical measures of the concepts the study discusses.

- Often a **rate** is a more valid measure than a **count.**

- Validity is simple for measurements of physical properties such as length, weight, and time. When we want to measure human personality and other vague properties, **predictive validity** is the most useful way to say whether our measures are valid.

- Also ask if there are **errors in measurement** that reduce the value of the data. You can think about errors in measurement like this:

measured value = true value + bias + random error

- Some ways of measuring are **biased,** or systematically wrong in the same direction.
- To reduce bias, you must use a better **instrument** to make the measurements.
- Other measuring processes lack **reliability,** so that measuring the same individuals again would give quite different results due to **random error.**
- A reliable measuring process will have a small **variance** of the measurements. You can improve the reliability of a measurement by repeating it several times and using the **average** result.

This chapter summary will help you evaluate the Case Study.

Link It

In reasoning from data to a conclusion, we start with the data. In statistics, data are ultimately represented by numbers. The planning of the production of data through a sample or experiment does not by itself produce these numbers. The extent to which these numbers represent the characteristics we wish to study affect the quality and relevance of our conclusions. When you work with data or read about a statistical study, ask exactly how the variables are defined and whether they leave out some things you want to know. This chapter presents several ideas one should think about in assessing the variables measured and hence the conclusions based on these measurements.

Case Study Evaluated

Use what you have learned in this chapter to evaluate the Case Study that opened the chapter. Start by reviewing the Chapter Summary. Then communicate clearly enough for any of your classmates to understand what you are saying. The Case Study that opened the chapter is motivated by research conducted in 2017 by van der Lindenan, Dunkel, and Madison.

1. Brain size was measured "as the sum of the total amount of gray and white matter, identical to the procedure used by Burgaleta et al. (2012). Further, the gray/white matter ratio was calculated and direct measures of intracranial volume were also included in the analyses." Do you think this is a valid measure of brain size? Do you think it is reliable? Do you think it is biased?

2. The researchers used a combination of several cognitive tests to measure general intelligence. Assume other researchers used the same combination of these cognitive tests. Is this a reliable measure of intelligence?

3. The researchers found some evidence that brain size and intelligence are related. However, the study described in Example 9 did not. Discuss the differences in the two studies.

In this chapter you:

- Learned the basics of measurement.
- Saw how to question whether measurements are valid or accurate.
- Considered how we can improve reliability and reduce bias.

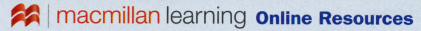 **macmillan** learning **Online Resources**

■ *LearningCurve* has good questions to check your understanding of the concepts.

Check the Basics

For Exercise 8.1, see page 166; for Exercise 8.2, see page 173.

8.3 Comparing marijuana use. According to the 2015–2016 National Survey on Drug Use and Health, among young adults aged 18–25 years, approximately 28.7% of young adults in Arizona, approximately 34.6% of young adults in Michigan, and approximately 47.5% of young adults in Colorado used marijuana in the past year. (*Note:* At the time of the survey, Colorado was the only state of the three with legalized marijuana.) Which of the following is true?

(a) We cannot use these percentages to compare marijuana use by young adults in these states because we do not know how many young adults used marijuana in the past year in each state.

(b) We cannot use these percentages to compare marijuana use by young adults in these states because we do not know how many young adults live in each state.

(c) We can use these percentages to compare marijuana use by young adults in these states because we are given rates.

(d) None of the above is true.

8.4 What is the instrument of measurement? A college president is interested in student satisfaction with recreational facilities on campus. A questionnaire is sent to all students and asks them to rate their satisfaction on a scale of 1 to 5 (with 5 being the best). The instrument of measurement is

(a) a student.

(b) the questionnaire.

(c) the rating on the scale.

(d) satisfaction.

8.5 Weight at the doctor's office. When you visit the doctor's office, several measurements may be taken, one of which is your weight. A doctor's office encourages patients to keep their shoes on to be weighed and promises to subtract 2 pounds for the weight of the patient's shoes. Which of the following is true once the patient's weight is adjusted by 2 pounds?

(a) This will be a reliable measure of the patient's weight.

(b) This will be a valid measure of the patient's weight.

(c) This will be an unbiased measure of the patient's weight.

(d) None of the above is true.

8.6 Measuring athletic ability. Which of the following is *not* a valid measurement of athletic ability?

(a) Time (in seconds) to run a 100-meter dash.

(b) Number of times a person goes to the gym per week.

(c) Maximum weight (in pounds) a person can bench press.

(d) Number of sit-ups a person can do in 1 minute.

8.7 Comparing teaching assistants (TAs). Professor Holmes has two teaching assistants who grade homework for a Statistics 101 course. Professor Holmes gives each of the two teaching assistants a rubric (a clear scoring guide) for the TAs to use when they grade the assignments. Holmes gives each TA the same student's paper to grade and has each TA grade the paper according to the rubric. Professor Holmes is doing this to try to guarantee the scores given by the TAs are

(a) not biased.

(b) predictive.

(c) reliable.

(d) valid.

Chapter 8 Exercises

8.8 Counting the unemployed? We could measure the extent of unemployment by a count (the number of people who are unemployed) or by a rate (the percentage of the labor force that is unemployed). The number of people in the labor force grew from 122 million in September 1988, to 138 million in September 1998, to 155 million in September 2008, to 162 million in September 2018. Use these facts to explain why the count of unemployed people is not a valid measure of the extent of unemployment.

8.9 Measuring a healthy lifestyle. You want to measure the "healthiness" of college students' lifestyles. Give an example of a clearly invalid way to measure healthiness. Then briefly describe a measurement process that you think is valid.

8.10 Rates versus counts. Customers returned 40 cell phones to Verizon this spring, and only 15 to Best Buy next door. Verizon sold 800 cell phones this spring, while Best Buy sold 200.

(a) Verizon had a greater number of cell phones returned. Why does this not show that Verizon's cell phone customers were less satisfied than those of Best Buy?

(b) What is the rate of returns (percentage of cell phones returned) at each of the stores?

(c) Use the rates of returns that you calculated to explain to a friend which store you would suggest your friend go to, to purchase a cell phone.

8.11 Seat belt safety. The National Highway Traffic Safety Administration reports that in 2016, 11,282 occupants of motor vehicles who were wearing a restraint died in motor vehicle accidents and 10,428 who were not wearing a restraint died. These numbers suggest that not using a restraining device is safer than using one. The *counts* aren't fully convincing, however. What *rates* would you like to know to compare the safety of using a restraint with not using one?

8.12 Tough course? A friend tells you, "In the 7:30 A.M. lecture for Statistics 101, 9 students failed, but 20 students failed in the 1:30 P.M. lecture. The 1:30 P.M. prof is a tougher grader than the 7:30 A.M. prof." Explain why the conclusion may not be true. What additional information would you need to compare the classes?

8.13 Obesity. A 2017 Organisation for Economic Co-operation and Development (OECD) report provided the prevalence of obesity among adults in 36 countries. Based on information in the report, only 3.7% of adults in Japan are obese, compared to 22.3% in Turkey and 38.2% in the United States. Do these numbers make a convincing case that Turkey and the United States have a more substantial problem with obesity than Japan?

8.14 Capital punishment. Between 1977 and 2017, 1465 convicted criminals were put to death in the United States. Here are data on the number of executions in several states during those years, as well as the estimated June 1, 2017, population of these states:

State	Population (thousands)	Executions
Alabama	4875	61
Arkansas	3004	31
Delaware	962	16
Florida	20,984	95
Indiana	6667	20
Nevada	2998	12
Oklahoma	3931	112
Texas	28,305	545

Texas is the leader in executions. Because Texas is a large state, we might expect it to have many executions. Find the *rate* of executions for each of the states listed above, in executions per million population. Because population is given in thousands, you can find the rate per million as

$$\text{rate per million} = \frac{\text{executions}}{\text{population in thousands}} \times 1000$$

Arrange the states in order of the number of executions relative to population. Is Texas still high by this measure? Do any other states stand out when you examine the rates?

8.15 Measuring intelligence. One way "intelligence" can be interpreted is as "general problem-solving ability." Explain why it is *not* valid to measure intelligence by a test that asks questions such as

Who wrote "The Star-Spangled Banner"?

Who won the last soccer World Cup?

8.16 National well-being in the UK. Established in 2010, the "Measuring National Well-being" (MNW) program administered by the Office for National Statistics in the United Kingdom attempts to measure "how the UK is doing." The MNW has questions that collect "both objective data (for example, unemployment rate) and subjective data (for example, satisfaction with job) to provide a more complete view of the nation's progress than economic measures such as gross domestic product (GDP) can do alone." How would you measure job satisfaction and its impact on the quality of life?

8.17 Measuring pain. There are 9 million enrollees in the Department of Veterans Affairs health care system. It wants doctors and nurses to treat pain as a "fifth vital sign," to be recorded along with blood pressure, pulse, temperature, and breathing rate. Help out the VA: How would you measure a patient's pain? (*Note*: There is not one correct answer for this question.)

8.18 Fighting cancer. Congress wants the medical establishment to show that progress is being made in fighting cancer. Here are some variables that might be used:

Variable 1. Total deaths from cancer. These have risen sharply over time, from 331,000 in 1970, to 505,000 in 1990, and to 572,000 in 2011.

Variable 2. The percentage of all Americans who die from cancer. The percentage of deaths due to cancer rose steadily, from 17.2% in 1970 to 23.5% in 1990, then leveled off around 23.2% in 2007.

Variable 3. The percentage of cancer patients who survive for 5 years from the time the disease was discovered. These rates are rising slowly. The 5-year survival rate was 50% in the 1975 to 1977 period and 66.5% from 2005 to 2011.

None of these variables is fully valid as a measure of the effectiveness of cancer treatment. Explain why Variables 1 and 2 could increase even if treatment is getting more effective, and why Variable 3 could increase even if treatment is getting less effective.

8.19 Testing job applicants. The law requires that tests given to job applicants must be shown to be directly job related. The Department of Labor believes that an employment test called the General Aptitude Test Battery (GATB) is valid for a broad range of jobs. As in the case of the SAT, blacks and Hispanics get lower average scores on the GATB than do whites. Describe briefly what must be done to establish that the GATB has predictive validity as a measure of future performance on the job.

8.20 Validity, bias, reliability. This winter I went to a local pharmacy to have my weight and blood pressure measured using a sophisticated electronic machine at the front of the store next to the checkout counter. Will the measurement of my weight be biased? Reliable? Valid? Explain your answer.

8.21 An activity on bias. Let's study bias in an intuitive measurement. Figure 8.3 is a drawing of a tilted glass. Reproduce this drawing on 10 sheets of paper. Choose 10 people: 5 men and 5 women. Explain that the drawing represents a tilted glass of water. Ask each subject to draw the water level when this tilted glass is holding as much water as it can.

Figure 8.3

The correct level is horizontal (straight back from the lower lip of the glass). Many people make large errors in estimating the level. Use a protractor to measure the angle of each subject's error. Were your subjects systematically wrong in the same direction? How large was the average error? Was there a clear difference between the average errors made by men and by women?

8.22 An activity on bias and reliability. Cut 5 pieces of string having these lengths in inches:

2.9 9.5 5.7 4.2 7.6

(a) Show the pieces to another student one at a time, asking the subject to estimate the length to the nearest tenth of an inch by eye. The error your subject makes is measured value minus true value and can be either positive or negative. What is the average of the 5 errors? Explain why this average would be close to 0 if there were no bias and we used many pieces of string rather than just 5.

(b) The following day, ask the subject to again estimate the length of each piece of string. (Present them in a different order on the second day.) Explain why the 5 differences between the first and second guesses would all be 0 if your subject were a perfectly reliable measurer of length. The bigger the differences, the less reliable your subject is. What is the average difference (ignoring whether they are positive or negative) for your subject?

8.23 More on bias and reliability. The previous exercise gives five true values for lengths. A subject measures each length twice by eye. Make up a set of results from this activity that matches each of the descriptions below. For simplicity, assume that bias means the same fixed error every time rather than an "on the average" error in many measurements.

(a) The subject has a bias of 0.5 inch too long and is perfectly reliable.

(b) The subject has no bias but is not perfectly reliable, so that the average difference in repeated measurements is 0.5 inch.

8.24 Even more on bias and reliability. Exercise 8.22 gives five true values for lengths. A subject measures the first length (true length = 2.9 inches) four times by eye. His measurements are

3.0 2.9 3.1 3.0

Suppose his measurements have a bias of +0.1 inches.

(a) What are the four random errors for his measurements?

(b) What is the variance of his four measurements?

8.25 Does job training work? To measure the effectiveness of government training programs, it is usual to compare workers' pay before and after training. But many workers sign up for training when their pay drops or they are laid off. So the "before" pay is unusually low and the pay gain looks large.

(a) Is this bias or random error in measuring the effect of training on pay? Why?

(b) How would you measure the success of training programs?

8.26 A recipe for poor reliability. Every month, the government releases data on "personal savings." This number tells us how many dollars individuals saved the previous month. Savings are calculated by subtracting personal spending (an enormously large number) from personal income (another enormous number). The result is one of the government's least reliable statistics.

Give a numerical example to show that small percentage changes in two very large numbers can produce a big percentage change in the difference between those numbers. A variable that is the difference between two big numbers is usually not very reliable.

8.27 Measuring crime. Crime data make headlines. We measure the amount of crime by the number of crimes committed or (better) by crime rates (crimes per 100,000 population). The FBI publishes data on crime in the United States by compiling crimes reported to police departments. The FBI data are recorded in the Uniform Crime Reporting Program and are based on reports from more than 18,000 law enforcement agencies across the United States. The National Crime Victimization Survey publishes data about crimes based on a national probability sample of about 90,000 households per year. The victim survey shows almost two times as many crimes as the FBI report. Explain why the FBI report has a large downward bias for many types of crime. (Here is a case in which bias in producing data leads to bias in measurement.)

8.28 Measuring crime. Twice each year, the National Crime Victimization Survey asks a random sample of households whether they have been victims of crime and, if so, the details. In all, nearly 160,000 people in about 90,000 households answer these questions per year. If other people in a household are in the room while one person is answering questions, the measurement of, for example, rape and other sexual assaults could be seriously biased. Why? Would the presence of other people lead to overreporting or under-reporting of sexual assaults?

8.29 Measuring pulse rate. You want to measure your resting pulse rate. You might count the number of beats in 5 seconds and multiply by 12 to get beats per minute.

(a) Consider counting the number of beats in 15 seconds and multiplying by 4 to get beats per minute. In what way will this improve the reliability of your measurement?

(b) Why are the first two measurement methods less reliable than actually measuring the number of beats in a minute?

8.30 Testing job applicants. A company used to give IQ tests to all job applicants. This is now illegal because IQ is not related to the performance of workers in all the company's jobs.

Does the reason for the policy change involve the *reliability*, the *bias*, or the *validity* of IQ tests as a measure of future job performance? Explain your answer.

8.31 The best earphones. You are writing an article for a consumer magazine based on a survey of the magazine's readers that asked about satisfaction with mid-priced earphones for the iPod and iPhone. Of 1648 readers who reported owning the Apple in-ear headphone with remote and mic, 347 gave it an outstanding rating. Only 69 outstanding ratings were given by the 134 readers who owned Klipsch Image S4i earphones with microphone. Describe an appropriate variable, which can be computed from these counts, to measure high satisfaction with a make of earphone. Compute the values of this variable for the Apple and Klipsch earphones. Which brand has the better high-satisfaction rating?

8.32 Where to work? Each year, *Forbes* magazine ranks the 2000 largest metropolitan areas in the United States in an article on the best places for businesses and careers. First place in 2017 went to Portland, Oregon. Portland was ranked 5th in 2016. Fourth place in 2017 went to Denver, Colorado. Denver was ranked 1st in 2016. At the other end, York, Pennsylvania, ranked 179th in 2017, up 15 spots from 194th in 2016! Are these facts evidence that *Forbes*'s ratings are invalid, biased, or unreliable? Explain your choice.

8.33 Validity, bias, reliability. Give your own example of a measurement process that is valid but has large bias. Then give your own example of a measurement process that is invalid but highly reliable.

Exploring the Web

Access these exercises on the text website: macmillanlearning.com/scc10e.

Do the Numbers Make Sense?

9

Case Study: Estimating Crowd Size

In this chapter you will:

- Discover what questions to ask about data you encounter.
- Learn how to check for consistency in data.
- Learn how to check for plausibility of the data.

Crowd science is used to estimate the number of people in a large crowd. Crowd scientists may have a background in census work, geospatial analysis, remote sensing, or cartography, among other qualifications. Individuals not trained in crowd science are not good at estimating the size of a crowd.

Consider the January 20, 2017, inauguration of Donald Trump. Was President Trump correct in estimating the number of people in attendance as somewhere between 1 and 1.5 million people? Estimates for the number of attendees at the Women's March in Washington held the day after the 2017 inauguration had estimates ranging between 500,000 and 1 million people. Side-by-side photos of the two events show more attendees at the Women's March than at the inauguration.

Both President Trump and the organizers of the Women's March wanted large numbers of people at their events. How do we get accurate data on the number of attendees at each event?

Business data, advertising claims, debate on public issues—we are assailed daily by numbers intended to prove a point, buttress an argument, or assure us that all is well. Sometimes we are fed false data. Sometimes, people who use data to argue a cause care more for the cause than about the accuracy of the data. Others simply lack the skills needed to employ numbers carefully. We know that we should always ask:

- How were the data produced?

- What do we want to measure?

- What exactly was measured?

We also know quite a bit about what good answers to these questions sound like. That's wonderful, but it isn't enough. We also need "number sense," the habit of asking if numbers make sense. Developing number sense is the purpose of this chapter. To help develop number sense, we will look at how bad data, or good data used wrongly, can mislead the unwary.

What Didn't They Tell Us?

The most common way to mislead with data is to cite correct numbers that don't quite mean what they appear to say because we aren't told the full story. The numbers are not made up, so the fact that the information is a bit incomplete may be an innocent oversight. Here are some examples. You decide how innocent they are.

Example 1
Snow! Snow! Snow!

Crested Butte attracts skiers by advertising that it has the highest average snowfall of any ski town in Colorado. That's true. But skiers want snow on the ski slopes, not in the town—and many other Colorado resorts get more snow on the slopes.

Example 2
Yet more snow

News reports of snowstorms say things like "A winter storm spread snow across the area, causing 28 minor traffic accidents." Eric Meyer, a reporter in Milwaukee, Wisconsin, says he often called the sheriff to gather such numbers. One day, he decided to ask the sheriff how many minor accidents are typical in good weather: about 48, said the sheriff. Perhaps, says Meyer, the news should say, "Today's winter storm prevented 20 minor traffic accidents."

Dmitry Kalinovsky/Shutterstock

Example 3

We attract really good students

Colleges know that many prospective students look at popular guidebooks to decide where to apply for admission. The guidebooks print information supplied by the colleges themselves. Surely no college would simply lie about, say, the average SAT score of its entering students or admission rates. But we do want our scores to look good and admission standards to appear high. What would happen if SAT scores were optional when applying for admission? Students with low scores will tend to not include them as part of their application so that average scores increase. In addition, the number of applicants increases, admittance rates decrease, and a college appears more selective.

Some argue that a test-optional policy (TOP) allows colleges to admit a more diverse group of students. How might this happen? In the past, "criticisms of the test's cultural and racial biases, lack of predictive power, and inability to measure an individual student's intelligence" are problematic for schools that are trying to increase diversity. Way back in 1969, Bowdoin College was the first school to adopt a TOP, claiming that standardized tests "tend to work in favor of the more advantaged elements of our society, while handicapping others." More recently, Hobart and William Smith Colleges adopted an SAT-optional policy for fall 2006, and their reported average SAT scores jumped 20 points. At the same time, national average SAT scores declined.

The point of these examples is that numbers have a context. If you don't know the context, the lonely, isolated, naked number doesn't tell you much.

Are the Numbers Consistent with Each Other?

Here is an example.

Example 4

The case of the missing vans

Auto manufacturers lend their dealers money to help them keep vehicles on their lots. The loans are repaid when the vehicles are sold. A Long Island auto dealer named John McNamara borrowed more than $6 billion from General Motors between 1985 and 1991. In December 1990 alone, Mr. McNamara borrowed $425 million to buy 17,000 GM vans customized by an Indiana company, allegedly for sale overseas. GM happily lent McNamara the money because he always repaid the loans.

Let's pause to consider the numbers, as GM should have done but didn't. At the time GM made this loan, the entire van-customizing industry produced only about 17,000 customized vans a month. So McNamara was claiming to buy an entire month's production. These large, luxurious, and gas-guzzling vehicles are designed for U.S. interstate highways. The recreational vehicle trade association says that only 1.35% (not quite 2800 vans) were exported in 1990. It's not plausible to claim that 17,000 vans in a single month are being bought for export. McNamara's claimed purchases were large even when compared with total production of vans. Chevrolet, for example, produced 100,067 full-sized vans in all of 1990.

Having looked at the numbers, you can guess the rest. McNamara admitted in federal court in 1992 that he was defrauding GM on a massive scale. The Indiana company was a shell company set up by McNamara, its invoices were phony, and the vans didn't exist. McNamara borrowed vastly from GM, used most of each loan to pay off the previous loan (thus establishing a record as a good credit risk), and skimmed off a bit for himself. The bit he skimmed amounted to more than $400 million. GM set aside $275 million to cover its losses. Two executives, who should have looked at the numbers relevant to their business, were fired.

John McNamara fooled General Motors because GM didn't compare his numbers with others. No one asked how a dealer could buy 17,000 vans in a single month for export when the entire custom van industry produces just 17,000 vans a month and only a bit over 1% are exported. Here's another example in which the numbers don't line up with each other.

Example 5   *Are unemployment numbers trustworthy?*

As mentioned in Example 8.3, the Bureau of Labor Statistics (BLS) has used the same formula to calculate the U.S. unemployment rate since 1994. Interestingly, prior to becoming president, Donald Trump criticized the unemployment rate reported by the BLS, even calling the calculated rate "total fiction" as recently as December 2016.

In October 2012, Trump said that a 7.8% unemployment figure was incorrect and claimed that unemployment was in reality at least 15%. Three years later, he said that the reported 5.4% unemployment rate was more like 40 or 42%. According to one *Washington Post* article, Trump called unemployment numbers "phony" at least ten times between 2012 and 2016. However, in March 2017, President Trump stated that the unemployment rate "may have been phony in the past, but it's very real now." There was a lack of consistency in the numbers that the BLS and President Trump reported over the years. If the BLS has been using the same formula since 1994, how can the unemployment rate be real and correct now?

During his campaign for president, Donald Trump counted everyone who didn't have a job, including those who are retired, are students, or are stay-at-home parents, as unemployed. However, the BLS considers these groups to be not in the labor force and thus does not include them in the calculation of the unemployment rate. If these groups were included in the calculation, the unemployment rate would skyrocket. This explains the inconsistency between President Trump's numbers and those reported by the BLS. Perhaps it was a lack of understanding of what goes into the calculation that led to this inconsistency prior to Donald Trump becoming president. Now that he is President Trump, he understands why the BLS calculation is correct. Any of us can come up with inconsistent numbers if we do not know how the official numbers are calculated.

Here's an example where we *know* something is wrong because the numbers don't agree. This is part of an article on a cancer researcher at the Sloan-Kettering Institute, which was accused of committing the ultimate scientific sin, falsifying data.

Example 6

Check the numbers

When reporting data about the Minnesota mouse experiments, the cancer researcher at Sloan-Kettering reported percentages of successes for groups of 20 mice as 53, 58, 63, 46, 48, and 67. This is impossible because any percentage of 20 must be a multiple of 5!

Are the Numbers Plausible?

You can often detect dubious numbers simply because they don't seem plausible. Sometimes, you can check an implausible number against data in reliable sources such as the annual *Statistical Abstract of the United States*. Sometimes, as the next example illustrates, you can do a calculation to show that a number isn't realistic.

Example 7

Now that's relief!

Hurricane Katrina struck the Gulf Coast in August 2005 and caused massive destruction. In September 2005, Senators Mary Landrieu (Democrat) and David Vitter (Republican) of Louisiana introduced the Hurricane Katrina Disaster Relief and Economic Recovery Act in Congress.

This bill sought a total of $250 billion in federal funds to provide long-term relief and assistance to the people of New Orleans and the Gulf Coast. Not all of this was to be spent on New Orleans alone, and the money was not meant to be distributed directly to residents affected by the hurricane. However, at the time, several people noticed that if you were one of the 484,674 residents of New Orleans, $250 billion in federal funds was the equivalent of

Jose Gil/Shutterstock

$$\text{dollars per resident} = \frac{250,000,000,000}{484,674} = 515,810.6$$

This would mean that a family of four would receive about $2,063,240!

Now it's your turn

9.1 New Year's Eve in Times Square. Ahead of the 2018 New Year's Eve celebration, Mayor Bill de Blasio said that New York City expected "up to 2 million people in Times Square itself" for the festivities. According to the Associated Press, "crowd-size experts scoff at those mammoth figures . . . saying it's impossible to squeeze that many of even the skinniest revelers into such a relatively small space." Is de Blasio's 2 million figure accurate? You can use the fact that the official website of Times Square (www.timessquarenyc.org) reported just under 420,000 people walked through Times Square on its busiest weekday during December 2018.

Are the Numbers Too Good to Be True?

In Example 6, lack of consistency led to the suspicion that the data were phony. *Too much precision or regularity* can lead to the same suspicion, as when a student's lab report contains data that are exactly as the theory predicts. The laboratory instructor knows that the accuracy of the equipment and the student's laboratory technique are not good enough to give such perfect results. He suspects that the student made them up. Here are just two examples about fraud in medical research.

Example 8

Fraud in medical research

In January 2018, a stem cell scientist at Kyoto University was found guilty of falsifying data for a paper that was published in *Stem Cell Reports*. Kyoto University asked that the paper be retracted from the journal. How did the scientist falsify data? An "investigation concluded that the lead author of the paper, Kohei Yamamizu, had fabricated all six main figures of the study, which were pivotal in the conclusions the author drew.'"

Example 9

Fraud in academic research

A classic example of false data happened in 1986. As an up-and-coming radiologist at the University of California, San Diego, Robert Slutsky appeared to publish many articles. However, when a reviewer was asked to write a letter for Slutsky's promotion, the reviewer discovered that two articles written by Slutsky had the exact same data, but with different numbers of subjects. Needless to say, Slutsky did not get a promotion—in fact, he resigned.

In the Slutsky case, suspicious regularity (exact same data) combined with inconsistency (different numbers of subjects) led a careful reader to suspect fraud.

Is the Arithmetic Right?

Conclusions that are wrong or just incomprehensible are often the result of small arithmetic errors. Rates and percentages cause particular trouble.

Statistics in Your World

Just a little arithmetic mistake In 1994, an investment club of grandmotherly women wrote a best-seller, *The Beardstown Ladies' Common-Sense Investment Guide: How We Beat the Stock Market—and How You Can, Too.* On the book cover and in their many TV appearances, the down-home authors claimed a 23.4% annual return, beating the market and most professionals. Four years later, a skeptic discovered that the club treasurer had entered data incorrectly. The Beardstown ladies' true return was only 9.1%, far short of the overall stock market return of 14.9% in the same period. We all make mistakes, but most of them don't earn as much money as this one did.

Example 10 — *Oh, those percents*

Here are a couple of examples involving percents. During the December 4, 2009, episode of the TV show *Fox & Friends,* a graphic was displayed with the question heading: "Did scientists falsify research to support their own theories on global warming?" The results, attributed to a Rasmussen Reports Poll on global warming, indicated that 59% of people believed this was "somewhat likely," 35% thought it was "very likely," and 26% considered it "not very likely." That adds up to a whopping 120% of those polled! Turns out that *Fox & Friends* misquoted the actual Rasmussen Reports Poll results but didn't notice the error.

Even smart people have problems with percentages. A newsletter for female university teachers asked, "Does it matter that women are 550% (five and a half times) less likely than men to be appointed to a professional grade?" Now 100% of something is all there is. If you take away 100%, there is nothing left. We have no idea what "550% less likely" might mean. Although we can't be sure, it is possible that the newsletter meant that the likelihood for women is the likelihood for men divided by 5.5. In this case, the percentage decrease would be

$$\text{percentage decrease} = \frac{\text{likelihood for men} - \text{likelihood for women}}{\text{likelihood for men}} \times 100\%$$

$$= \frac{\text{likelihood for men} - (\text{likelihood for men}/5.5)}{\text{likelihood for men}} \times 100\%$$

$$= \frac{1 - (1/5.5)}{1} \times 100\%$$

$$= \frac{4.5}{5.5} \times 100\% = 81.8\%$$

Arithmetic is a skill that is easily lost if not used regularly. Fortunately, those who continue to do arithmetic are less likely to be taken in by meaningless numbers. A little thought and a calculator go a long way.

Example 11 — *Summertime is burglary time*

An advertisement for a home security system says, "When you go on vacation, burglars go to work. According to FBI statistics, over 26% of home burglaries take place between Memorial Day and Labor Day."

This is supposed to convince us that burglars are more active in the summer vacation period. Look at your calendar. There are 14 weeks between Memorial Day and Labor Day. As a percentage of the 52 weeks in the year, this is

$$\frac{14}{52} = 0.269 \quad \text{(that is, 26.9\%)}$$

So the ad claims that 26% of burglaries occur in 27% of the year. This seems to make sense, so there is no cause for concern.

Example 12

The old folks are coming

A writer in *Science* claimed in 1976 that "people over 65, now numbering 10 million, will number 30 million by the year 2000, and will constitute an unprecedented 25 percent of the population." Sound the alarm: the elderly were going to triple in a quarter century to become a fourth of the population.

Let's check the arithmetic. Thirty million is 25% of 120 million, because

$$\frac{30}{120} = 0.25$$

interstid/Shutterstock

So the writer's numbers make sense only if the population in 2000 is 120 million. The U.S. population in 1975 was already 216 million. Something is wrong.

Thus alerted, we can check the *Statistical Abstract of the United States* to learn the truth. In 1975, there were 22.4 million people over age 65, not 10 million. That's more than 10% of the total population. The estimate of 30 million by the year 2000 was only about 11% of the population of 281 million for that year. Looking back, we now know that people at least 65 years old were 12% of the total U.S. population. As people live longer, the numbers of the elderly are growing. But growth from 10% to 12% over 25 years is far slower than the *Science* writer claimed.

Calculating the percentage increase or decrease in some quantity seems particularly prone to mistakes. The percentage change in a quantity is found by

$$\text{percentage change} = \frac{\text{amount of change}}{\text{starting value}} \times 100\%$$

Example 13

Stocks go up, stocks go down

On March 26, 2018, the NASDAQ composite index of stock prices closed at 7220.54 with its ninth largest point increase of 227.87. The NASDAQ composite index closed at 6992.60 the previous day. What percentage increase was this?

$$\text{percentage change} = \frac{\text{amount of change}}{\text{starting value}} \times 100\%$$

$$= \frac{7220.54 - 6992.80}{6992.80} \times 100\%$$

$$= \frac{227.87}{6992.80} \times 100\% = 0.033 \times 100\% = 3.3\%$$

Of course, stock prices go down as well as up. The very next day, the NASDAQ composite index of stock prices closed at 7008.81 with its ninth largest point *decrease* of 211.73. According to NASDAQ's stock market news, "Fears of an ensuing trade war between the U.S. and China led to grievous losses for the benchmarks on Friday. . . . Further, technology stocks, which have been driving the markets this year as well as in 2017, took a massive hit."

That's a percentage decrease of

$$\frac{\text{amount of change}}{\text{starting value}} \times 100\% = \frac{-211.73}{7220.54} \times 100\% = -2.9\%$$

Remember to always use the *starting* value, not the smaller value, in the denominator of your fraction. Also remember that you are interested in the *percentage* change, not the actual change (here, the −2.9% is the correct value to report instead of the −211.73 difference).

Now it's your turn

9.2 Percentage increase and decrease. Dejah earned $100,000 last year at her startup and wants to prepare for potential lean times. She plans to take a 20% salary decrease next year and will then take a 20% increase in the second year. One of her friends says, "That's not bad, Dejah. You will be back to your original salary in two years." Help Dejah explain to her friend that she will not be making her original salary in two years.

A quantity can increase by any amount—a 100% increase just means it has doubled. But nothing can go down more than 100%—it has then lost 100% of its value, and 100% is all there is.

Is There a Hidden Agenda?

Lots of people feel strongly about various issues, so strongly that they would like the numbers to support their feelings. Often, they can find support in numbers by choosing carefully which numbers to report or by working hard to squeeze the numbers into the shape they prefer. Here are two examples.

Example 14

Heart disease in women

A highway billboard says simply, "Half of all heart disease victims are women." What might be the agenda behind this true statement? Perhaps the billboard sponsors just want to make women aware that they do face risks from heart disease. (Surveys show that many women underestimate the risk of heart disease.)

On the other hand, perhaps the sponsors want to fight what some people see as an overemphasis on male heart disease. In that case, we might want to know that although half of heart disease victims are women, they are on the average much older than male victims. Roughly 50,000 women under age 65 and 100,000 men under age 65 die from heart disease each year. The American Heart Association says, "Risk of death due to coronary heart disease in women is roughly similar to that of men 10 years younger."

Example 15

Income inequality

During the economic boom of the 1980s and 1990s in the United States, the gap between the highest and lowest earners widened. In 1980, the bottom fifth of households received 4.3% of all income, and the top fifth received 43.7%. By 1998, the share of the bottom fifth had fallen to 3.6% of all income, and the share of the top fifth of households had risen to 49.2%. That is, the top fifth's share was almost 14 times the bottom fifth's share.

Can we massage the numbers to make it appear that the income gap is smaller than it actually is? An article in *Forbes* (an American business magazine) tried. First, according to data from the Current Population Survey, household income tends to be larger for larger households, so let's change to income per person. The rich pay more taxes, so look at income after taxes. The poor receive food stamps and other assistance, so let's count that. Finally, high earners work more hours than low earners, so we should adjust for hours worked. After all this, the share of the top fifth is only three times that of the bottom fifth. Of course, hours worked are reduced by illness, disability, care of children and aged parents, and so on. If *Forbes*'s hidden agenda is to show that income inequality isn't important, we may not agree.

Yet other adjustments are possible. Income, in these U.S. Census Bureau figures, does not include capital gains from, for example, selling stocks that have gone up. Almost all capital gains go to the rich, so including them would widen the income gap. *Forbes* didn't make this adjustment. Making every imaginable adjustment in the meaning of "income," says the U.S. Census Bureau, gives the bottom fifth of households 4.7% of total income in 1998 and the top fifth 45.8%.

The gap between the highest and lowest earners continues to widen. In 2017, according to the U.S. Census Bureau, the bottom fifth of households received 3.1% of all income and the top fifth received 51.5%, with the top 5 percent of households receiving 22.6% of all income.

Chapter 9: Statistics in Summary

- Pay attention to self-reported statistics. Ask exactly what a number measures and decide if it is a **valid measure.**

- Look for the context of the numbers and ask if there is important **missing information.**

- Look for **inconsistencies,** numbers that don't agree as they should, and check for **incorrect arithmetic.**

- Compare numbers that are **implausible**—surprisingly large or small—with numbers you know are right.

- Be suspicious when numbers are **too regular or agree too well** with what their author would like to see.

- Look with special care if you suspect the numbers are put forward in support of some **hidden agenda.**

This chapter summary will help you evaluate the Case Study.

Link It

In Chapters 1 through 8, we have seen that we should always ask how the data were produced and exactly what was measured. Both affect the quality of any conclusions drawn. The goal is to gain insight by means of the numbers that make up our data. Numbers are most likely to yield insights to those who examine them properly. We need to develop "number sense," the habit of asking if the numbers make sense. To assist you, in this chapter we have provided you with examples of bad data and of good data used wrongly. If you form the habit of looking at numbers properly, your friends will soon think that you are brilliant. They might even be right.

Case Study Evaluated

Use what you have learned in this chapter to evaluate the Case Study that opened the chapter. Start by reviewing the Chapter Summary. Then communicate clearly enough for any of your classmates to understand what you are saying.

How were the crowd sizes for the 2017 U.S. Presidential Inauguration and the Women's March on Washington calculated? We don't know how the estimates were made by either President Trump or the organizers of the Women's March. However, crowd scientists estimate that the crowd at the "women's march in Washington was roughly three times the size of the audience at President Trump's inauguration." By analyzing photographs and video, crowd scientists estimate the actual attendances as about 170,000 at the inauguration and about 460,000 at the Women's March. For more information, use the Internet to find out more about crowd science.

In this chapter you:

- Discovered what questions to ask about data you encounter.
- Learned how to check for consistency in data.
- Learned how to check for plausibility of the data.

macmillan learning Online Resources

LearningCurve has good questions to check your understanding of the concepts.

Check the Basics

For Exercise 9.1, see page 190; for Exercise 9.2, see page 194.

9.3 Academic good standing. To be in good standing at college, a student needs to maintain a certain grade point average (GPA). Typically, a student who falls below 2.0 is put on academic probation for at least a semester. At a local college, administrators wanted to increase the GPA to remain in good standing from 2.0 and reported that the anticipated increase of students on probation would be from 3% to 5% of the student body. Relatively speaking, the increase in the percentage of students at the college who would be on academic probation would be

(a) 2% over the current figure.

(b) 40% over the current figure.

(c) 67% over the current figure.

(d) We cannot determine the percentage increase because we do not know how many students attend the college.

9.4 HIV infection rates. According to a June 2015 report by the Centers for Disease Control and Prevention (CDC), one in eight Americans with HIV do not know that they are infected with the virus. The night the report was released, a news station reported "1 in 8 Americans do not know that they are infected with HIV." What is wrong with the news station's reporting? [*Note: More than one answer is possible.*]

(a) Nothing is wrong. The news station reported exactly the information it should have.

(b) The news station's report implies that one in eight Americans is infected with HIV.

(c) The news station's report implies that one in eight Americans who have not been tested for HIV actually is infected with HIV.

(d) The news station's report omitted the important piece of information that the one in eight figure is for those who are infected with HIV.

9.5 **Decreasing household trash.** A large city has started a new recycling program that aims to reduce household trash by 25% by educating residents about more household items that can be recycled. A friend of yours says, "This is great! The amount of recycling will increase by 25%. That's a great start!" What is wrong with your friend's comment?

(a) Your friend assumed that the amount of trash and the amount of recycling was equal to begin with.

(b) We know that a 25% reduction in household trash will result in less than a 25% increase in recycling.

(c) We know that a 25% reduction in household trash will result in more than a 25% increase in recycling.

(d) None of the above.

9.6 **Increasing test scores.** The average score on an exam in organic chemistry was 60. The professor gave a retest and reported that the average score on the exam increased by 25%. This means that the average score on the retest was

(a) 45.

(b) 75.

(c) 85.

(d) It is impossible to know without having all of the scores from the retest.

9.7 **Classes on Fridays.** A professor is concerned that not enough courses at his college hold class meetings on Fridays. After speaking with administrators, the professor finds out that 15% of courses have class meetings on Fridays. At this school, classes meet on either Monday-Wednesday-Friday or Tuesday-Thursday and do not meet on Saturday or Sunday.

(a) The 15% figure is about right because Friday is one of seven days of the week.

(b) The 15% figure is low because it is less than 20%, and Friday is one of five days on which classes meet.

(c) We cannot make a decision about the 15% figure because we do not know how courses are divided among teaching days (e.g., Monday-Wednesday-Friday or Tuesday-Thursday).

(d) The 15% figure is high because students tend to skip classes on Fridays.

Chapter 9 Exercises

9.8 **Drunk driving.** A newspaper article on drunk driving cited data on traffic deaths in Rhode Island: "Forty-two percent of all fatalities occurred on Friday, Saturday, and Sunday, apparently because of increased drinking on the weekends." What percent of the week do Friday, Saturday, and Sunday make up? Are you surprised that 42% of fatalities occur on those days?

9.9 **Advertising painkillers.** Shortly after its release, an advertisement for the pain reliever Tylenol was headlined "Why Doctors Recommend Tylenol More Than All Leading Aspirin Brands Combined." The makers of Bayer Aspirin, in a reply headlined "Makers of Tylenol, Shame on You!" accused Tylenol of misleading by giving the truth but not the whole truth. You be the detective. How was Tylenol's claim misleading even if true?

9.10 **Play fantasy sports and win.** FanDuel and DraftKings are two popular fantasy sports providers that allow players to bet and (possibly) win money. Prior to a 2016 lawsuit against both providers, commercials about how you could win a lot of money with FanDuel or DraftKings were rampant. According to a *Fortune* article, the ads "implied that all fantasy sports players, regardless of resources and experience, had the same chance of winning—and winning big." The article went on to say that "in reality, just a small percentage of professional players with research, software, and large bankrolls won the majority of the jackpot." If FanDuel and Draft-Kings continued advertising as they were without disclosing their terms, conditions, expected winnings, and fees, how might people watching the commercials be misled?

9.11 Deer in the suburbs. Westchester County is a suburban area covering 438 square miles immediately north of New York City. A garden magazine claimed that the county is home to 800,000 deer. Do a calculation that shows this claim to be implausible.

9.12 Suicides among Vietnam veterans. Did the horrors of fighting in Vietnam drive many veterans of that war to suicide? A figure of 150,000 suicides among Vietnam veterans in the 20 years following the end of the war has been widely quoted. Explain why this number is not plausible. To help you, here are some facts: about 20,000 to 25,000 American men commit suicide each year; about 3 million men served in Southeast Asia during the Vietnam War; there were roughly 93 million adult men in the United States 20 years after the war.

9.13 Trash at sea? A report on the problem of vacation cruise ships polluting the sea by dumping garbage overboard said:

On a seven-day cruise, a medium-size ship (about 1,000 passengers) might accumulate 222,000 coffee cups, 72,000 soda cans, 40,000 beer cans and bottles, and 11,000 wine bottles.

Are these numbers plausible? Do some arithmetic to back up your conclusion. Suppose, for example, that the crew is as large as the passenger list. How many cups of coffee must each person drink every day?

9.14 Funny numbers. Here's a quotation from a book review in a scientific journal:

. . . a set of 20 studies with 57 percent reporting significant results, of which 42 percent agree on one conclusion while the remaining 15 percent favor another conclusion, often the opposite one.

Do the numbers given in this quotation make sense? Can you determine how many of the 20 studies agreed on "one conclusion," how many favored another conclusion, and how many did not report significant results?

9.15 Airport delays. An article in a midwestern newspaper about flight delays at major airports said:

According to a Gannett News Service study of U.S. airlines' performance during the past five months, Chicago's O'Hare Field scheduled 114,370 flights. Nearly 10 percent, 1,136, were canceled.

Check the newspaper's arithmetic. What percent of scheduled flights from O'Hare were actually canceled?

9.16 How many miles do we drive? Here is an excerpt from Robert Sullivan's "A Slow-Road Movement?" in the Sunday magazine section of the *New York Times* on June 25, 2006:

In 1956, Americans drove 628 million miles; in 2002, 2.8 billion. . . . In 1997, according to the Department of Transportation, the Interstate System handled more than 1 trillion ton-miles of stuff, a feat executed by 21 million truckers driving approximately 412 billion miles.

(a) There were at least 100 million drivers in the United States in 2002. How many miles per driver per year is 2.8 billion miles? Does this seem plausible?

(b) According to the report, on average, how many miles per year do truckers drive? Does this seem plausible?

(c) Check the most recent *Statistical Abstract of the United States* at www.census.gov and determine how many miles per year Americans actually drive.

9.17 Battered women? A letter to the editor of the *New York Times* complained about a *Times* editorial that said "an American woman is beaten by her husband or boyfriend every 15 seconds." The writer of the letter claimed that "at that rate, 21 million women would be beaten by their husbands or boyfriends every year. That is simply not the case." He cited the National Crime Victimization Survey, which estimated 56,000 cases of violence against women by their husbands and 198,000 by boyfriends or former boyfriends. The survey showed 2.2 million assaults against women in all, most by strangers or someone the woman knew who was not her past or present husband or boyfriend.

(a) First do the arithmetic. Every 15 seconds is 4 per minute. At that rate, how many beatings would take place in an hour? In a day? In a year? Is the letter writer's arithmetic correct?

(b) Is the letter writer correct to claim that the *Times* overstated the number of cases of domestic violence against women?

9.18 We can read, but can we count? The U.S. Census Bureau once gave a simple test of literacy in English to a random sample of 3400 people. The *New York Times* printed some of the questions under the headline "113% of Adults in U.S. Failed This Test." Why is the percent in the headline clearly wrong?

9.19 Stocks go down. On February 5, 2018, the Dow Jones Industrial Average dropped 1175 points from its opening level of 25,521. This was the biggest one-day point decline ever. By what percentage did the Dow drop that day? On October 19, 1987, the Dow Jones Industrial Average dropped 508 points from its opening level of 2245. By what percentage did the Dow drop that day? This was the biggest one-day percentage drop ever.

9.20 Poverty. The number of Americans living below the official poverty line increased from 35,574,000 to 39,698,000 in the 20 years between 1997 and 2017. What percentage increase was this? You should not conclude from these values alone that poverty became more common during this time period. Why not?

9.21 Reducing CO_2 emissions. An online article reported the following:

In order to limit warming to two degrees Celsius and permit development, several developed countries would have to reduce their CO_2 emissions by more than 100 percent.

Explain carefully why it is impossible to reduce anything by more than 100%.

9.22 Are men more promiscuous? Are men more promiscuous by nature than women? Surveys seem to bear this out, with men reporting more sexual partners than women.

On August 12, 2007, the *New York Times* reported the following.

One survey, recently reported by the U.S. government, concluded that men had a median of seven female sex partners. Women had a median of four male sex partners. Another study, by British researchers, stated that men had 12.7 heterosexual partners in their lifetimes and women had 6.5.

There is one mathematical problem with these results. What is this problem?

9.23 Don't dare to drive? In early 2019 several media outlets reported new statistics from the National Safety Council that revealed that, for the first time in history, Americans are more likely to die from an accidental opioid overdose than from motor vehicle crashes. According to the NSC,

The odds of dying accidentally from an opioid overdose have risen to one in 96, eclipsing the odds of dying in a motor vehicle crash (one in 103).

Should you be worried about your chance of dying in a motor vehicle accident this year? To answer this question, use the facts that there are about 327 million people in the United States and there are about 48,000 motor vehicle deaths per year. What is the chance a typical person will die in a motor vehicle accident this year?

9.24 How many miles of highways? *Organic Gardening* magazine once said that "the U.S. Interstate Highway System spans 3.9 million miles and is wearing out 50% faster than it can be fixed. Continuous road deterioration adds $7 billion yearly in fuel costs to motorists." The distance from the East Coast to the West Coast of the United States is about 3000 miles. How many separate highways across the continent would be needed to account for 3.9 million miles of roads? What do you conclude about the number of miles in the interstate system?

9.25 In the garden. *Organic Gardening* magazine, describing how to improve your garden's soil, said, "Since a 6-inch layer of soil in a 100-square-foot plot weighs about 45,000 pounds, adding 230 pounds of compost will give you an instant 5% organic matter."

(a) What percent of 45,000 is 230?

(b) Water weighs about 62 pounds per cubic foot. There are 50 cubic feet in a garden layer 100 square feet in area and 6 inches deep. What would 50 cubic feet of water weigh? Is it plausible that 50 cubic feet of soil weighs 45,000 pounds?

(c) It appears from part (b) that the 45,000 pounds isn't right. In fact, soil weighs about 75 pounds per cubic foot. If we use the correct weight, is the "5% organic matter" conclusion roughly correct?

9.26 No eligible men? A news report quotes a sociologist as saying that for every 233 unmarried women in their 40s in the United States, there are only 100 unmarried men in their 40s. These numbers point to an unpleasant social situation for women of that age. Are the numbers plausible? (*Optional:* The *Statistical Abstract of the United States* has a table titled "Marital status of the population by age and sex" that gives the actual counts.)

9.27 Too good to be true? The late English psychologist Cyril Burt was known for his studies of the IQ scores of identical twins who were raised apart. The high correlation between the IQs of separated twins in Burt's studies pointed to heredity as a major factor in IQ. ("Correlation" measures how closely two variables are connected. We will meet correlation in Chapter 14.) Burt wrote several accounts of his work, adding more pairs of twins over time. Here are his reported correlations as he published them:

Publication date	Twins reared apart	Twins reared together
1955	0.771 (21 pairs)	0.944 (83 pairs)
1966	0.771 (53 pairs)	0.944 (95 pairs)

What is suspicious here?

9.28 Where you start matters. When comparing numbers over time, you can slant the comparison by choosing your starting point. Say the Chicago Cubs lose five games, then win four, then lose one. You can truthfully say that the Cubs have lost 6 of their last 10 games (sounds bad) or that they have won 4 of their last 5 (sounds good).

The following example can also be used to make numbers sound bad or good. The median income of American households (in dollars of 2017 buying power) was $56,533 in 1997, $59,534 in 2007, and $61,372 in 2017. All three values are in dollars of 2017 buying power, which allows us to compare the numbers directly. By what percentage did household income increase between 1997 and 2017? Between 2007 and 2017? You see that you can make the income trend sound bad or good by choosing your starting point.

9.29 Being on top also matters. The previous exercise noted that median household income increased slightly between 2007 and 2017. The top 5% of households earned $209,773 or more in 2007 and $237,034 or more in 2017. It is important to note that these amounts are both in dollars of 2017 buying power, which allows us to compare the numbers directly.

By what percentage did the income of top earners increase between 2007 and 2017? How does this compare with the percentage increase in median household income between 2007 and 2017?

9.30 Boating safety. Data on accidents in recreational boating in the U.S. Coast Guard's *Recreational Boating Statistics Report* show that the number of deaths has dropped from 685 in 2007 to 658 in 2017. The number of injuries reported also fell from 3673 in 2007 to 2629 in 2017. Why does it make sense that the number of deaths in these data is less than number of injuries? Which count (deaths or injuries) is probably more accurate?

9.31 Obesity and income. An article in the November 3, 2009, issue of the *Guardian* reported, "A separate opinion poll yesterday suggested that 50% of obese people earn less than the national average income." Income has a distribution that is such that more than 50% of all workers would earn less than the national average. Is this evidence that obese people tend to earn less than other workers?

Exploring the Web

Access these exercises on the text website: macmillanlearning.com/scc10e.

PART I Review

The first and most important question to ask about any statistical study is "Where do the data come from?" Chapter 1 addressed this question. The distinction between observational and experimental data is a key part of the answer. Good statistics starts with good designs for producing data. Chapters 2, 3, and 4 discussed sampling, the art of choosing part of a population to represent the whole. Figure I.1 summarizes the big idea of a simple random sample. Chapters 5 and 6 dealt with the statistical aspects of designing experiments, studies that impose some treatment in order to learn about the response. The big idea is the randomized comparative experiment. Figure I.2 outlines the simplest design.

Random sampling and randomized comparative experiments are perhaps the most important statistical inventions of the twentieth century. Both were slow to gain acceptance, and you will still see many voluntary response samples and uncontrolled experiments. Both random samples and randomized experiments involve the deliberate use of chance to eliminate bias and produce a regular pattern of outcomes. The regular pattern allows us to give margins of error, make confidence statements, and assess the statistical significance of conclusions based on samples or experiments.

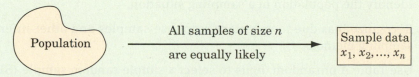

Figure I.1 The idea of a simple random sample.

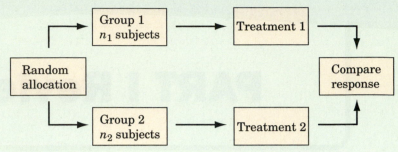

Figure I.2 The idea of a randomized comparative experiment.

When we collect data about people, ethical issues can be important. Chapter 7 discussed these issues and introduced three principles that apply to any study with human subjects. The last step in producing data is to measure the characteristics of interest to produce numbers we can work with. Measurement was the subject of Chapter 8. "Where do the data come from?" is the first question we should ask about a study, and "Do the numbers make sense?" is the second. Chapter 9 encouraged the valuable habit of looking skeptically at numbers before accepting what they seem to say.

PART I SUMMARY

Here are the most important skills you should have acquired after reading Chapters 1 to 9.

A. DATA

1. Recognize the individuals and variables in a statistical study.

2. Distinguish observational from experimental studies.

3. Identify sample surveys, censuses, and experiments.

B. SAMPLING

1. Identify the population in a sampling situation.

2. Recognize bias due to voluntary response samples and other inferior sampling methods.

3. Use Table A of random digits to select a simple random sample (SRS) from a population.

4. Explain how sample surveys deal with bias and variability in their conclusions. Explain in simple language what the margin of error for a sample survey result tells us and what "95% confidence" means.

5. Use the quick method to get an approximate margin of error for 95% confidence.

6. Understand the distinction between sampling errors and nonsampling errors. Recognize the presence of undercoverage and nonresponse as sources of error in a sample survey. Recognize the effect of the wording of questions on the responses.

7. Use random digits to select a stratified random sample from a population when the strata are identified.

C. EXPERIMENTS

1. Explain the differences between observational studies and experiments.

2. Identify the explanatory variables, treatments, response variables, and subjects in an experiment.

3. Recognize bias due to confounding of explanatory variables with lurking variables in either an observational study or an experiment.

4. Outline the design of a completely randomized experiment using a diagram like that in Figure I.2. Such a diagram should show the sizes of the groups, the specific treatments, and the response variable.

5. Use random digits to carry out the random assignment of subjects to groups in a completely randomized experiment.

6. Make use of matched pairs or block designs when appropriate.

7. Recognize the placebo effect.

8. Recognize when the double-blind technique should be used.

9. Be aware of weaknesses in an experiment, especially in the ability to generalize its conclusions.

10. Explain why a randomized comparative experiment can give good evidence for cause-and-effect relationships.

11. Explain the meaning of statistical significance.

D. OTHER TOPICS

1. Explain the three first principles of data ethics (protect subjects from harm, informed consent, and confidentiality). Discuss how they might apply in specific settings.

2. Explain how measuring leads to clearly defined variables in specific settings.

3. Evaluate the validity of a variable as a measure of a given characteristic, including predictive validity.

4. Explain how to reduce bias and improve reliability in measurement.

5. Recognize inconsistent numbers, implausible numbers, numbers so good they are suspicious, and arithmetic mistakes.

6. Calculate percentage increase or decrease correctly.

PART I REVIEW EXERCISES

Review exercises are short and straightforward exercises that help you solidify the basic ideas and skills in each part of this book. We have provided "hints" that indicate where you can find the relevant material for the odd-numbered problems.

I.1 Know these terms. A friend who knows no statistics has encountered some statistical terms in reading for her psychology course. Explain each of the following terms in one or two simple sentences.
(a) Simple random sample (*Hint*: See pages 24–29.)
(b) 95% confidence (*Hint*: See pages 45–46.)
(c) Sampling error (*Hint*: See page 61.)
(d) Statistically significant (*Hint*: See page 97.)

I.2 Know these terms. A friend who knows no statistics has encountered some statistical terms in her biology course. Explain each of the following terms in one or two simple sentences.

(a) Observational study
(b) Double-blind
(c) Nonsampling error
(d) Block design

I.3 A Twitter poll. In July 2018, *CSPAN* ran a poll on Twitter in which they asked, "Do you SUPPORT or OPPOSE the nomination of Judge Brett Kavanaugh to the Supreme Court?" The final result was 54% SUPPORT, 39% OPPOSE, and 7% UNDECIDED. The number of votes received was 42, 145. Explain why these sample results are almost certainly biased. (*Hint*: See pages 22–24.)

I.4 Select an SRS. A student at a large university wants to study the responses that students receive when calling an academic department for information. She selects an SRS of four departments from the following list for her study. Use the *Simple Random Sample* applet, other software, or Table A, starting at line 115 to do this.

Accounting	Electrical Engineering	Natural Resources
Architecture	Elementary Education	Nursing
Art	English	Pharmacy
Biology	Foreign Languages	Philosophy
Business Administration	History	Physics
Chemistry	Horticulture	Political Science
Communication	International Studies	Pre-Med
Computer Science	Marketing	Psychology
Dance	Mathematics	Sociology
Economics	Music	Veterinary Science

I.5 Select an SRS. An outside review team is visiting the history department to assess the quality of its undergraduate major. The department chair wants to select four senior undergraduate majors to meet with the review team. She decides to select these four students by taking an SRS of size 4 from the list of all 30 senior undergraduate history majors. Use the *Simple Random Sample* applet, other software, or Table A, starting at line 128 to draw an SRS of size 4 from the following 30 undergraduate history majors.

Anis	Frazier	Lehman	Murphy	Utlaut
Atlas	Gardner	Leonard	Pagolu	Valente
Bailey	Guthrie	Mee	Ramsey	Weese
Banks	Kowalski	Michelson	Ray	Wendelberger
Edwards	Kupka	Morgan	Sall	Woodall
Foyston	Laperriere	Mosienko	Sawchuk	Yzerman

(*Hint*: See pages 24–29.)

I.6 Errors in surveys. Give an example of a source of nonsampling error in a sample survey. Then give an example of a source of sampling error.

I.7 Errors in surveys. An overnight opinion poll calls randomly selected telephone numbers. This polling method misses all people without a phone. Is this a source of nonsampling error or of sampling error? Does the poll's announced margin of error take this source of error into account? (*Hint*: See pages 64–65.)

I.8 Errors in surveys. A college chooses an SRS of 100 students from the registrar's list of all undergraduates to interview about student life. If it selected two SRSs of 100 students at the same time, the two samples would give somewhat different results. Is this variation a source of sampling error or of nonsampling error? Does the survey's announced margin of error take this source of error into account?

I.9 Errors in surveys. Exercises I.7 and I.8 each mention a source of error in a sample survey. Would each kind of error be reduced by doubling the size of the sample with no other changes in the survey procedures? Explain your answers. (*Hint*: See pages 45–48 and pages 61–65.)

I.10 Errors in surveys. A Gallup Poll found that 46% of American smartphone users agree with the statement, "I can't imagine my life without my smartphone." The Gallup press release says:

> *Results of attitudes and behaviors of smartphone usage are based on 15,747 members of the Gallup Panel who have smartphones . . . For results based on this sample, one can say that the margin of sampling error is ±1 percentage point at the 95% confidence level.*

The release also points out that this margin of error is due only to sampling error. Give one example of a source of error in the poll result that is *not* included in this margin of error.

I.11 Find the margin of error. *Accelerating Acceptance 2018* was published by GLAAD with the results of surveys conducted by the Harris Poll in November 2017. The Accelerating Acceptance report is the "first-of-its-kind index to measure American attitudes toward lesbian, gay, bisexual, transgender, and queer (LGBTQ) people and issues." For the first time since its original publication four years before, GLAAD found an increase in non-LGBTQ people who were uncomfortable around their LGBTQ counterparts in certain situations. Of the 1897 randomly sampled non-LGBT adults, 31% reported they were "very" or "somewhat" uncomfortable seeing a same-sex couple holding hands.
(a) What is the population for this 2017 survey? (*Hint*: See pages 8–11.)
(b) Assuming the 2017 survey used random sampling, use the quick and approximate method to find a margin of error. Then give a complete confidence statement for a conclusion about the population. (*Hint*: See pages 43–44.)

I.12 Find the margin of error. The Accelerating Acceptance report mentioned in Exercise I.11 also found an increase in discrimination against LGBTQ people. A majority, 55%, of LGBTQ persons reported having experienced discrimination based on sexual orientation or gender identity. This was based on 263 randomly selected LGBTQ people.
(a) What is the population for this sample survey?
(b) Assuming the sample was a random sample, use the quick and approximate method to find a margin of error. Make a confidence statement about the opinion of the population.
(c) Explain why there is a difference in the size of the margin of error you calculated in this exercise when compared to the margin of error from Exercise I.11.

I.13 What kind of sample? At a party, there are 30 students over age 21 and 20 students under

age 21. You choose at random 6 of those over 21 and separately choose at random 4 of those under 21 to interview about attitudes toward alcohol. You have given every student at the party the same chance to be interviewed: What is that chance? Why is your sample not an SRS? What is this kind of sample called? (*Hint*: See pages 24–29 and pages 68–72.)

I.14 Design an experiment. A university's Department of Statistics wants to attract more majors. It prepares two advertising brochures. Brochure A stresses the intellectual excitement of statistics. Brochure B stresses how much money statisticians make. Which will be more attractive to first-year students? You have a questionnaire to measure interest in majoring in statistics, and you have 50 first-year students to work with. Outline the design of an experiment to decide which brochure works better.

I.15 Design an experiment. Gary and Greg disagree on which burger chain makes the best fries. Gary likes McDonald's fries but Greg prefers Burger King's. Gary and Greg recruit 20 friends who are willing to participate in a taste test. Outline the design of an experiment to decide which burger chain makes the best fries. (*Hint*: See pages 92–94.)

Exercises I.16 to I.19 are based on an article in the *Lancet* that investigates the use of aspirin to prevent cardiovascular events. This study is related to the Physicians' Health Study discussed in Exercise 5.11. The clinical trial discussed in the article was randomized, double-blind, and placebo controlled and was administered in a total of seven countries between 2007 and 2016. A total of 12,546 participants (at least age 55 for men and at least age 60 for women) were involved in the trial. Subjects were randomized into one of two groups, with 6270 subjects in the aspirin group and 6276 subjects in the placebo group. According to the Lancet article, "The overall incidence rate of serious adverse events was similar in both treatment groups, 20.19% in the aspirin group versus 20.89% in the placebo group. The overall incidence of adverse events was similar in both treatment groups, 82.01% in the aspirin group versus 81.72% in the placebo group."

I.16 Know these terms. Explain in one sentence each what "randomized," "double-blind," "placebo-controlled," and "clinical trial" mean in the description of the design of the study.

I.17 Experiment basics. Identify the subjects, the explanatory variable, and several response variables for this study. (*Hint*: See pages 89–93.)

I.18 Design an experiment. Use a diagram to outline the design of the experiment in this medical study.

I.19 Ethics. What are the three first principles of data ethics? Explain briefly what the medical study must do to apply each of these principles. (*Hint*: See pages 135–137.)

I.20 Measuring. Joni wants to measure the degree to which male college students belong to the political left. She decides simply to measure the length of their hair—longer hair will mean more left-wing.
(a) Is this method likely to be reliable? Why?
(b) This measurement appears to be invalid. Why?
(c) Nevertheless, it is possible that measuring politics by hair length might have some predictive validity. Explain how this could happen.

I.21 Reliability. You are laboring through a chemistry laboratory assignment in which you measure the conductivity of a solution. What does it mean for your measurement to be reliable? How can you improve the reliability of your final result? (*Hint*: See pages 171–174.)

I.22 Observation or experiment? The Nurses' Health Study has queried a sample of more than 100,000 female registered nurses every two years since 1976. Beginning in 1980, the study asked questions about diet, including alcohol consumption. The researchers concluded that "light-to-moderate drinkers had a significantly lower risk of death" than either nondrinkers or heavy drinkers.
(a) Is the Nurses' Health Study an observational study or an experiment? Why?
(b) What does "significant" mean in a statistical report?
(c) Suggest some lurking variables that might explain why moderate drinkers have lower death rates than nondrinkers. (The study did adjust for these variables.)

I.23 Observation or experiment? Can medical cannabis combat the opioid epidemic? An online article in *Leafly* describes a study conducted by scientists at the University of New Mexico. The study compared opioid use by a group of patients with chronic back pain. Some of these patients had been enrolled in New Mexico's Medical Cannabis Program. Others were not. Researchers found that 84% of patients who received access to medical cannabis reduced their opioid prescriptions, compared to 45% in the comparison group. Notably, 41% of the cannabis users stopped using opioids altogether.
(a) Is this an observational study or an experiment? Why? (*Hint*: See pages 7–8 and 12–13.)
(b) We cannot conclude that medical cannabis causes a reduction in opioid use. Suggest other relationships among these variables and perhaps lurking variables that might explain the reduction in opioid use by cannabis users. (*Hint*: See pages 4–6, 7–8, and 90.)

I.24 Percents up and down. Between October 14, 2017, and October 14, 2018, the average price of regular gasoline increased from $2.46 per gallon to $2.90 per gallon.

(a) Verify that this is almost an 18% increase in price.

(b) If the price of gasoline decreases by 18% from its October 14, 2018, level of $2.90 per gallon, what would be the new price? Notice that an 18% increase followed by an 18% decrease does not take us back to the starting point.

I.25 Percentage decrease. On Monday, September 10, 2001 (the day before the September 11 attacks), the NASDAQ stock index closed the day at 1695. By the end of Monday, September 17, 2001 (the first full day of trading after the attacks), the NASDAQ stock index had dropped to 1580. By what percentage did the index drop? (*Hint*: See pages 191–194.)

I.26 An implausible number? *Newsweek* once said in a story that a woman who is not currently married at age 40 has a better chance of being killed by a terrorist than of getting married. Do you think this is plausible? What kind of data would help you check this claim?

PART PROJECTS

Projects are longer exercises that require gathering information or producing data and that emphasize writing a short essay to describe your work. Many are suitable for teams of students.

Access these projects on the text website: **macmillanlearning.com/scc10e.**

PART II

Organizing Data

Words alone don't tell a story. A writer organizes words into sentences and organizes the sentences into a story line. If the words are badly organized, the story isn't clear. Data also need organizing if they are to tell a clear story. Too many words obscure a subject rather than illuminate it. Vast amounts of data are even harder to digest—we often need a brief summary to highlight essential facts. How to organize, summarize, and present data are our topics in the second part of this book.

Organizing and summarizing a large body of facts opens the door to distortions, both unintentional and deliberate. This is no less (but also no more) the case when the facts take the form of numbers rather than words. We will point out some of the traps that data presentations can set for the unwary. Those who picture statistics as primarily a piece of the liar's art concentrate on the part of statistics that deals with summarizing and presenting data. We claim that misleading summaries and selective presentations go back to that after-the-apple conversation among Adam, Eve, and God. Don't blame statistics. Do remember the saying "Figures won't lie, but liars will figure," and beware.

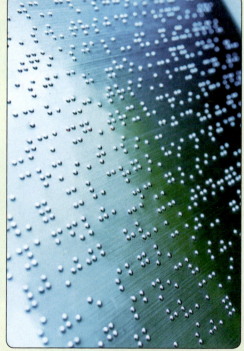

cozyta/Deposit Photos

10

Graphs, Good and Bad

Case Study: Leading Causes of Death

In this chapter, you will:

- Learn some basic methods for displaying data.

- Learn how to assess the quality of the graphics you see in the media.

The National Center for Health Statistics (NCHS) produces a report each year summarizing leading causes of death, changes in mortality over time, and other characteristics for mortality in the United States. From its November 2017 brief "Death: Final Data for 2015," along with its report on the "10 Leading Causes of Death by Age Group–2015," Figure 10.1 was constructed to display the top five causes of death for four age categories. A clear graphical display can tell us a story about a data set and can show us how multiple variables are related. It can also make it easy for policy- and decision-makers to decide how to allocate resources for prevention or intervention. For instance, a quick glance at the pie chart for ages 1–24 shows that a large portion, 15% of deaths in that age range, is attributed to suicide. A clear display of this staggering number makes a strong argument for allocation of resources.

Statistics deals with data, and we use tables and graphs to present data. Tables and graphs help us see what the data say. But not all tables and graphs do so accurately or clearly. By the end of the chapter, you will be able to determine whether Figure 10.1 is a good or a bad graphic.

Data Tables

Take a look at the *Statistical Abstract of the United States,* an annual volume packed with almost every variety of numerical information. Has the number of private elementary and secondary schools grown over time? What about

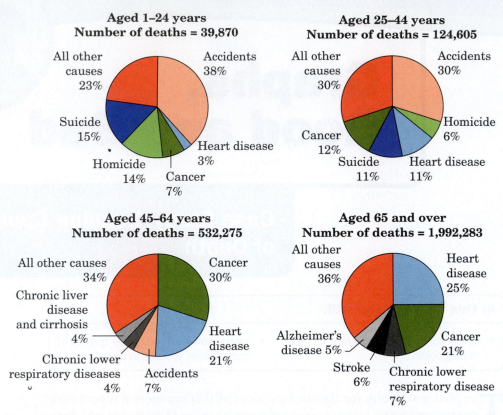

Figure 10.1 Top five causes of death compared across four age categories. (Data from Murphy, SL, Xu, J, Kochanek, KD, et al. Deaths: Final data for 2015. National vital statistics reports; vol 66 no 6. Hyattsville, MD: National Center for Health Statistics. 2017. https://www.cdc.gov/nchs/data/nvsr/nvsr66/nvsr66_06.pdf)

minority enrollments in these schools? How many college degrees were given in each of the past several years, and how were these degrees divided among fields of study and by the age, race, and sex of the students? You can find all this and more in the education section of the *Statistical Abstract*. The tables *summarize* data. We don't want to see information on every college degree individually, only the counts in categories of interest to us.

Example 1

What makes a clear table?

How well educated are adults? Table 10.1 presents the data for people aged 25 years and over. This table illustrates some good practices for data tables. It is clearly *labeled* so that we can see the subject of the data at once. The main heading describes the general subject of the data and gives the date because these data will change over time. Labels within the table identify the variables and state the *units* in which they are measured. Notice, for example, that the counts are in thousands. The *source* of the data appears at the foot of the table. This Census Bureau publication, in fact, presents data from our old friend, the Current Population Survey.

Table 10.1 Education of people aged 25 years and over in 2017

Level of education	Number of persons (thousands)	Percent
Less than high school	22,540	10.4
High school graduate	62,512	28.8
Some college, no degree	35,455	16.4
Associate's degree	22,310	10.3
Bachelor's degree	46,262	21.3
Advanced degree	27,841	12.8
Total	216,921	100.0

Data from Census Bureau, *Educational Attainment in the United States: 2017*.

Table 10.1 starts with the *counts* of people aged 25 years and over who have attained each level of education. *Rates* (percentages or proportions) are often clearer than counts—it is more helpful to hear that 10.4% of this age group did not finish high school than to hear that there are 22,540,000 such people. The percentages also appear in Table 10.1. The last two columns of the table present the *distribution* of the variable "level of education" in two alternate forms. Each of these columns gives information about what values the variable takes and how often it takes each value.

> ## Key Terms
>
> The **distribution** of a variable tells us what values it takes and how often it takes these values.

Example 2   *Roundoff errors*

Did you check Table 10.1 for consistency? The total number of people should be

$$22,540 + 62,512 + 35,455 + 22,310 + 46,262 + 27,841 = 216,920 \text{ (thousands)}$$

But the table gives the total as 216,921. What happened? Each entry is rounded to the nearest thousand. The rounded entries don't quite add to the total, which is rounded separately. Such **roundoff errors** will be with us from now on as we do more arithmetic.

Key Terms

A **categorical variable** places an individual into one of several groups or categories.

A **quantitative variable** takes numerical values for which arithmetic operations such as adding and averaging make sense. A quantitative variable is also sometimes referred to as a numerical variable.

It is not uncommon to see roundoff errors in tables. For example, when table entries are percentages, the total may sum to a value slightly different from 100%, often to 99.9 or 100.1%. When table entries are proportions, the total may sum to a value slightly different from 1, such as 0.99 or 1.01.

Types of Variables

When we think about graphs, it is helpful to distinguish between variables that place individuals into categories (such as gender, occupation, or education level) and those whose values have a meaningful numerical scale (such as height in centimeters or SAT scores).

Pie Charts and Bar Graphs

The variable displayed in Table 10.1, "level of education," is a categorical variable. There are six possible values of the variable. To picture this distribution in a graph, we might use a **pie chart.** Figure 10.2 is a pie chart of the level of education of people aged 25 years and over. Pie charts show how a whole is divided into parts. To make a pie chart, first draw a circle. The circle represents the whole, in this case all people aged 25 years and over. Wedges within the circle represent the parts, with the angle spanned by

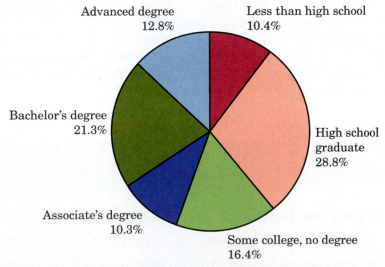

Figure 10.2 Pie chart of the distribution of highest level of education among persons aged 25 years and over in 2017.

each wedge in proportion to the size of that part. For example, 21.3% of those in this age group and over have a bachelor's degree but not an advanced degree. Because there are 360 degrees in a circle, the "bachelor's degree" wedge spans an angle of

$$0.213 \times 360 \text{ degrees} = 76.68 \text{ degrees}$$

Pie charts force us to see that the parts do make a whole. However, it is much easier for our eyes to compare the heights of the bars on a bar graph than it is to compare the size of the angles on a pie chart.

Figure 10.3 is a **bar graph** of the same data. There is a single bar for each education level. The height of each bar shows the percentage of people aged 25 years and over who have attained that highest level of education. Notice that each bar has the same width—this is always the case with a bar graph. Also, there is a space between the bars. It is obvious on either the pie chart or bar graph that the largest category is "High school graduate." However, smaller differences are much more subtle on a pie chart. It is very clear on the bar graph that "Bachelor's degree" is the second largest category. This is not as clear on the pie chart. Bar graphs are better for making comparisons of the sizes of categories. In addition, if there is a natural ordering of the variable, such as how much education a person has, this order can be displayed along the horizontal axis of the bar graph but cannot be displayed in an obvious way in a pie chart.

Pie charts and bar graphs can show the distribution (either counts or percentages) of a categorical variable such as highest level of education. A pie chart usually displays the percentage for each category (rather than the count) and only works if you have all the categories (the percentages

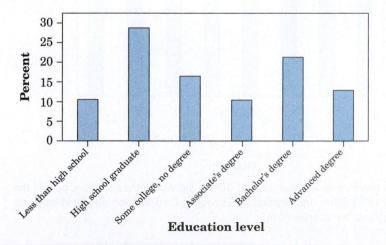

Figure 10.3 Bar graph of the distribution of highest level of education among persons aged 25 years and over in 2017.

add to 100%). A bar graph can display an entire distribution or can be used to display only a few categories. A bar graph can also compare the size of categories that are not parts of one whole. If you have one number to represent each category, you can use a bar graph.

Example 3

High taxes?

Figure 10.4 compares the level of taxation in eight democratic nations. Each democratic nation is a category and each category is described by one value, so a bar graph is a good choice for a graphical display of these data. The heights of the bars show the percentages of each nation's gross domestic product (GDP, the total value of all goods and services produced) that is taken in taxes. Americans accustomed to complaining about high taxes may be surprised to see that the United States, at 25.4% of GDP, is at the bottom of the group. Notice that a pie chart is not possible for these data since we are displaying eight separate quantities, not the parts of a whole.

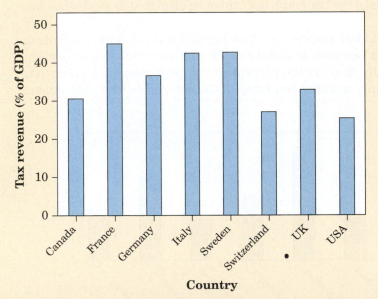

Figure 10.4 Total tax revenue as a percentage of GDP in eight countries that are part of the OECD, Example 3. (Data from the Organization for Economic Co-operation and Development, https://data.oecd.org/tax/tax-revenue.htm.)

Now it's your turn

10.1 Taxes. There are currently seven states in the United States that do not collect income tax. The table below contains the combined state and average local sales tax in 2018 for these seven states as computed by the Tax Foundation.

State	State and average local sales tax (%)
Alaska	1.76
Florida	6.80
Nevada	8.14
South Dakota	6.40
Texas	8.17
Washington	9.18
Wyoming	5.46

Is the variable "state" categorical or quantitative? Should you use a pie chart or a bar graph to display the state and average local sales tax for these states? Why?

Example 4

Government tax revenue breakdown

Governments generate tax revenue through various means. In its 2018 report, the Tax Foundation analyzed data for 2015 from the Organization for Economic Cooperation and Development (OECD). The United States' tax revenue is comprised of individual income tax (40.5%), social insurance tax (23.7%), consumption tax (17.0%), property tax (10.3%), and corporate income tax (8.5%). A bar graph is appropriate to display these data since we have one value to explain the size of each tax category. Notice, a pie graph is appropriate as well since these are all the parts of a whole. What if we wanted to compare the distribution of tax revenue for the United States to the average for all the other OECD countries? Pie charts are not good for comparisons. Figure 10.5 is a bar graph for the distribution of government tax revenue for the United States with a second set of bars representing the average for all other OECD nations adjacent to the bars for the United States. This is called a side-by-side bar graph. A side-by-side bar graph is useful for making comparisons. It is now clear that the United States relies more heavily on individual income tax, while the other countries rely more heavily on consumption taxes.

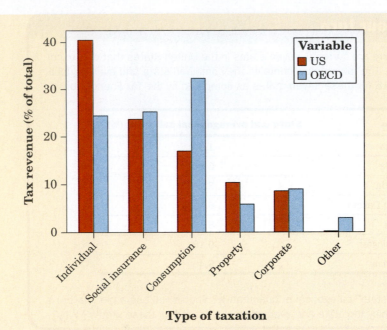

Figure 10.5 Side-by-side bar graph comparing the distribution for sources of government tax revenue in 2015 for the United States to the average of all other countries belonging to the Organization for Economic Co-operation and Development, Example 4. (Data from OECD, StatExtrats, http:stats.oecd.org.)

Beware the Pictogram

Bar graphs compare several quantities by means of the differing heights of bars that represent the quantities. Our eyes, however, react to the *area* of the bars as well as to their height. When all bars have the same width, the area (width × height) varies in proportion to the height and our eyes receive the right impression. When you draw a bar graph, make the bars equally wide. Artistically speaking, bar graphs are a bit dull. It is tempting to replace the bars with pictures for greater eye appeal.

Example 5   *A misleading graph*

Figure 10.6 is a **pictogram.** It is a bar graph in which pictures replace the bars. The graph is aimed at consumers shopping for a vacuum. It claims that a Dyson vacuum has more than twice the suction of any other vacuum. Although not clearly labeled, type of vacuum is shown along the x axis or horizontal axis, and the y axis or vertical axis, gives the calculated number of air watts for each vacuum. We see the Dyson vacuum

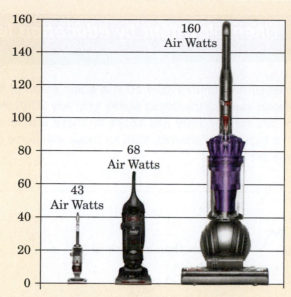

Figure 10.6 Advertisement found on Dyson's website for vacuum cleaners claiming a Dyson vacuum has more than twice the suction of any other vacuum, Example 5. (From *For All Practical Purposes: Mathematical Literacy in Today's World*, 10e by COMAP. Copyright 2016 W.H. Freeman and Company. All rights reserved. Used by permission of the publisher Macmillan Learning.)

(far right) has roughly 4 times the air watts (160 is roughly 4 times 43) of the vacuum on the far left. However, this graphic makes it appear as though there is a much larger difference. Why is this?

To magnify a picture, the artist must increase *both* height and width to avoid distortion of the image. By increasing both the height and width of the Dyson vacuum, it appears to be $4 \times 4 = 16$ times larger. Remember a bar graph should have bars of equal width—only the heights of the bars should vary. Replacing the bars on a bar graph with pictures is tempting, but it is difficult to keep the "bar" width equal without distorting the picture.

Change Over Time: Line Graphs

Many quantitative variables are measured at intervals over time. We might, for example, measure the height of a growing child or the price of a stock at the end of each month. In these examples, our main interest is change over time. To display how a quantitative variable changes over time, make a *line graph*.

When constructing a line graph, make sure to use equally spaced time intervals on the horizontal axis to avoid distortion. Line graphs can also be used to show how a quantitative variable changes over time, broken down according to some other categorical variable. Use a separate line for each category.

Key Terms

A **line graph** of a quantitative variable plots each observation against the time at which it was measured. Always put time on the horizontal axis of your plot and the variable you are measuring on the vertical axis. Connect the data points by lines to display the change over time.

Example 6

Unemployment by education level

How has unemployment changed over time? Figure 10.7 is a line graph of the monthly national unemployment rate, by education level, for the United States from July 2008 through July 2018. For example, the unemployment rate for July 2009 was 15.3% for those with less than a high school diploma, 9.5% for those with a high school diploma, 8.0% for those with some college, and 4.8% for those with a bachelor's degree or higher.

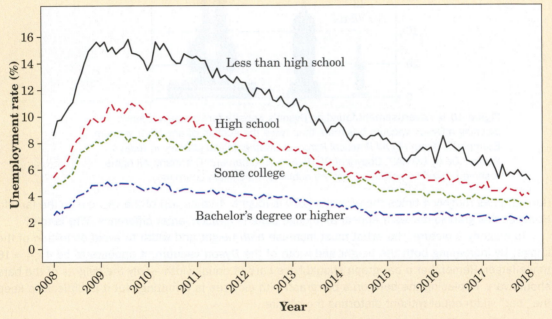

Figure 10.7 Monthly unemployment rate for four education categories from July 2008 through July 2018 reported by the Bureau of Labor Statistics, Example 6. (Data from https://www.bls.gov/charts /employment-situation/civilian-unemployment-rate.htm#.)

It would be difficult to see patterns in a long table of numbers. Figure 10.7 makes the patterns clear. What should we look for?

- First, look for an **overall pattern.** For example, a **trend** is a long-term upward or downward movement over time. Unemployment was at its highest for all education groups in 2009 and 2010, due to the Great Recession.

Since then, the overall trend is showing a decrease in the unemployment rate for all education levels (each line is generally decreasing).

- Next, look for striking **deviations** from the overall pattern. There is a noticeable increase from 2008 to the beginning of 2009. This was a side effect of the recession economy. Unemployment hovered around these record highs through 2010, when the rates finally started their decline to the current levels. There is a striking deviation in the unemployment rate for those with less than a high school degree around mid-2010. As the overall pattern of generally decreasing begins, there is a much more drastic dip in mid-2010 for those with less than a high school degree.

- Change over time often has a regular pattern of **seasonal variation** that repeats each year. Calculating the unemployment rate depends on the size of the workforce and the number of those in the workforce who are working. For example, the unemployment rate rises every year in January as holiday sales jobs end and outdoor work slows in the north due to winter weather. It would cause confusion (and perhaps political trouble) if the government's official unemployment rate jumped every January. These are regular and predictable changes in unemployment. As such, the Bureau of Labor Statistics makes seasonal adjustments to the monthly unemployment rate before reporting it to the general public.

Key Terms

A pattern that repeats itself at known regular intervals of time is called **seasonal variation.** Many series of regular measurements over time are **seasonally adjusted.** That is, the expected seasonal variation is removed before the data are published.

The line graph of Figure 10.7 shows the unemployment rate for four different education levels. A line graph may have only one line or more than one, as was the case in Figure 10.7. Picturing the unemployment rate for all four groups over this time period gives an additional message. We see that unemployment is always lower with more education—a strong argument for a college education! Also, the increase in unemployment was much more drastic during the Great Recession for the lower education levels. It appears that the unemployment rate is much more stable for those with higher education and much more variable for those with lower education levels.

Now it's your turn

10.2 Gasoline prices. The U.S. Energy Information Administration (EIA) keeps track of weekly and monthly average gasoline and diesel prices across the United States. The table below gives the average regular-grade gasoline prices for the month of August, from 1996 to 2018.

Date	Price ($)	Date	Price ($)
1996	1.21	2008	2.78
1997	1.22	2009	2.68
1998	1.03	2010	2.73
1999	1.22	2011	3.64
2000	1.47	2012	3.72
2001	1.42	2013	3.75
2002	1.40	2014	3.49
2003	1.62	2015	2.64
2004	1.88	2016	2.18
2005	2.49	2017	2.38
2006	2.95	2018	2.84
2007	2.78		

Make a line graph of these data and comment on any patterns you observe.

Watch Those Scales!

Because graphs speak so strongly, they can mislead the unwary. The careful reader of a line graph looks closely at the *scales* marked off on the axes.

Example 7   *Living together*

The number of unmarried couples living together has increased in recent years, to the point that some people say that cohabitation is delaying or even replacing marriage. Figure 10.8 presents two line graphs of the number of unmarried-couple households in the United States. The data once again come from the Current Population Survey. The graph on the left suggests a steady but moderate increase. The right-hand graph says that cohabitation is thundering upward.

The secret is in the scales. You can transform the left-hand graph into the right-hand graph by stretching the vertical scale, squeezing the horizontal scale, and cutting off the vertical scale

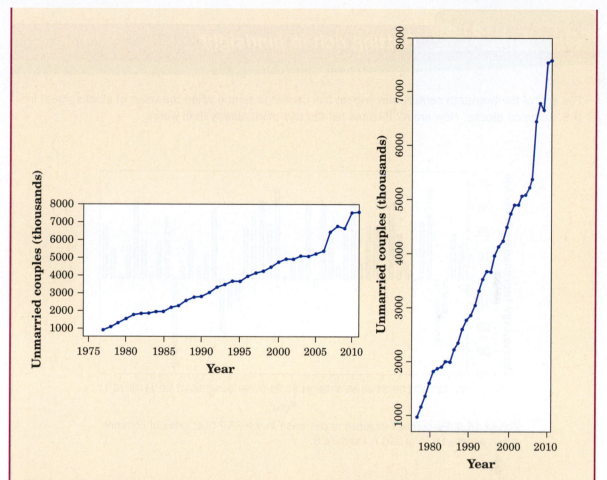

Figure 10.8 The effect of changing the scales in a line graph, Example 7. Both graphs plot the same data, but the right-hand graph makes the increase appear much more rapid.

just above and below the values to be plotted. Now you know how to either exaggerate or play down a trend in a line graph.

Which graph is correct? Both are accurate graphs of the data, but both have scales chosen to create a specific effect. Because there is no one "right" scale for a line graph, correct graphs can give different impressions by their choices of scale. Watch those scales!

Another important issue concerning scales is the following. When examining the change in the price or value of an item over time, plotting the actual increase can be misleading. It is often better to plot the percentage increase from the previous period.

Example 8

Getting rich in hindsight

The end of the twentieth century saw a great bull market (a period when the value of stocks rises) in U.S. common stocks. How great? Pictures tell the tale more clearly than words.

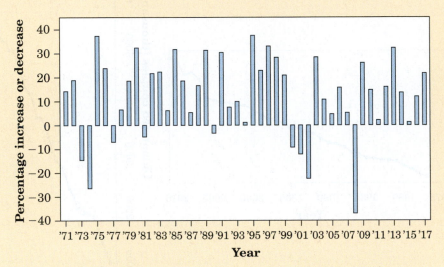

Figure 10.9 Percentage increase or decrease in the S&P 500 index of common stock prices, 1971 to 2017, Example 8.

Look first at Figure 10.9. This shows the percentage increase or decrease in stocks (measured by the Standard & Poor's 500 index) in each year from 1971 to 2017. Until 1982, stock prices bounce up and down. Sometimes they go down a lot—stocks lost 14.7% of their value in 1973 and another 26.5% in 1974. But starting in 1982, stocks go up in 17 of the next 18 years, often by a lot. From 2000 to 2009 stocks again bounce up and down, with a large loss of 37% in the recession that began in 2008.

Figure 10.10 shows, in hindsight, how you could have become rich in the period from 1971 to 1999. If you had invested $1000 in stocks at the end of 1970, the graph shows how much money you would have had at the end of each of the following years. After 1974, your $1000 was down to $853, and at the end of 1981, it had grown to only $2145. That's an increase of only 7.2% a year. Then the great bull market begins its work. By the end of 1999, it would have turned your $1000 into $36,108. Over the next 18 years as a whole, stocks again bounced up and down, and by the end of 2017, your $1000 would have turned into $117,290.

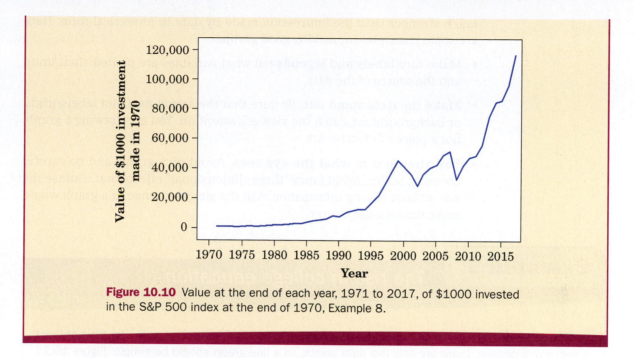

Figure 10.10 Value at the end of each year, 1971 to 2017, of $1000 invested in the S&P 500 index at the end of 1970, Example 8.

Figures 10.9 and 10.10 are instructive. For example, Figure 10.10 might give the impression that increases between 1970 and 1980 were negligible but that increases between 1995 and 1999 were dramatic. While it is true that the *actual value* of our investment increased much more between 1995 and 1999 than it did between 1970 and 1980, it would be incorrect to conclude that investments in general increased much more dramatically between 1995 and 1999 than in any of the years between 1970 and 1980.

Figure 10.9 tells a different, and more accurate, story. For example, the percentage increase in 1975 (approximately 37%) rivaled that in any of the years between 1995 and 1999. However, in 1975 the actual value of our investment was relatively small ($1170) and a 37% increase in such a small amount is nearly imperceptible on the scale used in Figure 10.10. By 1995, the actual value of our investment was about $14,000, and a 37% increase appears much more striking.

This might be a good place to read the "What's the verdict?" story on page 241, and to answer the questions. These questions ask you to assess a graph that appeared in a Reuters News story.

Making Good Graphs

Graphs are the most effective way to communicate using data. A good graph will reveal facts about the data that would be difficult or impossible to detect from a table. What is more, the immediate visual impression of a graph is much stronger than the impression made by data in numerical form. Here are some principles for making good graphs:

- Make sure **labels and legends** tell what variables are plotted, their units, and the source of the data.

- **Make the data stand out.** Be sure that the actual data, not labels, grids, or background art, catch the viewer's attention. You are drawing a graph, not a piece of creative art.

- **Pay attention to what the eye sees.** Avoid pictograms and be careful choosing scales. Avoid fancy "three-dimensional" effects that confuse the eye without adding information. Ask if a simple change in a graph would make the message clearer.

Example 9   *The rise in college education*

Figure 10.11 shows the rise in the percentage of women aged 25 years and over who have at least a bachelor's degree. There are only five data points, so a line graph should be simple. Figure 10.11 isn't simple. The artist couldn't resist a nice background sketch and also cluttered the graph with grid lines. We find it harder than it should be to see the data. Using pictures of the female graduate at each time point distorts the values in the same way as a pictogram. Grid lines on a graph serve no purpose—if your audience must know the actual numbers, give them a table along with the graph. A good graph uses no more ink than is needed to present the data clearly. In statistical graphics, decoration is always a distraction from the data, and sometimes an actual distortion of the data.

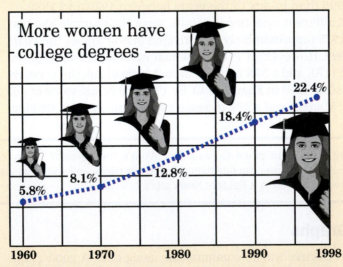

Figure 10.11 Chart junk: This graph is so cluttered with unnecessary ink that it is hard to see the data, Example 9.

Example 10

High taxes, reconsidered

Figure 10.4 is a respectable bar graph comparing taxes as a percentage of gross domestic product in eight countries. The countries are arranged alphabetically. Wouldn't it be clearer to arrange them in order of their tax burdens? Figure 10.12 does this. This simple change improves the graph by making it clearer where each country stands in the group of eight countries. Figure 10.12 also demonstrates the ability to display a bar graph horizontally.

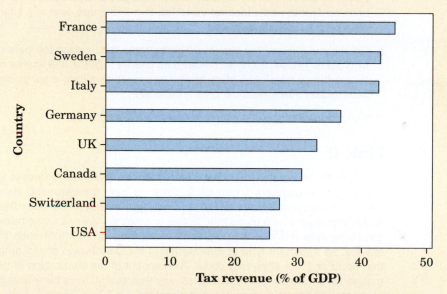

Figure 10.12 Total tax revenue as a percentage of the GDP for eight countries that are part of the OECD, Example 10. Changing the order of the bars has improved the graph in Figure 10.4. (Data from https://data.oecd.org/tax/tax-revenue.htm.)

Chapter 10: Statistics in Summary

- To see what data say, start with graphs.

- The choice of graph depends on the type of data. Do you have a **categorical variable,** such as level of education or occupation, which puts individuals into categories? Or do you have a **quantitative variable** measured in meaningful numerical units?

- Check data presented in a table for **roundoff errors.**

- The **distribution** of a variable tells us what values it takes and how often it takes those values.

- To display the **distribution** of a categorical variable, use a **pie chart** or a **bar graph.** Pie charts always show the parts of some whole, but bar graphs can compare any set of numbers measured in the same units. Bar graphs are better for comparisons. Bar graphs can be displayed vertically or horizontally.

- To show how a quantitative variable changes over time, use a **line graph** that plots values of the variable (vertical scale) against time (horizontal scale). If you have values of the variable for different categories, use a separate line for each category. Look for **trends** and **seasonal variation** in a line graph, and ask whether the data have been **seasonally adjusted.**

- Graphs can mislead the eye. Avoid **pictograms** that replace the bars of a bar graph by pictures whose height and width both change. Look at the scales of a line graph to see if they have been stretched or squeezed to create a particular impression. Avoid clutter that makes the data hard to see.

This chapter summary will help you evaluate the Case Study.

Link It

In reasoning from data to a conclusion, where the data come from is important. We studied this in Chapters 1 through 9. Once we have the data, and are satisfied that they were produced appropriately, we can begin to determine what the data tell us. Tables and graphs help us do this. In this chapter, we learned some basic methods for displaying data with tables and graphs. We learned what information these graphics provide. An important type of information is the distribution of the data—the values that occur and how often they occur. The concept of the distribution of data or the distribution of a variable is a fundamental way that statisticians think about data. We will encounter it again and again in future chapters.

Data that are produced badly can mislead us. Likewise, graphs that are produced badly can mislead us. In this chapter we learned how to recognize bad graphics. Developing "graphic sense," the habit of asking if a graphic accurately and clearly displays our data, is as important as developing "number sense," discussed in Chapter 9.

Case Study Evaluated

Use what you have learned in this chapter to evaluate the Case Study that opened the chapter. Start by reviewing the Chapter Summary. Then answer each of the following questions in complete sentences. Be sure to communicate clearly enough for any of your classmates to understand what you are saying.

Is Figure 10.1 the best graphical representation of the top five leading causes of death across four major age categories? What are the drawbacks to the current graphical display? Discuss which graphical display would be better and why. If you have access to statistical software, create the graphical display you think is better.

In this chapter you:

- Learned some basic methods for displaying data.
- Learned how to assess the quality of the graphics you see in the media.

macmillan learning Online Resources

- The Snapshots video *Visualizing and Summarizing Categorical Data* discusses categorical data and describes pie charts and bar graphs to summarize categorical data in the context of data from a NASA program.

- The StatClips video *Summaries and Pictures for Categorical Data* discusses how to draw a pie chart and a bar graph using two examples.

- The StatClips Examples video *Summaries and Pictures for Categorical Data Example A* discusses pie charts and bar graphs in the context of data about the choice of field of study by first-year students.

- The StatClips Examples video *Summaries and Pictures for Categorical Data Example B* discusses bar graphs in the context of data from a survey about high-tech devices.

- The StatClips Examples video *Summaries and Pictures for Categorical Data Example C* discusses the use of pie charts and bar graphs in the context of data from the Arbitron ratings.

Check the Basics

For Exercise 10.1, see page 219; for Exercise 10.2, see page 224.

10.3 Categorical and quantitative variables A survey was conducted and respondents were asked what color car they drive and how many miles they travel per day. The variable type for "car color" and "miles per day" are

(a) both categorical.

(b) both quantitative.

(c) car color is quantitative, miles per day is categorical.

(d) car color is categorical, miles per day is quantitative.

10.4 Line Graph Which of the following is not acceptable for a line graph?

(a) Having lines that cross

(b) Having equally spaced time intervals

(c) Having time intervals that are not equally spaced

(d) Including seasonal variation

10.5 Distributions A recent report on the religious affiliation of Hispanics says 55% are Catholic, 22% Protestant, 18% unafilliated, and 4% other. This adds up to 99%. You conclude

(a) the remaining 1% are not accounted for.

(b) there was a calculation or data entry error.

(c) this is due to roundoff error.

(d) there is a religion missing from the survey.

10.6 Which graph? You have the average SAT score of entering freshman for five universities. The best graphical display for these data would be a

(a) pie chart.

(b) bar graph.

(c) line graph.

(d) side-by-side bar graph.

10.7 Which graph? You want to show how the price of cable television has changed over the years. You should use a

(a) pie chart.

(b) bar graph.

(c) line graph.

(d) side-by-side bar graph.

10.8 Bar graph For a bar graph to be accurate

(a) the bars should be vertical rather than horizontal.

(b) the bars should have equal height.

(c) the bars must touch each other.

(d) the bars should have equal width.

Chapter 10 Exercises

10.9 Lottery sales. States sell lots of lottery tickets. Table 10.2 shows where the money comes from in the state of Illinois. Make a bar graph that shows the distribution of lottery sales by type of game. Is it also proper to make a pie chart of these data? Explain.

Table 10.2 Illinois State Lottery sales by type of game, fiscal year 2010

Game	Sales ($)
Pick Three	301,417,049
Pick Four	191,038,518
Lotto	111,158,528
Little Lotto	119,634,946
Mega Millions	221,809,484
Megaplier5	848,077
Pick N Play	1,549,252
Raffle	19,999,460
Powerball	43,269,461
Power Play	8,469,680
Instants	1,190,109,917
Total	2,209,304,371

Data from Illinois Lottery Fiscal Year 2010 Financial Release.

10.10 Consistency? Table 10.2 shows how Illinois State Lottery sales are divided among different types of games. What is the sum of the amounts spent on the 11 types of games? Why is this sum not exactly equal to the total given in the table?

10.11 Marital status. In the U.S. Census Bureau document *America's Families and Living Arrangements: 2017,* we find these data on the marital status of American women aged 15 years and older as of 2017:

Current marital status	Count (thousands)
Never married	39,087
Married	65,193
Widowed	11,642
Divorced	14,591

(a) How many women were not married in 2017?

(b) Make a bar graph to show the distribution of marital status, and describe what you see in this bar graph.

(c) Would it also be correct to use a pie chart? Explain.

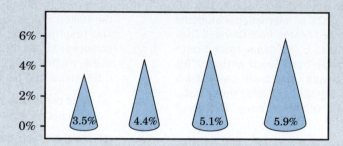

Figure 10.13 Comparing interest rates, Exercise 10.12.

10.12 **We pay high interest.** Figure 10.13 shows a graph taken from an advertisement for an investment that promises to pay a higher interest rate than bank accounts and other competing investments. Is this graph a correct comparison of the four interest rates? Explain your answer.

10.13 **Attitudes on same-sex marriage.** Attitudes on same-sex marriage have changed over time, but they also differ according to age. Figure 10.14 shows change in the attitudes

on same-sex marriage for four generational cohorts. The y axis in this figure shows the proportion in each generational cohort who favor same-sex marriage. Comment on the overall trend in change in attitude. Explain how attitude on same-sex marriage differs according to generation.

10.14 **International students.** The Institute of International Education (IIE) collects data each year about international students who attend college or university in the United States.

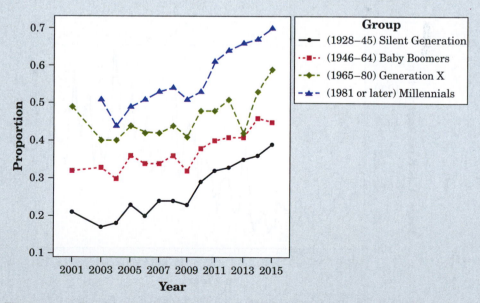

Figure 10.14 Changing attitudes on same-sex marriage by generational cohort, Exercise 10.13. Data from Pew Research Center.

In their 2018 *Open Doors* report, the IIE reported that 33.2% of international students came to the United States from China, 17.9% came from India, 5.0% came from South Korea, 4.1% came from Saudi Arabia, 2.4% came from Canada, and 1.4% came from Mexico. Make a graph to display these data. Do you need an "other countries" category? Why?

10.15 The cost of imported oranges. Figure 10.15 is a line graph of the average cost of imported oranges each month from July 1995 to April 2012. These data are the price in U.S. dollars per metric ton.

(a) The graph shows strong seasonal variation. How is this visible in the graph? Why would you expect the price of oranges to show seasonal variation?

(b) What is the overall trend in orange prices during this period, after we take into account seasonal variation?

10.16 College majors. A report by the National Center for Education Statistics presented data on the percentage of bachelor's degrees conferred by postsecondary institutions in the United States, by field of study, in 2016. The results: 5.9%, biology and biomedical sciences; 19.4%, business; 4.5%, education; 5.6%, engineering; 6.1%, psychology; and 4.8%, visual and performing arts.

(a) What percentage of bachelor's degrees were conferred in fields other than those listed?

(b) Make a graph that compares the percentages of bachelor's degrees conferred in different fields of study.

10.17 Order matters. A common statistic considered by administration at universities is credit hour production. This is calculated by taking the credit hours for a course taught times the student enrollment. For example, a faculty member teaching two 3-credit courses with 50 students each would have produced $2 \times 3 \times 50 = 300$ credit hours. Suppose you are the chair for the Department of Statistics and you give Figure 10.16 to your dean. The dean asks you to explain why the bars decrease in height from left to right. How would you respond?

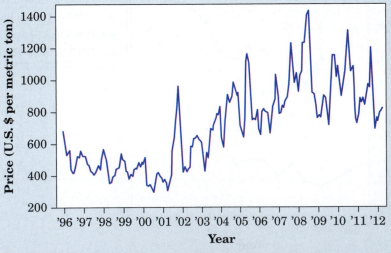

Figure 10.15 The price of oranges, July 1995 to June 2010, Exercise 10.15.

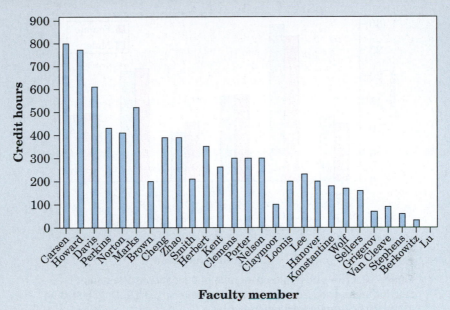

Figure 10.16 Bar graph depicting credit hours produced for each faculty member in a department, Exercise 10.17.

10.18 Civil disorders. The years around 1970 brought unrest to many U.S. cities. Here are government data on the number of civil disturbances in each three-month period during the years 1968 to 1972.

(a) Make a line graph of these data.

(b) The data show both a longer-term trend and seasonal variation within years. Describe the nature of both patterns. Can you suggest an explanation for the seasonal variation in civil disorders?

Period	Count	Period	Count
1968, Jan.–Mar.	6	1970, July–Sept.	20
Apr.–June	46	Oct.–Dec.	6
July–Sept.	25	1971, Jan.–Mar.	12
Oct.–Dec.	3	Apr.–June	21
1969, Jan.–Mar.	5	July–Sept.	5
Apr.–June	27	Oct.–Dec.	1
July–Sept.	19	1972, Jan.–Mar.	3
Oct.–Dec.	6	Apr.–June	8
1970, Jan.–Mar.	26	July–Sept.	5
Apr.–June	24	Oct.–Dec.	5

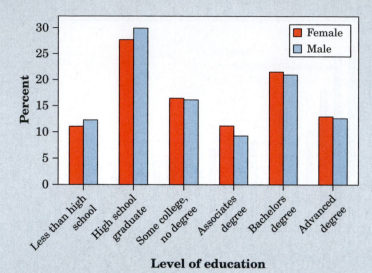

Level of education

Figure 10.17 A side-by-side bar graph of educational attainment by sex, for those aged 25 years and older, Exercise 10.19. Data collected in the 2017 Current Population Survey.

10.19 Educational attainment by sex. Figure 10.17 is a side-by-side bar graph of educational attainment, by sex, for those aged 25 years and older. These data were collected in the 2017 Current Population Survey. Compare educational attainment for males and females. Comment on any patterns you see.

10.20 A bad graph? Figure 10.18 shows a graph that appeared in the *Lexington (Ky.) Herald-Leader* on October 5, 1975. Discuss the correctness of this graph.

10.21 Seasonal variation. You examine the average temperature in Chicago each month for many years. Do you expect a line graph of the data to show seasonal variation? Describe the kind of seasonal variation you expect to see.

10.22 Sales are up. The sales at your new gift shop in December are double the November value. Should you conclude that your shop is growing

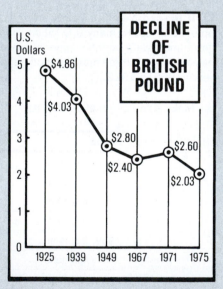

Figure 10.18 A newspaper's graph of the value of the British pound, Exercise 10.20.

more popular and will soon make you rich? Explain your answer.

10.23 Counting employed people. Consider a news article that states more Americans were working in June than any other month. The article goes on to say, "The report that employment plunged in June, with nonfarm payrolls declining by 117,000, helped to persuade the Federal Reserve to cut interest rates yet again." In conclusion, the article notes that "457,000 more people were employed in June than in May." What explains the difference between the fact that employment went up by 457,000 and the official report that employment went down by 117,000?

10.24 The sunspot cycle. Some plots against time show **cycles** of up-and-down movements. Figure 10.19 is a line graph of the average number of sunspots on the sun's visible face for each month from 1900 to 2011. What is the approximate length of the sunspot cycle? That is, how many years are there between the successive valleys in the graph? Is there any overall trend in the number of sunspots?

10.25 Trucks versus cars. Do consumers prefer trucks, SUVs, and minivans to passenger cars? Here are data on sales and leases of new cars and trucks in the United States from 1996 to 2010. (The definition of "truck" includes SUVs and minivans.) Plot two line graphs on the same axes to compare the change in car and truck sales over time. Describe the trend that you see.

Year:	1996	1997	1998	1999
Cars (1000s):	10,550	10,510	10,990	11,410
Trucks (1000s):	8,130	8,430	9,080	11,010

Year:	2000	2001	2002	2003
Cars (1000s):	11,710	11,060	10,250	9,860
Trucks (1000s):	10,990	10,750	10,498	10,212

Year:	2004	2005	2006	2007
Cars (1000s):	10,100	9,942	10,118	9,943
Trucks (1000s):	10,194	10,546	10,060	10,200

Year:	2008	2009	2010
Cars (1000s):	8,833	7,193	7,530
Trucks (1000s):	7,482	5,860	7,020

10.26 Who sells cars? Figure 10.20 is a pie chart of the percentage of passenger car sales in 2016 by various manufacturers. The artist has tried to make the graph attractive by using the wheel of a car for the "pie." Is the graph still a correct display of the data? Explain your answer.

10.27 Who sells cars? Make a bar graph of the data in Exercise 10.26. What advantage does your new graph have over the pie chart in Figure 10.20?

10.28 Crime in New York City. Here are the numbers of robberies in New York City, according to the New York Police Department (NYPD), for 2000 through 2017. Display these data in a graph. What are the most important facts that the data show?

Year:	2000	2001	2002	2003	2004
Count:	32,562	28,202	27,229	25,989	24,373

Year:	2005	2006	2007	2008	2009
Count:	24,722	23,739	21,809	22,401	18,601

Year:	2010	2011	2012	2013
Count:	19,486	19,717	20,144	19,128

Year:	2014	2015	2016	2017
Count:	16,539	16,931	15,500	13,956

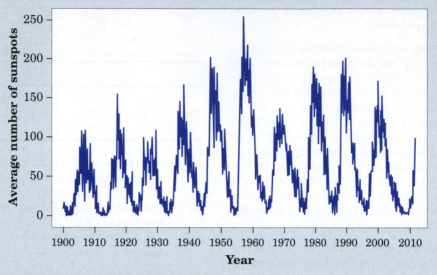

Figure 10.19 The sunspot cycle, Exercise 10.24. This is a line graph of the average number of sunspots per month for the years 1900–2011. Data from the National Oceanic and Atmospheric Administration.

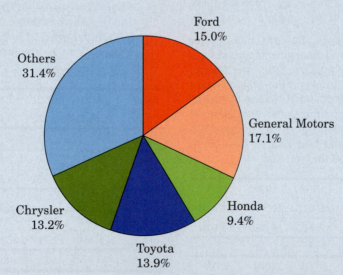

Figure 10.20 Passenger car sales by several manufacturers in 2016, Exercise 10.26.

10.29 Bad habits. According to the National Household Survey on Drug Use and Health, when asked in 2016, 23.5% of those aged 18 to 25 years used cigarettes in the past month, 5.2% used smokeless tobacco, 23.2% used illicit drugs, and 38.4% engaged in binge alcohol drinking. Explain why it is *not* correct to display these data in a pie chart.

10.30 Accidental deaths. In 2015, there were 146,571 deaths from unintentional injury in the United States. Among these were 47,478 deaths from poisoning, 37,757 from motor vehicle accidents, 33,381 from falls, 6,914 from suffocation, and 3,602 from drowning.

(a) Find the percentage of accidental deaths from each of these causes, rounded to the nearest percent. What percentage of accidental deaths were due to other causes?

(b) Make a well-labeled graph of the distribution of causes of accidental deaths.

10.31 Yields of money market funds. Many people invest in money market funds. These are mutual funds that attempt to maintain a constant price of $1 per share while paying monthly interest. Table 10.3 gives the average annual interest rates (in percent) paid by all taxable money market funds from 1973 (the first full year in which such funds were available) to 2008.

(a) Make a line graph of the interest paid by money market funds for these years.

(b) Interest rates, like many economic variables, show **cycles,** clear but repeating up-and-down movements. In which years did the interest rate cycle reach temporary peaks?

(c) A plot against time may show a consistent **trend** underneath cycles. When did interest rates reach their overall peak during these years? Describe the general trend downward since that year.

Table 10.3 Average annual interest rates (in percent) paid by money market funds, 1973–2008

Year	Rate	Year	Rate	Year	Rate	Year	Rate
1973	7.60	1982	12.23	1991	5.71	2000	5.89
1974	10.79	1983	8.58	1992	3.36	2001	3.67
1975	6.39	1984	10.04	1993	2.70	2002	1.29
1976	5.11	1985	7.71	1994	3.75	2003	0.64
1977	4.92	1986	6.26	1995	5.48	2004	0.82
1978	7.25	1987	6.12	1996	4.95	2005	2.66
1979	10.92	1988	7.11	1997	5.10	2006	4.51
1980	12.68	1989	8.87	1998	5.04	2007	4.70
1981	16.82	1990	7.82	1999	4.64	2008	2.05

Data from Albert J. Fredman, "A closer look at money market funds," *American Association of Individual Investors* Journal, February 1997, pp. 22–27; and the 2010 *Statistical Abstract of the United States.*

Table 10.4 Women's winning times (minutes) in the Boston Marathon, 1972–2018

Year	Time	Year	Time	Year	Time	Year	Time
1972	190	1985	154	1998	143	2011	143
1973	186	1986	145	1999	143	2012	152
1974	167	1987	146	2000	146	2013	146
1975	162	1988	145	2001	144	2014	139
1976	167	1989	144	2002	141	2015	145
1977	168	1990	145	2003	145	2016	149
1978	165	1991	144	2004	144	2017	142
1979	155	1992	144	2005	145	2018	160
1980	154	1993	145	2006	144		
1981	147	1994	142	2007	149		
1982	150	1995	145	2008	145		
1983	143	1996	147	2009	152		
1984	149	1997	146	2010	146		

Data from the website en.wikipedia.org/wiki/List_of_winners_of_the_Boston_Marathon/.

10.32 The Boston Marathon. Women were allowed to enter the Boston Marathon in 1972. The time (in minutes, rounded to the nearest minute) for each winning woman from 1972 to 2018 appears in Table 10.4.

(a) Make a graph of the winning times.

(b) Give a brief description of the pattern of Boston Marathon winning times over these years. Have times stopped improving in recent years?

Exploring the Web

Access these exercises on the text website: macmillanlearning.com/scc10e.

What's the Verdict?

The following "What's the verdict" story asks you to examine a widely shared graph that can be found online. You will use what you have learned in this chapter to assess the quality of this graph.

Go online and look at the graph that can be found at

http://graphics.thomsonreuters.com/14/02/US-FLORIDA0214.gif

This widely shared graph by Reuters News shows Florida gun deaths and displays what happened before and after Florida's "Stand Your Ground" law was passed. This law allows people to shoot others who threaten their homes or families without consequences for the shooting.

Questions

WTV10.1. What is the variable that is being measured? Is it categorical or quantitative?

WTV10.2. What type of graph is this?

WTV10.3. How would you describe the trend of this graph, particularly before and after 2005, if you had to summarize it for a caption?

WTV10.4. Take a closer look at the y axis, if you haven't already. What is unusual here? Does this change your interpretation of the trend in the question above?

WTV10.5. What are some possible explanations for why the graph's creator, Christine Chan, made the graph with this orientation?

What's the verdict? Pay close attention to your axes, and do not be distracted by artistic effects.

Displaying Distributions with Graphs

11

Case Study: Apartment Rental Prices

In this chapter you will:

- Learn how to construct two types of graphics—histograms and stemplots—for displaying quantitative data.

- Learn how to interpret histograms and stemplots.

For some students, the cost of living can be a deciding factor when it comes to choosing a college or university to attend. Students who want to live off-campus, in particular, might be especially interested in knowing how much they will need to spend per month on housing. Which cities across the United States offer the most and least expensive apartment rental prices?

In 2018, the website abodo.com published its 2017 *Annual Rent Report*. The report included a list of the average rental prices of one-bedroom apartments in 85 U.S. cities. Topping the list, with an average rental price of $3,333, was San Francisco, California. The least expensive city on the list was Fort Wayne, Indiana, with an average rental price of $526. Although some characteristics of a data set can be discerned from examining a list of numbers, it can be difficult to see patterns in large lists. As we saw in Chapter 10, graphs are a powerful way to make sense of large collections of numbers. By the end of this chapter, you will know how to make a graph to examine the distribution of average apartment rental prices and what to look for when examining this graph.

Histograms

Categorical variables record group membership, such as the marital status of a man or the race of a college student. We can use a pie chart or bar graph to display the distribution of categorical variables because they have relatively few values. What about quantitative variables such as the SAT scores of students admitted to a college or the income of families? These variables take so many values that a graph of the distribution is clearer if nearby values are grouped together. The most common graph of the distribution of a quantitative variable is a **histogram.**

Example 1 — How to make a histogram

Table 11.1 presents the percentage of residents aged 65 years and older in each of the 50 states. To make a histogram of this distribution, proceed as follows.

Step 1. Divide the range of the data into classes of equal width. The data in Table 11.1 range from 7.7 to 17.3, so we choose as our classes

$$7.0 \leq \text{percentage of residents 65 and older} < 8.0$$
$$8.0 \leq \text{percentage of residents 65 and older} < 9.0$$
$$\vdots$$
$$17.0 \leq \text{percentage of residents 65 and older} < 18.0$$

Table 11.1 Percentage of residents aged 65 and older in the states, 2010

State	Percent	State	Percent	State	Percent
Alabama	13.8	Louisiana	12.3	Ohio	14.1
Alaska	7.7	Maine	15.9	Oklahoma	13.5
Arizona	13.8	Maryland	12.3	Oregon	13.9
Arkansas	14.4	Massachusetts	13.8	Pennsylvania	15.4
California	11.4	Michigan	13.8	Rhode Island	14.4
Colorado	10.9	Minnesota	12.9	South Carolina	13.7
Connecticut	14.2	Mississippi	12.8	South Dakota	14.3
Delaware	14.4	Missouri	14.0	Tennessee	13.4
Florida	17.3	Montana	14.8	Texas	10.3
Georgia	10.7	Nebraska	13.5	Utah	9.0
Hawaii	14.3	Nevada	12.0	Vermont	14.6
Idaho	12.4	New Hampshire	13.5	Virginia	12.2
Illinois	12.5	New Jersey	13.5	Washington	12.3
Indiana	13.0	New Mexico	13.2	West Virginia	16.0
Iowa	14.9	New York	13.5	Wisconsin	13.8
Kansas	13.2	North Carolina	12.9	Wyoming	12.4
Kentucky	13.3	North Dakota	14.5		

Data from *Age and Sex Composition: 2010 Census Briefs;* available online at https://www.census.gov/prod/cen2010/briefs/c2010br-03.pdf.

Be sure to specify the classes precisely so that each individual falls into exactly one class. In other words, be sure that the classes are *exclusive* (no individual is in more than one class) and *exhaustive* (every individual appears in some class). The individuals in this example are states. A state with 7.9% of its residents aged 65 and older would fall into the first class, whereas a state with 8.0% of its residents aged 65 and older would fall into the second class.

Step 2. Count the number of individuals in each class. Here are the counts of states in each class:

Class	Count	Class	Count	Class	Count
7.0 to 7.9	1	11.0 to 11.9	1	15.0 to 15.9	2
8.0 to 8.9	0	12.0 to 12.9	11	16.0 to 16.9	1
9.0 to 9.9	1	13.0 to 13.9	17	17.0 to 17.9	1
10.0 to 10.9	3	14.0 to 14.9	12		

Step 3. Draw the histogram. Mark on the horizontal axis the scale for the variable whose distribution you are displaying. That's "percentage of residents aged 65 and older" in this example. The scale runs from 5 to 20 because that range spans the classes we chose. The vertical axis contains the scale of counts. Each bar represents a class. The base of the bar covers the class, and the bar height is the class count. There is no horizontal space between the bars unless a class is empty, so that its bar has height zero. Figure 11.1 is our histogram.

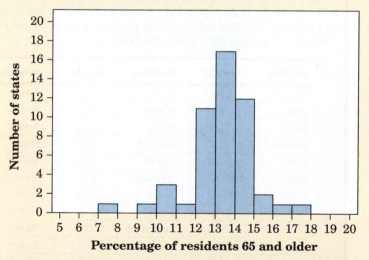

Figure 11.1 Histogram of the percentages of residents aged 65 and older in the 50 states, Example 1. Note the outlier.

Just as with bar graphs, our eyes respond to the area of the bars in a histogram. Be sure that the classes for a histogram have equal widths. There is no one right choice for the number of classes. Some people recommend between 10 and 20 classes but suggest using fewer when the

Statistics in Your World

What the eye really sees We make the bars in bar graphs and histograms equal in width because the eye responds to their area. That's roughly true. Careful study by statistician William Cleveland shows that our eyes "see" the size of a bar in proportion to the 0.7 power of its area. Suppose, for example, that one figure in a pictogram is both twice as high and twice as wide as another. The area of the bigger figure is 4 times that of the smaller. But we perceive the bigger figure as only 2.6 times the size of the smaller because 2.6 is 4 to the 0.7 power.

size of the data set is small. Too few classes will give a "skyscraper" histogram, with all values in a few classes with tall bars. Too many classes will produce a "pancake" graph, with most classes having one or no observations. Neither choice will give a good picture of the shape of the distribution. You must use your judgment in choosing classes to display the shape. Statistics software will choose the classes for you and may use slightly different rules than those we have discussed. The computer's choice is usually a good one, but you can change it if you want. When using statistical software, it is good practice to check what rules are used to determine the classes.

Now it's your turn

11.1 Personal record for weightlifting. Bodyshop Athletics keeps a dry erase board for members to keep track of their personal record for various events. The "dead lift" is a weightlifting maneuver where a barbell is lifted from the floor to the hips. The data below are the personal records for members at Bodyshop Athletics in pounds lifted during the dead lift.

Member	Weight	Member	Weight	Member	Weight
Baker, B.	175	G.T.C.	250	Pender	205
Baker, T.	100	Harper	155	Porth	215
Birnie	325	Horel	215	Ross	115
Bonner	155	Hureau	285	Stapp	190
Brown	235	Ingram	165	Stokes	305
Burton	155	Johnson	175	Taylor, A.	165
Coffey, L.	135	Jones, J.	195	Taylor, Z.	305
Coffey, S.	275	Jones, L.	205	Thompson	285
Collins, C.	215	LaMonica	235	Trent	135
Collins, E.	95	Lee	165	Tucker	245
Dalick, B.	225	Lord	405	Watson	155
Dalick, K.	335	McCurry	165	Wind, J.	350
Edens	255	Moore	145	Wind, K.	185
Flowers	205	Morrison	145		

Make a histogram of this distribution following the three steps described in Example 1. Create your classes using $75 \leq$ weight < 125, then $125 \leq$ weight < 175, then $175 \leq$ weight < 225, and so on.

Interpreting Histograms

Making a statistical graph is not an end in itself. The purpose of the graph is to help us understand the data. After you (or your computer) make a graph, always ask, "What do I see?" Here is a general strategy for looking at graphs.

We have already applied this strategy to line graphs. Trend and seasonal variation are common overall patterns in a line graph. The more drastic decrease in unemployment in mid-2010 for those with less than a high school degree, seen in Figure 10.7 (page 222), is deviating from the overall pattern of small dips and increases apparent throughout the rest of the line. This is an example of a striking deviation from the general pattern that one observes for the rest of the time period. In the case of the histogram of Figure 11.1, it is easiest to begin with deviations from the over-all pattern of the histogram. One state stands out as separated from the main body of the histogram. You can find this in the table once the histogram has called attention to it. Alaska has 7.7% of its residents over age 65. This state is a clear *outlier*.

> ## Key Terms
>
> In any graph of data, look for an **overall pattern** and also for striking **deviations** from that pattern.

> ## Key Terms
>
> An **outlier** in any graph of data is an individual observation that falls outside the overall pattern of the graph.

Is Utah, with 9.0% of its population over 65, an outlier? Whether an observation is an outlier is to some extent a matter of judgment, although statisticians have developed some objective criteria for identifying possible outliers. Utah is the smallest of the main body of observations and, unlike Alaska, is not separated from the general pattern. We would not call Utah an outlier. Once you have spotted outliers, look for an explanation. Many outliers are due to mistakes, such as typing 4.0 as 40. Other outliers point to the special nature of some observations. Explaining outliers usually requires some background information. It is not surprising that Alaska, the northern frontier, has few residents 65 and older.

To see the *overall pattern* of a histogram, ignore any outliers. This is not to suggest that outliers should be discarded or that we should pretend those values do not exist. If any outliers are present, we want to acknowledge their existence and think about how they impact the distribution. Here is a simple way to organize your thinking.

We will learn how to describe center and variability numerically in Chapter 12. For now, we can describe the center of a distribution by its *midpoint,* the value at roughly the middle of all the values in the distribution. We can describe the variability of a distribution by giving the *smallest and largest values,* ignoring any outliers. We can then mention the presence of possible outliers in an effort to more completely describe the distribution.

> ## Key Terms
>
> To describe the overall pattern of a distribution:
>
> - Describe the **center** and the **variability**.
> - Describe the **shape** of the histogram in a few words.

Example 2

Describing distributions

Look again at the histogram in Figure 11.1. **Shape:** The distribution has a *single peak*. It is roughly *symmetric*—that is, the pattern is similar on both sides of the peak. **Center:** The midpoint of the distribution is close to the single peak, at about 13%. **Variability:** The variability is about 9% to 18% if we ignore the outlier. We have one outlying value, at 7.7%, that falls away from the overall pattern.

Example 3

Tuition and fees in Illinois

There are 114 colleges and universities in Illinois. Their tuition and fees for the 2016–2017 school year range from $2095 at Kaskaskia College to $53,649 at the University of Chicago. Figure 11.2 is a histogram of the tuition and fees charged by all 114 Illinois colleges and universities. We see that many (mostly community colleges) charge less than $6000. The distribution extends out to the right. At the upper extreme, two colleges charge between $48,000 and $54,000.

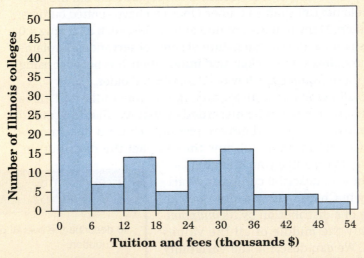

Figure 11.2 Histogram of the tuition and fees charged by 114 Illinois colleges and universities in the 2016–2017 academic year, Example 3. Data from www.isac.org.

The distribution of tuition and fees at Illinois colleges, shown in Figure 11.2, has a quite different **shape** from the distribution in Figure 11.1. There is a strong *peak* in the lowest cost class. Most colleges charge less than $12,000, but there is a long right tail extending up to almost $54,000. We call a distribution with a long tail on one side *skewed*. The **center** is roughly $12,000 since close to half the colleges charge less than this. The **variability** is large, from $2095 to more than $53,000. According to what we can see in the histogram, there appear to be no outliers—the colleges with the highest tuitions just continue the long right tail that is part of the overall pattern.

When you describe a distribution, concentrate on the main features. Look for major peaks, not for minor ups and downs in the bars of the histogram like those in Figure 11.2. Look for gaps and clear outliers that fall away from the overall pattern, not just for the smallest and largest observations. Look for rough *symmetry* or clear *skewness*.

A distribution is symmetric if the two sides of a figure like a histogram are exact mirror images of each other. Data are almost never exactly symmetric, so we are willing to call histograms like that in Figure 11.1 roughly symmetric as an overall description. The tuition distribution in Figure 11.2, on the other hand, is clearly skewed to the right. Here are more examples.

Key Terms

A distribution is **symmetric** if the right and left sides of the histogram are approximately mirror images of each other.

A distribution is **skewed to the right** if the right side of the histogram (containing the half of the observations with larger values) extends much farther out than the left side. It is **skewed to the left** if the left side of the histogram extends much farther out than the right side.

Example 4
Lake elevation levels

Lake Murray is a man-made reservoir located in South Carolina. It is used mainly for recreation, such as boating, fishing, and water sports. It is also used to provide backup hydroelectric power for South Carolina Electric and Gas. The lake levels fluctuate with the highest levels in summer (for safe boating and good fishing) and the lowest levels in winter (for water quality). Water can be released at the dam in the case of heavy rains or to let water out to maintain winter levels. The U.S. Geological Survey (USGS) monitors

Red_Rock_Gal/Shutterstock

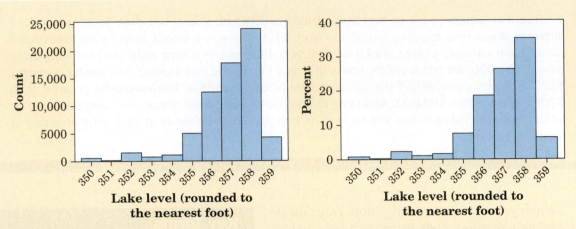

Figure 11.3 U.S. Geological Survey (USGS) reported hourly lake levels from November 1, 2007, through August 11, 2015, for Lake Murray, SC. There are 67,810 lake level observations. The histogram on the left is showing the number of times the lake was at each level. The histogram on the right shows this same data as the percent of times Lake Murray reached each level.

water levels in Lake Murray. The histograms in Figure 11.3 were created using 67,810 hourly elevation levels from November 1, 2007, through August 11, 2015.

The two histograms of lake levels were made from the same data set, and the histograms look identical in shape. The shape of the distribution of lake levels is skewed left, since the left side of the histogram is longer. The minimum lake level is 350 feet and the maximum is 359 feet. Using the histogram on the right, by adding the height of the bar for 358 and 359 feet elevations, we see the lake level is at 358 or 359 roughly 40% of the time. Using this information, it appears that a lake level of 357 feet is the midpoint of the distribution.

Let's examine the difference in the two histograms. The histogram on the left puts the count of observations on the y axis (this is called a frequency histogram), while the histogram on the right uses the percent of times the lake reaches a certain level (this is called a relative frequency histogram). The frequency histogram tells us the lake reached an elevation of 358 feet approximately 24,000 times (24,041 to be exact!). If a fisherman considering a move to Lake Murray cares about how often the lake reaches a certain level, it is more illustrative to use the relative frequency histogram on the right, which reports the percent of times the lake reached 358 feet. The height of the bar for 358 feet is 35, so the fisherman would know the lake is at the 358-foot elevation roughly 35% of the time.

It is not uncommon in the current world to have very large data sets or "big data." Google uses big data to rank web pages and provide the best search results. Banks use big data to analyze spending patterns and learn when to flag your debit or credit card for fraudulent use. Large firms use big data to analyze market patterns and adapt marketing strategies accordingly. Our data set of size 67,810 is actually small in the realm of "big data," but still big enough to see that it is almost always better to use a relative frequency histogram when sample sizes grow large. A relative frequency histogram is also a better choice if one wants to make comparisons between two distributions.

Example 5

Shakespeare's words

Figure 11.4 shows the distribution of lengths of words used in Shakespeare's plays. This distribution has a single peak and is somewhat skewed to the right. There are many short words (3 and 4 letters) and few very long words (10, 11, or 12 letters), so that the right tail of the histogram extends out farther than the left tail. The center of the distribution is about 4. That is, about half of Shakespeare's words have four or fewer letters. The variability is from 1 letter to 12 letters.

Notice that the vertical scale in Figure 11.4 is not the *count* of words but the *percentage* of all of Shakespeare's words that have each length. A histogram of percentages rather than counts is convenient since this was a large data set. Different kinds of writing have different distributions of word lengths, but all are skewed to the right because short words are common and very long words are rare.

claudiodivizia/Deposit Photos

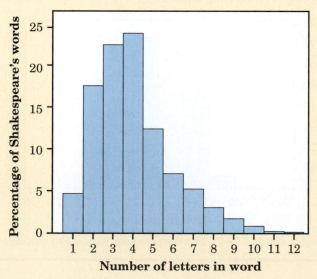

Figure 11.4 The distribution of word lengths used by Shakespeare in his plays, Example 5. This distribution is skewed to the right. Data from C. B. Williams, *Style and Vocabulary: Numerical Studies*, Griffin, 1970.

The overall shape of a distribution is important information about a variable. Some types of data regularly produce distributions that are symmetric or skewed. For example, the sizes of living things of the same species (like lengths of crickets) tend to be symmetric. Data on incomes (whether of individuals, companies, or nations) are usually strongly skewed to the right. There are many moderate incomes, some large incomes, and a few very large incomes. It is very common for data to be skewed to the right when they have a strict minimum value (often 0). Income and the lengths of Shakespeare's words are examples. Likewise, data that have a strict maximum value (such as 100, as in student test scores) are often skewed to the left. Do remember that many distributions have shapes that are neither symmetric nor skewed. Some data show other patterns. Scores on an exam, for example, may have a cluster near the top of the scale if many students did well. Or they may show two distinct peaks if a tough problem divided the class into those who did and didn't solve it. Use your eyes and describe what you see.

Now it's your turn

11.2 Height distribution. Height distributions generally have a predictable pattern. In a large introductory statistics class, students were asked to report their height. The histogram in Figure 11.5 displays the distribution of heights, in inches, for 266 females from this class. Describe the shape, center, and variability of this distribution. Are there any outliers?

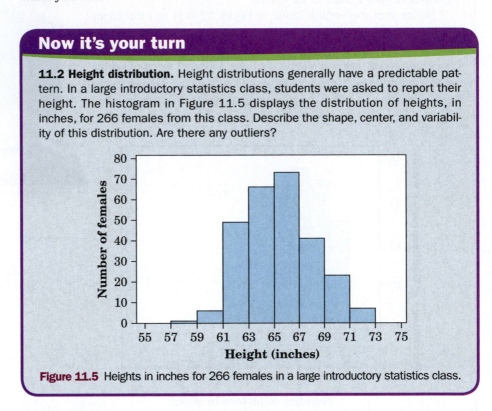

Figure 11.5 Heights in inches for 266 females in a large introductory statistics class.

Stemplots

Histograms are not the only graphical display of distributions. For small data sets, a *stemplot* (sometimes called a *stem-and-leaf plot*) is quicker to make and presents more detailed information.

A stemplot looks like a histogram turned on end. The stemplot in Figure 11.6 is just like the histogram in Figure 11.1 because the classes chosen for the histogram are the same as the stems in the stemplot. Figure 11.7 is a stemplot of the Illinois tuition data discussed in Example 3. This stemplot has almost six times as many classes as the histogram of the

Example 6

Stemplot of the "65 and older" data

For the "65-and-older" percentages in Table 11.1, the whole-number part of the observation is the stem, and the final digit (tenths) is the leaf. The Alabama entry, 13.8, has stem 13 and leaf 8. Stems can have as many digits as needed, but each leaf must consist of only a single digit. Figure 11.6 shows the steps in making a stemplot for the data in Table 11.1. First, write the stems. Then go through the table adding a leaf for each observation. Finally, arrange the leaves in increasing order out from each stem.

stem	leaf	stem	leaf		stem	leaf
7		7	7		7	7
8		8			8	
9		9	0		9	0
10		10	973		10	379
11		11	4		11	4
12		12	45339809234		12	02333445899
13		13	88023885552559748		13	02234555557888889
14		14	424390851436		14	012334445689
15		15	94		15	49
16		16	0		16	0
17		17	3		17	3

Key: A stem of 7 and a leaf of 7 means 7.7%

Figure 11.6 Making a stemplot of the 65-and-older data in Table 11.1. Whole percents form the stems, and tenths of a percent form the leaves.

same data in Figure 11.2. We interpret stemplots as we do histograms, looking for the overall pattern and for any outliers.

You can choose the classes in a histogram. The classes (the stems) of a stemplot are given to you. You can get more flexibility by **rounding** the data so that the final digit after rounding is suitable as a leaf. Do this when the data have too many digits. For example, the tuition charges of Illinois colleges look like

$4,500 $9,948 $13,587 $25,088 . . .

A stemplot would have too many stems if we took all but the final digit as the stem and the final digit as the leaf. To make the stemplot in Figure 11.7, we round these data to the nearest hundred dollars:

45 99 136 251 . . .

These values appear in Figure 11.7 on the stems 4, 9, 13, and 25.

Stemplots

To make a **stemplot:**

1. Separate each observation into a **stem** consisting of all but the final (rightmost) digit and a **leaf,** the final digit. Stems may have as many digits as needed, but each leaf contains only a single digit. Do not include commas or decimal points with your leaves.

2. Write the stems in a vertical column with the smallest at the top, and draw a vertical line at the right of this column.

3. Write each leaf in the row to the right of its stem, in increasing order out from the stem.

Figure 11.7
Stemplot of the
Illinois tuition and fee
data in Example 3.
Data from
www.isac.org.

```
stem   leaf
  1 |
  2 | 15889
  3 | 0111122344455555555666678899
  4 | 01111112224445
  5 | 14
  6 |
  7 |
  8 |
  9 | 99
 10 | 5
 11 | 0245
 12 | 08
 13 | 0256
 14 | 138
 15 | 78
 16 | 7
 17 | 67
 18 |
 19 | 5
 20 | 578
 21 | 25
 22 | 8
 23 |
 24 |
 25 | 11788
 26 | 37
 27 | 19
 28 | 12
 29 | 19
 30 | 16688
 31 | 7
 32 | 03
 33 | 89
 34 | 0
 35 | 345
 36 |
 37 | 0
 38 | 5
 39 | 1
 40 | 7
 41 |
 42 |
 43 | 3
 44 | 11
 45 | 2
 46 |
 47 |
 48 | 6
 49 |
 50 |
 51 |
 52 |
 53 | 6
```
Key: A stem of 2 and a leaf of 1 means $2100.

The chief advantage of a stemplot is that it displays the actual values of the observations. We can see from the stemplot in Figure 11.7, but not from the histogram in Figure 11.2, that Illinois's most expensive college charges $53,600 (rounded to the nearest hundred dollars). Stemplots are also faster to draw than histograms. A stemplot requires that we use the first digit or digits as stems. This amounts to an automatic choice of classes and can give a poor picture of the distribution. Stemplots do not work well with large data sets because the stems then have too many leaves.

Chapter 11: Statistics in Summary

■ The **distribution** of a variable tells us what values the variable takes and how often it takes each value.

■ To display the distribution of a quantitative variable, use a **histogram** or a **stemplot.** We usually favor stemplots when we have a small number of observations and histograms for larger data sets. Make sure to choose the appropriate number of classes so that the distribution shape is displayed accurately. For really large data sets, use a histogram of percents (relative frequency histogram).

■ When you look at a graph, look for an **overall pattern** and for **deviations** from that pattern, such as **outliers.**

■ We can characterize the overall pattern of a histogram or stemplot by describing its **shape, center,** and **variability.** Some distributions have simple shapes such as **symmetric**, **skewed to the left,** or **skewed to the right,** but others are too irregular to describe by a simple shape.

This chapter summary will help you evaluate the Case Study.

Link It

In Chapter 10, we learned how to use tables and graphs to see what data tell us. In this chapter, we looked at two additional graphs, histograms and stemplots, that help us make sense of large collections of numbers. These graphics are pictures of the distribution of a single quantitative variable. Although the bar graphs in Chapter 10 look much like histograms, the difference is that bar graphs are used to display the distribution of a categorical variable, while histograms display the distribution of a quantitative variable. The overall pattern (shape, center, and variability) and deviations from this pattern (outliers) are important features of the distribution of a variable. We will look more carefully at these features in future chapters. In addition, these features will figure prominently in some of the conclusions that we will draw about a variable from data.

Case Study Evaluated

Use what you have learned in this chapter to evaluate the Case Study that opened the chapter. Start by reviewing the Chapter Summary. Then answer each of the following questions in complete sentences. Be sure to communicate clearly enough for any of your classmates to understand what you are saying.

Figure 11.8 is a histogram of the distribution of average rental prices of one-bedroom apartments from a sample of 85 U.S. cities.

1. Describe the overall shape, the center, and the variability of the distribution.

2. Are there any outliers?

3. In Dallas, Texas, the average rental price of a one-bedroom apartment is $1,190. Is this a low, high, or typical amount? Please explain.

4. The data that inspired this case study can be found on the following website: https://www.abodo.com/blog/2017-annual-rent-report/. Go to this website and scroll down to the bottom of the website to see a table of data. Find the 2017 average one-bedroom rental price for the location where you live, or for a location close to where you live. Is this average a low, high, or typical amount? Please explain.

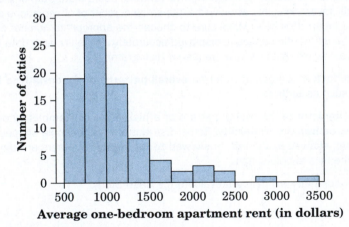

Figure 11.8 Histogram of the average rental prices (in dollars) of 85 one-bedroom apartments in 2017.

In this chapter you:

- Learned how to construct two types of graphics–histograms and stemplots–for displaying quantitative data.
- Learned how to interpret histograms and stemplots.

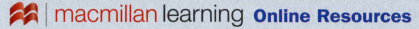

macmillan learning **Online Resources**

- The StatBoards video *Creating and Interpreting a Histogram* discusses how to construct a histogram.

- The Snapshots video *Visualizing Quantitative Data* discusses stemplots and histograms in the setting of two real examples.

Check the Basics

For Exercise 11.1, see page 246; for Exercise 11.2, see page 252.

11.3 Histograms Use a histogram when

(a) the number of observations is small.

(b) you want to look at the distribution of a quantitative variable.

(c) you want to look at the distribution of a categorical variable.

(d) you want to show the actual observations.

11.4 Histograms The heights of the bars on a relative frequency histogram displaying the lengths of rivers will add to

(a) 100%.

(b) the sample size.

(c) the sum of all the river lengths.

(d) the midpoint of the distribution.

11.5 Stemplots An advantage of a stemplot over a histogram is

(a) they are good for really large data sets.

(b) they are horizontal.

(c) one can recover the actual observations from the display.

(d) the classes are chosen for you.

11.6 Shape of distributions Figure 11.9 contains exam scores for 75 students. What is the shape of the exam score distribution?

(a) Symmetric

(b) Skewed to the right

(c) Skewed to the left

(d) None of the above

11.7 Shape of distributions In a certain town, most haircuts are between $10 and $20, but a few salons cater to high-end clients and charge $30 to $60. The distribution of haircut prices is

(a) Symmetric

(b) Skewed to the right

(c) Skewed to the left

(d) None of the above

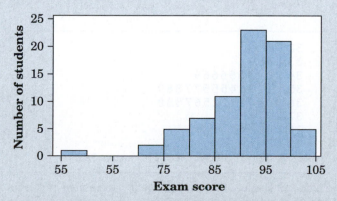

Figure 11.9 Histogram of the exam scores for 75 students.

Chapter 11 Exercises

11.8 Fast-food. If you go to the website for any fast-food chain, you can generally find information about the nutritional content of menu items. In February 2019, a sample of 130 menu items was selected from 13 popular fast-food chains, and the amount of carbohydrates (measured in grams) was recorded for each of these items. Figure 11.10 shows the distribution of grams of carbohydrates from the 130 menu items. Describe the shape, center, and variability of this distribution. Are there any outliers?

11.9 Where do 18- to 44-year-olds live? Figure 11.11 is a stemplot of the percentage of residents aged 18 to 44 in each of the 50 states in 2010. As in Figure 11.6 (page 253) for older residents, the stems are whole percents and the leaves are tenths of a percent.

(a) It turns out that Utah has the largest percentage of residents aged 18 to 44 years. What is the percentage for this state?

(b) Describe the shape, center, and variability of this distribution.

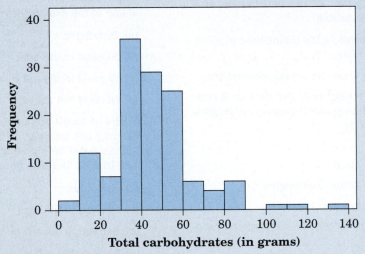

Figure 11.10 Histogram of the amount of carbohydrates (in grams) from 130 fast-food items, Exercise 11.8.

```
32 | 5
33 | 489
34 | 044556669
35 | 1234455577889
36 | 0001122567888
37 | 01457
38 | 12247
39 | 7
Key: A stem of 32 and a leaf of 5 means 32.5%.
```

Figure 11.11 Stemplot of the percentage of each state's residents who are 18 to 44 years old, Exercise 11.9.

(c) Is the distribution for residents aged 18 to 44 years more or less variable than the distribution in Figure 11.6 for residents aged 65 and older?

(d) From an examination of the stemplot, would it be possible to determine what percentage of residents aged 18 to 44 years lived in Arizona in 2010? Please explain why or why not.

11.10 The Super Bowl MVP. When the New England Patriots won the Super Bowl in 2017, Tom Brady, at 39 years of age, became the oldest player in National Football League (NFL) history to be named a Super Bowl Most Valuable Player (or MVP). Figure 11.12 is a histogram of the ages of players who have been named MVP for the first 53 Super Bowl games.

(a) Briefly describe the shape, center, and variability of this distribution.

(b) Does it appear that Tom Brady's age would be an outlier in this distribution?

11.11 Returns on common stocks. The total return on a stock is the change in its market price plus any dividend payments made. Total return is usually expressed as a percentage of the beginning price. Figure 11.13 is a histogram of the distribution of total returns for all 1528 common stocks listed on the New York Stock Exchange in one year.

(a) Describe the overall shape of the distribution of total returns.

(b) What is the approximate center of this distribution? Approximately what were the smallest and largest total returns? (This describes the variability of the distribution.)

(c) A return less than zero means that owners of the stock lost money. About what percentage of all stocks lost money?

(d) Explain why we prefer a histogram to a stemplot for describing the returns on 1528 common stocks.

11.12 The Kentucky Derby. On the first Saturday in May, the Kentucky Derby takes place at Churchill Downs in Louisville, Kentucky. A total of 20 horses race in the Kentucky Derby, and they race around a track that is 1.25 miles long. Figure 11.14 shows the speeds (in miles per hour, or mph) for all 123 horses who have won the Kentucky Derby from 1896 through 2018.

(a) Describe the shape, center, and variability of the distribution of speeds.

(b) The 2018 winner of the Kentucky Derby was Justify. Justify had a speed of 36.23 mph.

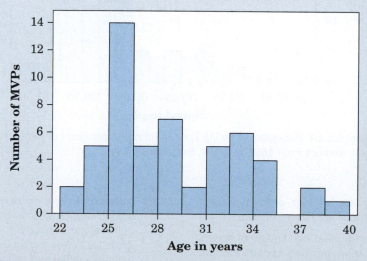

Figure 11.12 Histogram of the ages of Super Bowl MVPs through 2017, Exercise 11.10.

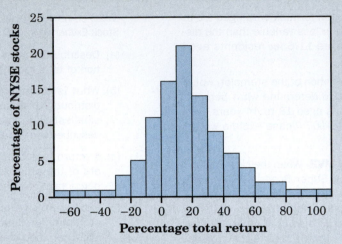

Figure 11.13 The distribution of total returns for all New York Stock Exchange common stocks in one year, Exercise 11.11. Data from J. K. Ford, "Diversification: How Many Stocks Will Suffice?" *American Association of Individual Investors Journal*, January 1990, pp. 14–16.

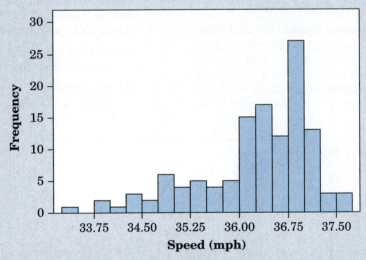

Figure 11.14 Histogram of racing speeds (in miles per hour) of Kentucky Derby winners from 1896 to 2018, Exercise 11.12.

Is 36.23 mph a low, high, or typical amount in this distribution? Please explain.

(c) Approximately what percentage of horses in this distribution have speeds that are greater than or equal to 37 mph?

11.13 Automobile fuel economy. Government regulations require automakers to give the city and highway gas mileages for each model of car. Table 11.2 gives the combined highway and city mileages (miles per gallon) for 31 model year 2015 sedans. Make a stemplot of the combined

Table 11.2 Combined city/highway gas mileage for model year 2015 midsize cars

Model	mpg	Model	mpg
Acura RLX	24	Hyundai Sonata	37
Audi A8	22	Infiniti Q40	22
BMW 528i	27	Jaguar XF	18
Bentley Mulsanne	13	Kia Forte	30
Buick LaCrosse	21	Lexus ES 350	24
Buick Regal	22	Lincoln MKZ	25
Cadillac CTS	14	Mazda 6	30
Cadillac CTS AWD	22	Mercedes-Benz B-Class	84
Chevrolet Malibu	29	Mercedez-Benz E350	23
Chevrolet Sonic	31	Nissan Altima	31
Dodge Challenger	18	Nissan Leaf	114
Ford C-max Energi Plug-in Hybrid	88	Rolls Royce Wraith	15
Ford Fusion Energi Plug-in Hybrid	88	Toyota Prius	50
Ford Fusion Hybrid	42	Toyota Prius Plug-in Hybrid	95
Honda Accord	31	Volvo S80	22
Honda Accord Hybrid	47		

Data from www.fueleconomy.gov/feg/byclass/Midsize_Cars2015.shtml.

gas mileages of these cars. What can you say about the overall shape of the distribution? Where is the center (the value such that half the cars have better gas mileage and half have worse gas mileage)? Some of these cars are electric and the reported mileage is the electric equivalent. These cars have far higher mileage. How many electric cars are in this data set?

11.14 Commute times. How long does it take to commute to work? Table 11.3 gives the mean travel time to work, in minutes, in each of the 50 states in 2016. Display the distribution in a graph and briefly describe its shape, center, and variability.

11.15 Twins money. Table 11.4 gives the salaries of the players on the Minnesota Twins baseball team for the 2018 season. Make a histogram of these data. Is the distribution roughly symmetric, skewed to the right, or skewed to the left? Explain.

11.16 The statistics of writing style. Numerical data can distinguish different types of writing, and sometimes even individual authors.

Here are data collected by students on the percentages of words of 1 to 15 letters used in articles in *Popular Science* magazine:

Length:	1	2	3	4	5
Percent:	3.6	14.8	18.7	16.0	12.5

Length:	6	7	8	9	10
Percent:	8.2	8.1	5.9	4.4	3.6

Length:	11	12	13	14	15
Percent:	2.1	0.9	0.6	0.4	0.2

(a) Make a histogram of this distribution. Describe its shape, center, and variability.

(b) How does the distribution of lengths of words used in *Popular Science* compare with the similar distribution in Figure 11.4 (page 251) for Shakespeare's plays? Look in particular at short words (2, 3, and 4 letters) and very long words (more than 10 letters).

Table 11.3 Mean travel time to work in minutes, 2016

State	Minutes	State	Minutes	State	Minutes
Alabama	24.5	Louisiana	25.2	Ohio	23.3
Alaska	18.8	Maine	23.7	Oklahoma	21.4
Arizona	24.9	Maryland	32.4	Oregon	23.2
Arkansas	21.6	Massachusetts	29.0	Pennsylvania	26.5
California	28.4	Michigan	24.3	Rhode Island	24.4
Colorado	24.9	Minnesota	23.2	South Carolina	24.1
Connecticut	25.7	Mississippi	24.2	South Dakota	16.9
Delaware	25.7	Missouri	23.4	Tennessee	24.7
Florida	26.7	Montana	17.9	Texas	25.9
Georgia	27.7	Nebraska	18.3	Utah	21.6
Hawaii	27.2	Nevada	23.9	Vermont	22.7
Idaho	20.4	New Hampshire	26.9	Virginia	28.1
Illinois	28.5	New Jersey	31.2	Washington	26.7
Indiana	23.4	New Mexico	21.7	West Virginia	25.6
Iowa	18.9	New York	32.6	Wisconsin	21.9
Kansas	19.2	North Carolina	24.1	Wyoming	18.1
Kentucky	23.0	North Dakota	17.3		

Data from 2012–2016 American Community Survey Five-Year Estimates, www.census.gov/programs-surveys/acs/.

Table 11.4 Salaries of the Minnesota Twins, 2018

Player	Salary ($)	Player	Salary ($)
Joe Mauer	23,000,000	Ryan Pressly	1,600,000
Ervin Santana	13,500,000	Ehire Adrianza	1,000,000
Phil Hughes	13,200,000	Trevor May	650,000
Lance Lynn	12,000,000	Eddie Rosario	602,500
Brian Dozier	9,000,000	Miguel Sano	602,500
Addison Reed	8,250,000	Max Kepler	587,500
Jason Castro	8,000,000	Ryan LaMarre	585,000
Logan Morrison	6,500,000	Byron Buxton	580,000
Jake Odorizzi	6,300,000	Jorge Polanco	575,000
Eduardo Escobar	4,850,000	Jose Berrios	570,000
Fernando Rodney	4,500,000	Taylor Rogers	565,000
Kyle Gibson	4,200,000	Trevor Hildenberger	555,000
Zach Duke	2,150,000	Mitch Garver	547,500
Michael Pineda	2,000,000	Gabriel Moya	545,000
Robbie Grossman	2,000,000	Tyler Kinley	545,000

Data from https://www.usatoday.com/sports/mlb/twins/salaries/2018/player/all/.

11.17 What's my shape? There are 30 teams in the National Basketball Association (NBA), and each team has a team payroll. The team payroll consists of the total amount of money available to pay all players on the team. As an example, in the 2018–19 season, the LA Lakers had a team payroll of $107,020,840. Suppose you were able to look at the distribution of team payrolls for all 30 teams in the NBA during the 2018–19 season. Would you expect this distribution to be roughly symmetric, clearly skewed to the right, or clearly skewed to the left? Why?

11.18 The Asian population in the eastern states. When the 2010 census was published, it was reported that the Asian population grew faster than any other race group in the United States during the decade from 2000 to 2010. Here are the percentages of the population who are of Asian origin in each state east of the Mississippi River in 2010:

State	Percent	State	Percent	State	Percent
Alabama	1.1	Connecticut	3.8	Delaware	3.2
Florida	2.4	Georgia	3.2	Illinois	4.6
Indiana	1.6	Kentucky	1.1	Maine	1.0
Maryland	5.5	Massachusetts	5.3	Michigan	2.4
Mississippi	0.9	New Hampshire	2.2	New Jersey	8.3
New York	7.3	North Carolina	2.2	Ohio	1.7
Pennsylvania	2.7	Rhode Island	2.9	South Carolina	1.3
Tennessee	1.4	Vermont	1.3	Virginia	5.5
West Virginia	0.7	Wisconsin	2.3		

Make a stemplot of these data. Describe the overall pattern of the distribution. Are there any outliers?

11.19 How many calories does a hot dog have? *Consumer Reports* magazine presented the following data on the number of calories in a hot dog for each of 17 brands of meat hot dogs:

173 191 182 190 172 147 146 139 175
136 179 153 107 195 135 140 138

Make a stemplot of the distribution of calories in meat hot dogs and briefly describe the shape of the distribution. Most brands of meat hot dogs contain a mixture of beef and pork, with up to 15% poultry allowed by government regulations. The only brand with a different makeup was Eat Slim Veal Hot Dogs. Which point on your stemplot do you think represents this brand?

11.20 The changing age distribution of the United States. The distribution of the ages of a nation's population has a strong influence on economic and social conditions. Table 11.5 shows the age distribution of U.S. residents in 1950 and 2050, in millions of persons. The 1950 data come from that year's census. The 2050 data are projections made by the Census Bureau.

(a) Because the total population in 2050 is much larger than the 1950 population, comparing percentages in each age group is clearer than comparing counts. Make a table of the percentage of the total population in each age group for both 1950 and 2050.

(b) Make a histogram of the 1950 age distribution (in percents). Then describe the main features of the distribution. In particular, look at the percentage of children relative to the rest of the population.

(c) Make a histogram of the projected age distribution for the year 2050. Use the same scales as in (b) for easy comparison. What are the most important changes in the U.S. age distribution projected for the 100-year period between 1950 and 2050?

11.21 Babe Ruth's home runs. Here are the numbers of home runs that Babe Ruth hit in his 15 years with the New York Yankees, 1920 to 1934:

54 59 35 41 46 25 47 60
54 46 49 46 41 34 22

Make a stemplot of these data. Is the distribution roughly symmetric, clearly skewed, or neither? About how many home runs did Ruth hit in a typical year? Is his famous 60 home runs in 1927 an outlier?

11.22 Back-to-back stemplot. The current major league single-season home run record is held by

Barry Bonds of the San Francisco Giants. Here are Bonds's home run counts for 1986 to 2007:

16	25	24	19	33	25	34	46
37	33	42	40	37	34	49	73
46	45	45	5	26	28		

Table 11.5 Age distribution in the United States, 1950 and 2050 (in millions of persons)

Age group	1950	2050
Under 10 years	29.3	53.6
10 to 19 years	21.8	53.1
20 to 29 years	24.0	50.4
30 to 39 years	22.8	48.9
40 to 49 years	19.3	45.1
50 to 59 years	15.5	43.3
60 to 69 years	11.0	39.4
70 to 79 years	5.5	29.8
80 to 89 years	1.6	21.2
90 to 99 years	0.1	8.2
100 years and older	—	0.8
Total	151.1	310.6

Data from https://www.census.gov/prod/2002pubs/censr-4.pdf and https://www.census.gov/prod/1/pop/p25-1130/p251130.pdf.

A **back-to-back stemplot** helps us compare two distributions. Write the stems as usual, but with a vertical line both to their left and to their right. On the right, put leaves for Ruth (Exercise 11.21). On the left, put leaves for Bonds. Arrange the leaves on each stem in increasing order out from the stem. Now write a brief comparison of Ruth and Bonds as home run hitters.

11.23 When it rains, it pours. On July 25 to 26, 1979, 42.00 inches of rain fell on Alvin, Texas. That's the most rain ever recorded in Texas for a 24-hour period. Table 11.6 gives the maximum precipitation ever recorded in 24 hours (through 2010) at any weather station in each state. The record amount varies a great deal from state to state—hurricanes bring extreme rains on the Atlantic coast, and the mountain West is generally dry.

(a) Make a graph to display the distribution of records for the states. Mark where your state lies in this distribution. Briefly describe the distribution in terms of shape, center, and variability.

(b) In April 2018, it was reported that 49.69 inches of rain fell in a 24-hour period

Table 11.6 Record 24-hour precipitation amounts (inches) by state

State	Precip.	State	Precip.	State	Precip.
Alabama	32.52	Louisiana	22.00	Ohio	10.75
Alaska	15.20	Maine	13.32	Oklahoma	15.68
Arizona	11.40	Maryland	14.75	Oregon	11.77
Arkansas	14.06	Massachusetts	18.15	Pennsylvania	13.50
California	25.83	Michigan	9.78	Rhode Island	12.13
Colorado	11.08	Minnesota	15.10	South Carolina	14.80
Connecticut	12.77	Mississippi	15.68	South Dakota	8.74
Delaware	8.50	Missouri	18.18	Tennessee	13.60
Florida	23.28	Montana	11.50	Texas	42.00
Georgia	21.10	Nebraska	13.15	Utah	5.08
Hawaii	38.00	Nevada	7.78	Vermont	9.92
Idaho	7.17	New Hampshire	11.07	Virginia	14.28
Illinois	16.91	New Jersey	14.81	Washington	14.26
Indiana	10.50	New Mexico	11.28	West Virginia	12.02
Iowa	13.18	New York	11.15	Wisconsin	11.72
Kansas	13.53	North Carolina	22.22	Wyoming	6.06
Kentucky	10.40	North Dakota	8.10		

Data from National Oceanic and Atmospheric Administration, www.noaa.gov.

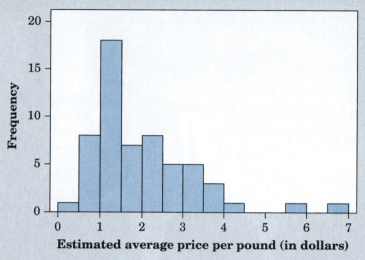

Figure 11.15 Histogram of the average estimated prices (in dollars) of 58 fruits and vegetables in 2016.

on the island of Kauai, in Hawaii. If we change the precipitation amount for Hawaii, in Table 11.6, from 38.00 inches to 49.69 inches, how does this impact the distribution of records for the states? In other words, would your description of shape, center, and variability from part (b) need to be modified? Please explain.

11.24 What's wrong? The Economic Research Service (ERS) from the U.S. Department of Agriculture (USDA) uses information from different retail establishments to provide estimates of the average price per pound (in dollars) of different fruits and vegetables. These prices are used when publishing the *Dietary Guidelines for Americans* in order to show individuals how much it will cost them to maintain a healthy diet. Figure 11.15 shows the distribution of the estimated average prices per pound, in 2016, of 58 fruits and vegetables. Each of the following descriptions of this distribution

is incorrect. Explain what is wrong with each description.

(a) The distribution is skewed to the left since most of the estimated average prices are on the left side of the histogram.

(b) A majority of the estimated average prices are between $1.00 and $1.50.

(c) We can see, by looking along the x axis, that estimated average prices start low at the beginning of 2016 and get higher later in 2016.

(d) There are five fruits and vegetables with estimated average prices of exactly $3.00 per pound.

Exploring the Web

Access these exercises on the text website: macmillanlearning.com/scc10e.

12

Describing Distributions with Numbers

Case Study: Education and Income

In this chapter you will:

- Learn new methods to summarize large data sets with a few numbers.
- Understand how to interpret these new summary numbers.
- Learn a simple graphical method for summarizing important features of large data sets.

Does education pay? We are told that people with more education earn more on average than people with less education. How much more? How can we answer this question?

Data on income can be found at the Census Bureau website. The data are estimates, for the year 2017, of the total incomes of 219,830,000 people aged 25 and older with earnings, and are based on the results of the Current Population Survey in 2018. The website gives the income distribution for each of several education categories. In particular, it gives the number of people in each of several education categories who earned between $1 and $2499, between $2500 and $4999, up to between $97,500 and $99,999, and $100,000 and over. That is a lot of information. A histogram could be used to display the data, but are there simple ways to summarize the information with just a few numbers that allow us to make sensible comparisons?

By the end of this chapter, using simple methods for summarizing large data sets, you will be able to provide an answer to whether education really pays.

Baseball has a rich tradition of using statistics to summarize and characterize the performance of players. In recent years, statistics have been used to

```
stem │ leaf
   0 │ 5
   1 │ 69
   2 │ 45568
   3 │ 334477
   4 │ 0255669
   5 │
   6 │
   7 │ 3
Key: A stem of 1 and a leaf of 6 means 16 home runs.
```

Figure 12.1 Stemplot of the number of home runs hit by Barry Bonds in his 22-year career.

evaluate player performance and potential to guide decisions about trades, when to replace the starting pitcher, and other coaching decisions. We begin by investigating ways to summarize the performance of the greatest home run hitters of all time.

In the summer of 2007, Barry Bonds shattered the career home run record, breaking the previous record set by Hank Aaron. Here are his home run counts for the years 1986 (his rookie year) to 2007 (his final season):

1986	1987	1988	1989	1990	1991	1992	1993	1994	1995	1996
16	25	24	19	33	25	34	46	37	33	42

1997	1998	1999	2000	2001	2002	2003	2004	2005	2006	2007
40	37	34	49	73	46	45	45	5	26	28

The stemplot in Figure 12.1 displays the data. The shape of the distribution is a bit irregular, but we see that it has one high outlier, and if we ignore this outlier, we might describe it as slightly skewed to the left with a single peak. The outlier is, of course, Bonds's record season in 2001.

A graph and a few words give a good description of Barry Bonds's home run career. But words are less adequate to describe, for example, the incomes of people with a high school education. We need *numbers* that summarize the center and variability of a distribution.

Median and Quartiles

A simple and effective way to describe center and variability is to give the **median** and the **quartiles.** The median, denoted M, is the midpoint, the value that separates the smaller half of the observations from the larger half. The middle 50% of the observations fall between the first and third quartiles. The quartiles get their name because with the median they divide the observations into quarters—one-quarter of the observations lie below the first quartile, half lie below the median, and three-quarters lie below the third quartile. That's the idea. To actually get numbers, we need a rule that makes the idea exact.

Example 1   *Finding the median*

We might compare Bonds's career with that of Hank Aaron's, the previous holder of the career record. Here are Aaron's home run counts for his 23 years in baseball.

13 27 26 44 30 39 40 34 45 44 24 32
44 39 29 44 38 47 34 40 20 12 10

To find the median, first arrange them in order from smallest to largest:

10 12 13 20 24 26 27 29 30 32 34 **34**
38 39 39 40 40 44 44 44 44 45 47

The bold 34 is the center observation, with 11 observations to its left and 11 to its right. When the number of observations n is odd (here $n = 23$), there is always one observation in the center of the ordered list. This is the median, $M = 34$.

How does this compare with Bonds's record? Here are Bonds's 22 home run counts, arranged in order from smallest to largest:

5 16 19 24 25 25 26 28 33 33 **34 34** 37 37 40 42 45 45 46 46 49 73

When n is even, there is no one middle observation. But there is a middle pair—the bold 34 and 34 have 10 observations on either side. We take the median to be halfway between this middle pair. So Bonds's median is

$$M = \frac{34 + 34}{2} = \frac{68}{2} = 34$$

Norm Hall/EPA/Shutterstock

There is a fast way to locate the median in an ordered list: count up $(n+1)/2$ places from the beginning of the list. Try it. For Aaron, $n = 23$ and $(23+1)/2 = 12$, so the median is the 12th entry in the ordered list. For Bonds, $n = 22$ and $(22+1)/2 = 11.5$. This means "halfway between the 11th and 12th" entries, so M is the average of these two entries. This "$(n+1)/2$ rule" is especially handy when you have many observations. The median of $n = 46,940$ incomes is halfway between the 23,470th and 23,471st in the ordered list. Be sure to note that $(n+1)/2$ does *not* give the median M, just its position in the ordered list of observations.

The median *M*

The **median *M*** is the midpoint of a distribution, the number such that half the observations are smaller and the other half are larger. To find the median of a distribution:

1. Arrange all observations in order of size, from smallest to largest.
2. If the number of observations *n* is odd, the median *M* is the observation that lies in the center of the ordered list. Find the location of the median by counting $(n+1)/2$ observations up from the bottom of the list.
3. If the number of observations *n* is even, the median *M* is the average of the two center observations in the ordered list. The location of the median is again $(n+1)/2$ from the bottom of the list.

The Census Bureau website provides data on income inequality. For example, it tells us that in 2017, the median income of Hispanic households was $50,486. That's helpful but incomplete. Do most Hispanic households earn close to this amount, or are the incomes widely variable? The simplest useful description of a distribution consists of both a measure of *center* and a measure of *variability*. If we choose the median (the midpoint) to describe center, the quartiles (in particular, the difference between the quartiles) provide natural descriptions of variability. Again, the idea is clear: find the points one-quarter and three-quarters up the ordered list of observations. Again, we need a rule to make the idea precise. The rule for calculating the quartiles uses the rule for the median.

The quartiles Q_1 and Q_3

To calculate the **quartiles:**

1. Arrange the observations in increasing order and locate the median *M* in the ordered list of observations. This "overall" median *M* will either be one of the observations in the ordered list (if there are an odd number of observations) or a value between two of the observations in the ordered list (if there are an even number of observations).
2. The **first quartile Q_1** is the median of the observations whose position in the ordered list is to the left of the location of the overall median. The overall median is not included in the observations considered to be to the left of the overall median.
3. The **third quartile Q_3** is the median of the observations whose position in the ordered list is to the right of the location of the overall median. The overall median is not included in the observations considered to be to the right of the overall median.

Example 2

Finding the quartiles

Hank Aaron's 23 home run counts are

10 12 13 20 24 26 27 29 30 32 34 **34** 38 39 39 40 40 44 44 44 44 45 47

$\quad\quad\quad\quad\quad\uparrow\quad\quad\quad\quad\quad\quad\quad\quad\quad\quad\uparrow\quad\quad\quad\quad\quad\quad\quad\quad\quad\quad\quad\uparrow$

$\quad\quad\quad\quad\quad Q_1\quad\quad\quad\quad\quad\quad\quad\quad\quad\quad M\quad\quad\quad\quad\quad\quad\quad\quad\quad\quad Q_3$

There is an odd number of observations, so the median is the one in the middle, the bold 34 in the list. To find the quartiles, ignore this central observation. The first quartile is the median of the 11 observations to the left of the bold 34 in the list. That's the sixth, so $Q_1 = 26$. The third quartile is the median of the 11 observations to the right of the bold 34. It is $Q_3 = 44$.

Barry Bonds's 22 home run counts are

5 16 19 24 25 25 26 28 33 33 **34 34** 37 37 40 42 45 45 46 46 49 73

$\quad\quad\quad\quad\quad\quad\uparrow\quad\quad\quad\quad\quad\quad\quad\quad\quad\uparrow\quad\quad\quad\quad\quad\quad\quad\quad\uparrow$

$\quad\quad\quad\quad\quad\quad Q_1\quad\quad\quad\quad\quad\quad\quad\quad M\quad\quad\quad\quad\quad\quad\quad\quad Q_3$

The median lies halfway between the middle pair. There are 11 observations to the left of this location. The first quartile is the median of these 11 numbers. That's the sixth, so $Q_1 = 25$. The third quartile is the median of the 11 observations to the right of the overall median's location, $Q_3 = 45$.

Now it's your turn

12.1 Babe Ruth. Prior to Hank Aaron, Babe Ruth was the holder of the career record. Here are Ruth's home run counts for his 22 years in Major League Baseball, arranged in order from smallest to largest:

0   2   3   4   6   11   22   25   29   34   35
41   41   46   46   46   47   49   54   54   59   60

Find the median, first quartile, and third quartile of these counts.

You can use the $(n+1)/2$ rule to locate the quartiles when there are many observations. The Census Bureau website tells us that there were 17,318,000 (rounded off to the nearest 1000) Hispanic households in the United States in 2017. If we ignore the roundoff, the median of these 17,318,000 incomes is halfway between the 8,659,000th and 8,659,001st in the list arranged in order from smallest to largest. So the first quartile is the median of the 8,659,000 incomes below this point in the list. Use the $(n+1)/2$ rule with $n = 8,659,000$ to locate the quartile:

$$\frac{n+1}{2} = \frac{8,659,000+1}{2} = 4,329,500.5$$

The average of the 4,329,500th and 4,329,501st incomes in the ordered list falls in the range $25,000 to $34,999 and we estimate the first quartile to be $26,681.

The third quartile is the median of the 8,659,000 incomes above the median. By the $(n+1)/2$ rule with 8,659,000, this will be the average of the 4,329,500th and 4,329,501st incomes above the median in the ordered list. We find that this falls in the range $75,000 to $99,999 and we estimate the third quartile to be $89,880.

In practice, people use statistical software to compute quartiles. Software can give results that differ from those you will obtain using the method described here. In fact, different software packages use slightly different rules for deciding how to divide the space between two adjacent values between which the quartile is believed to lie. We have chosen to select the point halfway between them, but other rules exist. Two different software packages can give two slightly different answers, depending on the rule employed.

The Five-Number Summary and Boxplots

The smallest and largest observations tell us little about the distribution as a whole, but they give information about the tails of the distribution that is missing if we know only the median and the quartiles. To get a quick summary of both center and variability, combine all five numbers.

These five numbers offer a reasonably complete description of center and variability. The five-number summaries of home run counts are

10   26   34   44   47

for Aaron and

5   25   34   45   73

for Bonds. The five-number summary of a distribution leads to a new graph, the *boxplot*. Figure 12.2 shows boxplots for the home run comparison.

You can draw boxplots either horizontally or vertically. Be sure to include a numerical scale in the graph. When you look at a boxplot, first locate the median, which marks the center of the distribution. Then look at the variability. The quartiles

> ### Key Terms
>
> The **five-number summary** of a distribution consists of the smallest observation, the first quartile, the median, the third quartile, and the largest observation, written in order from smallest to largest. In symbols, the five-number summary is
>
> Minimum Q_1 M Q_3 Maximum

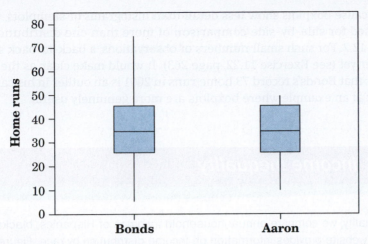

Figure 12.2 Boxplots comparing the yearly home run production of Barry Bonds and Hank Aaron.

(more precisely, the difference between the two quartiles) show the variability of the middle half of the data, and the extremes (the smallest and largest observations) indicate the variability of the entire data set. We see from Figure 12.2 that Bonds's usual performance, as indicated by the median and the box that marks the middle half of the distribution, is similar to that of Aaron. We also see that the distribution for Aaron is less variable than the distribution for Bonds.

Key Terms

A **boxplot** is a graph of the five-number summary.

- A central box spans the quartiles.
- A line in the box marks the median.
- Lines extend from the box out to the smallest and largest observations.

Now it's your turn

12.2 Babe Ruth. Here are Babe Ruth's home run counts for his 22 years in Major League Baseball, arranged in order from smallest to largest:

$$0 \quad 2 \quad 3 \quad 4 \quad 6 \quad 11 \quad 22 \quad 25 \quad 29 \quad 34 \quad 35$$
$$41 \quad 41 \quad 46 \quad 46 \quad 46 \quad 47 \quad 49 \quad 54 \quad 54 \quad 59 \quad 60$$

Draw a boxplot of this distribution. How does it compare with those of Barry Bonds and Hank Aaron in Figure 12.2?

Because boxplots show less detail than histograms or stemplots, they are best used for side-by-side comparison of more than one distribution, as in Figure 12.2. For such small numbers of observations, a back-to-back stemplot is better yet (see Exercise 11.22, page 263). It would make clear, as the boxplot cannot, that Bonds's record 73 home runs in 2001 is an outlier in his career. Let us look at an example where boxplots are more genuinely useful.

Example 3   *Income inequality*

To investigate income inequality, we compare annual household incomes of Hispanics, blacks, and whites. The Census Bureau website provides information on income distribution by race. Figure 12.3 compares the income distributions for Hispanics, blacks, and whites in 2017. This figure is a variation on the boxplot idea. The largest income among several million people will surely be very large. Figure 12.3 uses the 95% points (the values representing where the top 5% of annual incomes start) in the distributions instead of the single largest incomes. So, for example, the line above the box for the Hispanic group extends only to $180,012 rather than to the highest income. Many statistical software packages allow you to produce boxplots that suppress extreme values, but the rules for what constitutes an extreme value usually do not use the 95% point in the distribution instead of the single largest value.

Figure 12.3 gives us a clear and simple visual comparison. We see that the median and middle half are slightly greater for Hispanics than for blacks and that for whites the median and middle half are greater than for both blacks and Hispanics. The income of the bottom 5% stays small because there are

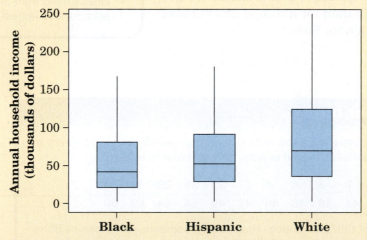

Figure 12.3 Boxplots comparing the distributions of income among Hispanics, blacks, and whites. The ends of each plot are at 0 and at the 95% point in the distribution.

some people in each group with no income or even negative income, perhaps due to illness or disability. The 95% point, marking off the top 5% of incomes, is greater for whites than for either blacks or Hispanics, and the 95% point of incomes for Hispanics is greater than for blacks. Overall, incomes for whites tend to be larger than those for Hispanics and blacks, highlighting racial inequities in income.

Figure 12.3 also illustrates how boxplots often indicate the symmetry or skewness of a distribution. In a symmetric distribution, the first and third quartiles are equally distant from the median. In most distributions that are skewed to the right, on the other hand, the third quartile will be farther above the median than the first quartile is below it. The extremes behave the same way. Even with the top 5% not present, we can see the right-skewness of incomes for all three races.

STATISTICAL CONTROVERSIES

Income Inequality

During the prosperous 1980s and 1990s, the incomes of all American households went up, but the gap between rich and poor grew. Figures 12.4 and 12.5 give two views of increasing inequality. Figure 12.4 is a line graph of household income, in dollars adjusted to have the same buying power every year. The lines show the 20th and 80th percentiles of income, which mark off the bottom fifth and the top fifth of households. The 80th percentile (up 66% between 1967 and 2017) is pulling away from the 20th percentile (up about 28%).

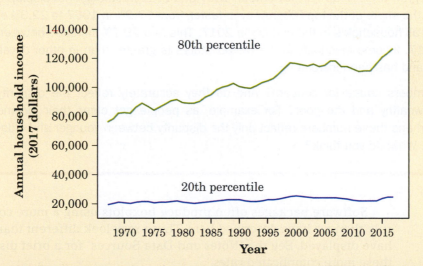

Figure 12.4 The change over time of two points in the distribution of incomes for American households. Eighty percent of households have incomes below the 80th percentile, and 20% have incomes below the 20th percentile. In 2017, the 20th percentile was $24,638, and the 80th percentile was $126,855.

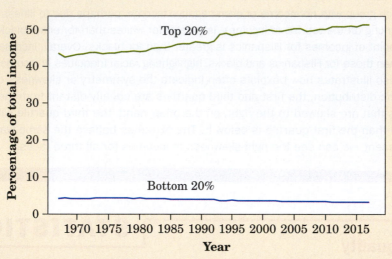

Figure 12.5 The change over time of the shares of total household income that go to the highest-income 20% and to the lowest-income 20% of households. In 2017, the top 20% of households received 52% of all income.

Figure 12.5 looks at the *share* of all income that goes to the top fifth and the bottom fifth. The bottom fifth's share has drifted down, to 3.1% of all income in 2017. The share of the top fifth grew to 52% (up 18.1% between 1967 and 2017). Although not displayed in the figures, the share of the top 5% grew even faster, from 17.2% in 1967 to 22.3% of the income of all households in the country in 2017. This is a 29.7% increase between 1967 and 2017. Income inequality in the United States is greater than in other developed nations and has been increasing.

Are these numbers cause for concern? And do they accurately reflect the disparity between the wealthy and the poor? For example, as people get older, their income increases. Perhaps these numbers reflect only the disparity between younger and older wage earners. What do you think?

Software packages often produce boxplots using a more complicated rule than we have described here. These may look different than those we have displayed. See the "Notes and Data Sources" for a brief discussion of these more complicated rules.

Mean and Standard Deviation

The five-number summary is not the most common numerical description of a distribution. That distinction belongs to the combination of the *mean* to measure center and the *standard deviation* to measure variability. The mean is familiar—it is the ordinary average of the observations. The idea of the standard deviation is to give the average distance of observations from the mean. The "average distance" in the standard deviation is found in a rather obscure way. We will give the details, but you may want to just think of the standard deviation as "average distance from the mean" and leave the details to your calculator.

Mean and standard deviation

The **mean $\bar{x}$** (pronounced "x-bar") of a set of observations is their arithmetic average. To find the mean of n observations, add the values and divide by n:

$$\bar{x} = \frac{\text{sum of the observations}}{n}$$

The **standard deviation s** measures the average distance of the observations from their mean. It is calculated by finding an average of the squared distances and then taking the square root. To find the standard deviation of n observations:

1. Find the distance of each observation from the mean and square each of these distances.
2. Average the squared distances by dividing their sum by $n-1$. This average squared distance is called the **variance.**
3. The standard deviation s is the square root of this average squared distance.

Example 4

Finding the mean and standard deviation

The numbers of home runs Barry Bonds hit in his 22 major league seasons are

16 25 24 19 33 25 34 46 37 33 42
40 37 34 49 73 46 45 45 5 26 28

To find the mean of these observations,

$$\bar{x} = \frac{\text{sum of observations}}{n}$$
$$= \frac{16 + 25 + \cdots + 28}{22}$$
$$= \frac{762}{22} = 34.6$$

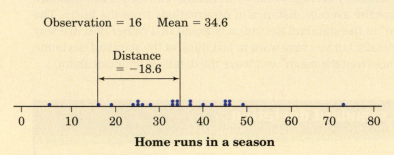

Figure 12.6 Barry Bonds's home run counts, Example 4, with their mean and the distance of one observation from the mean indicated. Think of the standard deviation as an average of these distances.

Figure 12.6 displays the data as points above the number line, with their mean marked by a vertical line. The arrow shows one of the distances from the mean. The idea behind the standard deviation s is to average the 22 distances. To find the standard deviation by hand, you can use a table layout:

Observation	Squared distance from mean
16	$(16 - 34.6)^2 = (-18.6)^2 = 345.96$
25	$(25 - 34.6)^2 = (-9.6)^2 = 92.16$
⋮	
28	$(28 - 34.6)^2 = (-6.6)^2 = 43.56$
	sum $= 4139.12$

The average is

$$\frac{4139.12}{21} = 197.1$$

Notice that we "average" by dividing by *one less* than the number of observations. Finally, the standard deviation is the square root of this number:

$$s = \sqrt{197.1} = 14.04$$

In practice, you can key the data into your calculator and hit the mean key and the standard deviation key. Or you can enter the data into a spreadsheet or other software to find $\overline{x}$ and s. It is usual, for good but somewhat technical reasons, to average the squared distances by dividing their total by $n-1$ rather than by n. Many calculators have two standard deviation buttons, giving you a choice between dividing by n and dividing by $n-1$. Be sure to choose $n-1$.

Now it's your turn

12.3 Hank Aaron. Here are Aaron's home run counts for his 23 years in baseball.

13 27 26 44 30 39 40 34 45 44 24 32
44 39 29 44 38 47 34 40 20 12 10

Find the mean and standard deviation of the number of home runs Aaron hit in each season of his career. How do the mean and median compare?

More important than the details of the calculation are the properties that show how the standard deviation measures variability.

Properties of the standard deviation s

- s measures variability about the mean $\overline{x}$. Use s to describe the variability of a distribution only when you use $\overline{x}$ to describe the center.
- $s = 0$ only when there is *no variability*. This happens only when all observations have the same value. So standard deviation zero means no variability at all. Otherwise, $s > 0$. As the observations become more variable about their mean, s gets larger.

Example 5

Investing 101

We have discussed examples about income. Here is an example about what to do with it once you've earned it. One of the first principles of investing is that taking more risk brings higher returns, at least on the average in the long run. People who work in finance define risk as the variability of returns from an investment (greater variability means higher risk) and measure risk by how unpredictable the return on an investment is. A bank account that is insured by the government and has a fixed rate of interest has no risk—its return is known exactly. Stock in a new company may soar one week and plunge the next. It has high risk because you can't predict what it will be worth when you want to sell.

bloomua/Deposit Photos

Investors should think statistically. You can assess an investment by thinking about the distribution of (say) yearly returns. That means asking about both the center and the variability of the pattern of returns. Only naive investors look for a high average return without asking about risk, that is, about how variable the returns are. Financial experts use the mean and standard deviation to describe returns on investments. The standard deviation was long considered too complicated to mention to the public, but now you will find standard deviations appearing regularly in mutual funds reports.

Here by way of illustration are the means and standard deviations of the yearly returns on three investments over the second half of the twentieth century (the 50 years from 1950 to 1999):

Investment	Mean return	Standard deviation
Treasury bills	5.34%	2.96%
Treasury bonds	6.12%	10.73%
Common stocks	14.62%	16.32%

You can see that risk (variability) goes up as the mean return goes up, just as financial theory claims. Treasury bills and bonds are ways of loaning money to the U.S. government. Treasury bills are paid back in 1 year, so their return changes from year to year depending on interest rates. Bonds are 30-year loans. They are riskier because the value of a bond you own will drop if interest rates go up. Stocks are even riskier. They give higher returns (on the average in the long run) but at the cost of lots of sharp ups and downs along the way. As the stemplot in Figure 12.7 shows, stocks went up by as much as 50% and down by as much as 26% in 1 year during the 50 years covered by our data.

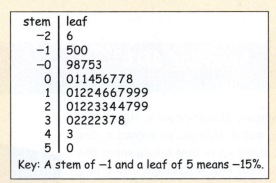

```
stem  leaf
 -2   6
 -1   500
 -0   98753
  0   011456778
  1   01224667999
  2   01223344799
  3   02222378
  4   3
  5   0
Key: A stem of -1 and a leaf of 5 means -15%.
```

Figure 12.7 Stemplot of the yearly returns on common stocks for the 50 years 1950–1999, Example 5. The returns are rounded to the nearest whole percent. The stems are 10s of percents and the leaves are single percents.

Choosing Numerical Descriptions

The five-number summary is easy to understand and is the best short description for most distributions. The mean and standard deviation are harder to understand but are more common. How can we decide which of these two descriptions of center and variability to use? Let's start by comparing the mean and the median. "Midpoint" and "arithmetic average" are both reasonable ideas for describing the center of a set of data, but they are different ideas with different uses. The most important distinction is that the mean (the average) is strongly influenced by a few extreme observations and the median (the midpoint) is not.

Example 6
Mean versus median

Table 12.1 gives the approximate salaries (in millions of dollars) of the 18 members of the Los Angeles Lakers basketball team for the 2018–2019 season. You can calculate that the mean is $\bar{x} = \$6.0$ million and that the median is $M = \$2.1$ million. No wonder professional basketball players have big houses.

Why is the mean so much higher than the median? Figure 12.8 is a stemplot of the salaries, with millions as stems. The distribution is skewed to the right and there is one very high outlier. The very high salary of LeBron James pulls up the sum of the salaries and so pulls up the mean. If we drop the outlier, the mean for the other 17 players is only \$4.2 million. The median doesn't change nearly as much: it drops from \$2.1 million to \$1.8 million.

We can make the mean as large as we like by just increasing LeBron James's salary. The mean will follow one outlier up and up. But to the median, LeBron's salary just counts as one observation at the upper end of the distribution. Moving it from \$35.7 million to \$357 million would not change the median at all.

DARREN ABATE/EFE/Alamy
Live News

Table 12.1 Salaries of the Los Angeles Lakers, 2018–2019 season

Player	Salary ($)	Player	Salary ($)
LeBron James	35.7 million	Moritz Wagner	1.8 million
Luoi Deng	18.0 million	Kyle Kuzma	1.7 million
Kentavious Caldwell-Pope	12.0 million	Josh Hart	1.7 million
Rajon Rondo	9.0 million	Ivica Zubac	1.5 million
Lonzo Ball	7.5 million	Isaac Bonga	1.0 million
Brandom Ingram	5.8 million	Joel Berry II	0.8 million
Lance Stephenson	4.4 million	Jeffrey Carroll	0.8 million
Michael Beasley	3.5 million	Alex Caruso	0.1 million
JaVale McGee	2.4 million	Travis Wear	0.1 million

Data from www.spotrac.com/nba/rankings/base/los-angeles-lakers.

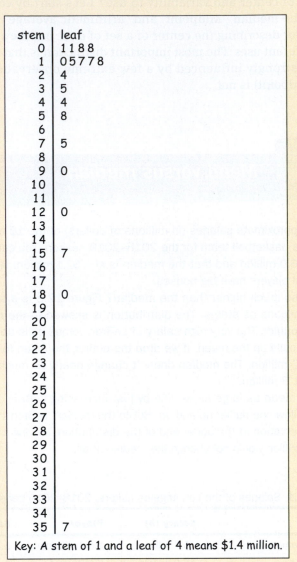

Key: A stem of 1 and a leaf of 4 means $1.4 million.

Figure 12.8 Stemplot of the salaries of Los Angeles Lakers players, from Table 12.1.

The mean and median of a symmetric distribution are close to each other. In fact, $\bar{x}$ and M are exactly equal if the distribution is exactly symmetric. In skewed distributions, however, the mean runs away from the median toward the long tail. Many distributions of monetary values—incomes, house prices, wealth—are strongly skewed to the right. The mean may be much larger than the median. For example, we saw in

Example 3 that the distribution of incomes for blacks, Hispanics, and whites is skewed to the right. The Census Bureau website gives the mean incomes for 2017 as $58,593 for blacks, $68,315 for Hispanics, and $93,453 for whites. Compare these with the corresponding medians of $40,258, $50,486, and $68,145. Because monetary data often have a few extremely high observations, descriptions of these distributions usually employ the median.

You should think about more than symmetry versus skewness when choosing between the mean and the median. The distribution of selling prices for homes in Middletown is no doubt skewed to the right—but if the Middletown City Council wants to estimate the total market value of all houses in order to set tax rates, the mean and not the median helps them out because the mean will be larger. (The total market value is just the number of houses times the mean market value and has no connection with the median.)

The standard deviation is pulled up by outliers or the long tail of a skewed distribution even more strongly than the mean. The standard deviation of the Lakers' salaries is $s = \$8.8$ million for all 18 players and only $s = \$4.9$ million when the outlier is removed. The quartiles are much less sensitive to a few extreme observations. There is another reason to avoid the standard deviation in describing skewed distributions. Because the two sides of a strongly skewed distribution have different amounts of variability, no single number such as s describes the variability well. The five-number summary, with its two quartiles and two extremes, does a better job. In most situations, it is wise to use $\bar{x}$ and s only for distributions that are roughly symmetric.

Why do we bother with the standard deviation at all? One answer appears in the next chapter: the mean and standard deviation are the natural measures of center and variability for an important kind of symmetric distribution, called the Normal distribution.

Remember that a graph gives the best overall picture of a distribution. Numerical measures of center and variability report specific facts about a distribution, but they do not describe its entire shape. Numerical summaries do not disclose the presence of multiple peaks or gaps, for example. *Always start with a graph of your data*.

Statistics in Your World

Poor New York? Is New York a rich state? New York's mean income per person ranks sixth among the states, right up there with its rich neighbors Connecticut and New Jersey, which rank second and fourth. But while Connecticut and New Jersey rank ninth and eighth in median household income, New York stands 22nd. What's going on? Just another example of mean versus median. New York has many very highly paid people, who pull up its mean income per person. But it also has a higher proportion of poor households than do Connecticut and New Jersey, and this brings the median down. New York is not a rich state—it's a state with extremes of wealth and poverty.

Choosing a summary

The mean and standard deviation are strongly affected by outliers or by the long tail of a skewed distribution. The median and quartiles are less affected.

The five-number summary is usually better than the mean and standard deviation for describing a skewed distribution or a distribution with outliers. Use $\bar{x}$ and s only for reasonably symmetric distributions that are free of outliers.

Chapter 12: Statistics in Summary

■ If we have data on a single quantitative variable, we start with a histogram or stemplot to display the distribution. Then we include numbers to describe the **center and variability** of the distribution.

■ There are two common descriptions of center and variability: the **five-number summary** and the **mean and standard deviation.**

■ The five-number summary consists of the **median** M, the midpoint of the observations, to measure the center, the two **quartiles** Q_1 and Q_3, and the smallest and largest observations. The difference between Q_1 and Q_3 and the difference between the smallest and largest observations describes variability.

■ A **boxplot** is a graph of the five-number summary.

■ The **mean** $\bar{x}$ is the average of the observations.

■ The **standard deviation** s measures variability as a kind of average distance from the mean, so use it only with the mean. The **variance** is the square of the standard deviation.

■ The mean and standard deviation can be changed a lot by a few outliers. The mean and median are the same for symmetric distributions, but the mean moves farther toward the long tail of a skewed distribution.

■ In general, use the five-number summary to describe most distributions and the mean and standard deviation only for roughly symmetric distributions.

This chapter summary will help you evaluate the Case Study.

Link It

In Chapter 11, we discussed histograms and stemplots as graphical displays of the distribution of a single quantitative variable. We were interested in the shape, center, and variability of the distribution. In this chapter, we introduce numbers to describe the center and variability. For symmetric distributions, the mean and standard deviation are used to describe the center and variability. For distributions that are not roughly symmetric, we use the five-number summary to describe the center and variability.

In most of the examples, we used graphical displays and numbers to describe the distribution of data on a single quantitative variable. These data are typically a sample from some population. Thus, the numbers that describe features of the distribution are statistics as discussed in Chapter 3. In the next chapter, we begin to think about distributions of populations. Thus, the numbers that describe features of these distributions are parameters. In later chapters, we will use statistics to draw conclusions, or make inferences, about parameters. Drawing conclusions about parameters that describe the center of a distribution of a single quantitative variable will be an important type of inference.

Case Study Evaluated

Use what you have learned in this chapter to evaluate the Case Study that opened the chapter. Start by reviewing the Chapter Summary. Then answer each of the following questions in complete sentences. Be sure to communicate clearly enough for any of your classmates to understand what you are saying.

   Find the data on income by education at the Census Bureau website listed in the Notes and Data Sources section at the end of the book.

1. What are the median incomes for people 25 years old and older who are high school graduates only, have some college but no degree, have a bachelor's degree, have a master's degree, and have a doctorate degree? At the bottom of the table, you will find median earnings in dollars.

2. From the distribution given in the tables, can you find the (approximately) first and third quartiles?

3. Do people with more education earn more than people with less education? Discuss.

In this chapter you have:

- Learned new methods to summarize large data sets with a few numbers such as the median, quartiles, mean, and standard deviation.

- Understood how to interpret these new summary numbers as measures of the center and the variability of a distribution.

- Learned a simple graphical method, the boxplot, for summarizing important features of large data sets.

macmillan learning Online Resources

- The StatClips Examples videos, *Summaries of Quantitative Data Example A, Example B, and Example C*, describe how to compute the mean, standard deviation, and median of data.

- The StatClips Examples video, *Exploratory Pictures for Quantitative Data Example C*, describes how to construct boxplots.

- The Snapshots Video, *Summarizing Quantitative Data*, discusses the mean, standard deviation, and median of data, as well as stemplots, in the context of a real example.

Check the Basics

For Exercise 12.1, see page 271; for Exercise 12.2, see page 273; for Exercise 12.3, see page 279.

12.4 Mean The mean of the three numbers, 1, 2, and 15, is equal to

(a) 1.

(b) 2.

(c) 5.

(d) 6.

12.5 Median The median of the three numbers, 1, 2, and 15, is equal to

(a) 1.

(b) 2.

(c) 5.

(d) 6.

12.6 Standard deviation Which of the following statements is true of the standard deviation?

(a) Removing an outlier will decrease the standard deviation.

(b) Removing an outlier will increase the standard deviation.

(c) It is the difference between the first and third quartile.

(d) It is the difference between the minimum and maximum values.

12.7 The five-number summary Which of the following is a graph of the five-number summary?

(a) A histogram

(b) A stemplot

(c) A boxplot

(d) A bar graph

12.8 Describing distributions Which of the following should you use to describe a distribution that is skewed?

(a) The five-number summary

(b) The mean, the first quartile, and the third quartile

(c) The mean and standard deviation

(d) The median and standard deviation

Chapter 12 Exercises

12.9 Median income. You read that the median income of U.S. households in 2017 was $68,145. Explain in plain language what "the median income" is.

12.10 What's the average? The Census Bureau website gives several choices for "average income" in its historical income data. In 2017, the median income of American households was $68,145. The mean household income was $93,453. The median income of families was $75,938, and the mean family income was $100,400. The Census Bureau says, "Households consist of all people who occupy a housing unit. The term 'family' refers to a group of two or more people related by birth, marriage, or adoption who reside together." Explain carefully why mean incomes are higher than median incomes and why family incomes are higher than household incomes.

12.11 Rich magazine readers. *Seattle Magazine* reports that the average income of its readers is $240,000. Is the median wealth of these readers greater or less than $240,000? Why?

12.12 College tuition. Figure 11.7 (page 254) is a stemplot of the tuition charged by 114 colleges in Illinois. The stems are thousands of dollars and the leaves are hundreds of dollars. For example, the highest tuition is $53,600 and appears as leaf 6 on stem 53.

(a) Find the five-number summary of Illinois college tuitions. You see that the stemplot already arranges the data in order.

(b) Would the mean tuition be clearly smaller than the median, about the same as the median, or clearly larger than the median? Why?

12.13 Where are the young more likely to live? Figure 11.11 (page 258) is a stemplot of the percentage of residents aged 18 to 34 in each of the 50 states. The stems are whole percents and the leaves are tenths of a percent.

(a) The shape of the distribution suggests that the mean will be about the same as the median. Why?

(b) Find the mean and median of these data and verify that the mean and the median are similar.

12.14 Gas mileage. Table 11.2 (page 261) gives the highway gas mileages for some model year 2015 midsized cars.

(a) Make a stemplot of these data if you did not do so in Exercise 11.13.

(b) Find the five-number summary of gas mileages. Which cars are in the bottom quarter of gas mileages?

(c) The stemplot shows a fact about the overall shape of the distribution that the five-number summary cannot describe. What is it?

12.15 Twins money. Table 11.4 (page 262) gives the salaries of the Minnesota Twins baseball team. What shape do you expect the distribution to have? Do you expect the mean salary to be close to the median, clearly higher, or clearly lower? Verify your choices by making a graph and calculating the mean and median.

12.16 The richest 5%. The distribution of individual incomes in the United States is strongly skewed to the right. In 2016, if we only look at the incomes of the top 5% of Americans, the mean and median incomes of the individuals in the top 5% were $215,000 and $375,000. Which of these numbers is the mean and which is the median? Explain your reasoning.

12.17 How many calories does a hot dog have? *Consumer Reports* magazine presented the following data on the number of calories in a hot dog for each of 17 brands of meat hot dogs:

173 191 182 190 172 147 146 139 175
136 179 153 107 195 135 140 138

Make a stemplot if you did not already do so in Exercise 11.19 (page 263), and find the five-number summary. The stemplot shows important facts about the distribution that the numerical summary does not tell us. What are these facts?

12.18 Returns on common stocks. Example 5 informs us that financial theory uses the mean and standard deviation to describe the returns on investments. Figure 11.13 (page 260) is a histogram of the returns of all New York Stock Exchange common stocks in one year. Are the mean and standard deviation suitable as a brief description of this distribution? Why?

12.19 The Super Bowl MVP. Figure 11.12 (page 259) is a histogram of the ages of players who have been named a Super Bowl Most Valuable Player (MVP) for the first 53 Super Bowl Games. The classes for Figure 11.12 are 22–23.5, 23.6–25, and so on.

(a) What is the position of each number in the five-number summary in a list of 53 observations arranged from smallest to largest?

(b) Even without the actual data, you can use your answer to (a) and the histogram to give the five-number summary approximately. Do this. About how old must an MVP be in order to be in the top quarter?

12.20 The statistics of writing style. Here are data on the percentages of words of 1 to 15 letters used in articles in *Popular Science* magazine. Exercise 11.16 (page 261) asked you to make a histogram of these data.

Length:	1	2	3	4	5	6	7	8
Percent:	3.6	14.8	18.7	16.0	12.5	8.2	8.1	5.9

Length:	9	10	11	12	13	14	15
Percent:	4.4	3.6	2.1	0.9	0.6	0.4	0.2

Find the five-number summary of the distribution of word lengths from this table.

12.21 Immigrants in the eastern states. Here are the number of legal immigrants (in thousands) who settled in each state east of the Mississippi River in 2017:

Alabama	3.8	New Hampshire	2.3
Connecticut	11.9	New Jersey	54.4
Delaware	2.2	New York	139.4
Florida	127.6	North Carolina	21.1
Georgia	26.2	Ohio	16.9
Illinois	40.5	Pennsylvania	27.8
Indiana	10.1	Rhode Island	4.1
Kentucky	7.5	South Carolina	5.0
Maine	1.6	Tennessee	9.8
Maryland	25.1	Vermont	0.8
Massachusetts	37.0	Virginia	29.5
Michigan	18.9	West Virginia	0.8
Mississippi	1.8	Wisconsin	6.7

Make a graph of the distribution. Describe its overall shape and any outliers. Then choose and calculate a suitable numerical summary.

12.22 Immigrants in the eastern states. New York and Florida are high outliers in the distribution of the previous exercise. Find the mean and the median for these data with and without New York and Florida. Which measure changes more when we omit the outliers?

12.23 State SAT scores. Figure 12.9 is a histogram of the average scores on the SAT Mathematics exam for college-bound senior students in the 50 states and the District of Columbia in 2018. The distinctive overall shape of this distribution implies that a single measure of center such as the mean or the median is of little value in describing the distribution. Explain why this is true.

12.24 Highly paid athletes. A news article reported that of the 446 players on National Basketball Association rosters in 2015, only 146 made more than $5 million. The article also stated that the average NBA salary in 2015 was $5.2 million. Was $5.2 million the mean or median salary for NBA players? How do you know?

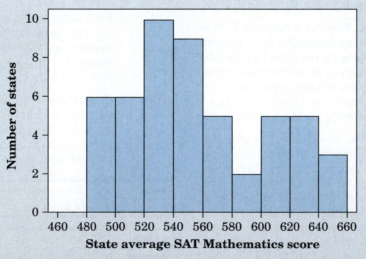

Figure 12.9 Histogram of the average scores on the SAT Mathematics exam for college-bound senior students in the 50 states and the District of Columbia in 2018, Exercise 12.23.

12.25 Mean or median? Which measure of center, the mean or the median, should you use in each of the following situations? Why?

(a) Middletown is considering imposing an income tax on citizens. The city government wants to know the average income of citizens so that it can estimate the total tax base.

(b) In a study of the standard of living of typical families in Middletown, a sociologist estimates the average family income in that city.

12.26 Mean or median? You are planning a party and want to know how many cans of soda to buy. A genie offers to tell you either the mean number of cans guests will drink or the median number of cans. Which measure of center should you ask for? Why? To make your answer concrete, suppose there will be 30 guests and the genie will tell you either $\bar{x} = 5$ cans or $M = 3$ cans. Which of these two measures would best help you determine how many cans you should have on hand?

12.27 State SAT scores. We want to compare the distributions of average SAT Math and Evidence-Based Reading and Writing (ERW) scores for the states and the District of Columbia. We enter these data into a computer with the names SATM for Math scores and SATERW for Evidence-Based Reading and Writing scores. Below is output from the statistical software package Minitab. (Other software produces similar output. Some software uses rules for finding the quartiles that differ slightly from ours. So software may not give exactly the answer you would get by hand.)

Use this output to make boxplots of SAT Math and Evidence-Based Reading and Writing scores for the states. Briefly compare the two distributions in words.

12.28 Do SUVs waste gas? Table 11.2 (page 261) gives the highway fuel consumption (in miles per gallon) for 31 model year 2015 midsized cars. You found the five-number summary for these data in Exercise 12.14. Here are the highway gas mileages for 26 four-wheel-drive model year 2015 sport utility vehicles:

Model	mpg
BMW X5 xdrive35i	27
Chevrolet Tahoe K1500	22
Chevrolet Traverse	23
Dodge Durango	24
Ford Expedition	20
Ford Explorer	23
GMC Acadia	23
GMC Yukon	22
Infiniti QX80	19
Jeep Grand Cherokee	20
Land Rover LR4	19
Land Rover Range Rover	23
Land Rover Range Rover Sport	19
Lexus GX460	20
Lexus LX570	17
Lincoln Navigator	20
Lincoln MKT	23
Mercedes-Benz ML250 Bluetec 4matic	29
Mercedes-Benz G63 AMG	14
Nissan Armada	18
Nissan Pathfinder Hybrid	27
Porsche Cayenne S	24
Porsche Cayenne Turbo	21
Toyota Highlander	24
Toyota Land Cruiser Wagon	18
Toyota 4Runner	21

Minitab output for Exercise 12.27

Variable	N	Mean	Median	StDev	Minimum	Maximum	Q1	Q3
SATM	51	557.25	547.00	48.89	480.00	655.00	521.00	606.00
SATERW	51	567.29	552.00	45.32	497.00	643.00	535.00	618.00

(a) Give a graphical and numerical description of highway fuel consumption for SUVs. What are the main features of the distribution?

(b) Make boxplots to compare the highway fuel consumption of the midsize cars in Table 11.2 and SUVs. What are the most important differences between the two distributions?

12.29 How many calories in a hot dog? Some people worry about how many calories they consume. *Consumer Reports* magazine, in a story on hot dogs, measured the calories in 20 brands of beef hot dogs, 17 brands of meat hot dogs, and 17 brands of poultry hot dogs. Here is computer output describing the beef hot dogs,

```
Mean = 156.8  Standard deviation = 22.64
Min = 111  Max = 190  N = 20  Median = 152.5
Quartiles = 140, 178.5
```

the meat hot dogs,

```
Mean = 158.7  Standard deviation = 25.24
Min = 107  Max = 195  N = 17  Median = 153
Quartiles = 139, 179
```

and the poultry hot dogs,

```
Mean = 122.5  Standard deviation = 25.48
Min = 87  Max = 170  N = 17  Median = 129
Quartiles = 102, 143
```

(Some software uses rules for finding the quartiles that differ slightly from ours. So software may not give exactly the answer you would get by hand.) Use this information to make boxplots of the calorie counts for the three types of hot dogs. Write a brief comparison of the distributions. Will eating poultry hot dogs usually lower your calorie consumption compared with eating beef or meat hot dogs? Explain.

12.30 Finding the standard deviation. The level of various substances in the blood influences our health. Here are measurements of the level of phosphate in the blood of a patient, in milligrams of phosphate per deciliter of blood, made on six consecutive visits to a clinic:

5.6   5.2   4.6   4.9   5.7   6.4

A graph of only six observations gives little information, so we proceed to compute the mean and standard deviation.

(a) Find the mean from its definition. That is, find the sum of the six observations and divide by 6.

(b) Find the standard deviation from its definition. That is, find the distance of each observation from the mean, square the distances, then calculate the standard deviation. Example 4 shows the method.

(c) Now enter the data into your calculator and use the mean and standard deviation keys to obtain $\bar{x}$ and s. Do the results agree with your hand calculations?

12.31 What s measures. Use a calculator to find the mean and standard deviation of these two sets of numbers:

(a) 4 2 4 2 4 2

(b) 5 5 5 1 1 1

Which data set is more variable?

12.32 What s measures. Add 2 to each of the numbers in data set (a) in the previous exercise. The data are now 6 4 6 4 6 4.

(a) Use a calculator to find the mean and standard deviation and compare your answers with those for data set part (a) in the previous exercise. How does adding 2 to each number change the mean? How does it change the standard deviation?

(b) Without doing the calculation, what would happen to $\bar{x}$ and s if we added 10 to each value in data set part (a) of the previous exercise? (This exercise demonstrates that the standard deviation measures only variability about the mean and ignores changes in where the data are centered.)

12.33 Cars and SUVs. Use the mean and standard deviation to compare the gas mileages of midsize cars (Table 11.2, page 261) and SUVs (Exercise 12.28). Do these numbers catch the main points of your more detailed comparison in Exercise 12.28?

12.34 A contest. This is a standard deviation contest. You must choose four numbers from the whole numbers 0 to 9, with repeats allowed.

(a) Choose four numbers that have the smallest possible standard deviation.

(b) Choose four numbers that have the largest possible standard deviation.

(c) Is more than one choice correct in either (a) or (b)? Explain.

12.35 $\bar{x}$ and s are not enough. The mean $\bar{x}$ and standard deviation s measure center and variability but are not a complete description of a distribution. Data sets with different shapes can have the same mean and standard deviation. To demonstrate this fact, use your calculator to find $\bar{x}$ and s for these two small data sets. Then make a stemplot of each and comment on the shape of each distribution.

| Data A: | 9.14 | 8.14 | 8.74 | 8.77 | 9.26 | 8.10 |
| Data B: | 6.58 | 5.76 | 7.71 | 8.84 | 8.47 | 7.04 |

| Data A: | 6.13 | 3.10 | 9.13 | 7.26 | 4.74 |
| Data B: | 5.25 | 5.56 | 7.91 | 6.89 | 12.50 |

12.36 Raising pay. A school system employs teachers at salaries between $40,000 and $70,000. The teachers' union and the school board are negotiating the form of next year's increase in the salary schedule. Suppose that every teacher is given a flat $3000 raise.

(a) How much will the mean salary increase? The median salary?

(b) Will a flat $3000 raise increase the variability as measured by the distance between the quartiles? Explain.

(c) Will a flat $3000 raise increase the variability as measured by the standard deviation of the salaries? Explain.

12.37 Raising pay. Suppose that the teachers in the previous exercise each receive a 5% raise. The amount of the raise will vary from $2000 to $3500, depending on present salary. Will a 5% across-the-board raise increase the variability of the distribution as measured by the distance between the quartiles? Do you think it will increase the standard deviation? Explain your reasoning.

12.38 Making colleges look good. Colleges announce an "average" SAT score for their entering freshmen. Usually the college would like this "average" to be as high as possible. A *New York Times* article noted, "Private colleges that buy lots of top students with merit scholarships prefer the mean, while open-enrollment public institutions like medians." Use what you know about the behavior of means and medians to explain these preferences.

12.39 What graph to draw? We now understand three kinds of graphs to display distributions of quantitative variables: histograms, stemplots, and boxplots. Give an example (just words, no data) of a situation in which you would prefer that graphing method.

Exploring the Web

Access these exercises on the text website: macmillanlearning.com/scc10e.

Normal Distributions

Case Study: Applying the Normal Distribution to Data

In this chapter you will:

- Discover the special properties of Normal curves.

- Understand how to use Normal curves to answer questions about underlying distributions that cannot easily be determined from the histograms.

Bar graphs and histograms are definitely old technology. Using bars to display data goes back to William Playfair (1759–1823), an English economist who was an early pioneer of data graphics. Histograms require that we choose classes, and their appearance can change with different choices. Surely, modern software offers a better way to picture distributions?

Software can replace the separate bars of a histogram with a smooth curve that represents the overall shape of a distribution. Look at Figure 13.1. The data are the number of text messages reported as being sent on a particular day in 2017 by a random sample of 447 high school seniors. This particular random sample was generated from the U.S. Census at School project of the American Statistical Association. The curve superimposed on the histogram is generated from statistical software as a way of replacing the histogram. The software doesn't start from the histogram—it starts with the actual observations and cleverly draws a curve to describe their distribution.

In Figure 13.1, the software has caught the overall shape and shows the ripples in the long right tail more effectively than does the histogram. It struggles a bit with the peak: it has extended the curve beyond zero in an attempt to smooth out the sharp peak. In Figure 13.2, we apply the same software to a set of data with a more regularly shaped distribution. These are

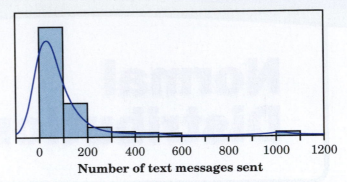

Figure 13.1 A histogram and a computer-drawn curve. Both picture the distribution of number of text messages reported as being sent on a particular day in 2017 by a random sample of 447 high school seniors. This distribution is skewed to the right. (This figure was created using the JMP software package.)

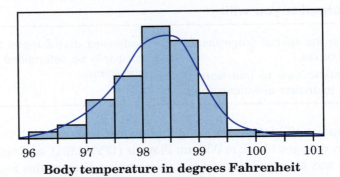

Figure 13.2 A histogram and a computer-drawn curve. Both picture the distribution of body temperatures from a sample of 130 healthy adults. This distribution is quite symmetric. (This figure was created using the JMP software package.)

the body temperatures from a sample of 130 healthy adults. The software draws a curve that shows a distinctive symmetric, single-peaked, bell shape.

For the irregular distribution in Figure 13.1, we can't do better. In the case of the very symmetric and bell-shaped distribution in Figure 13.2, however, there is another way to get a smooth curve. The distribution can be described by a specific kind of smooth curve called a *Normal curve*. Figure 13.3 shows the Normal curve for these data. The curve looks a lot like the one in Figure 13.2, but a close look shows that it is smoother. The Normal curve is much easier to work with and does not require clever software.

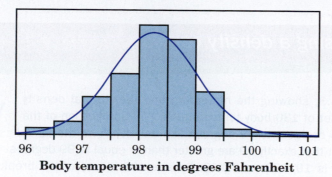

Figure 13.3 A perfectly symmetric Normal curve used to describe the distribution of body temperatures. (This figure was created using the JMP software package.)

We now have a kit of graphical and numerical tools for describing distributions. What is more, we have a clear strategy for exploring data on a single quantitative variable:

1. Always plot your data: make a graph, usually a histogram or a stemplot.
2. Look for the overall pattern (shape, center, variability) and for striking deviations such as outliers.
3. Choose either the five-number summary or the mean and standard deviation to briefly describe center and variability in numbers.

Here is one more step to add to this strategy:

4. Sometimes the overall pattern of a large number of observations is so regular that we can describe it by a smooth curve.

Density Curves

Figures 13.1 and 13.2 show curves used in place of histograms to picture the overall shape of a distribution of data. You can think of drawing a curve through the tops of the bars in a histogram, smoothing out the irregular ups and downs of the bars. There are two important distinctions between histograms and these curves. First, most histograms show the *counts* of observations in each class by the heights of their bars and therefore by the areas of the bars. We set up curves to show the *proportion* of observations in any region by areas under the curve. To do that, we choose the scale so that the total area under the curve is exactly 1. We then have a **density curve.** Second, a histogram is a plot of data obtained from a sample. We use this histogram to understand the actual distribution of the population from which the sample was selected. The density curve is intended to reflect the idealized shape of the population distribution.

Example 1

Using a density curve

Figure 13.4 copies Figure 13.3, showing the histogram and the Normal density curve that describe this data set of 130 body temperatures. What proportion of the temperatures are greater than or equal to 99 degrees Fahrenheit? From the actual 130 observations, we can count that exactly 19 are greater than or equal to 99 degrees Fahrenheit. So the proportion is 19/130, or 0.146. Because 99 is one of the break points between the classes in the histogram, the area of the shaded bars in Figure 13.4(a) makes up 0.146 of the total area of all the bars.

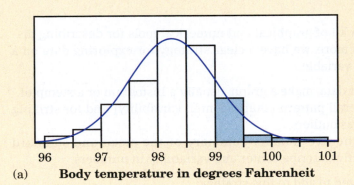

(a)   **Body temperature in degrees Fahrenheit**

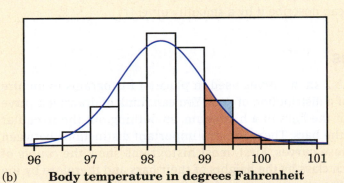

(b)   **Body temperature in degrees Fahrenheit**

Figure 13.4 A histogram and a Normal density curve, for Example 1. **(a)** The area of the shaded bars in the histogram represents observations greater than 99 degrees Fahrenheit. These make up 19 of the 130 observations. **(b)** The shaded area under the Normal curve represents the proportion of observations greater than 99 degrees Fahrenheit. This area is 0.1587. (This figure was created using the JMP software package.)

Now concentrate on the density curve drawn through the histogram. The total area under this curve is 1, and the shaded area in Figure 13.4(b) represents the proportion of observations that are greater than or equal to 99 degrees Fahrenheit. This area is 0.1587. You can see that the density curve is a quite good approximation—0.1587 is close to 0.146.

The area under the density curve in Example 1 is not exactly equal to the true proportion because the curve is an idealized picture of the distribution. For example, the curve is exactly symmetric but the actual data are only approximately symmetric. Because density curves are smoothed-out idealized pictures of the overall shapes of distributions, they are most useful for describing large numbers of observations.

The Center and Variability of a Density Curve

Density curves help us better understand our measures of center and variability. Areas under a density curve represent proportions of the total number of observations. The median and quartiles are easy. The median is the point with half the observations on either side. So *the median of a density curve is the equal-areas point,* the point with half the area under the curve to its left and the remaining half of the area to its right. The quartiles divide the area under the curve into quarters. One-fourth of the area under the curve is to the left of the first quartile, and three-fourths of the area is to the left of the third quartile. You can roughly locate the median and quartiles of any density curve by eye by dividing the area under the curve into four equal parts.

Because density curves are idealized patterns, a symmetric density curve is exactly symmetric. The median of a symmetric density curve is therefore at its center. Figure 13.5(a) shows the median of a symmetric curve. We can roughly locate the equal-areas point on a skewed curve like that in Figure 13.5(b) by eye.

What about the mean? The mean of a set of observations is their arithmetic average. If we think of the observations as weights stacked on a seesaw, the mean is the point at which the seesaw would balance. This fact is also true of density curves. *The mean is the point at which the curve would balance if made of solid material.* Figure 13.6 illustrates this fact about the mean. A symmetric curve balances at its center because the two sides are identical. *The mean and median of a symmetric density curve are equal,* as in Figure 13.5(a).

Key Terms

The **median** of a density curve is the equal-areas point, the point that divides the area under the curve in half.

The **mean** of a density curve is the balance point, or center of gravity, at which the curve would balance if made of solid material.

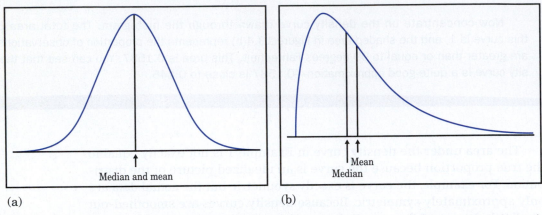

(a)   (b)

Figure 13.5 The median and mean for two density curves: **(a)** a symmetric Normal curve and **(b)** a curve that is skewed to the right.

Figure 13.6 The mean of a density curve is the point at which it would balance.

The mean of a skewed distribution is pulled toward the long tail. Figure 13.5(b) shows how the mean of a skewed density curve is pulled toward the long tail more than is the median.

Normal Distributions

The density curves in Figures 13.3 and 13.4 belong to a particularly important family: the Normal curves. (We capitalize "Normal" to remind you that these curves are special.) Figure 13.7 presents two more Normal density curves. Normal curves are symmetric, single-peaked, and bell-shaped. Their tails fall off quickly, so that we do not expect outliers. Because Normal distributions are symmetric, the mean and median lie together at the peak in the center of the curve.

Normal curves also have the special property that we can locate the standard deviation of the distribution by eye on the curve. This isn't true for most other density curves. Here's how to do it. Imagine that you

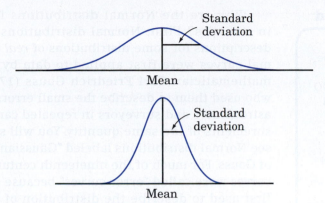

Figure 13.7 Two Normal curves. The standard deviation determines the variability of a Normal curve.

are skiing down a mountain that has the shape of a Normal curve. At first, you descend at an ever-steeper angle as you go out from the peak:

Fortunately, before you find yourself going straight down, the slope begins to grow flatter rather than steeper as you go out and down:

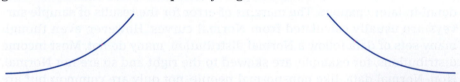

The points at which this change of curvature takes place are located one standard deviation on either side of the mean. The standard deviations are marked on the two curves in Figure 13.7. You can feel the change as you run a pencil along a Normal curve, and so find the standard deviation.

Normal curves have the special property that giving the mean and the standard deviation completely specifies the curve. The mean determines the center of the curve, and the standard deviation determines its shape. Changing the mean of a Normal distribution does *not* change its shape, only its location on the axis. Changing the standard deviation *does* change the shape of a Normal curve, as Figure 13.7 illustrates. The distribution with the smaller standard deviation is less variable and more sharply peaked. The box at the right summarizes basic facts about Normal curves.

Normal density curves

The **Normal curves** are symmetric, bell-shaped curves that have these properties:

- A specific Normal curve is completely described by giving its mean and its standard deviation.
- The mean determines the center of the distribution. It is located at the center of symmetry of the curve.
- The standard deviation determines the shape of the curve. It is the distance from the mean to the change-of-curvature points on either side.

Statistics in Your World

The bell curve? Does the distribution of human intelligence follow the "bell curve" of a Normal distribution? Scores on IQ tests do roughly follow a Normal distribution. That is because a test score is calculated from a person's answers in a way that is designed to produce a Normal distribution. To conclude that intelligence follows a bell curve, we must agree that the test scores directly measure intelligence. Many psychologists don't think there is one human characteristic that we can call "intelligence" and can measure by a single test score.

Why are the Normal distributions important in statistics? First, Normal distributions are good descriptions for some distributions of *real data*. Normal curves were first applied to data by the great mathematician Carl Friedrich Gauss (1777–1855), who used them to describe the small errors made by astronomers and surveyors in repeated careful measurements of the same quantity. You will sometimes see Normal distributions labeled "Gaussian" in honor of Gauss. For much of the nineteenth century, Normal curves were called "error curves" because they were first used to describe the distribution of measurement errors. As it became clear that the distributions of some biological and psychological variables were at least roughly Normal, the "error curve" terminology was dropped. The curves were first called "Normal" by Francis Galton in 1889. Galton, a cousin of Charles Darwin, pioneered the statistical study of biological inheritance.

Normal curves also describe the distribution of statistics such as sample proportions (when the sample size is large and the value of the proportion is moderate) and sample means (when we take many samples from the same population). We will explore distributions of statistics in more detail in later chapters. The margins of error for the results of sample surveys are usually calculated from Normal curves. However, even though many sets of data follow a Normal distribution, many do not. Most income distributions, for example, are skewed to the right and so are not Normal. Non-Normal data, like non-normal people, not only are common but are sometimes more interesting than their normal counterparts.

The 68–95–99.7 rule

In any Normal distribution, approximately

- **68%** of the observations fall within one standard deviation of the mean.
- **95%** of the observations fall within two standard deviations of the mean.
- **99.7%** of the observations fall within three standard deviations of the mean.

The 68–95–99.7 Rule

There are many Normal curves, each described by its mean and standard deviation. All Normal curves share many properties. In particular, the standard deviation is the natural unit of measurement for Normal distributions. This fact is described in the box at the left.

Figure 13.8 illustrates the 68–95–99.7 rule. By remembering these three numbers, you can think about Normal distributions without constantly making detailed calculations. Remember also, though, that no set of data is exactly described by a Normal curve. The 68–95–99.7 rule will be only approximately true for SAT scores or the lengths of crickets since these distributions are approximately Normal.

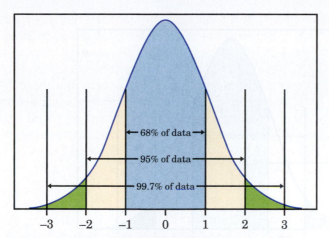

Figure 13.8 The 68–95–99.7 rule for Normal distributions. Note the x-axis, or the horizontal axis, displays the number of standard deviations from the mean.

Example 2

Heights of women

Dean Drobot/Shutterstock

The distribution of heights of women is approximately Normal with mean 63.7 inches and standard deviation 2.5 inches. To use the 68–95–99.7 rule, always start by drawing a picture of the Normal curve. Figure 13.9 shows what the rule says about women's heights.

Half of the observations in any Normal distribution lie above the mean, so half of all young women are taller than 63.7 inches.

The central 68% of any Normal distribution lies within one standard deviation of the mean. Half of this central 68%, or 34%, lies above the mean. So 34% of women are between 63.7 inches and 66.2 inches tall. Adding the 50% who are shorter than 63.7 inches, we see that 84% of women have heights less than 66.2 inches. That leaves 16% who are taller than 66.2 inches.

The central 95% of any Normal distribution lies within two standard deviations of the mean. Two standard deviations is 5 inches here, so the middle 95% of women's heights are between 58.7 inches (that's $63.7 - 5$) and 68.7 inches (that's $63.7 + 5$).

The other 5% of women have heights outside the range from 58.7 to 68.7 inches. Because the Normal distributions are symmetric, half of these women are on the short side. The shortest 2.5% of women are less than 58.7 inches (just under 5 feet) tall.

Almost all (99.7%) of the observations in any Normal distribution lie within three standard deviations of the mean. Almost all young women are between 56.2 and 71.2 inches tall.

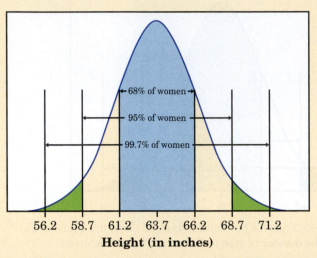

Figure 13.9 The 68–95–99.7 rule for heights of young women, for Example 2. This Normal distribution has mean 63.7 inches and standard deviation 2.5 inches.

Now it's your turn

13.1 Heights of men. The distribution of heights of men is approximately Normal with mean 69.2 inches and standard deviation 2.5 inches. Between which heights do the middle 95% of men fall?

Standard Scores

Madison scored 600 on the SAT Mathematics college entrance exam. How good a score is this? That depends on where a score of 600 lies in the distribution of all scores. The SAT exams are scaled so that scores should roughly follow the Normal distribution with mean 500 and standard deviation 100. Madison's 600 is one standard deviation above the mean. The 68–95–99.7 rule now tells us just where she stands (Figure 13.10). Half of all scores are below 500, and another 34% are between 500 and 600. So Madison did better than 84% of the students who took the SAT. Her score report not only will say she scored 600 but will add that this is at the "84th percentile." That's statistics speak for "You did better than 84% of those who took the test."

Because the standard deviation is the natural unit of measurement for Normal distributions, we restated Madison's score of 600 as "one standard deviation above the mean." Observations expressed in standard deviations above or below the mean of a distribution are called *standard scores*. Standard scores are also sometimes referred to as z-scores.

Key Terms

The **standard score** for any observation is

$$\text{standard score} = \frac{\text{observation} - \text{mean}}{\text{standard deviation}}$$

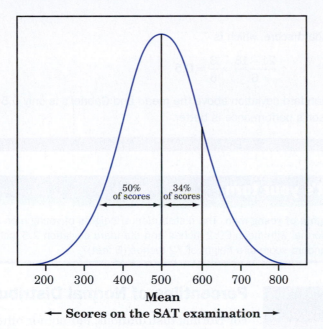

Figure 13.10 The 68–95–99.7 rule shows that 84% of any Normal distribution lies to the left of the point one standard deviation above the mean. Here, this fact is applied to SAT scores.

A standard score of 1 says that the observation in question lies one standard deviation above the mean. An observation with standard score -2 is two standard deviations below the mean. Standard scores can be used to compare values in different distributions. Of course, you should not use standard scores unless you are willing to use the standard deviation to describe the variability of the distributions. That requires that the distributions be at least roughly symmetric.

Example 3 — *ACT versus SAT scores*

Madison scored 600 on the SAT Mathematics exam. Her friend Gabriel took the American College Testing (ACT) test and scored 21 on the math part. ACT scores are Normally distributed with mean 18 and standard deviation 6. Assuming that both tests measure the same kind of ability, who has the higher score?

Madison's standard score is

$$\frac{600 - 500}{100} = \frac{100}{100} = 1.0$$

smolaw/Shutterstock

Compare this with Gabriel's standard score, which is

$$\frac{21-18}{6} = \frac{3}{6} = 0.5$$

Because Madison's score is 1 standard deviation above the mean and Gabriel's is only 0.5 standard deviation above the mean, Madison's performance is better.

Now it's your turn

13.2 Heights of young men. The distribution of heights of young men is approximately Normal with mean 69.2 inches and standard deviation 2.5 inches. What is the standard score of a height of 72 inches (6 feet)?

Key Terms

The **cth percentile** of a distribution is a value such that c percent of the observations lie below it and the rest lie above.

Percentiles of Normal Distributions*

For Normal distributions, but not for other distributions, standard scores translate directly into *percentiles*.

The median of any distribution is the 50th percentile, and the quartiles are the 25th and 75th percentiles. In any Normal distribution, the point one standard deviation above the mean (standard score 1) is the 84th percentile. Figure 13.10 shows why. Every standard score for a Normal distribution translates into a specific percentile, which is the same no matter what the mean and standard deviation of the original Normal distribution are. Table B at the back of this book gives the percentiles corresponding to various standard scores. This table enables us to do calculations in greater detail than does the 68–95–99.7 rule. Many technology tools, including some graphing calculators, will give you percentiles for standard scores and with more precision. In this text, we have chosen to use Table B with only one decimal point for the standard score because we are focused more on conceptual understanding than on precision.

Example 4

Percentiles for college entrance exams

Madison's score of 600 on the SAT translates into a standard score of 1.0. We saw that the 68–95–99.7 rule says that this is the 84th percentile. Table B is a bit more precise: it says that standard score 1 is the 84.13 percentile of a Normal distribution. Gabriel's 21 on the ACT is a standard score

---

*This material is not needed to read the rest of the book.

of 0.5. Table B says that this is the 69.15 percentile. Gabriel did well, but not as well as Madison. The percentile is easier to understand than either the raw score or the standard score. That's why reports of exams such as the SAT usually give both the score and the percentile.

Example 5

Finding the observation that matches a percentile

How high must a student score on the SAT to fall in the top 5% of all scores? That requires a score at or above the 95th percentile. Look in the body of Table B for the percentiles closest to 95. You see that standard score 1.6 is the 94.52 percentile and standard score 1.7 is the 95.54 percentile. We can average these two standard scores to get 1.65. The standard score of 1.65 is approximately the 95th percentile *of any Normal distribution*.

To go from the standard score back to the scale of SAT scores, "undo" the standard score calculation as follows:

$$\text{observation} = \text{mean} + \text{standard score} \times \text{standard deviation}$$
$$= 500 + (1.65)(100) = 665$$

A score of 665 or higher will be in the top 5%.

Example 6

Finding the percentage below a particular value

Returning again to the distribution of SAT scores, if Ting obtains a score of 430 on this exam, what percentage of individuals score the same as or lower than Ting? To answer this question, we must first convert Ting's score to a standard score. This standard score is

$$\frac{430 - 500}{100} = \frac{-70}{100} = -0.7$$

From Table B, we can see that the Standard score of -0.7 corresponds to a percentile of 24.20. Therefore, 24.20% of those individuals who took this exam scored at or below Ting's score of 430.

Example 7

Finding the percentage above a particular value

If Jordan scores 725 on the SAT Mathematics exam, what percentage of individuals would score at least as high or higher than Jordan? To answer this question, we must again begin by converting Jordan's score to a standard score. This standard score is

$$\frac{725 - 500}{100} = \frac{225}{100} = 2.25$$

To use Table B, we must first round 2.25 to 2.3. From Table B, we can see that the Standard score of 2.3 corresponds to a Percentile of 98.93. We might be tempted to assume our final answer will be 98.93%. If we were attempting to determine the percentage of individuals who scored 725 or lower, our final answer would be 98.93%. We want to know, however, the percentage who score 725 or higher. To find this percentage, we need to remember that the total area below the Normal curve is 1. Expressed as a percentage, the total area is 100%. To find the percentage of values that fall at or above 725, we subtract the percentage of values that fall below 725 from 100%. This gives us 100% − 98.93% = 1.07%. Our final answer is that 1.07% of individuals who take the SAT Mathematics exam score at or above Jordan's score of 725.

Example 8

Finding the percentage between two particular values

What percentage of individuals would score between Ting's and Jordan's score on the SAT Mathematics exam? From Example 6, we know that 24.20% of individuals score below Ting. Similarly, from Example 7, we know that 98.93% of individuals score below Jordan. To find the percentage of individuals who score *between* Ting's and Jordan's scores, we subtract 24.20% from 98.93% to get 98.93% − 24.20% = 74.74%. We find that 74.74% of individuals who take the SAT Mathematics exam have scores between Ting and Jordan.

Now it's your turn

13.3 SAT scores. How high must a student score on the SAT Mathematics exam to fall in the top 25% of all scores? What percentage of students obtain scores at or below 475? What percentage of students obtain scores at or above 580? What percentage of students obtain scores between 475 and 580?

Chapter 13: Statistics in Summary

- Stemplots, histograms, and boxplots all describe the distributions of quantitative variables.

- **Density curves** also describe distributions. A density curve is a curve with area exactly 1 underneath it whose shape describes the overall pattern of a distribution.

- An area under the curve gives the proportion of the observations that fall in an interval of values.

- You can roughly locate the median (equal-areas point) and the mean (balance point) by eye on a density curve.

- Stemplots, histograms, and boxplots are created from samples. Density curves are intended to display the idealized shape of the distribution of the population from which the samples are taken.

- **Normal curves** are a special kind of density curve that describes the overall pattern of some sets of data. Normal curves are symmetric and bell-shaped. A specific Normal curve is completely described by its mean and standard deviation. You can locate the mean (center point) and the standard deviation (distance from the mean to the change-of-curvature points) on a Normal curve. All Normal distributions obey the **68–95–99.7 rule.**

- **Standard scores** express observations in standard deviation units about the mean, which has standard score 0. A given standard score corresponds to the same **percentile** in any Normal distribution. Table B gives percentiles of Normal distributions.

This chapter summary will help you evaluate the Case Study.

Link It

Chapters 10, 11, and 12 provide us with a strategy for exploring data on a single quantitative variable.

- Make a graph, usually a histogram or stemplot.

- Look for the overall pattern (shape, center, variability) and striking deviations from the pattern.

- Choose the five-number summary or the mean and standard deviation to briefly describe the center and variability in numbers.

   In this chapter, we added another step: sometimes the overall pattern of a large number of observations is so regular that we can describe it by a smooth density curve, such as the Normal curve. This step also allows us to identify "a large number of observations" as a population and use density curves to describe the distribution of a population. We did precisely this when we used the Normal distribution to describe the distribution of the heights of all women or the scores of all students on the SAT exam.

   Using a density curve to describe the distribution of a population is a convenient summary, allowing us to determine percentiles of the distribution without having to see a list of all the values in the population. It also suggests the nature of the conclusions we might draw about a single quantitative variable. Use statistics that describe the distribution of the sample to draw conclusions about parameters that describe the distribution of a population. We will explore this in future chapters.

Case Study Evaluated

Use what you have learned in this chapter to evaluate the Case Study that opened the chapter. Start by reviewing the Chapter Summary. Then answer each of the following questions in complete sentences. Be sure to communicate clearly enough for any of your classmates to understand what you are saying.

The Normal curve that best approximates the distribution of body temperatures in Figures 13.2 and 13.3 has mean 98.25 degrees Fahrenheit and standard deviation 0.73 degrees Fahrenheit.

1. According to the 68-95-99.7 rule, 68% of body temperatures fall between what two values? Between what two values do 95% of body temperatures fall?

2. There was a time when 98.6 degrees Fahrenheit was considered the average body temperature. Given what you know about the distribution of body temperatures given in Figures 13.2 and 13.3, what percentage of individuals would you expect to have body temperatures greater than 98.6 degrees Fahrenheit? What percentage would you expect to have body temperature less than 98.6 degrees Fahrenheit?

In this chapter you

- Discovered the special properties of Normal curves.
- Understood how to use Normal curves to answer questions about the underlying distributions represented in Figures 13.2 and 13.3 that cannot easily be determined from the histograms.

macmillan learning Online Resources

- *LearningCurve* has good questions to check your understanding of the concepts.

Check the Basics

For Exercise 13.1, see page 302; for Exercise 13.2, see page 304; for Exercise 13.3, see page 306.

13.4 Density curves. One characteristic of a density curve is that there is a specific total area under the curve. What is this area equal to?

(a) Exactly 1.

(b) Approximately 1.

(c) It depends on what is being measured.

(d) It depends on whether the distribution is Normal.

13.5 The mean and median. Which of the following is an incorrect statement?

(a) If a density curve is skewed to the right, the mean will be larger than the median.

(b) In a symmetric density curve, the mean is equal to the median.

(c) The median is the balance point in a density curve.

(d) The mean of a skewed distribution is pulled toward the long tail.

13.6 Pulse rates. Suppose that resting pulse rates for healthy adults are found to follow a Normal distribution, with a mean of 69 beats per minute and a standard deviation of 9.5 beats per minutes. If Bonnie has a pulse rate of 78.5 beats per minute, this means that

(a) Approximately 32% of adults have pulse rates higher than Bonnie's.

(b) Approximately 16% of adults have pulse rates higher than Bonnie's.

(c) Bonnie's pulse rate is two standard deviations above the mean.

(d) Bonnie's pulse rate, when converted to a standard score, would be 1.5.

13.7 More on pulse rates. Let's again assume that the resting pulse rates for healthy adults follow a Normal distribution with a mean of 69 beats per minute and a standard deviation of 9.5 beats per minute. When converted to a standard score, Adam's pulse rate becomes −0.5. How should this standard score be interpreted?

(a) Adam should see a doctor because his pulse rates is unusually low.

(b) Adam's pulse rate is above the mean.

(c) Exactly 5% of healthy adults have pulse rates less than Adam's pulse rate.

(d) Adam's pulse rates is about one half of a standard deviation below the mean.

13.8 Reading test scores. A standardized reading test is given to fifth-grade students. Scores on this test are Normally distributed, with a mean of 32 points and a standard deviation of 8 points. When Corey gets his test results, he is told that his score is at the 95th percentile. What does this mean?

(a) Corey's score is higher than 95% of the students who took this test.

(b) Corey's score is 48 points.

(c) Corey's score is exactly 2 standard deviations above the mean.

(d) All of the above.

Chapter 13 Exercises

13.9 Density curves

(a) Sketch a density curve that is strongly skewed to the left.

(b) Sketch a density curve that is symmetric but has a shape different from that of the Normal curves.

13.10 Mean and median. Figure 13.11 shows density curves of several shapes. Briefly describe the overall shape of each distribution. Two or more points are marked on each curve. The mean and the median are among these points. For each curve, which point is the median and which is the mean?

13.11 Random numbers. If you ask a computer to generate "random numbers" between 0 and 5, you will get observations from a *uniform distribution*. Figure 13.12 shows the density curve for a uniform distribution. This curve takes the constant value 0.2 between 0 and 5 and is zero outside that range. Use this density curve to answer these questions.

(a) Why is the total area under the curve equal to 1?

(b) The curve is symmetric. What is the value of the mean and median?

(c) What percentage of the observations lie between 4 and 5?

(d) What percentage of the observations lie between 1.5 and 3?

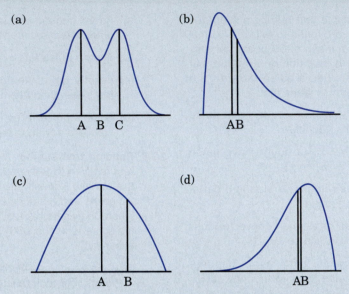

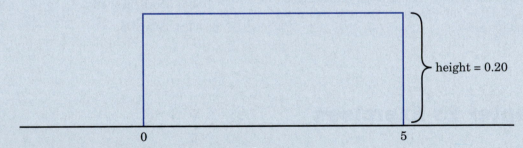

Figure 13.11 Four density curves of different shapes, for Exercise 13.10. In each case, the mean and the median are among the marked points.

height = 0.20

0   5

Figure 13.12 The density curve of a uniform distribution, for Exercise 13.11. Observations from this distribution are spread "at random" between 0 and 5.

IQ test scores. Figure 13.13 is a stemplot of the IQ scores that are consistent with the 2018 article "Flynn effect and its reversal are both environmentally caused." This distribution is very close to Normal with mean 100 and standard deviation 10. Use the Normal distribution with mean 100 and standard deviation 10 as a description of the IQ test scores of all adults. Use this distribution and the 68–95–99.7 rule to answer Exercises 13.12 to 13.14.

13.12 Between what values do the IQ scores of 68% of all adults lie?

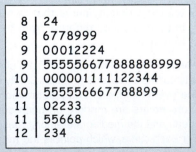

```
 8 | 24
 8 | 6778999
 9 | 00012224
 9 | 555556677888888999
10 | 000001111122344
10 | 555556667788899
11 | 02233
11 | 55668
12 | 234
```

Figure 13.13 Stemplot of the IQ scores of 80 adults, for Exercises 13.12 to 13.14. (Key: A stem of 8 and a leaf of 2 means 82.)

13.13 What percentage of IQ scores for all adults are more than 100?

13.14 What percentage of all students have IQ scores below 80? None of the 80 adults in our sample had scores this low. Are you surprised at this? Why?

13.15 Length of pregnancies. The length of human pregnancies from conception to birth varies according to a distribution that is approximately Normal with mean 266 days and standard deviation 16 days. Use the 68–95–99.7 rule to answer the following questions.

(a) Almost all (99.7%) pregnancies fall in what range of lengths?

(b) How long are the longest 2.5% of all pregnancies?

(c) How short are the shortest 16% of all pregnancies?

13.16 A Normal curve. What are the mean and standard deviation of the Normal curve in Figure 13.14?

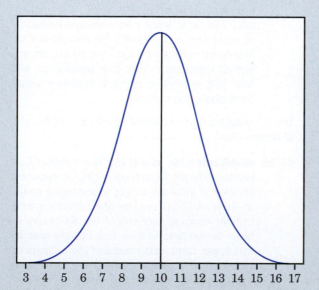

Figure 13.14 What are the mean and standard deviation of this Normal density curve? For Exercise 13.16.

13.17 Horse pregnancies. Bigger animals tend to carry their young longer before birth. The length of horse pregnancies from conception to birth varies according to a roughly Normal distribution with mean 336 days and standard deviation 3 days. Use the 68–95–99.7 rule to answer the following questions.

(a) Between what values do the lengths of the middle 95% of all horse pregnancies fall?

(b) What percentage of horse pregnancies are less than 333 days?

13.18 Great hitters then and now. Three landmarks of baseball achievement are Ty Cobb's batting average of .420 in 1911, Ted Williams's .406 in 1941, and George Brett's .390 in 1980. These batting averages cannot be compared directly because the distribution of major league batting averages has changed over the years. The distributions are quite symmetric and (except for outliers such as Cobb, Williams, and Brett) reasonably Normal. While the mean batting average has been held roughly constant by rule changes and the balance between hitting and pitching, the standard deviation has dropped over time. Here are the facts:

Decade	Mean	Std. dev.
1910s	.266	.0371
1940s	.267	.0326
1970s	.261	.0317

(a) Compute the standard scores for the batting averages of Cobb, Williams, and Brett to compare how far each stood above his peers.

(b) According to ESPN, the batting average leaders in 2018 were Mookie Betts of the Boston Red Sox for the American League (AL) and Christian Yelich of the Milwaukee Brewers for the National League (NL). AL batting averages had a mean of .262 with standard deviation .0315, while NL batting averages had a mean of .269 with standard deviation .0233. Use standard scores to compare the batting averages for Betts and Yelich.

Which player performed better relative to his peers?

13.19 Comparing IQ scores. The Wechsler Adult Intelligence Scale (WAIS) is an IQ test. Scores on the WAIS for the 20 to 34 age group are approximately Normally distributed with mean 110 and standard deviation 15. Scores for the 60 to 64 age group are approximately Normally distributed with mean 90 and standard deviation 15. Jessica, who is 28, scores 140 on the WAIS. Her mother, who is 63, takes the test and scores 120.

(a) Express both scores as standard scores that show where each woman stands within her own age group.

(b) Who scored higher relative to her age group, Jessica or her mother? Who has the higher absolute level of the variable measured by the test?

13.20 Men's heights. The distribution of heights of men is approximately Normal with mean 69.2 inches and standard deviation 2.5 inches. Sketch a Normal curve on which this mean and standard deviation are correctly located. (*Hint:* Draw the curve first, locate the points where the curvature changes, then mark the horizontal axis.)

13.21 More on men's heights. The distribution of heights of men is approximately Normal with mean 69.2 inches and standard deviation 2.5 inches. Use the 68–95–99.7 rule to answer the following questions.

(a) What percentage of men are shorter than 61.7 inches?

(b) Between what heights do the middle 95% of men fall?

(c) What percentage of men are taller than 66.7 inches?

13.22 Heights of men and women. The heights of women are approximately Normal with mean 63.7 inches and standard deviation 2.5 inches. The heights of men have mean 69.2 inches and standard deviation 2.5 inches. What percentage of women are taller than a man of average (mean) height?

13.23 Heights of adults. The mean height of men is about 69.2 inches. Women that age have a mean height of about 63.7 inches. Do you think that the distribution of heights for all adults is approximately Normal? Explain your answer.

13.24 Sleep. The distribution of hours of sleep per school night, among high school seniors, is found to be Normally distributed, with a mean of 6.6 hours and a standard deviation of 1.3 hours. Use this information and the 68–95–99.7 rule to answer the following questions.

(a) What percentage of high school seniors sleep more than 7.9 hours? Less than 4 hours?

(b) What range contains the middle 68% of hours slept per weeknight by high school seniors?

13.25 Cholesterol. Low-density lipoprotein, or LDL, is the main source of cholesterol buildup and blockage in the arteries. This is why LDL is known as "bad cholesterol." LDL is measured in milligrams per deciliter of blood, or mg/dL. In a population of adults at risk for cardiovascular problems, the distribution of LDL levels is Normal, with a mean of 123 mg/dL and a standard deviation of 41 mg/dL. If an individual's LDL is at least one standard deviation or more above the mean, the patient will be monitored carefully by a doctor. What percentage of individuals from this population will have LDL levels one or more standard deviations above the mean?

The following optional exercises require use of Table B of Normal distribution percentiles.

13.26 NCAA rules for athletes. The National Collegiate Athletic Association (NCAA) requires Division II athletes to get a combined score of at least 820 on the Mathematics and Critical Reading sections of the SAT exam in order to compete in their first college year. In 2018, the combined scores of the millions of college-bound seniors taking the SATs were approximately Normal with mean 1068 and standard deviation approximately 204. What percentage of all college-bound seniors had scores less than 820?

13.27 More NCAA rules. For Division I athletes the NCAA uses a sliding scale, based on both core GPA and the combined Mathematics and Critical Reading SAT score, to determine eligibility to compete in the first year of college. For athletes with a core GPA of 3.0, a score of at least 620 on the combined Mathematics and Critical Reading sections of the SAT exam is required. Use the information in the previous exercise to find the percentage of all SAT scores of college-bound seniors that are less than 620.

13.28 800 on the SAT. It is possible to score higher than 800 on the SAT, but scores above 800 are reported as 800. (That is, a student can get a reported score of 800 without a perfect paper.) In 2018, the scores of college-bound seniors SAT Math test followed a Normal distribution with mean 542 and standard deviation 123. What percentage of scores were above 800 (and so reported as 800)?

13.29 SAT ERW scores. In 2018, the average performance of college-bound seniors on the Evidence-Based Reading and Writing (ERW) portion of the SAT followed a Normal distribution with mean 522 and standard deviation 114. The mean for the SAT Math portion was 542. What percentage of scores on the SAT ERW portion were lower than the SAT math mean?

13.30 Are we getting smarter? When the Stanford-Binet IQ test came into use in 1932, it was adjusted so that scores for each age group of children followed roughly the Normal distribution with mean 100 and standard deviation 15. The test is readjusted from time to time to keep the mean at 100. If present-day American children took the 1932 Stanford-Binet test, their mean score would be about 120. The reasons for the increase in IQ over time are not known but probably include better childhood nutrition and more experience in taking tests.

(a) IQ scores above 130 are often called "very superior." What percentage of children had very superior scores in 1932?

(b) If present-day children took the 1932 test, what percentage would have very superior scores? (Assume that the standard deviation 15 does not change.)

13.31 Japanese IQ scores. The Wechsler Intelligence Scale for Children is used (in several languages) in the United States and Europe. Scores in each case are approximately Normally distributed with mean 100 and standard deviation 15. When the test was standardized in Japan, the mean was 111. To what percentile of the American-European distribution of scores does the Japanese mean correspond?

13.32 The stock market. The annual rate of return on stock indexes (which combine many individual stocks) is very roughly Normal. Since 1945, the Standard & Poor's 500 index has had a mean yearly return of 12.5%, with a standard deviation of 17.8%. Take this Normal distribution to be the distribution of yearly returns over a long period.

(a) In what range do the middle 95% of all yearly returns lie?

(b) The market is down for the year if the return on the index is less than zero. In what proportion of years is the market down?

(c) In what proportion of years does the index gain 25% or more?

13.33 Locating the quartiles. The quartiles of any distribution are the 25th and 75th percentiles. About how many standard deviations from the mean are the quartiles of any Normal distribution?

13.34 Women's heights. The heights of adult women are approximately Normal with mean 69.2 inches and standard deviation 2.5 inches. How tall are the tallest 10% of women? (Use the closest percentile that appears in Table B.)

13.35 High IQ scores. Scores on the Wechsler Adult Intelligence Scale for the 20 to 34 age group are approximately Normally distributed with mean 110 and standard deviation 15. How high must a person score to be in the top 5% of all scores?

Exploring the Web

Access these exercises on the text website: macmillanlearning.com/scc10e.

Describing Relationships: Scatterplots and Correlation

Case Study: SAT Scores

In this chapter you will:

* Learn how to examine relations among quantitative variables.

The news media have a weakness for lists. Best places to live, best colleges, healthiest foods, most popular dog breeds . . . a list of best or worst is sure to find a place in the news. When the state-by-state SAT scores come out each year, it's therefore no surprise that we find news articles ranking the states from best (Minnesota in 2017) to worst (District of Columbia in 2017) according to the average SAT Mathematics score achieved by their high school seniors. Unfortunately, such reports leave readers believing that schools in the District of Columbia must be much worse than those in Minnesota. Where does your home state rank? And do you believe the ranking reflects the quality of education you received?

The College Board, which sponsors the SAT exams, doesn't like this practice at all. In fact, the following caveat is included in the annual report the College Board prepares to summarize SAT results for each state: "The College Board strongly discourages using scores to compare or evaluate teachers, schools, districts, or states, because of differences in participation and test taker populations. Relationships between test scores and other background or contextual factors are complex and interdependent." To see why, let's look at the data.

Figure 14.1 shows the distribution of average scores on the SAT Mathematics exam for the 50 states and the District of Columbia. Minnesota leads at 651, and the District of Columbia trails at 468 on the SAT scale of 200 to 800. The distribution has one clear peak and is skewed to the right.

To understand one variable, such as SAT scores, we must look at how it is related to other variables. By the end of this chapter, you will be able to use what you have learned to appreciate why the College Board discourages ranking states on SAT scores alone.

A medical study finds that short women are more likely to have heart attacks than women of average height, while tall women have the fewest heart attacks. An insurance group reports that heavier cars are involved in fewer fatal accidents per 10,000 vehicles registered than are lighter cars. These and many other statistical studies look at the relationship between two variables. To understand such a relationship, we must often examine other variables as well. To conclude that shorter women have a higher risk of heart attacks, for example, the researchers had to eliminate the effect of other variables such as weight and exercise habits. Our topic in this and the following chapters is relationships between variables. One of our main themes is that the relationship between two quantitative variables can be strongly influenced by other variables that are lurking in the background.

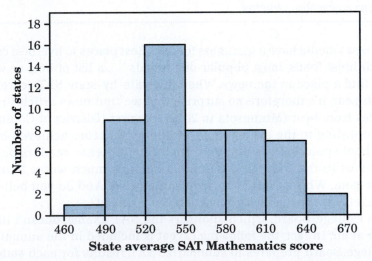

Figure 14.1 Histogram of the average scores of students in the 50 states and the District of Columbia on the SAT Mathematics exam.

Many statistical studies examine data on more than one quantitative variable. Fortunately, statistical analysis of several-variable data builds on the tools we used to examine individual variables. The principles that guide our work also remain the same:

- First plot the data, then add numerical summaries.

- Look for overall patterns and deviations from those patterns.

- When the overall pattern is quite regular, there is sometimes a way to describe it very briefly.

Scatterplots

The most common way to display the relation between two quantitative variables is a *scatterplot*.

Example 1

Healthy foods

For those who want to watch their weight and follow a nutritious diet, it isn't difficult to find guidelines and recommendations for healthy eating. One example is the *Dietary Guidelines for Americans*, which is a publication put out every five years by the U.S. Department of Health and Human Services (HHS) and the U.S. Department of Agriculture (USDA). In the 2015–2020 edition of this publication, it is recommended that individuals "consume a healthy eating pattern that accounts for all foods and beverages within an

ZoneCreative/Shutterstock

appropriate calorie level." A variety of vegetables, fruits, grains, fat-free and low-fat dairy products, proteins, and oils are included in this "healthy eating pattern."

Naturally, this begs the question of whether Americans are any good at identifying healthy foods. Are the foods nutrition experts consider to be "healthy" the same as the foods an average American adult would consider to be "healthy"? According to a 2016 article published in the *New York Times*, when a survey was conducted of a representative sample of American voters, the types of foods these voters considered to be healthy didn't always align with the types of foods a panel of nutritionists identified as being healthy. A list was prepared of 52 food items, and the list was shared with 2000 registered voters and 672 nutrition experts who were members of the American Society for Nutrition. Figure 14.2 is a scatterplot that shows how the percentage of nutritionists who identified food items as being healthy compared with the percentage of American voters who identified these items as being healthy. We think what nutritionists believe is "healthy" will help explain what American voters believe is "healthy." That is, the percentage of nutritionists who believe a food item is healthy is the *explanatory variable*, and the percentage of American voters who believe a food item is healthy is the *response variable*. We want to see how the percentage of American voters who identify food items as healthy changes as the percentage of nutritionists who label these items as healthy changes, so we put the percentage of agreement among nutritionists (the explanatory variable) on the horizontal axis. Each point on the plot represents one food item. We can see, for several of the food items, that as the percentage of agreement among nutritionists goes up, the percentage of agreement among American voters also goes up. We can also see that there isn't always agreement among nutritionists and American voters. For example, items such as frozen yogurt, granola, granola bars, and coconut oil were rated more often as healthy by American voters than by nutrition experts. Conversely, items like sushi, quinoa, and tofu were rated more often as healthy by nutrition experts than by American voters.

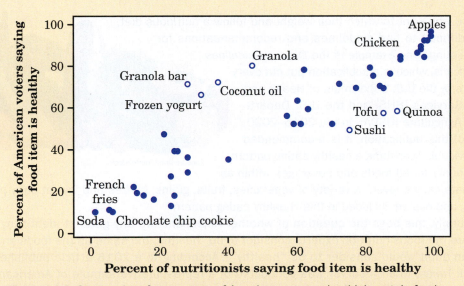

Figure 14.2 Scatterplot of percentage of American voters who think certain foods are healthy against percentage of nutritionists who think certain foods are healthy, Example 1.

Always plot the explanatory variable, if there is one, on the horizontal axis (the *x* axis) of a scatterplot. As a reminder, we usually call the explanatory variable *x* and the response variable *y*. If there is no explanatory-response distinction, either variable can go on the horizontal axis.

Key Terms

A **scatterplot** shows the relationship between two quantitative variables measured on the same individuals. The values of one variable appear on the horizontal axis, and the values of the other variable appear on the vertical axis. Each individual in the data appears as the point in the plot fixed by the values of both variables for that individual.

Example 2

Health and wealth

Figure 14.3 is a scatterplot of data from the World Bank for 2016. The individuals are all the world's nations for which data are available. The explanatory variable is a measure of how rich a country is: the gross domestic product (GDP) per capita. GDP is the total value of the goods and services produced in a country, converted into dollars. The response variable is life expectancy at birth.

We might expect people in richer countries to live longer. The overall pattern of the scatterplot does show this, but the relationship has an interesting shape. Life expectancy tends to rise very

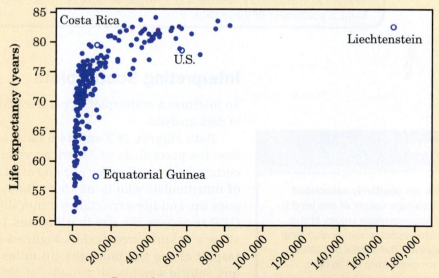

Figure 14.3 Scatterplot of the life expectancy of people in many nations against each nation's gross domestic product per person, for Example 2.

quickly as GDP increases, then levels off. People in very rich countries such as the United States typically live no longer than people in poorer but not extremely poor nations. Some of these countries, such as Costa Rica, do just about as well as the United States.

Two nations are outliers. In one, Equatorial Guinea, life expectancies are similar to those of its neighbors but its GDP is higher. Equatorial Guinea produces oil. It may be that income from mineral exports goes mainly to a few people and so pulls up GDP per capita without much effect on either the income or the life expectancy of ordinary citizens. That is, GDP per person is a mean, and we know that mean income can be much higher than median income.

The other outlier is Liechtenstein, a tiny nation bordering Switzerland and Austria. Liechtenstein has a strong financial sector and is considered a tax haven.

Now it's your turn

14.1 Brain size and intelligence. For centuries, people have associated intelligence with brain size. A recent study used magnetic resonance imaging to measure the brain size of several individuals. The IQ and brain size (in units of 10,000 pixels) of six individuals are as follows:

Brain size:	100	90	95	92	88	106
IQ:	140	90	100	135	80	103

Is there an explanatory variable and a response variable? If so, what are they? Make a scatterplot of these data.

Interpreting Scatterplots

To interpret a scatterplot, apply the usual strategies of data analysis.

Both Figures 14.2 and 14.3 have a clear *direction:* the percentage of American voters who think certain foods are healthy goes up as the percentage of nutritionists who think those foods are healthy goes up, and life expectancy generally goes up as GDP increases. We say that Figures 14.2 and 14.3 show a *positive association*. Figure 14.4 is a scatterplot of the gas mileages (in miles per gallon) and vehicle weights (in pounds) of 396 model-year 2017 cars. The response variable is gas mileage and the explanatory variable is vehicle weight. We see that gas mileage decreases as vehicle weight increases. We say that Figure 14.4 shows a *negative association*.

Key Terms

Two variables are **positively associated** when above-average values of one tend to accompany above-average values of the other and below-average values also tend to occur together. The scatterplot slopes upward as we move from left to right.

Two variables are **negatively associated** when above-average values of one tend to accompany below-average values of the other, and vice versa. The scatterplot slopes downward from left to right.

Examining a scatterplot

In any graph of data, look for the **overall pattern** and for striking **deviations** from that pattern.

You can describe the overall pattern of a scatterplot by the **direction, form,** and **strength** of the relationship.

An important kind of deviation is an **outlier,** an individual value that falls outside the overall pattern of the relationship.

Each of our scatterplots has a distinctive *form*. Figures 14.2 and 14.4 show a roughly straight-line trend, and Figure 14.3 shows a curved relationship. The *strength* of a relationship in a scatterplot is determined by how closely the points follow a clear form. The relationships in Figures 14.2 and 14.3 are not strong. American voters and nutrition experts are not always in agreement about which foods are healthy, and nations with similar GDPs can have quite different life expectancies. The relationship in Figure 14.4 is moderately strong. Here is an example of a stronger relationship with a simple form.

Statistics in Your World

After you plot your data, think!
Abraham Wald (1902–1950), like many statisticians, worked on war problems during World War II. Wald invented some statistical methods that were military secrets until the war ended. Here is one of his simpler ideas. Asked where extra armor should be added to airplanes, Wald studied the location of enemy bullet holes in planes returning from combat. He plotted the locations on an outline of the plane. As data accumulated, most of the outline filled up. Put the armor in the few spots with no bullet holes, said Wald. That's where bullets hit the planes that didn't make it back.

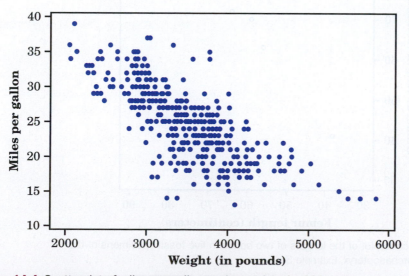

Figure 14.4 Scatterplot of miles per gallon against weight for 396 cars.

Example 3   *Classifying fossils*

Archaeopteryx is an extinct beast having feathers like a bird but teeth and a long bony tail like a reptile. Only six fossil specimens are known. Because these fossils differ greatly in size, some scientists think they are different species rather than individuals from the same species. We will examine data on the lengths in centimeters of the femur (a leg bone) and the humerus (a bone in the upper arm) for the five fossils that preserve both bones. Here are the data:

Femur:	38	56	59	64	74
Humerus:	41	63	70	72	84

Because there is no explanatory-response distinction, we can put either measurement on the *x* axis of a scatterplot. The plot appears in Figure 14.5.

Jason Edwards/National Geographic/ Getty Images

The plot shows a *strong, positive, straight-line association*. The straight-line form is important because it is common and simple. The association is strong because the points lie close to a line. It is positive because as the length of one bone increases, so does the length of the other bone. These data suggest that all five fossils belong to the same species and differ in size because some are younger than others. We might expect that a different species would have a different relationship between the lengths of the two bones, so that it would appear as an outlier.

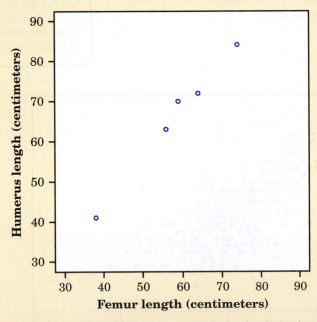

Figure 14.5 Scatterplot of the lengths of two bones in five fossil specimens of the extinct beast archaeopteryx, Example 3.

Now it's your turn

14.2 Brain size and intelligence. For centuries, people have associated intelligence with brain size. A recent study used magnetic resonance imaging to measure the brain size of several individuals. The IQ and brain size (in units of 10,000 pixels) of six individuals are as follows:

Brain size:	100	90	95	92	88	106
IQ:	140	90	100	135	80	103

Look carefully at the scatterplot you made of these data when you completed Exercise 14.1. What is the form, direction, and strength of the association? Are there any outliers?

This might be a good place to read the "What's the verdict?" story on page 341, and to answer the questions. These questions ask you to explore a scatterplot of exam scores in a statistics class.

Multiple Variables

Relationships between variables can be complicated, and, often, a single response can be explained not just by one explanatory variable but by a combination of explanatory variables. In addition to showing the relationship between two variables, a scatterplot can allow us to see how three or more variables are related. Recall Figure 14.4 where car weight was shown to have a negative association with gas mileage. What happens when we take the effect of a third variable into account, such as the number of cylinders in the car? Cylinders are the power units in a car's engine, and the number of cylinders is generally related to the size of the engine. There is a negative relationship between the weight of a vehicle and the gas mileage of the vehicle, but Figure 14.6 shows how this relationship is suddenly not as straightforward when we account for the number of cylinders in the vehicle. For instance, the relationship appears negative for vehicles with six or fewer cylinders but is either nonexistent or slightly positive for vehicles with eight or more cylinders.

More examples that illustrate the interplay among multiple variables can be found on the Gapminder website (http://www.gapminder.org/tools), where data are compiled from a variety of different sources. Figure 14.7 is a scatterplot from Gapminder that displays the relationship between fertility rate (or the average number of babies born per woman) and life expectancy (in years) for many countries around the world in 2014. The individuals are countries. Each dot in the scatterplot represents a different country. The size of the dot relates to the population size of the country, with larger dots representing more populated countries. The color of the dot refers to the region of the world in which the country is located. From an examination of this scatterplot, we can see an overall negative relationship; as

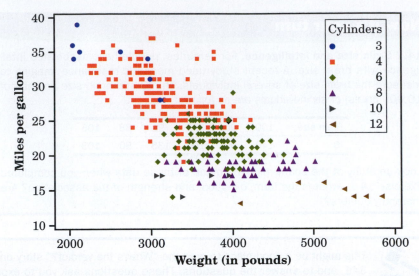

Figure 14.6 Scatterplot of miles per gallon against weight for 396 cars. Symbols show what happens when number of cylinders is taken into account.

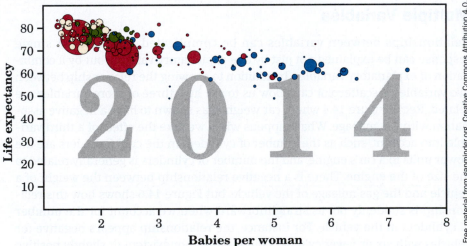

Figure 14.7 Scatterplot from www.gapminder.org of life expectancy against fertility rate for many countries. Dot size and color provide information about population size and region of the world, respectively.

fertility rate goes up, life expectancy goes down. Both larger and smaller countries vary in terms of fertility rate and life expectancy, but patterns clearly emerge when countries are categorized based on population size and region of the country. For example, the scatterplot allows us to see that the most populated countries have among the lowest fertility rates, but their life expectancies are not always higher when compared to less-populated countries. We can also see a striking difference in both fertility rate and life expectancy for countries in different regions of the world; the highest fertility rates and lowest life expectancies come from countries in Africa, as denoted by the blue dots, and the lowest fertility rates and highest life expectancies come from countries in Europe, as denoted by the yellow dots. Scatterplots can be manipulated on the Gapminder website to display relationships among many different variables, and each scatterplot includes a "play" button that can allow the user to see how relationships among these variables have changed over time.

Correlation

A scatterplot displays the direction, form, and strength of the relationship between two quantitative variables. Straight-line relations are particularly important because a straight line is a simple pattern that is quite common. A straight-line relation is strong if the points lie close to a straight line and weak if they are widely scattered about a line. Our eyes are not good judges of how strong a relationship is. The two scatterplots in Figure 14.8 depict the same data, but the right-hand plot is drawn smaller in a large field. The right-hand plot seems to show a stronger straight-line relationship. Our eyes can be fooled by changing the plotting scales or the amount of blank space around the cloud of points in a scatterplot. We need to follow our strategy for data analysis by using a numerical measure to supplement the graph. *Correlation* is the measure we use.

> **Key Terms**
>
> The **correlation** describes the direction and strength of a straight-line relationship between two quantitative variables. Correlation is usually written as *r*.

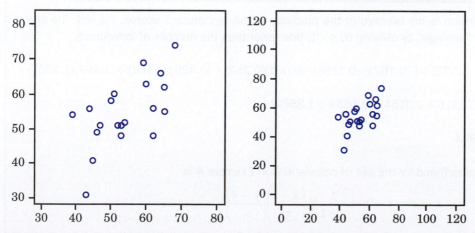

Figure 14.8 Two scatterplots of the same data. The right-hand plot suggests a stronger relationship between the variables because of the surrounding space.

Calculating a correlation takes a bit of work. You can usually think of r as the result of pushing a calculator button or giving a command in software and concentrate on understanding its properties and use. Knowing how we obtain r from data, however, does help us understand how correlation works, so here we go.

Example 4

Calculating correlation

We have data on two variables, x and y, for n individuals. For the fossil data in Example 3, x is femur length, y is humerus length, and we have data for $n = 5$ fossils.

Step 1. Find the mean and standard deviation for both x and y. For the fossil data, a calculator tells us that

Femur:	$\bar{x} = 58.2$ cm	$s_x = 13.20$ cm
Humerus:	$\bar{y} = 66.0$ cm	$s_y = 15.89$ cm

We use s_x and s_y to remind ourselves that there are two standard deviations, one for the values of x and the other for the values of y.

Step 2. Using the means and standard deviations from Step 1, find the standard scores for each x-value and for each y-value:

Value of x	Standard score $(x - \bar{x})/s_x$	Value of y	Standard score $(y - \bar{y})/s_y$
38	$(38 - 58.2)/13.20 = -1.530$	41	$(41 - 66.0)/15.89 = -1.573$
56	$(56 - 58.2)/13.20 = -0.167$	63	$(63 - 66.0)/15.89 = -0.189$
59	$(59 - 58.2)/13.20 = 0.061$	70	$(70 - 66.0)/15.89 = 0.252$
64	$(64 - 58.2)/13.20 = 0.439$	72	$(72 - 66.0)/15.89 = 0.378$
74	$(74 - 58.2)/13.20 = 1.197$	84	$(84 - 66.0)/15.89 = 1.133$

Step 3. The correlation is the average of the products of these standard scores. As with the standard deviation, we "average" by dividing by $n-1$, one fewer than the number of individuals:

$$r = \frac{1}{4}[(-1.530)(-1.573) + (-0.167)(-0.189) + (0.061)(0.252) + (0.439)(0.378) + (1.197)(1.133)]$$

$$= \frac{1}{4}(2.4067 + 0.0316 + 0.0154 + 0.1659 + 1.3562)$$

$$= \frac{3.9758}{4} = 0.994$$

The algebraic shorthand for the set of calculations in Example 4 is

$$r = \frac{1}{n-1}\sum\left(\frac{x - \bar{x}}{s_x}\right)\left(\frac{y - \bar{y}}{s_y}\right)$$

The symbol $\sum$, called "sigma," means "add them all up."

Understanding Correlation

More important than calculating r (a task for technology) is understanding how correlation measures association. Here are the facts:

- **Positive r indicates positive association between the variables, and negative r indicates negative association.** The scatterplot in Figure 14.5 shows strong positive association between femur length and humerus length. In three fossils, both bones are longer than their average values, so their standard scores are positive for both x and y. In the other two fossils, the bones are shorter than their averages, so both standard scores are negative. The products are all positive, giving a positive r.

- **The correlation r always falls between -1 and 1.** Values of r near 0 indicate a very weak straight-line relationship. The strength of the relationship increases as r moves away from 0 toward either -1 or 1. Values of r close to -1 or 1 indicate that the points lie close to a straight line. The extreme values $r = -1$ and $r = 1$ occur only when the points in a scatterplot lie exactly along a straight line.

  The result $r = 0.994$ in Example 4 reflects the strong positive straight-line pattern in Figure 14.5. The scatterplots in Figure 14.9 illustrate how r measures both the direction and the strength of a straight-line relationship. Study them carefully. Note that the sign of r matches the direction of the slope in each plot, and that r approaches -1 or 1 as the pattern of the plot comes closer to a straight line.

- Because r uses the standard scores for the observations, **the correlation does not change when we change the units of measurement** of x, y, or both. Measuring length in inches rather than centimeters in Example 4 would not change the correlation $r = 0.994$.

  Our descriptive measures for one variable all share the same units as the original observations. If we measure length in centimeters, the median, quartiles, mean, and standard deviation are all in centimeters. The correlation between two variables, however, has no unit of measurement; it is just a number between -1 and 1.

- **Correlation ignores the distinction between explanatory and response variables.** If we reverse our choice of which variable to call x and which to call y, the correlation does not change.

- **Correlation measures the strength of only straight-line association between two variables.** Correlation does not describe curved relationships between variables, no matter how strong they are.

- Like the mean and standard deviation, **the correlation is strongly affected by a few outlying observations.** Use r with caution when outliers appear in the scatterplot. Look, for example, at Figure 14.10. We changed the femur length of the first fossil from 38 to 60 centimeters. Rather than falling in line with the other fossils, the first is now an outlier. The correlation drops from $r = 0.994$ for the original data to $r = 0.640$.

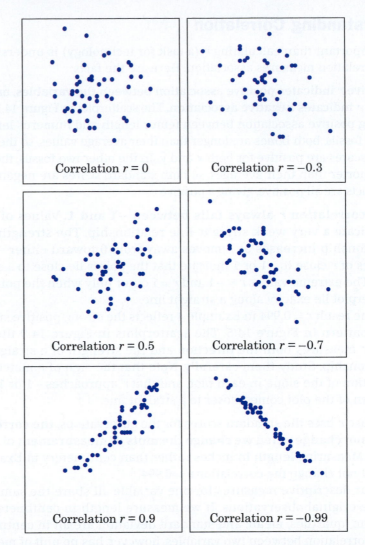

Figure 14.9 How correlation measures the strength of a straight-line relationship. Patterns closer to a straight line have correlations closer to 1 or −1.

Now it's your turn

14.3 Brain size and intelligence. For centuries, people have associated intelligence with brain size. A recent study used magnetic resonance imaging to measure the brain size of several individuals. The IQ and brain size (in units of 10,000 pixels) of six individuals are as follows:

Brain size:	100	90	95	92	88	106
IQ:	140	90	100	135	80	103

Review the scatterplot you made of this data when you completed Exercise 14.1. Compare your plot with those in Figure 14.8. What would you estimate the correlation r to be?

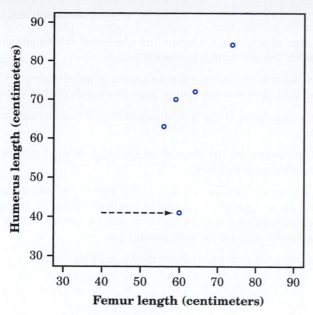

Figure 14.10 Moving one point reduces the correlation from $r = 0.994$ to $r = 0.640$.

There are many kinds of relationships between variables and many ways to measure them. Although correlation is very common, remember its limitations. Correlation makes sense only for quantitative variables—we can speak of the relationship between the sex of voters and the political party they prefer, but not of the correlation between these variables. Even for quantitative variables such as the length of bones, correlation measures only straight-line association.

Remember also that correlation is not a complete description of two-variable data, even when there is a straight-line relationship between the variables. You should give the means and standard deviations of both x and y along with the correlation. Because the formula for correlation uses the means and standard deviations, these measures are the proper choice to accompany a correlation.

Chapter 14: Statistics in Summary

- A **scatterplot** is a graph of the relationship between two quantitative variables. If you have an explanatory and a response variable, put the explanatory variable on the x (horizontal) axis of the scatterplot.

- When you examine a scatterplot, look for the **direction, form,** and **strength** of the relationship and also for possible **outliers.**

- If there is a clear direction, is it positive (the scatterplot slopes upward from left to right) or negative (the plot slopes downward)?

- Is the form straight or curved? Are there clusters of observations? Is the relationship strong (a tight pattern in the plot) or weak (the points scatter widely)?

- How might our description of the relationship change if multiple variables are taken into account?

- The **correlation r** measures the direction and strength of a straight-line relationship between two quantitative variables.

- Correlation is a number between -1 and 1. The sign of r shows whether the association is positive or negative. The value of r gets closer to -1 or 1 as the points cluster more tightly about a straight line. The extreme values -1 and 1 occur only when the scatterplot shows a perfectly straight line.

This chapter summary will help you evaluate the Case Study.

Link It

Chapters 11, 12, and 13 discussed graphical and numerical summaries suitable for a single quantitative variable. In practice, most statistical studies examine relationships between two or more variables. In this chapter, we learn about scatterplots, a type of graph that displays the relationship between two quantitative variables, and correlation, a number that measures the direction and strength of a straight-line relationship between two quantitative variables.

As with other graphics and numbers that summarize data, scatterplots and correlations help us see what the data are telling us, in this case about the possible relationship between two quantitative variables. The goal is to draw conclusions about whether the relationships observed in our data are true in general. In the next chapter, we discuss this in detail.

Case Study Evaluated

Use what you have learned in this chapter to evaluate the Case Study that opened the chapter. Start by reviewing the Chapter Summary. Then answer each of the following questions in complete sentences. Be sure to communicate clearly enough for any of your classmates to understand what you are saying.

Figure 14.11 is a scatterplot of each state's average SAT Mathematics score and the proportion of that state's high school seniors who took the SAT exam. SAT score is the response variable, and the proportion of a state's high school seniors who took the SAT exam is the explanatory variable.

1. Describe the overall pattern in words. Is the association positive or negative? Is the relationship strong?

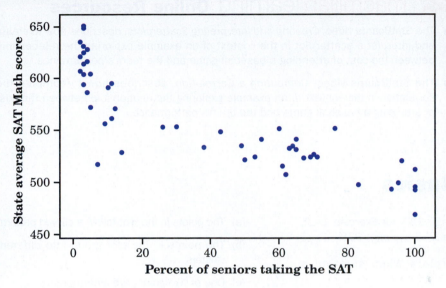

Figure 14.11 Scatterplot of average SAT Mathematics score for each state against the proportion of the state's high school seniors who took the SAT.

2. The plot shows two groups of states. In one group, fewer than 20% took the SAT. In the other, at least 26% took the exam and the average scores tend to be lower. There are two common college entrance exams, the SAT and the ACT. In ACT states, only students applying to selective colleges take the SAT. Which group of states in the plot corresponds to states in which most students take the ACT exam?

3. Write a paragraph, in language that someone who knows no statistics would understand, explaining why comparing states on the basis of average SAT scores alone would be misleading as a way of comparing the quality of education in the states.

In this chapter you:

- Learned how to examine relations among quantitative variables.

macmillan learning **Online Resources**

- The StatBoards video, *Creating and Interpreting Scatterplots*, describes how to create and interpret a scatterplot in the context of an example exploring the relationship between the cost of attending a baseball game and the team's performance.

- The StatBoards video, *Computing a Correlation*, describes how to compute the correlation in the context of an example exploring the relationship between the cost of attending a baseball game and the team's performance.

Check the Basics

For Exercise 14.1, see page 320; for Exercise 14.2, see page 323; for Exercise 14.3, see page 328.

14.4 Creating scatterplots When creating a scatterplot,

(a) always put the categorical variable on the horizontal axis.

(b) always put the categorical variable on the vertical axis.

(c) if you have an explanatory variable, put it on the horizontal axis.

(d) if you have a response variable, put it on the horizontal axis.

14.5 Interpreting scatterplots If the points in a scatterplot of two variables slope downward from left to right, we say the direction of the relationship between the variables is

(a) positive.

(b) negative.

(c) strong.

(d) weak.

14.6 Interpreting scatterplots Which of the following patterns might one observe in a scatterplot?

(a) The points in the plot follow a curved pattern.

(b) The points in the plot group into different clusters.

(c) One or two points are clear outliers.

(d) All of the above.

14.7 Correlation Which of the following is true of the correlation r?

(a) It cannot be greater than 1 or less than -1.

(b) It measures the strength of the *straight-line* relationship between two quantitative variables.

(c) A correlation of $+1$ or -1 can only happen if there is a perfect straight-line relationship between two quantitative variables.

(d) All of the above.

14.8 Correlation and scatterplots If the points in a scatterplot are very tightly clustered around a straight line, the correlation must be

(a) close to 0.

(b) close to $+1$.

(c) close to -1.

(d) close to either $+1$ or -1.

Chapter 14 Exercises

14.9 What number can I be?

(a) What are all the values that a correlation *r* can possibly take?

(b) What are all the values that a standard deviation *s* can possibly take?

(c) What are all the values that a mean $\bar{x}$ can possibly take?

14.10 Measuring mice. For a biology project, you measure the tail length (millimeters) and weight (grams) of 10 mice.

(a) Explain why you expect the correlation between tail length and weight to be positive.

(b) If you measured tail length in centimeters rather than millimeters, would you expect the correlation to change? Please explain.

14.11 Living on campus. A February 2, 2008, article in the *Columbus Dispatch* reported a study on the distances students lived from campus and average grade point average (GPA). Here is a summary of the results:

Residence	Avg. GPA
Residence hall	3.33
Walking distance	3.16
Near campus, long walk or short drive	3.12
Within the county, not near campus	2.97
Outside the county	2.94

Based on these data, is the association between the distance a student lives from campus and average GPA positive, negative, or near 0?

14.12 The endangered manatee. Manatees are large, gentle, slow-moving creatures found along the coast of Florida. Many manatees are injured or killed by boats. Figure 14.12 is a scatterplot of the number of manatee deaths by boats versus the number of boats

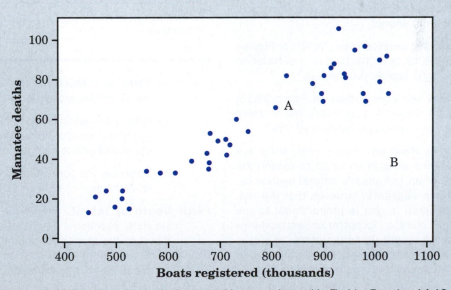

Figure 14.12 Manatee deaths by boats and boats registered in Florida, Exercise 14.12.

registered in Florida (in thousands) for the years between 1977 and 2016.

(a) Describe the overall pattern of the relationship in words.

(b) About what are the number of boats registered and manatee deaths for point A?

(c) Suppose there was a point near B. Would this be an outlier. If so, say how it is unusual (for example, "a moderately high number of deaths but a low number of boats registered").

14.13 Calories and salt in hot dogs. Figure 14.13 (page 335) shows the calories and sodium content in 17 brands of hot dogs. Describe the overall pattern of these data. In what way is the point marked A unusual?

14.14 The endangered manatee. Is the correlation r for the data in Figure 14.12 near -1, clearly negative but not near -1, near 0, clearly positive but not near 1, or near 1? Explain your answer.

14.15 Calories and salt in hot dogs. Is the correlation r for the data in Figure 14.13 near -1, clearly negative but not near -1, near 0, clearly positive but not near 1, or near 1? Explain your answer.

14.16 Comparing correlations. Which of Figures 14.2, 14.12, and 14.13 has a correlation closer to 0? Explain your answer.

14.17 Outliers and correlation. In Figure 14.13, the point marked A is an outlier. Will removing the outlier *increase* or *decrease* r? Why?

14.18 Long eyelashes. False eyelashes are sometimes worn in an effort to extend the length of an individual's natural eyelashes. Research suggests, however, that our natural eyelash length is proportional to the width of our eye. To better understand how eye width length (in centimeters, or cm) is related to eyelash length (in centimeters, or cm), measurements were obtained from 22

mammals. Those measurements are presented here.

Mammal	Eye width (cm)	Eyelash length (cm)
Armadillo	0.57	0.13
Hedgehog	0.39	0.23
Anteater	0.74	0.23
Elephant shrew	0.58	0.41
Lemur	1.08	0.43
Opposum	0.72	0.24
Porcupine	0.70	0.36
Red panda	1.52	0.32
Jack rabbit	0.92	0.61
Badger	1.75	0.42
Panda	1.89	0.60
Cat	1.98	0.38
Chimpanzee	1.98	0.47
Wild sheep	2.02	0.95
Camel	2.20	0.76
Snow leopard	2.41	0.89
Cougar	2.54	0.86
Tapir	2.60	1.21
African wild dog	2.64	0.76
Wild boar	2.92	0.80
Red kangaroo	3.09	0.67
Giraffe	4.08	1.66

(a) Make a scatterplot. (Which variable should be the explanatory variable?)

(b) Is the association between these variables positive or negative? Explain why you expect the relationship to have this direction.

(c) Describe the form and strength of the relationship.

14.19 Death by intent. Homicide and suicide are both intentional means of ending a life. However, the reason for committing a homicide is different than that for suicide, and we might expect homicide and suicide rates to be uncorrelated. On the other hand, both can involve some degree of violence, so perhaps we might expect some level of correlation in the rates. Table 14.1 gives data from 2008–2010 for 26 counties in Ohio. Rates are per 100,000 people. The data also indicate that the homicide rates for some counties should be treated with caution because

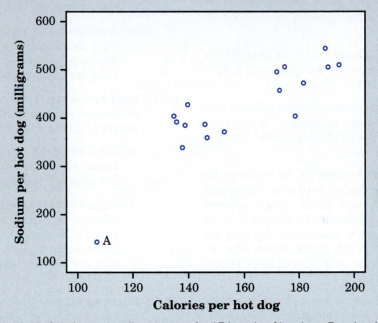

Figure 14.13 Calories and sodium content for 17 brands of hot dogs, Exercise 14.13.

Table 14.1 Homicide and suicide rates per 100,000 people

County	Homicide rate	Suicide rate	Caution	County	Homicide rate	Suicide rate	Caution
Allen	4.2	9.2	Y	Lorrain	3.1	11.0	Y
Ashtabula	1.8	15.5	Y	Lucas	7.4	13.3	N
Butler	2.6	12.7	Y	Mahoning	10.9	12.4	N
Clermont	1.0	16.0	Y	Medina	0.5	10.0	Y
Clark	5.6	14.5	N	Miami	2.6	9.2	Y
Columbiana	3.5	16.6	N	Montgomery	9.5	15.2	N
Cuyahoga	9.2	9.5	N	Portage	1.6	9.6	Y
Delaware	0.8	7.6	Y	Stark	4.7	13.5	N
Franklin	8.7	11.4	N	Summit	4.9	11.5	N
Greene	2.7	12.8	Y	Trumbull	5.8	16.6	N
Hamilton	8.9	10.8	N	Warren	0.7	11.3	Y
Lake	1.8	11.3	Y	Wayne	1.8	8.9	Y
Licking	4.5	12.9	N	Wood	1.0	7.4	Y

of low counts (Y = Yes, treat with caution, and N = No, do not treat with caution).

(a) Does it matter in this example which variable is considered to be the explanatory variable and which variable is considered to be the response variable? Please explain.

(b) Make a scatterplot of the data for the counties for which the data do not need to be treated with caution.

(c) Is the association between these variables positive or negative? What is the form of the relationship? How strong is the relationship?

(d) Now add the data for the counties for which the data do need to be treated with caution to your graph, using a different color or a different plotting symbol. Does the pattern of the relationship that you observed in part (c) hold for the counties for which the data do need to be treated with caution also?

14.20 Marriage. Suppose that men always married women three years younger than themselves. Draw a scatterplot of the ages of six married couples, with the husband's age as the explanatory variable. What is the correlation r for your data? Why?

14.21 Stretching a scatterplot. Changing the units of measurement can greatly alter the appearance of a scatterplot. Return to the fossil data from Example 3:

Femur:	38	56	59	64	74
Humerus:	41	63	70	72	84

These measurements are in centimeters. Suppose a deranged scientist measured the femur in meters and the humerus in millimeters. The data would then be

Femur:	0.38	0.56	0.59	0.64	0.74
Humerus:	410	630	700	720	840

(a) Draw an x axis extending from 0 to 75 and a y axis extending from 0 to 850. Plot the original data on these axes. Then plot the new data on the same axes in a different color. The two plots look very different.

(b) Nonetheless, the correlation is exactly the same for the two sets of measurements. Why do you know that this is true without doing any calculations?

14.22 Long eyelashes. Exercise 14.18 gives data on the eye width and eyelash length for a sample of 22 mammals.

(a) Use a calculator to find the correlation r. Explain from looking at the scatterplot why this value of r is reasonable.

(b) Suppose that the eye widths had been recorded in inches rather than centimeters. For example, the eye width of the giraffe would be 1.61 inches rather than 4.08 centimeters. How would the value of r change?

14.23 Death by intent. Table 14.1 gives data on on homicide and suicide rates from 2008–2010 for 26 counties in Ohio. The homicide rates for 14 of the counties should be treated with caution because of low counts. You made a scatterplot of these data in Exercise 14.19.

(a) Do you think the correlation will be about the same for the counties for which the data do need to be treated with caution and for the counties for which the data do not need to be treated with caution, or quite different for the two groups? Why?

(b) Calculate r for the counties for which the data do need to be treated with caution alone and also for the counties for which the data do not need to be treated with caution alone. (Use your calculator.)

14.24 Strong association but no correlation. The gas mileage of an automobile first increases and then decreases as the speed increases. Suppose that this relationship is very regular, as shown by the following data on speed (miles per hour) and mileage (miles per gallon):

Speed:	25	35	45	55	65
Mileage:	20	24	26	24	20

Make a scatterplot of mileage versus speed. Use a calculator to show that the correlation between speed and mileage is $r = 0$. Explain why the correlation is 0 even though there is a strong relationship between speed and mileage.

14.25 Death by intent. The data in Table 14.1 are given in deaths per 100,000 people. If we changed the data from deaths per 100,000 people to deaths per 1,000 people, how would the rates change? How would the correlation between homicide and suicide rates change?

14.26 What are the units? How sensitive to changes in water temperature are coral reefs? To find out, measure the growth of corals in aquariums (where growth is the change in weight, in pounds, of the coral before and after the experiment) when the water temperature (in degrees Fahrenheit) is controlled at different levels. In what units are each of the following descriptive statistics measured?

(a) The mean growth of the coral

(b) The standard deviation of the growth of the coral

(c) The correlation between weight gain and temperature

(d) The median growth of the coral

14.27 Teaching and research. A college newspaper interviews a psychologist about student ratings of the teaching of faculty members. The psychologist says, "The evidence indicates that the correlation between the research productivity and teaching rating of faculty members is close to zero." The paper reports this as "Professor McDaniel said that good researchers tend to be poor teachers, and vice versa." Explain why the paper's report is wrong. Write a statement in plain language (don't use the word "correlation") to explain the psychologist's meaning.

14.28 Sloppy writing about correlation. Each of the following statements contains a blunder. Explain in each case what is wrong.

(a) "There is a high correlation between the manufacturer of a car and the gas mileage of the car."

(b) "We found a high correlation ($r = 1.09$) between the horsepower of a car and the gas mileage of the car."

(c) "The correlation between the weight of a car and the gas mileage of the car was found to be $r = 0.53$ mile per gallon."

14.29 Guess the correlation. Measurements in large samples show that the correlation

(a) between this semester's GPA and the previous semester's GPA of an upper-class student is about _____.

(b) between IQ and the scores on a test of the reading ability of seventh-grade students is about _____.

(c) between the number of hours a student spends studying per week and the average number of hours spent studying by his or her roommates is about _____.

The answers (in scrambled order) are

$$r = 0.2 \quad r = 0.5 \quad r = 0.8$$

Match the answers to the statements and explain your choice.

14.30 Guess the correlation. For each of the following pairs of variables, would you expect a substantial negative correlation, a substantial positive correlation, or a weak correlation?

(a) The cost of a cable TV service and the number of channels provided by the service.

(b) The weight of a road-racing bicycle and the cost of the bicycle.

(c) The number of hours a student spends on Facebook and the student's GPA.

(d) The heights and salaries of faculty members at your university.

14.31 Investment diversification. A mutual funds company's newsletter says, "A well-diversified portfolio includes assets with low correlations." The newsletter includes a table of correlations between the returns on various classes of investments. For example, the correlation between municipal bonds and large-cap stocks is 0.50, and the correlation between municipal bonds and small-cap stocks is 0.21.

(a) Rachel invests heavily in municipal bonds. She wants to diversify by adding an investment whose returns do not closely follow the returns on her bonds. Should she choose large-cap stocks or small-cap stocks for this purpose? Explain your answer.

(b) If Rachel wants an investment that tends to increase when the return on her bonds drops, what kind of correlation should she look for?

14.32 Take me out to the ball game. What is the relationship between the price charged for a hot dog and the price charged, per ounce, for beer in Major League Baseball stadiums? Table 14.2 gives some data. Make a scatterplot appropriate for showing how beer price helps explain hot dog price. Describe the relationship that you see. Are there any outliers?

14.33 When it rains, it pours. Figure 14.14 plots the highest *yearly* precipitation ever recorded in each state, as of 2010, against the highest *daily* precipitation ever recorded in that state. The points for Alaska (AK), Hawaii (HI), and Texas (TX) are marked on the scatterplot.

(a) About what are the highest daily and yearly precipitation values for Alaska?

(b) Alaska and Hawaii have very high yearly maximums relative to their daily maximums. Omit these two states as outliers. Describe the nature of the relationship for the other states. Would knowing a state's highest daily precipitation be a great help in predicting that state's highest yearly precipitation?

Table 14.2 2017 Hot dog and beer (per ounce) prices (in dollars) at some Major League Baseball stadiums

Team	Hot dog	Beer	Team	Hot dog	Beer	Team	Hot dog	Beer
Angels	5.00	0.38	Giants	6.25	0.59	Rays	5.00	0.42
Astros	5.25	0.43	Indians	4.00	0.42	Reds	5.25	0.45
Blue Jays	4.16	0.41	Marlins	6.25	0.50	Red Sox	5.25	0.67
Braves	4.25	0.42	Mets	6.50	0.53	Rockies	5.25	0.25
Brewers	5.50	0.44	Padres	5.25	0.42	Royals	5.75	0.33
Cardinals	5.00	0.42	Phillies	4.00	0.58	Tigers	4.75	0.42
Diamondbacks	4.00	0.29	Pirates	3.50	0.38	Twins	5.00	0.43
Dodgers	6.50	0.39	Rangers	6.00	0.38	White Sox	4.50	0.44

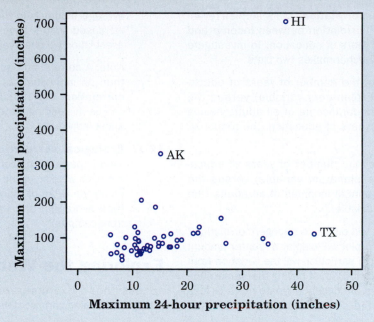

Figure 14.14 Record-high yearly precipitation recorded at any weather station in each state plotted against record-high daily precipitation for the state, Exercise 14.33.

14.34 How many corn plants are too many? How much corn per acre should a farmer plant to obtain the highest yield? To find the best planting rate, do an experiment: plant at different rates on several plots of ground and measure the harvest. Here are data from such an experiment:

Plants per acre	Yield (bushels per acre)			
12,000	150.1	113.0	118.4	142.6
16,000	166.9	120.7	135.2	149.8
20,000	165.3	130.1	139.6	149.9
24,000	134.7	138.4	156.1	
28,000	119.0	150.5		

(a) Is yield or planting rate the explanatory variable? Why?

(b) Make a scatterplot of yield and planting rate.

(c) Describe the overall pattern of the relationship. Is it a straight line? Is there a positive or negative association, or neither? Explain why increasing the number of plants per acre of ground has the effect that your graph shows.

14.35 Why so small? Make a scatterplot of the following data:

x	1	2	3	4	9	10
y	12	2	3	5	9	11

Use your calculator to show that the correlation is about 0.4. What feature of the data is responsible for reducing the correlation to this value despite a strong straight-line association between x and y in most of the observations?

14.36 Ecological correlation. Many studies reveal a positive correlation between income and number of years of education. To investigate this, a researcher makes two plots.

Plot 1: Plot the number of years of education (the explanatory variable) versus the average annual income of all adults having that many years of education (the response variable).

Plot 2: Plot the number of years of education (the explanatory variable) versus the individual annual incomes of all adults (the response variable).

Which plot will display a stronger correlation? (*Hint*: Which plot will display a greater amount of scatter? In particular, will the variation from individual to individual having the same number of years of education create more or less scatter in Plot 2 compared with plotting the average incomes in Plot 1? What effect will increased scatter have on the strength of the association we observe?)

Note: A correlation based on averages rather than on individuals is called an **ecological correlation.** Correlations based on averages can be misleading if they are interpreted to be about individuals.

14.37 Ecological correlation again. In Exercise 14.11 (page 333), would the association be stronger, weaker, or the same if the data given listed the GPAs of individual students (rather than averages) and the distance they lived from campus?

Exploring the Web

Access these exercises on the text website: macmillanlearning.com/scc10e.

What's the Verdict?

The following "What's the verdict?" is a true story and asks you to use what you have learned in this chapter to examine a scatterplot.

A statistics teacher plotted the Exam 1 and Exam 2 scores for her students. The plot is displayed in Figure 14.15.

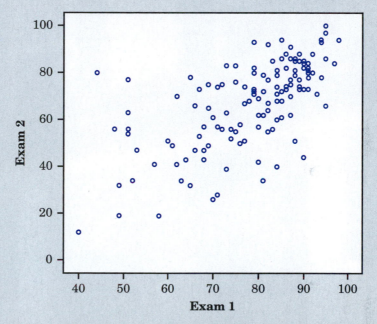

Figure 14.15 A scatterplot of the first and second exam scores of students in a statistics class. What's the verdict?

Questions

WTV14.1. What does each dot represent?

WTV14.2. What variables are being measured, and are they categorical or quantitative?

WTV14.3. How would you describe the form, direction, and strength of these data?

WTV14.4. Describe in terms of the story what the direction means.

WTV14.5. Do you notice any unusual points? What is the most surprising point? Why is it unusual?

We will continue to explore these data in Chapter 15.

Describing Relationships: Regression, Prediction, and Causation

15

Case Study: Housing Prices

In this chapter you will:

- Learn how to apply statistical methods in order to predict one variable from other variables.

- Learn how to distinguish between the ability to predict one variable from others and the issue of whether changes in one variable are caused by changes in others.

Imagine you are getting ready to purchase a house. You have searched high and low for the perfect dwelling, and you think you've found it, but if the price isn't right, you might not be able to afford it, or you might not feel that it's worth the investment. For many reasons, housing prices are important. They can have an impact on whether buyers are able to purchase houses in the first place, and they can also have an impact on whether homeowners will be motivated to sell their properties. Prices that are too high might prevent buyers from being able to own their own homes, whereas prices that are too low might be discouraging to sellers. If the value of houses in a particular area is low, for example, a homeowner who wants to move might be forced to wait until values rise again in order to avoid losing money.

What affects the price of a house? Variables such as the size of the house, the number of acres included on the property, the age of the house, the location of the house, the quality of nearby schools, and whether the house has particular amenities (e.g., multiple bathrooms, a fireplace, a newly remodeled kitchen, etc.) might all come to mind when thinking about what adds value to a house, but what about the number of Starbucks in the neighborhood? According to the results of a 2018 study from the Harvard Business School, in which data were analyzed from the restaurant review site Yelp and the U.S. Census Bureau, as the number of Starbucks in a particular ZIP code increases, so do the prices of houses in that ZIP code. Does this mean we should be concerned about the resale value of a house if there are no nearby Starbucks? By the end of this chapter, you will be able to critically evaluate the relationship between the number of Starbucks in a neighborhood and house prices.

Key Terms

A **regression line** is a straight line that describes how a response variable *y* changes as an explanatory variable *x* changes. We often use a regression line to predict the value of *y* for a given value of *x*.

Regression Lines

If a scatterplot shows a straight-line relationship between two quantitative variables, we would like to summarize this overall pattern by drawing a line on the graph. A *regression line* summarizes the relationship between two variables, but only in a specific setting: one of the variables helps explain or predict the other. That is, regression describes a relationship between an explanatory variable and a response variable.

Example 1   *Fossil bones*

In Examples 3 and 4 in Chapter 14, we saw that the lengths of two bones in fossils of the extinct beast archaeopteryx closely follow a straight-line pattern. Figure 15.1 plots the lengths for the five available fossils. The regression line on the plot gives a quick summary of the overall pattern.

Another archaeopteryx fossil is incomplete. Its femur is 50 centimeters long, but the humerus is missing. Can we predict how long the humerus is? The straight-line pattern connecting humerus length to femur length is so strong that we feel quite safe in using femur length to predict humerus length. Figure 15.1 shows how: starting at the femur length (50 cm),

Jason Edwards/National Geographic/ Getty Images

go up to the line, then over to the humerus length axis. We predict a length of about 56 cm. This is the length the humerus would have if this fossil's point lay exactly on the line. All the other points are close to the line, so we think the missing point would also be close to the line. That is, we think this prediction will be quite accurate.

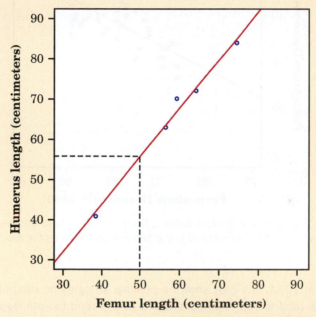

Figure 15.1 Using a straight-line pattern for prediction, Example 1. The data are the lengths of two bones in five fossils of the extinct beast archaeopteryx.

Example 2 — *Presidential elections, the Reagan years*

Republican Ronald Reagan was elected president twice, in 1980 and in 1984. His economic policy of tax cuts to stimulate the economy, eventually leading to increases in tax revenue, was still advocated by some Republican presidential candidates in 2015. Figure 15.2 plots the percentage of voters in each state who voted for Reagan's Democratic opponents: Jimmy Carter in 1980

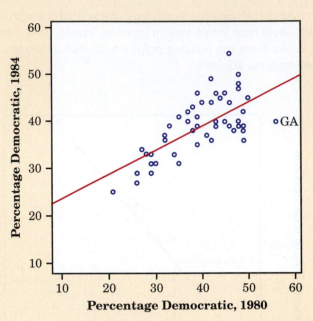

Figure 15.2 A weaker straight-line pattern, Example 2. The data are the percentage in each state who voted Democratic in the two Reagan presidential elections.

and Walter Mondale in 1984. The plot shows a positive straight-line relationship. We expect this because some states tend to vote Democratic and others tend to vote Republican. There is one outlier: Georgia, President Carter's home state, voted 56% for the Democrat Carter in 1980 but only 40% Democratic in 1984.

We could use the regression line drawn in Figure 15.2 to predict a state's 1984 vote from its 1980 vote. The points in this figure are more widely scattered about the line than are the points in the fossil bone plot in Figure 15.1. The correlations, which measure the strength of the straight-line relationships, are $r = 0.994$ for Figure 15.1 and $r = 0.704$ for Figure 15.2. The scatter of the points makes it clear that predictions of voting will be generally less accurate than predictions of bone length.

Regression Equations

When a plot shows a straight-line relationship as strong as that in Figure 15.1, it is easy to draw a line close to the points by eye. In Figure 15.2, however, different people might draw quite different lines by eye. Because we want to predict *y* from *x*, we want a line that is close to the points in the *vertical* (*y*) direction. It is hard to concentrate on just the vertical distances when drawing a line by eye. What is more, drawing by eye gives us a line on the graph but not an equation for the line. We need a way to find from the data the equation of the line that comes closest to the points in the vertical direction. There are many ways to make the collection of vertical distances "as small as possible." The most common is the *least-squares* method.

Figure 15.3 illustrates the least-squares idea. This figure magnifies the center part of Figure 15.1 to focus on three of the points. We see the vertical distances of these three points from the regression line. To find the least-squares line, look at these vertical distances (all five for the fossil data), square them, and move the line until the sum of the squares is the smallest it can be for any line. The lines drawn on the scatterplots in

Statistics in Your World

Regression toward the mean To "regress" means to go backward. Why are statistical methods for predicting a response from an explanatory variable called "regression"? Sir Francis Galton (1822–1911), who was the first to apply regression to biological and psychological data, looked at examples such as the heights of children versus the heights of their parents. He found that the taller-than-average parents tended to have children who were also taller than average, but not as tall as their parents. Galton called this fact "regression toward the mean," and the name came to be applied to the statistical method.

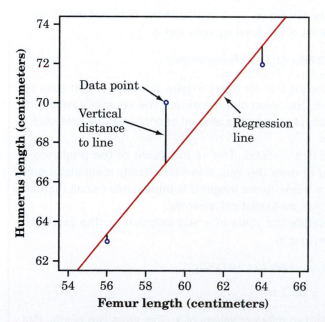

Figure 15.3 A regression line aims to predict *y* from *x*. So a good regression line makes the vertical distances from the data points to the line small.

Figures 15.1 and 15.2 are the least-squares regression lines. We won't give the formula for finding the least-squares line from data—that's a job for a calculator or computer. You should, however, be able to use the equation that the machine produces.

In writing the equation of a line, x stands as usual for the explanatory variable and y for the response variable. The equation of a line has the form

$$y = a + bx$$

The number b is the **slope** of the line, the amount by which y changes when x increases by one unit. The number a is the **intercept,** the value of y when $x = 0$. To use the equation for prediction, just substitute your x-value into the equation and calculate the resulting y-value.

Example 3

Using a regression equation

In Example 1, we used the "up-and-over" method in Figure 15.1 to predict the humerus length for a fossil whose femur length is 50 cm. The equation of the least-squares line is

$$\text{humerus length} = -3.66 + (1.197 \times \text{femur length})$$

The *slope* of this line is $b = 1.197$. This means that for these fossils, humerus length goes up by 1.197 cm when femur length goes up 1 cm. The slope of a regression line is usually important for understanding the data. The slope is the rate of change, the amount of change in the predicted y when x increases by 1.

The *intercept* of the least-squares line is $a = -3.66$. This is the value of the predicted y when $x = 0$. Although we need the intercept to draw the line, it is statistically meaningful only when x can actually take values close to zero. Here, femur length 0 is impossible (recall that the femur is a bone in the leg), so the intercept has no statistical meaning.

To use the equation for *prediction*, substitute the value of x and calculate y. The predicted humerus length for a fossil with a femur 50 cm long is

$$\text{humerus length} = -3.66 + (1.197)(50)$$

$$= 56.2 \text{ cm}$$

To *draw the line* on the scatterplot, predict y for two different values of x. This gives two points. Plot them and draw the line through them.

Now it's your turn

15.1 Fossil bones. Use the equation of the least-squares line

$$\text{humerus length} = -3.66 + (1.197 \times \text{femur length})$$

to predict the humerus length for a fossil with a femur 70 cm long.

Understanding Prediction

Computers make prediction easy and automatic, even from very large sets of data. Anything that can be done automatically is often done thoughtlessly. Regression software will happily fit a straight line to a curved relationship, for example. Also, the computer cannot decide which is the explanatory variable and which is the response variable. This is important because the same data give two different lines depending on which is the explanatory variable.

In practice, we often use several explanatory variables to predict a response. As part of its admissions process, a college might use SAT Math and Verbal scores and high school grades in English, math, and science (five explanatory variables) to predict first-year college grades. Although the details are messy, all statistical methods of predicting a response share some basic properties of least-squares regression lines.

- **Prediction is based on fitting some "model" to a set of data.** In Figures 15.1 and 15.2, our model is a straight line that we draw through the points in a scatterplot. Other prediction methods use more elaborate models.

- **Prediction works best when the model fits the data closely.** Compare again Figure 15.1, where the data closely follow a line, with Figure 15.2, where they do not. Prediction is more trustworthy in Figure 15.1. Also, it is not so easy to see patterns when there are many variables, but if the data do not have strong patterns, prediction may be very inaccurate.

- **Prediction outside the range of the available data is risky.** Suppose that you have data on a child's growth between 3 and 8 years of age. You find a strong straight-line relationship between age x and height y. If you fit a regression line to these data and use it to predict height at age 25 years, you will predict that the child will be 8 feet tall. Growth slows down and stops at maturity, so extending the straight line to adult ages is foolish. No one would make this mistake in predicting height. But almost all economic predictions try to tell us what will happen next quarter or next year. No wonder economic predictions are often wrong. Prediction outside the range of available data is referred to as extrapolation. Beware of **extrapolation**!

Example 4

Predicting the national deficit

The Congressional Budget Office is required to submit annual reports that predict the federal budget and its deficit or surplus for the next five years. These forecasts depend on future economic trends (unknown) and on what Congress will decide about taxes and spending (also unknown). Even the prediction of the state of the budget if current policies are not changed has been wildly inaccurate. The forecast made in January 2008 for 2012, for example, underestimated the deficit by nearly $1000 billion! The January 2009 forecast for 2013 underestimated the deficit by $423 billion, but the January 2010 forecast for 2014 under-

Jim Lo Scalzo/EPA/Shutterstock

estimated the deficit by only $8 billion. As former Senator Everett Dirksen once said, "A billion here and a billion there and pretty soon you are talking real money." In 1999, the Budget Office was predicting a surplus (ignoring Social Security) of $996 billion over the following 10 years. Politicians debated what to do with the money, but no one else believed the prediction (correctly, as it turned out). In 2015, there was a $439 billion deficit; in 2016, a $587 billion deficit; and in 2017, a $665 billion deficit. The forecast in January 2015 was for a $652 billion deficit in 2019. Was this forecast accurate?

Correlation and Regression

Correlation measures the direction and strength of a straight-line relationship. Regression draws a line to describe the relationship. Correlation and regression are closely connected, even though regression requires choosing an explanatory variable and correlation does not.

Both correlation and regression are strongly affected by outliers. Be wary if your scatterplot shows strong outliers. Figure 15.4 plots the record-high yearly precipitation in each state against that state's record-high 24-hour precipitation. Hawaii is a high outlier, with a yearly record of 704.83 inches of rain recorded at Kukui in 1982. The correlation for all 50 states in Figure 15.4 is 0.510. If we leave out Hawaii, the correlation drops to $r = 0.248$. The solid line in the figure is the least-squares line for predicting the annual record from the 24-hour record. If we leave out Hawaii, the least-squares line drops down to the dotted line. This line is nearly flat—there is little relation between yearly and 24-hour record precipitation once we decide to ignore Hawaii.

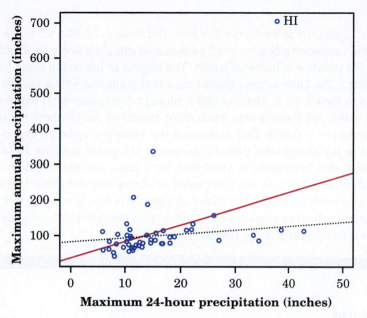

Figure 15.4 Least-squares regression lines are strongly influenced by outliers. The solid line is based on all 50 data points. The dotted line leaves out Hawaii.

The usefulness of the regression line for prediction depends on the strength of the association. That is, the usefulness of a regression line depends on the correlation between the variables. It turns out that the *square* of the correlation is the right measure.

The idea is that when there is a straight-line relationship, some of the variation in y is accounted for by the fact that as x changes it pulls y along with it.

Key Terms

The **square of the correlation, r^2,** is the proportion of the variation in the values of y that is explained by the least-squares regression of y on x.

Example 5

Using r^2

Look again at Figure 15.1. There is a lot of variation in the humerus lengths of these five fossils, from a low of 41 cm to a high of 84 cm. The scatterplot shows that we can explain almost all of this variation by looking at femur length and at the regression line. As femur length increases, it pulls humerus length up with it along the line. There is very little leftover variation in humerus length, which appears in the scatter of points about the line. Because $r = 0.994$ for these data, $r^2 = (0.994)^2 = 0.988$. So, the variation "along the line" as femur length pulls humerus length with it, accounts for 98.8% of all the variation in humerus length. The scatter of the points about the line accounts for only the remaining 1.2%. Little leftover scatter says that prediction will be accurate.

Contrast the voting data in Figure 15.2. There is still a straight-line relationship between the 1980 and 1984 Democratic votes, but there is also much more scatter of points about the regression line. Here, $r = 0.704$ and so $r^2 = 0.496$. Only about half the observed variation in the 1984 Democratic vote is explained by the straight-line pattern. You would still guess a higher 1984 Democratic vote for a state that was 45% Democratic in 1980 than for a state that was only 30% Democratic in 1980. But lots of variation remains in the 1984 votes of states with the same 1980 vote. That is the other half of the total variation among the states in 1984. It is due to other factors, such as differences in the main issues in the two elections and the fact that President Reagan's two Democratic opponents came from different parts of the country.

Statistics in Your World

Did the vote counters cheat?

Republican Bruce Marks was ahead of Democrat William Stinson when the voting-machine results were tallied in their 1993 Pennsylvania election. But Stinson was ahead after absentee ballots were counted by the Democrats, who controlled the election board. A court fight followed. The court called in a statistician, who used regression with data from past elections to predict the counts of absentee ballots from the voting-machine results. Marks's lead of 564 votes from the machines predicted that he would get 133 more absentee votes than Stinson. In fact, Stinson got 1025 more absentee votes than Marks. Did the vote counters cheat?

In reporting a regression, it is usual to give r^2 as a measure of how successful the regression was in explaining the response. When you see a correlation, square it to get a better feel for the strength of the association. Perfect correlation ($r = -1$ or $r = 1$) means the points lie exactly on a line. Then, $r^2 = 1$ and all of the variation in one variable is accounted for by the straight-line relationship with the other variable. If $r = -0.7$ or $r = 0.7$, $r^2 = 0.49$ and about half the variation is accounted for by the straight-line relationship. In the r^2 scale, correlation ± 0.7 is about halfway between 0 and ± 1.

Now it's your turn

15.2 At the ballpark. Table 14.2 (page 338) gives data on the prices charged for beer (per ounce) and for a hot dog at Major League Baseball stadiums. The correlation between the prices is $r = 0.21$. What proportion of the variation in hot dog prices is explained by the least-squares regression of hot dog prices on beer prices (per ounce)?

The Question of Causation

There is a strong relationship between cigarette smoking and death rate from lung cancer. Does smoking cigarettes *cause* lung cancer? There is a strong association between the availability of handguns in a nation and that nation's homicide rate from guns. Does easy access to handguns *cause* more murders? It says right on the pack that cigarettes cause lung cancer. Whether more guns cause more murders is hotly debated. Why is the evidence for cigarettes and cancer better than the evidence for guns and homicide?

We already know three big facts about statistical evidence for cause and effect.

Statistics and causation

1. A strong relationship between two variables does not always mean that changes in one variable cause changes in the other.
2. The relationship between two variables is often influenced by other variables lurking in the background.
3. The best evidence for causation comes from randomized comparative experiments.

Example 6

Does television extend life?

Measure the number of television sets per person *x* and the life expectancy *y* for the world's nations. There is a high positive correlation: nations with many TV sets have higher life expectancies.

The basic meaning of causation is that by changing *x* we can bring about a change in *y*. Could we lengthen the lives of people in Botswana by shipping them TV sets? No. Rich nations have more TV sets than poor nations. Rich nations also have longer life expectancies because they offer better nutrition, clean water, and better health care. There is no cause-and-effect tie between the number of TV sets and length of life.

goglik83/Deposit Photos

Example 6 illustrates our first two big facts. Correlations such as this are sometimes called "nonsense correlations." The correlation is real. What is nonsense is the conclusion that changing one of the variables causes changes in the other. A lurking variable—such as national wealth in Example 6—that influences both *x* and *y* can create a high correlation even though there is no direct connection between *x* and *y*. We might call this *common response*: both the explanatory and the response variable are responding to some lurking variable.

Example 7
Obesity in mothers and daughters

What causes obesity in children? Inheritance from parents, overeating, lack of physical activity, and too much television have all been named as explanatory variables.

The results of a study of Mexican-American girls aged 9 to 12 years are typical. Researchers measured body mass index (BMI), a measure of weight relative to height, for both the girls and their mothers. People with high BMI are overweight or obese. They also measured hours of television watched, minutes of physical activity, and intake of several kinds of food. The result: the girls' BMIs were weakly correlated with physical activity ($r = -0.18$), diet, and television. The strongest correlation ($r = 0.506$) was between the BMI of daughters and the BMI of their mothers.

Body type is in part determined by heredity. Daughters inherit half their genes from their mothers. There is, therefore, a direct causal link between the BMI of mothers and daughters. Of course, the causal link is far from perfect. The mothers' BMIs explain only 25.6% (that's r^2 again) of the variation among the daughters' BMIs. Other factors, some measured in the study and others not measured, also influence BMI. *Even when direct causation is present, it is rarely a complete explanation of an association between two variables.*

Can we use r or r^2 from Example 7 to say how much inheritance contributes to the daughters' BMIs? No. Remember *confounding*. It may well be that mothers who are overweight also set an example of little exercise, poor eating habits, and lots of television. Their daughters pick up these habits to some extent, so the influence of heredity is mixed up with influences from the girls' environment. We can't say how much of the correlation between mother and daughter BMIs is due to inheritance.

Figure 15.5 shows in outline form how a variety of underlying links between variables can explain association. The dashed line represents an observed association between the variables x and y. Some associations are explained by a *direct cause-and-effect* link between the variables. The first diagram in Figure 15.5 shows "x causes y" by an arrow running from x to y. The second diagram illustrates *common response*. The observed association between the variables x and y is explained by a lurking variable z. Both x and y change in response to changes in z. This common response creates an association even though there may be no direct causal link between x and y. The third diagram in Figure 15.5 illustrates *confounding*. Both the explanatory variable x and the lurking variable z may influence the response variable y. Variables x and z are themselves associated, so we cannot distinguish the influence of x from the influence of z. We cannot

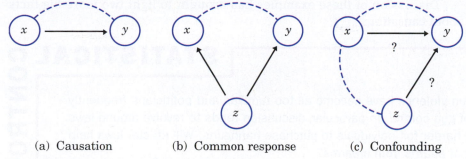

(a) Causation   (b) Common response   (c) Confounding

Figure 15.5 Some explanations for an observed association. A dashed line shows an association. An arrow shows a cause-and-effect link. Variable x is explanatory, y is a response variable, and z is a lurking variable.

say how strong the direct effect of x on y is. In fact, it can be hard to say if x influences y at all.

In Example 7, there is a causal link between the BMI of mothers and daughters. However, other factors, some measured in the study and some not measured, also influence the BMI of daughters. This is an example of confounding, illustrated in Figure 15.5(c). The x in the figure corresponds to the BMI of the mother, the z to one of the other factors, and the y to the BMI of the daughter.

Both common response and confounding involve the influence of a lurking variable or variables z on the response variable y. We won't belabor the distinction between the two kinds of relationships. Just remember that "beware the lurking variable" is good advice in thinking about relationships between variables. Here is another example of common response, in a setting where we want to do prediction.

Example 8
SAT scores and college grades

High scores on the SAT examinations in high school certainly do not *cause* high grades in college. The moderate association (r^2 is about 27%) is no doubt explained by common response variables such as academic ability, study habits, and staying sober. Figure 15.5(b) illustrates this. In the figure, z might correspond to academic ability, x to SAT scores, and y to grades in college.

The ability of SAT scores to partly predict college performance doesn't depend on causation. We need only believe that the relationship between SAT scores and college grades that we see in past years will continue to hold for this year's high school graduates. Think once more of our fossils, where femur length predicts humerus length very well. The strong relationship is explained by common response to the overall age and size of the beasts whose fossils we now examine. *Prediction doesn't require causation.*

Discussion of these examples has brought to light two more big facts about causation:

STATISTICAL CONTROVERSIES

Gun Control

Headlines about gun violence have become all too familiar, and politicians frequently debate the issue of gun control. In particular, discussion tends to revolve around laws aimed at making it harder for individuals to purchase handguns. Will stricter laws help reduce the number of deaths from firearms?

In 2016, a research team led by Dr. Bindu Kalesan published the results of a study that examined 31,672 firearm deaths that occurred in the United States in 2010. The purpose of this study was to assess the effect of different state-specific firearm laws on firearm mortality. Results indicated that three state laws—universal background checks for firearm purchases, ammunition background checks, and identification requirements—were most strongly associated with reduced overall firearm mortality. Kalesan and colleagues argued that federal adoption of these three laws across all states could drastically reduce firearm mortality.

Advanced regression methods were used in this study to determine the relationship between firearm deaths and whether or not certain state firearm laws were in effect after adjusting (or controlling) for other variables that could affect firearm death, such as unemployment rates, rates for non-firearm homicides, gun ownership, and firearm exports. Is reducing firearm mortality in the United States as simple as enacting three federal laws? Skeptics are not so sure.

Critics of this gun control study caution against drawing conclusions about cause and effect. They point out many important variables that could affect firearm mortality that were not included in this study, and they question whether we should be so quick to accept conclusions based on an examination of firearm deaths from just one year. What do you think? What variables do you think should have been controlled for in this study? What do you see as the weaknesses of this study based on what you have learned so far about statistics?

Now it's your turn

15.3 The endangered manatee. Exercise 14.12 (page 333) gives data on the number of boats registered in Florida (in thousands) and the number of manatee deaths for the years between 1977 and 2016. The correlation between these variables is $r = 0.94$. Do you think the observed relationship is due to direct causation, common response, confounding, or some combination of these? Explain your answer.

Evidence for Causation

Despite the difficulties, it is sometimes possible to build a strong case for causation in the absence of experiments. The evidence that smoking causes lung cancer is about as strong as nonexperimental evidence can be.

Doctors had long observed that most lung cancer patients were smokers. Observational studies comparing smokers and "similar" (in the sense of characteristics such as age, gender, and overall health) nonsmokers showed a strong association between smoking and death from lung cancer. Could the association be explained by lurking variables that the studies could not measure? Might there be, for example, a genetic factor that predisposes people both to nicotine addiction and to lung cancer? Smoking and lung cancer would then be positively associated even if smoking had no direct effect on the lungs. How were these objections overcome?

Let's answer this question in general terms. What are the criteria for establishing causation when we cannot do an experiment?

- **The association is strong.** The association between smoking and lung cancer is very strong.

- **The association is consistent.** Many studies of different kinds of people in many countries link smoking to lung cancer. That reduces the chance that a lurking variable specific to one group or one study explains the association.

- **Higher doses are associated with stronger responses.** People who smoke more cigarettes per day or who smoke over a longer period get lung cancer more often. People who stop smoking reduce their risk.

- **The alleged cause precedes the effect in time.** Lung cancer develops after years of smoking. The number of men dying of lung cancer rose as smoking became more common, with a lag of about 30 years. Lung cancer kills more men than any other form of cancer. Lung cancer was rare among women until women began to smoke. Lung cancer in women rose along with smoking, again with a lag of about 30 years, and has now passed breast cancer as the leading cause of cancer death among women.

- **The alleged cause is plausible.** Experiments with animals show that tars from cigarette smoke do cause cancer.

Medical authorities do not hesitate to say that smoking causes lung cancer. The U.S. Surgeon General has long stated that cigarette smoking is "the largest avoidable cause of death and disability in the United States." The evidence for causation is overwhelming—but it is not as strong as the evidence provided by well-designed experiments.

More about statistics and causation

4. The observed relationship between two variables may be due to **direct causation, common response,** or **confounding.** Two or more of these factors may be present together.

5. An observed relationship can, however, be used for prediction without worrying about causation as long as the patterns found in past data continue to hold true.

This might be a good place to read the "What's the verdict?" story on page 369, and to answer the questions. These questions ask you to continue exploring the scatterplot first introduced in the "What's the verdict?" story in Chapter 14.

Correlation, Prediction, and Big Data

In 2008, researchers at Google were able to track the spread of influenza across the United States much faster than the Centers for Disease Control and Prevention (CDC). By using computer algorithms to explore millions of online Internet searches, the researchers discovered a correlation between what people searched for online and whether they had flu symptoms. The researchers used this correlation to make their surprisingly accurate predictions.

Massive databases, or "big data," that are collected by Google, Facebook, credit card companies, and others, contain petabytes—or 10^{15} bytes—of data and continue to grow in size. Big data allow researchers, businesses, and industry to search for correlations and patterns in data that will enable them to make accurate predictions about public health, economic trends, or consumer behavior. Using big data to make predictions is increasingly common. Big data explored with clever algorithms opens exciting possibilities. Will the experience of Google become the norm?

Proponents for big data often make the following claims for its value. First, big data may include all members of a population, eliminating the need for statistical sampling. Second, there is no need to worry about causation because correlations are all we need to know for making accurate predictions. Third, scientific and statistical theory is unnecessary because, with enough data, the numbers speak for themselves.

Are these claims correct? First, as we saw in Chapter 3, it is true that sampling variability is reduced by increasing the sample size and will become negligible with a sufficiently large sample. It is also true that there is no sampling variability when one has information on the entire population of interest. However, sampling variability is not the only source of error in statistics computed from data. Bias is another source of error and is not eliminated because the sample size is extremely large. Big data are often enormous convenience samples, the result of recording huge numbers of web searches, credit card purchases, or mobile phones pinging the nearest phone tower. This is not equivalent to having information about the entire population of interest. For example, in principle, it is possible to record every message on Twitter and use these data to draw conclusions about public opinion. However, Twitter users are not representative of the population as a whole. According to the Pew Research Internet Project, in 2013, U.S.-based users were disproportionally young, urban or suburban, and black. In other words, the large amount of data generated by Twitter users is biased when the goal is to draw conclusions about public opinion of all adults in the United States.

Second, it is true that correlation can be exploited for purposes of prediction even if there is no causal relation between explanatory and response variables. However, if you have no idea what is behind a correlation you have no idea what might cause prediction to fail, especially when one exploits the correlation to extrapolate to new situations. For a few winters after their success in 2008, Google Flu Trends continued to accurately track the spread of influenza using the correlations they discovered. But during the 2012–2013 flu season, data from the CDC showed that Google's estimate of the spread of flu-like illnesses was overstated by almost a factor of two. A possible explanation was that the news was full of stories about the flu and this provoked Internet searches by people who were otherwise healthy. The failure to understand why search terms were correlated with the spread of flu resulted in incorrectly assuming previous correlations extrapolated into the future.

Adding to the perception of the infallibility of big data are news reports touting successes, with few reports of the failures. The claim that theory is unnecessary because the numbers speak for themselves is misleading when all the numbers concerning successes and failures of big data are not reported. Statistical theory has much to say that can prevent data analysts from making serious errors. Providing examples of where mistakes have been made and explaining how, with proper statistical understanding and tools, those mistakes could have been avoided, is an important contribution.

The era of big data is exciting and challenging and has opened incredible opportunities for researchers, businesses, and industry. But simply being big does not exempt big data from statistical pitfalls such as bias and extrapolation.

Chapter 15: Statistics in Summary

- **Regression** is the name for statistical methods that fit some model to data in order to predict a response variable from one or more explanatory variables.

- The simplest kind of regression fits a straight line on a scatterplot for use in predicting y from x. The most common way to fit a line is the **least-squares** method, which finds the line that makes the sum of the squared vertical distances of the data points from the line as small as possible.

- The **squared correlation r^2** tells us what fraction of the variation in the responses is explained by the straight-line tie between y and x.

- **Extrapolation,** or prediction outside the range of the data, is risky because the pattern may be different there. Beware of extrapolation!

- A strong relationship between two variables is not always evidence that changes in one variable **cause** changes in the other. Lurking variables can create relationships through **common response** or **confounding.**

- If we cannot do experiments, it is often difficult to get convincing evidence for causation.

This chapter summary will help you evaluate the Case Study.

Link It

In Chapter 14, we used scatterplots and the correlation to explore and describe the relationship between two quantitative variables. In this chapter, we looked carefully at fitting a straight line to data in a scatterplot when there appears to be a straight-line trend, and then we used this line to predict the response from the explanatory variable. In doing this, we have used data to draw conclusions. We assume that the straight line that we fit to our data describes the actual relationship between the response and the explanatory variable and, thus, that conclusions (predictions) about additional values of the response based on other values of the explanatory variable are valid.

Are these conclusions (predictions) justified? The squared correlation provides information about the likelihood of a successful prediction. Small values of the squared correlation suggest that our predictions are not likely to be accurate. Extrapolation is another setting in which our predictions are not likely to be accurate.

Finally, when there is a strong relationship between two variables, it is tempting to draw an additional conclusion: namely, that changes in one variable cause changes in another. However, the case for causation requires more than a strong relationship. Unless our data are produced by a proper experiment, the case for causation is difficult to prove.

Case Study Evaluated

Use what you have learned in this chapter to evaluate the Case Study that opened the chapter. Start by reviewing the Chapter Summary. Then answer each of the following questions in complete sentences. Be sure to communicate clearly enough for any of your classmates to understand what you are saying.

What should we conclude about the relationship between the number of Starbucks in a ZIP code and the prices of houses in that ZIP code? Write a paragraph explaining why the association between the number of Starbucks in a ZIP code and housing prices in that ZIP code is not surprising and why it would be incorrect to conclude that building more Starbucks in a neighborhood will cause housing prices to increase.

In this chapter you:

- Learned how to apply statistical methods in order to predict one variable from other variables.
- Learned how to distinguish between the ability to predict one variable from others and the issue of whether changes in one variable are caused by changes in others.

macmillan learning **Online Resources**

- The StatClips video, *Regression—Introduction and Motivation*, describes many of the topics in this chapter in the context of an example about hair growth.

- The StatBoards video, *Beware Extrapolation!*, discusses the dangers of extrapolation in the context of several examples.

Check the Basics

For Exercise 15.1, see page 349; for Exercise 15.2, see page 352; for Exercise 15.3, see page 356.

15.4 Least-squares. The least-squares regression line

(a) is the line that makes the sum of the vertical distances of the data points from the line as small as possible.

(b) is the line that makes the sum of the vertical distances of the data points from the line as large as possible.

(c) is the line that makes the sum of the squared vertical distances of the data points from the line as small as possible.

(d) is the line that makes the sum of the squared vertical distances of the data points from the line as large as possible.

15.5 Correlation. The quantity that tells us what fraction of the variation in the responses is explained by the straight-line tie between the response and explanatory variables is

(a) the correlation.

(b) the absolute value of the correlation.

(c) the square root of the correlation.

(d) the square of the correlation.

15.6 Extrapolation. Extrapolation, or prediction outside the range of the data, is risky

(a) because the pattern observed in the data may be different outside the range of the data.

(b) because correlation does not necessarily imply causation.

(c) unless the correlation is very close to 1.

(d) unless the square of the correlation is very close to 1.

15.7 Prediction. An observed relationship between two variables can be used for prediction

(a) as long as we know the relationship is due to direct causation.

(b) as long as the relationship is a straight-line relationship.

(c) as long as the patterns found in past data continue to hold true.

(d) in all of the above instances.

15.8 Causation. The best evidence that changes in one variable cause changes in another comes from

(a) randomized comparative experiments.

(b) data for which the square of the correlation is near 1.

(c) higher values of the explanatory variable are associated with stronger responses.

(d) there is a plausible theory for causation.

Chapter 15 Exercises

15.9 Obesity in mothers and daughters. The study in Example 7 found that the correlation between the body mass index of young girls and their hours of physical activity in a day was $r = -0.18$. Why might we expect this correlation to be negative? What percentage of the variation in BMI among the girls in the study can be explained by the straight-line relationship with hours of activity?

15.10 State SAT scores. Figure 14.11 (page 331) plots the average SAT Mathematics score of each state's high school seniors against the percentage of each state's seniors who took the exam. In addition to two clusters, the plot shows an overall roughly straight-line pattern. The least-squares regression line for predicting average SAT Math score from percentage taking is

$$\text{average Math SAT score} = 602.4 - (1.142 \times \text{percentage taking})$$

(a) What does the slope $b = -1.142$ tell us about the relationship between these variables?

(b) In Georgia, the percentage of high school seniors who took the SAT was 61%. Predict their average score. (The actual average score in Georgia was 515.)

(c) On page 349, we mention that using least-squares regression to do prediction outside the range of available data is risky. For what range of data is it reasonable to use the least-squares regression line for predicting average SAT Mathematics score from percentage taking?

15.11 The endangered manatee. Figure 14.12 (page 333) plots the number of manatee deaths by boats versus the number of boats registered in Florida (in thousands). There is a clear straight-line pattern with a modest amount of scatter. The correlation between these variables is $r = 0.94$. What percentage of the observed variation among the manatees' deaths by boats is explained by the straight-line relationship between manatee deaths and number of boats registered?

15.12 State SAT scores. The correlation between the average SAT Mathematics score in the states and the percent of high school seniors who take the SAT is $r = -0.86$.

(a) The correlation is negative. What does that tell us?

(b) How well does proportion taking predict average score? (Use r^2 in your answer.)

15.13 The endangered manatee. The least-squares line for predicting manatee deaths by boats from number of boats registered in Florida, based on the data plotted in Figure 14.12 (page 333), is

number of manatee deaths by boats $= -47.16$ $+ (0.136 \times$ number of boats registered in Florida)

Explain in words the meaning of the slope $b = 0.136$. Then predict the number of manatee deaths by boats when the number of boats registered in Florida is 1000.

15.14 Long eyelashes. Here are the eye widths (in centimeters) and eyelash lengths (in centimeters) for a sample of 22 mammals.

Mammal	Eye width	Eyelash length
Armadillo	0.57	0.13
Hedgehog	0.39	0.23
Anteater	0.74	0.23
Elephant shrew	0.58	0.41
Lemur	1.08	0.43
Opposum	0.72	0.24
Porcupine	0.70	0.36
Red panda	1.52	0.32
Jack rabbit	0.92	0.61
Badger	1.75	0.42
Panda	1.89	0.60
Cat	1.98	0.38
Chimpanzee	1.98	0.47
Wild sheep	2.02	0.95
Camel	2.20	0.76
Snow leopard	2.41	0.89
Cougar	2.54	0.86
Tapir	2.60	1.21
African wild dog	2.64	0.76
Wild boar	2.92	0.80
Red kangaroo	3.09	0.67
Giraffe	4.08	1.66

You made a scatterplot of these data in Exercise 14.18 (page 334). The least-squares regression line is

eyelash length $= 0.05 + (0.3102 \times$ eye width)

What would you predict the eyelash length to be for a mammal with an eye width of 1.25 centimeters? How accurate is your prediction?

15.15 Wine and heart disease. Drinking moderate amounts of wine may help prevent heart attacks. Let's look at data for entire nations. Table 15.1 gives data on yearly wine consumption (liters of alcohol from drinking wine, per person) and yearly deaths from heart disease (deaths per 100,000 people) in 19 developed countries in 2001.

Table 15.1 Wine consumption and heart disease

Country	Alcohol from wine[a]	Heart disease death rate[b]	Country	Alcohol from wine[a]	Heart disease death rate[b]
Australia	3.25	80	Italy	7.50	60
Austria	4.75	100	Netherlands	2.75	70
Belgium	2.75	60	New Zealand	2.50	100
Canada	1.50	80	Norway	1.75	80
Denmark	4.50	90	Spain	5.00	50
Finland	3.00	120	Sweden	2.50	90
France	8.50	40	Switzerland	6.00	70
Germany	3.75	90	United Kingdom	2.75	120
Iceland	1.25	110	United States	1.25	120
Ireland	2.00	130			

[a]Liters of alcohol from drinking wine, per person.
[b]Deaths per 100,000 people, ischemic heart disease.

(a) Make a scatterplot that shows how national wine consumption helps explain heart disease death rates.

(b) Describe in words the direction, form, and strength of the relationship.

(c) The correlation for these variables is $r = -0.645$. Why does this value agree with your description in part (b)?

15.16 The 2008 and 2012 presidential elections. Democrat Barack Obama was elected president in 2008 and 2012. Figure 15.6 plots the percentage who voted for Obama in 2008 and 2012 for each of the 50 states and the District of Columbia.

(a) Describe in words the direction, form, and strength of the relationship between the

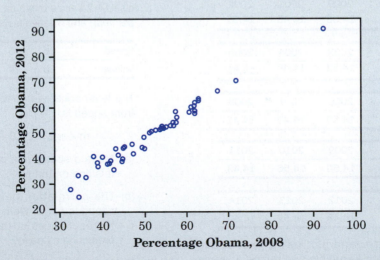

Figure 15.6 Scatterplot of the percentage who voted for Obama in 2008 and 2012 for each of the 50 states and the District of Columbia, Exercise 15.16.

percentage of votes for Obama in 2008 and the percentage in 2012. Are there any unusual features in the plot?

(b) The least-squares regression line is

percentage in 2012 = $-4.75 + (1.05 \times$ percentage in 2008)

Draw this line on a separate sheet of paper. (To draw the line, use the equation to predict y for $x = 90$ and for $x = 50$. Plot the two (x, y) points and draw the line through them.)

(c) The correlation between these variables is $r = 0.983$. What percentage of the observed variation in 2012 percentages can be explained by straight-line dependence on 2008 percentages?

15.17 Global warming. Here are annual average global temperatures for the years between 1994 and 2014 in degrees Celsius:

Year	1994	1995	1996
Temperature	14.23	14.35	14.22

Year	1997	1998	1999
Temperature	14.42	14.54	14.36

Year	2000	2001	2002
Temperature	14.33	14.45	14.51

Year	2003	2004	2005
Temperature	14.52	14.48	14.55

Year	2006	2007	2008
Temperature	14.50	14.49	14.41

Year	2009	2010	2011
Temperature	14.50	14.56	14.43

Year	2012	2013	2014
Temperature	14.48	14.52	14.59

(a) Make a scatterplot using year as the explanatory variable and temperature as the response variable.

(b) The least-squares regression line is

temperature = $-7.8 + (0.0111 \times$ year)

What would you predict for the annual average temperature for 2014 based on this line? How accurate is your prediction?

(c) The correlation between these variables is $r = 0.675$. What percentage of the variation in temperature can be explained by straight-line dependence on year?

15.18 Wine and heart disease. Table 15.1 gives data on wine consumption and heart disease death rates in 19 countries in 2001. A scatterplot (Exercise 15.15) shows a moderately strong relationship. The least-squares regression line for predicting heart disease death rate from wine consumption, calculated from the data in Table 15.1, is

$$y = 115.86 - 8.05x$$

Use this equation to predict the heart disease death rate in a country where adults average 1 liter of alcohol from wine each year and in a country that averages 8 liters per year. Use these two results to draw the least-squares line on your scatterplot.

15.19 Strong association but no correlation. Exercise 14.24 gives these data on the speed (miles per hour) and mileage (miles per gallon) of a car:

Speed:	25	35	45	55	65
Mileage:	20	24	26	24	20

The least-squares line for predicting mileage from speed is

mileage = $22.8 + (0 \times$ speed)

(a) Make a scatterplot of the data and draw this line on the plot.

(b) The correlation between mileage and speed is $r = 0$. What does this say about the usefulness of the regression line in predicting mileage?

15.20 Wine and heart disease. In Exercises 15.15 and 15.18, you examined data on wine consumption and heart disease deaths from Table 15.1. Suggest some differences among nations that may be confounded with wine-drinking habits. (*Note:* What is more, data about nations may tell us little about individual people. So these data alone are not evidence that you can lower your risk of heart disease by drinking more wine.)

15.21 Correlation and regression. If the correlation between two variables x and y is $r = 0$, there is no straight-line relationship between the variables. It turns out that the correlation is 0 exactly when the slope of the least-squares regression line is 0. Explain why slope 0 means that there is no straight-line relationship between x and y. Start by drawing a line with slope 0 and explaining why in this situation x has no value for predicting y.

15.22 Acid rain. Researchers studying acid rain measured the acidity of precipitation in a Colorado wilderness area for 150 consecutive weeks. Acidity is measured by pH. Lower pH values show higher acidity. The acid rain researchers observed a straight-line pattern over time. They reported that the least-squares regression line

$$pH = 5.43 - (0.0053 \times weeks)$$

fit the data well.

(a) Draw a graph of this line. Is the association positive or negative? Explain in plain language what this association means.

(b) According to the regression line, what was the pH at the beginning of the study (weeks = 1)? At the end (weeks = 150)?

(c) What is the slope of the regression line? Explain clearly what this slope says about the change in the pH of the precipitation in this wilderness area.

(d) Is it reasonable to use this least-squares regression line to predict the pH of precipitation after 200 weeks? Explain your answer.

15.23 Review of straight lines. Fred keeps his savings in his mattress. He began with $1000 from his mother and adds $250 each year.

His total savings y after x years are given by the equation

$$y = 1000 + 250x$$

(a) Draw a graph of this equation. (Choose two values of x, such as 0 and 10. Compute the corresponding values of y from the equation. Plot these two points on graph paper and draw the straight line joining them.)

(b) After 20 years, how much will Fred have in his mattress?

(c) If Fred had added $300 instead of $250 each year to his initial $1000, what is the equation that describes his savings after x years?

15.24 Review of straight lines. During the period after birth, a male white rat gains exactly 39 grams (g) per week. (This rat is unusually regular in his growth, but 39 g per week is a realistic rate.)

(a) If the rat weighed 110 g at birth, give an equation for his weight after x weeks. What is the slope of this line?

(b) Draw a graph of this line between birth and 10 weeks of age.

(c) Would you be willing to use this line to predict the rat's weight at age two years? Do the prediction and think about the reasonableness of the result. (There are 454 grams in a pound. A large cat weighs about 10 pounds.)

15.25 More on correlation and regression. In Exercises 15.11 and 15.13, the correlation and the slope of the least-squares line for the number of boats registered in Florida and the number of manatee deaths by boats are both positive. In Exercises 15.15 and 15.18, both the correlation and the slope for wine consumption and heart disease deaths are negative. Is it possible for these two quantities (the correlation and the slope) to have opposite signs? Explain your answer.

15.26 Always plot your data! Table 15.2 presents four sets of data prepared by the statistician Frank Anscombe to illustrate the dangers of

Table 15.2 Four data sets for exploring correlation and regression

Data Set A

x	10	8	13	9	11	14	6	4	12	7	5
y	8.04	6.95	7.58	8.81	8.33	9.96	7.24	4.26	10.84	4.82	5.68

Data Set B

x	10	8	13	9	11	14	6	4	12	7	5
y	9.14	8.14	8.74	8.77	9.26	8.10	6.13	3.10	9.13	7.26	4.74

Data Set C

x	10	8	13	9	11	14	6	4	12	7	5
y	7.46	6.77	12.74	7.11	7.81	8.84	6.08	5.39	8.15	6.42	5.73

Data Set D

x	8	8	8	8	8	8	8	8	8	8	19
y	6.58	5.76	7.71	8.84	8.47	7.04	5.25	5.56	7.91	6.89	12.50

Data from Frank J. Anscombe, "Graphs in Statistical Analysis," *The American Statistician* 27 (1973): 17–21.

calculating without first plotting the data. *All four sets have the same correlation and the same least-squares regression line to several decimal places.* The regression equation is

$$y = 3 + 0.5x$$

(a) Make a scatterplot for each of the four data sets and draw the regression line on each of the plots. (To draw the regression line, substitute $x = 5$ and $x = 10$ into the equation. Find the predicted y for each x. Plot these two points and draw the line through them on all four plots.)

(b) In which of the four cases would you be willing to use the regression line to predict y given that $x = 10$? Explain your answer in each case.

15.27 Going to class helps. A study of class attendance and grades among first-year students at a state university showed that in general, students who attended a higher percentage of their classes earned higher grades. Class attendance explained 25% of the variation in grade index among the students. What is the numerical value of the correlation between percentage of classes attended and grade index?

15.28 The average age of farm owners. The average age of American farm owners has risen steadily during the last 30 years. Here are data on the average age of farm owners (years) from 1982 to 2012:

Year:	1982	1987	1992	1997	2002	2007	2012
Average age:	50.5	52.0	53.3	54.3	55.3	57.1	58.3

(a) Make a scatterplot of these data. Draw a regression line by line for predicting a year's farm population.

(b) Extend your line to predict the average age of farm owners in 2100. Is this result reasonable? Why?

15.29 Lots of wine. Exercise 15.18 gives us the least-squares line for predicting heart disease deaths per 100,000 people from liters of alcohol from wine consumed, per person. The line is based on data from 19 rich countries. The equation is $y = 115.86 - 8.05x$. What is the predicted heart disease death rate for a country where wine consumption is 150 liters of alcohol per person? Explain why this result can't be true. Explain why using the regression line for this prediction is not intelligent.

15.30 Do emergency personnel make injuries worse? Someone says, "There is a strong positive correlation between the number of emergency personnel at the scene of an accident and the extent of injuries of those in the accident. So sending lots of emergency personnel just causes more severe injuries." Explain why this reasoning is wrong.

15.31 Facebook and grades. A September 2010 article on msnbc.com reported on a study that found that college students who are on Facebook while studying or doing homework wind up getting lower grades. Perhaps limiting time on Facebook will improve grades. Can you think of explanations for the association between time on Facebook and grades other than "time on Facebook causes a drop in grades"?

15.32 Healthy eating. In Example 14.1, you learned about a 2016 article from the *New York Times* in which the percentage of American voters who believed certain foods to be "healthy" was associated with the percentage of nutritionists who believed those food items to be "healthy." The correlation between the percentage of American voters who identify food items as "healthy" and the percentage of nutritionists who identify those items as "healthy" is $r = 0.89$. Do you think this observed relationship is due to direct causation, common response, confounding, or some combination of these? Explain your answer.

15.33 Health and wealth. An article entitled "The Health and Wealth of Nations" says, concerning the positive correlation between health and income per capita:

This correlation is commonly thought to reflect a causal link running from income to health . . . Recently, however, another intriguing possibility has emerged: that the health-income correlation is partly explained by a causal link running the other way—from health to income.

Explain how higher income in a nation can cause better health. Then explain how better health can cause higher income. There is no simple way to determine the direction of the link.

15.34 Is word length the key to success in college? Many prospective college students are asked to write an essay as part of the college application process. Do the lengths of the words used in the essay matter? According to researchers from The University of Texas at Austin, in a 2015 study that examined 50,000 admissions essays, the answer is "Yes." The researchers found that parts of speech such as articles and pronouns make a difference and "provide a better yardstick than long words for measuring a person's potential." Interestingly, the study determined that students relying on the word "I" in their essays tended to have lower grade point averages (GPAs), once enrolled, than those using the words "the" and "a."

What lurking variables might explain the association between the word lengths used in an essay and success in college? Explain why teaching students to use shorter words while writing (and to refrain from overreliance on the word "*I*") may have little effect on who succeeds in college.

15.35 Does low-calorie salad dressing cause weight gain? People who use low-calorie salad dressing in place of regular dressing tend to be heavier than people who use regular dressing. Does this mean that low-calorie salad dressings cause weight gain? Give a more plausible explanation for this association.

15.36 Internet use and school grades. Children who spend many hours on the Internet get lower grades in school, on average, than those who spend less time on the Internet. Suggest some lurking variables that may explain this relationship because they contribute to both heavy Internet use and poor grades.

15.37 Correlation again. The correlation between percentage voting Democrat in 1980 and percentage voting Democrat in 1984 (Example 15.2) is $r = 0.704$. The correlation between percentage of high school seniors taking the SAT and average SAT Mathematics score in the states (Exercise 15.12) is $r = -0.86$. Which of these two correlations indicates a stronger straight-line relationship? Explain your answer.

15.38 Religion is best for lasting joy. An August 2015 article in the *Washington Post* reported a study in which researchers looked at volunteering or working with a charity; taking educational courses; participating in religious organizations; and participating in a political or community organization. Of these, participating in religious organizations was the only social activity associated with "sustained happiness." What do you think of the claim that "joining a religious group" causes "sustained happiness"?

15.39 Living on campus. A February 2, 2008, article in the *Columbus Dispatch* reported a study on the distances students lived from campus and average grade point average (GPA). Here is a summary of the results:

Residence	Avg. GPA
Residence hall	3.33
Walking distance from campus	3.16
Near campus, long walk or short drive	3.12
Within the county, not near campus	2.97
Outside the county	2.94

Based on these data, the association between the distance a student lives from campus and GPA is negative. Many universities require freshmen to live on campus, but these data have prompted some to suggest that sophomores should also be required to live on campus in order to improve grades. Do these data imply that living closer to campus improves grades? Why?

15.40 Calculating the least-squares line. Like to know the details when you study something? Here is the formula for the least-squares regression line for predicting y from x. Start with the means $\bar{x}$ and $\bar{y}$ and the standard deviations s_x and s_y of the two variables and the correlation r between them. The least-squares line has equation $y = a + bx$ with

$$\text{slope:} \quad b = r\frac{s_y}{s_x} \qquad \text{intercept:} \quad a = \bar{y} - b\bar{x}$$

Example 4 in Chapter 14 (page 326) gives the means, standard deviations, and correlation for the fossil bone length data. Use these

values in the formulas just given to verify the equation of the least-squares line given on page 347:

$$\text{humerus length} = -3.66 + (1.197 \times \text{femur length})$$

The remaining exercises require a two-variable statistics calculator or software that will calculate the least-squares regression line from data.

15.41 Long eyelashes. Return to the eye width and eyelash data in Exercise 15.14.

(a) Verify the equation given for the least-squares line in that exercise.

(b) Suppose you were told only that the eyelash length of a mammal was 0.75 centimeters. You now want to "predict" the eye width of the mammal. Find the equation of the least-squares regression line that is appropriate for this purpose. What is your prediction?

(c) The two lines in parts (a) and (b) are different. Explain clearly why there are two different regression lines.

15.42 Is wine good for your heart? Table 15.1 gives data on wine consumption and heart disease death rates in 19 countries. Verify the equation of the least-squares line given in Exercise 15.18.

15.43 Always plot your data! A skeptic might wonder if the four very different data sets in Table 15.2 really do have the same correlation and least-squares line. Verify that (to a close approximation) the least-squares line is $y = 3 + 0.5x$, as given in Exercise 15.26.

Exploring the Web

Access these exercises on the text website: macmillanlearning.com/scc10e.

What's the Verdict?

The following "What's the verdict" is a continuation of the "What's the verdict" story in Chapter 14. You will use what you have learned in this chapter to further explore a scatterplot.

A statistics teacher plotted the Exam 1 and Exam 2 scores for her students. The plot is displayed in Figure 15.7.

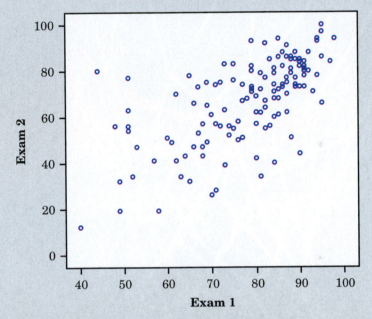

Figure 15.7 A scatterplot of the first and second exam scores of students in a statistics class. What's the verdict?

Questions

WTV15.1. Does doing well on Exam 1 *cause* a student to do well on Exam 2?

WTV15.2. (Repeated from Chapter 14). Do you notice any unusual points? What is the most surprising point? Why is it unusual?

WTV15.3. If you were the teacher of this class, what would you think about the student who has this unusual point?

Here is the rest of the story. The teacher recognized the student belonging to this dot as somebody who had been sleeping through most of the lectures between Exam 1 and Exam 2, but he also sat next to his girlfriend in class, who got a similar score on Exam 2. Cheating on Exam 2 could not be proven beyond a reasonable doubt, but the proctors kept a close eye on him, and he was later sent to the dean after being seen cheating on the Final Exam.

What's the verdict? Outliers can be quite interesting, especially for detective work.

16

The Consumer Price Index and Government Statistics

Case Study: Who's the Greatest Basketball Player of All Time?

In this chapter you will:

- Learn about the Consumer Price Index and other index numbers.

- Study methods to compare the buying power of the dollar across different years.

- Understand how the government collects and uses statistics.

Sportscasters and fans alike have been debating whether LeBron James or Michael Jordan is the greatest basketball player of all time. We pay for what we value, so one way to compare LeBron James and Michael Jordan is with their salaries.

Strictly speaking, Michael Jordan's 1997–1998 salary of $33,140,000 is less than the 2018–2019 salary of $35,654,150 for LeBron James. Does this mean that LeBron James is the best? We know that a dollar in 1997 bought more than a dollar in 2018, but how much more? Maybe in terms of buying power, Michael Jordan is the best.

We all notice the high salaries paid to professional athletes. In the National Basketball Association (NBA), for example, the mean salary rose from \$2,160,000 in 1997 to \$7,430,431 in 2018. That's a big jump. Not as big as it first appears, however. *A dollar in 2018 did not buy as much as a dollar in 1997, so 1997 salaries cannot be directly compared with 2018 salaries.* The hard fact that the dollar has steadily lost buying power over time means that we must make an adjustment whenever we compare dollar values from different years. The adjustment is easy. What is not easy is measuring the changing buying power of the dollar. The government's Consumer Price Index (CPI) is the tool we need.

Index Numbers

The CPI is one kind of numerical description, an *index number*. We can attach an index number to any quantitative variable that we measure repeatedly over time. The idea of the index number is to give a comparison of changes in a variable over time. For example, "The average cost of tuition and fees at public four-year colleges rose 103.7% between 1998–1999 and 2018–2019." That is, an index number reflects the change in a variable when compared to a *base period*.

Key Terms

An **index number** measures the value of a variable relative to its value at a **base period**. To find the index number for any value of the variable:

$$\text{index number} = \frac{\text{value}}{\text{base value}} \times 100$$

Example 1

Calculating an index number

A gallon of unleaded regular gasoline cost \$0.997 during the first week of November 1998 and \$2.670 during the first week of November 2018. (These are national average prices calculated by the U.S. Department of Energy.) The gasoline price index number for the first week in November 2018, with the first week in November 1998 as the base period, is

$$\begin{aligned}\text{index number} &= \frac{\text{value}}{\text{base value}} \times 100 \\ &= \frac{2.670}{0.997} \times 100 = 267.8\end{aligned}$$

The gasoline price index number for the base period, November 1998, is

$$\text{index number} = \frac{0.997}{0.997} \times 100 = 100$$

© Wave/Corbis

Knowing the base period is essential to making sense of an index number. Because the index number for the base period is always 100, it is usual to identify the base period as 1998 by writing "1998 = 100." In news reports concerning the CPI, you will notice the mysterious equation "1982–84 = 100." That's shorthand for the fact that the years 1982 to 1984 are the base period for the CPI. An index number just gives the current value as a percentage of the base value. Index number 271.1 means 271.1% of the base value, or a 171.1% increase from the base value. Index number 57 means that the current value is 57% of the base, a 43% decrease.

Fixed Market Basket Price Indexes

It may seem that index numbers are little more than a plot to disguise simple statements in complex language. Why say, "The Consumer Price Index (1982–84 = 100) stood at 252.9 in October 2018," instead of "Consumer prices rose 152.9% between the 1982–84 average and October 2018"? In fact, the term "index number" usually means more than a measure of change relative to a base. It also tells us the kind of variable whose change we measure. That variable is a weighted average of several quantities, with fixed weights. Let's illustrate the idea by a simple price index.

Example 2
The mountain man price index

Andrew Johnson lives in a cabin in the mountains and strives for self-sufficiency. He buys only salt, kerosene, and the services of a professional welder. Here are Andrew's purchases in 2000, the base period. His cost, in the last column, is the price per unit multiplied by the number of units he purchased.

welcomia/Deposit Photos

Good or service	2000 quantity	2000 price	2000 cost
Salt	100 pounds	$0.55/pound	$55.00
Kerosene	50 gallons	0.979/gallon	48.95
Welding	10 hours	13.75/hour	137.50
		Total cost =	$241.45

The total cost of Andrew's collection of goods and services in 2000 was $241.45. To find the "Mountain Man Price Index" for 2018, we use 2018 prices to calculate the 2018 cost of this *same* collection of goods and services. Here is the calculation:

Good or service	2000 quantity	2018 price	2018 cost
Salt	100 pounds	$0.675/pound	$67.50
Kerosene	50 gallons	2.209/gallon	110.45
Welding	10 hours	22.34/hour	223.40
		Total cost =	$401.35

The same goods and services that cost \$241.45 in 2000 cost \$401.35 in 2018. So the Mountain Man Price Index (2000 = 100) for 2018 is

$$\text{index number} = \frac{401.35}{241.45} \times 100 = 166.3$$

Key Terms

A **fixed market basket price index** is an index number for the total cost of a fixed collection of goods and services.

The point of Example 2 is that we follow the cost of the *same* collection of goods and services over time. It may be that Andrew refused to hire the welder in 2018 because he could not afford him. No matter— the index number uses the 2000 quantities, ignoring any changes in Andrew's purchases between 2000 and 2018. We call the collection of goods and services whose total cost we follow a *market basket*. The index number is then a *fixed market basket price index*.

The basic idea of a fixed market basket price index is that the weight given to each component (salt, kerosene, welding) remains fixed over time. The CPI is in essence a fixed market basket price index, with several hundred items that represent all consumer purchases. Holding the market basket fixed allows a legitimate comparison of prices because we compare the prices of exactly the same items at each time. As we will see, it also poses severe problems for the CPI.

Using the CPI

For now, think of the CPI as an index number for the cost of everything that American consumers buy. That the CPI for October 2018 was 252.9 means that we must spend \$252.9 in October 2018 to buy goods and services that cost \$100 in the 1982 to 1984 base period. An index number for "the cost of everything" lets us compare dollar amounts from different years by converting all the amounts into dollars of the same year. You will find tables in the *Statistical Abstract of the United States,* for example, with headings such as "Median Household Income (in 2017 dollars)." That table has restated all incomes in dollars that will buy as much as the dollar would buy in 2017. Watch for the term *constant dollars* and for phrases like *real income* or *real terms*. They mean that all dollar amounts represent the same buying power even though they may describe different years.

Table 16.1 gives the annual CPI for the years from 1915 to 2018. Figure 16.1 is a line graph of the *annual percentage increase* in CPI values. It shows that the periods from 1915 to 1920, the 1940s, and from 1975 to 1985 experienced high inflation. Although there is considerable variation, the annual percentage increase is positive in most years. In general, the twentieth century was a time of steep inflation. The twenty-first century has inflation too, but it is less extreme than the previous century.

Table 16.1 Annual Average Consumer Price Index, 1982–1984 = 100

Year	CPI	Year	CPI	Year	CPI
1915	10.1	1986	109.6	2004	188.9
1920	20.0	1987	113.6	2005	195.3
1925	17.5	1988	118.3	2006	201.6
1930	16.7	1989	124.0	2007	207.3
1935	13.7	1990	130.7	2008	215.3
1940	14.0	1991	136.2	2009	214.5
1945	18.0	1992	140.3	2010	218.1
1950	24.1	1993	144.5	2011	224.9
1955	26.8	1994	148.2	2012	229.6
1960	29.6	1995	152.4	2013	233.0
1965	31.5	1996	156.9	2014	236.7
1970	38.8	1997	160.5	2015	237.0
1975	53.8	1998	163.0	2016	240.0
1980	82.4	1999	166.6	2017	245.1
1981	90.9	2000	172.2	2018	251.1
1983	99.6	2001	177.1		
1984	103.9	2002	179.9		
1985	107.6	2003	184.0		

Data from Bureau of Labor Statistics.

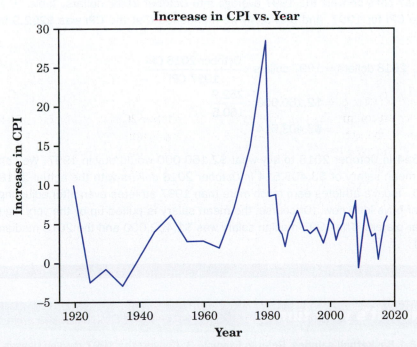

Figure 16.1 The increase in the Consumer Price Index (1982–84 = 100) for selected years from 1915 to 2018. Percentage increases are generally positive with the exception of the early twentieth century and twenty-first century between 2008 and 2009.

Faced with the depressing fact of inflation, it would be foolish to think about dollars without adjusting for their decline in buying power. Here is the recipe for converting dollars of one year into dollars of another year.

Adjusting for changes in buying power

To convert an amount in dollars at time A to the amount with the same buying power at time B:

$$\text{dollars at time B} = \text{dollars at time A} \times \frac{\text{CPI at time B}}{\text{CPI at time A}}$$

Notice that the CPI for the time you are *going to* appears on the top in the ratio of CPIs in this formula. Here are some examples.

Example 3
Salaries of professional athletes

The mean salary of NBA players rose from $2,160,000 in 1997 to $7,430,431 in 2018. How big was the increase in real terms? Let's convert the 1997 average into October 2018 dollars. Table 16.1 gives the annual average CPI for 1997, and previously we mentioned that the CPI was $252.9 in October 2018.

$$\text{2018 dollars} = \text{1997 dollars} \times \frac{\text{October 2018 CPI}}{\text{1997 CPI}}$$

$$= \$2{,}160{,}000 \times \frac{252.9}{160.5}$$

$$= \$3{,}403{,}514$$

That is, it took $3,403,514 in October 2018 to buy what $2,160,000 would buy in 1997. We can now compare the 1997 mean salary of $3,403,514 *in October 2018 dollars* with the actual 2018 mean salary, $7,430,431. Today's athletes earn much more than 1997 athletes even after adjusting for the fact that the dollar buys less now. (Of course, the mean salary is pulled up by the very high salaries paid to a few star players. The 1997 median salary was $1,402,000 and the 2018 median salary was $3,627,842.)

Now it's your turn

16.1 Basketball salaries. Refer to Example 3. Convert the 1997 median basketball salary into 2018 dollars. How do the values compare?

Example 4

Rising incomes?

For a more serious example, let's leave the pampered world of professional athletes and look at the incomes of ordinary people. The median annual income of all American households was $22,415 in 1984. By 2017 (the most recent year for which data were available at the time of this writing) the median income had risen to $61,372. Dollar income more than doubled, but we know that much of that rise is an illusion because of the dollar's declining buying power. To compare these incomes, we must express them in dollars of the same year. Let's express the 1984 median household income in 2017 dollars:

$$\text{2017 dollars} = \$22,415 \times \frac{245.1}{103.9} = \$52,786$$

Real household incomes rose from $52,876 to $61,372 in the 33 years between 1984 and 2017. That's a 16.1% increase.

The picture is different at the top. The 5% of households with the highest incomes earned $68,500 or more in 1984. In 2017 dollars, this is

$$\text{2017 dollars} = \$68,500 \times \frac{245.1}{103.9} = \$161,591$$

In fact, the top 5% of households earned $237,034 or more in 2017. That is, the real income of the highest earners increased by 46.7%.

Finally, let's look at the bottom. The 20% of households with the lowest incomes earned $9,500 or less in 1984. In 2017 dollars, this is

$$\text{2013 dollars} = \$9,500 \times \frac{245.1}{103.9} = \$22,410$$

In fact, the bottom 20% of households earned $24,638 or less in 2017. That is, the real income of the lowest earners increased by only 9.9%.

Now it's your turn

16.2 Production workers. Let's look at hourly paid workers in the United States. Their median annual earnings were $24,856 in 2007 and $30,430 in 2017. Compute the 2007 earnings in 2017 dollars. By what percentage did the real earnings of hourly paid workers change between 2007 and 2017?

Statistics in Your World

So you think that's inflation?

Americans were unhappy when oil price increases in 1973 set off a round of inflation that saw the Consumer Price Index almost double in the following decade. That's nothing—in Argentina, prices rose 127% in a single month, July 1989. The Turkish lira went from 14 to the dollar in 1970 to 579,000 to the dollar in 2000. The Zimbabwe dollar went from 253 to the dollar in September 2007 to 60,623 to the dollar by December 2008. There were 65 German marks to a dollar in January 1920, and 4,200,000,000,000 marks to a dollar in November 1923. Now *that's* inflation.

Example 4 illustrates how using the CPI to compare dollar amounts from different years brings out truths that are otherwise hidden. In this case, the truth is that the fruits of prosperity since the 1980s went mostly to those at the top of the income distribution and that very little real progress was made by those at the bottom. Put another way, people with the highest pay (usually those with skills and education) did much better than people with the lowest pay (usually those who lack special skills and college educations). Economists suggest several reasons: the "new economy" that rewards knowledge, high immigration leading to competition for less-skilled jobs, more competition from abroad, and so on. Exactly why incomes at the top have increased so much and what we should do about the stagnant incomes of those at the bottom are controversial questions.

Understanding the CPI

The idea of the CPI is that it is an index number for the cost of everything American consumers buy. That idea needs lots of adjusting to be practical. Much of the fiddling uses the results of large sample surveys.

Who is covered? The official name for the common version of the CPI (there are others, but we will ignore them) is the Consumer Price Index for All Urban Consumers. The CPI market basket represents the purchases of people living or working in urban areas. The official definition of "urban" is broad, so that about 80% of the U.S. population is covered. But if you live on a farm, the CPI doesn't apply to you.

How is the market basket chosen? Different households buy different things, so how can we get a single market basket? From a sample survey—the Consumer Expenditure Survey gathers detailed data on the spending of more than 30,000 households. The Bureau of Labor Statistics (BLS) breaks spending into categories such as "fresh fruits and vegetables," "new and used motor vehicles," and "hospital and related services." Then it chooses specific items, such as "fresh oranges," to represent each group in the market basket. The items in the market basket get weights that represent their category's proportion of all spending. The weights, and even the specific market basket items, are updated regularly to keep up with changing buying habits. So the market basket isn't actually fixed.

How are the prices determined? From more sample surveys. The BLS must discover the price of "fresh oranges" every month. That price differs from city to city and from store to store in the same city. Each month, the BLS records 80,000 prices in 87 cities at a sample of stores. The Point of

Purchase Survey of 16,800 households keeps the BLS up-to-date on where consumers shop for each category of goods and services (supermarkets, convenience stores, discount stores, and so on).

Does the CPI measure changes in the cost of living? A fixed market basket price index measures the cost of *living the same* over time, as Example 2 illustrated. In fact, we don't keep buying the same market basket of goods and services over time. We switch from LP records to cassette tapes and CDs and then to digital music downloads. We don't buy new 1998 cars in 2008 or 2018. As prices change, we may change what we buy. For example, if beef becomes expensive, we may buy less beef and more chicken or more tofu. A fixed market basket price index can't accurately measure changes in the cost of living when the economy itself is changing.

The BLS tries hard to keep its market basket up-to-date and to compensate for changes in quality. Every year, for example, the BLS must decide how much of the increase in new-car prices is paying for better quality. Only what's left counts as a genuine price increase in calculating the CPI. Between December 1967 and December 1994, actual car prices went up 313.4%, but the new-car price in the CPI went up only 172.1%. In 1995, adjustments for better quality reduced the overall rise in the prices of goods and services from 4.7% to only 2.2%. Prices of goods and services make up about 70% of the CPI. Most of the rest is the cost of shelter—renting an apartment or buying a house. House prices are another problem for the BLS. People buy houses partly to live in and partly because they think owning a house is a good investment. If we pay more for a house because we think it's a good investment, the full price should not go into the CPI.

By now it is clear that the CPI is *not* a fixed market basket price index, though that is the best way to start thinking about it. The BLS must constantly change the market basket as new products appear and our buying habits change. It must adjust the prices its sample surveys record to take account of better quality and the investment component of house prices. Yet the CPI still does not measure all changes in our cost of living. It leaves out taxes, for example, which are certainly part of our cost of living.

Even if we agree that the CPI should look only at the goods and services we buy, it doesn't perfectly measure changes in our cost of living. In principle, a true "cost-of-living index" would measure the cost of the *same standard of living* over time. That's why we start with a fixed market basket price index, which also measures the cost of living the same over time but takes the simple view that "the same" means buying exactly the same things. If we are just as satisfied after switching from beef to tofu to avoid paying more for beef, our standard of living hasn't changed and a cost-of-living index should ignore the higher price of beef. If we are willing to pay more for products that keep our environment clean, we are paying for a higher standard of living and the index should treat this just like an improvement in the quality of a new car. The BLS says that it would like

Does the CPI Overstate Inflation?

In 1995, Federal Reserve Chairman Alan Greenspan estimated that the CPI overstates inflation by somewhere between 0.5% and 1.5% per year. Mr. Greenspan was unhappy about this because increases in the CPI automatically drive up federal spending. At the end of 1996, a group of outside experts appointed by the Senate Finance Committee estimated that the CPI had in the past overstated the rate of inflation by about 1.1% per year. The Bureau of Labor Statistics (BLS) agreed that the CPI overstates inflation but thought that the experts' guess of 1.1% per year was too high.

RTimages/Shutterstock

The reasons the CPI shows the value of a dollar falling faster than is true are due partly to the nature of the CPI and partly to limits on how quickly the BLS can adjust the details of the enormous machine that lies behind the CPI. Think first about the details. The prices of new products, such as digital cameras and flat-screen televisions, often start high and drop rapidly. The CPI market basket changes too slowly to capture the drop in price. Discount stores with lower prices also enter the CPI sample slowly. Although the BLS tries hard to adjust for better product quality, the outside experts thought these adjustments were often too little and too late. The BLS has made many improvements in these details (most recently, in February 2015). The improved CPI would have grown about 0.5% per year more slowly than the actual CPI between 1978 and 1998.

The wider issue is the nature of the CPI as essentially a fixed market basket index. What sort of bias does such an index create? Does it produce an upward bias—that is, does it overstate the cost of living? Or does it create a downward bias—that is, does it understate the cost of living? Why?

the CPI to track changes in the cost of living, but that a true cost-of-living index isn't possible in the real world.

The Place of Government Statistics

Modern nations run on statistics. Economic data, in particular, guide government policy and inform the decisions of private business and individuals. Price indexes and unemployment rates, along with many other, less-publicized series of data, are produced by government statistical offices.

Some countries have a single statistical office, such as Statistics Canada (www.statcan.ca), the Australian Bureau of Statistics (www.abs.gov.au), and Statistics Sweden (www.scb.se/en). Others attach smaller offices to various branches of government. The United States is an extreme case: there are 13 principal statistical agencies with another 89 federal statistical offices. The U.S. Census Bureau and the Bureau of Labor Statistics are the largest, but you may at times use the products of the Bureau of Economic Analysis, the National Center for Health Statistics, the Bureau of Justice Statistics, or others in the federal government's collection of statistical agencies.

A 1993 ranking of government statistical agencies by the heads of these agencies in several nations put Canada at the top, with the United States tied with Britain and Germany for sixth place. The top spots generally went to countries with a single, independent statistical office. In 1996, Britain combined its main statistical agencies to form a new Office for National Statistics (www.gov.uk/government/statistics). U.S. government statistical agencies remain fragmented.

What do citizens need from their government statistical agencies? First of all, they need data that are *accurate* and *timely* and that *keep up with changes in society and the economy*. Producing accurate data quickly demands considerable resources. Think of the large-scale sample surveys that produce the unemployment rate and the CPI. The major U.S. statistical offices have a good reputation for accuracy and lead the world in getting data to the public quickly. Their record for keeping up with changes is less impressive. The struggle to adjust the CPI for changing buying habits and changing quality is one issue. Another is the failure of U.S. economic statistics to keep up with trends such as the shift from manufacturing to health care and technology as the major types of economic activity.

Much of the difficulty stems from lack of money. In the years after 1980, reducing federal spending was a political priority. Government statistical agencies lost staff and cut programs. Lower salaries made it hard to attract the best economists and statisticians to government. The level of government spending on data also depends on our view of what data the government should produce. In particular, should the government produce data that are used mainly by private business rather than by the government's own policymakers? Perhaps such data should be either compiled by private concerns or produced only for those who are willing to pay. This is a question of political philosophy rather than statistics, but it helps determine what level of government statistics we want to pay for.

Freedom from political influence is as important to government statistics as accuracy and timeliness. When a statistical office is part of a government ministry, it can be influenced by the needs and desires of that ministry. The U.S. Census Bureau is in the Department of Commerce, which serves business interests. The Bureau of Labor Statistics is in the Department of Labor. Thus, business and labor each have "their own" statistical office. The professionals in the statistical offices successfully resist direct political interference—a poor unemployment report is never delayed until after an election, for example. But indirect influence is clearly present. The BLS must compete with other Department of Labor activities for its budget, for example. Political interference with statistical work seems to be increasing, as when Congress refused to allow the U.S. Census Bureau to use sample surveys to correct for undercounting in the 2000 census. Such corrections may convince statisticians but do not necessarily convince the general public. Congress, not without justification, considers the public legitimacy of the census to be as important as its technical perfection. Thus, from Congress's point of view, political interference was justified even though the decision was disappointing to statisticians.

The 1996 reorganization of Britain's statistical offices was prompted in part by a widespread feeling that political influence was too strong. The details of how unemployment is measured in Britain were changed many times in the 1980s, for example, and almost all the changes had the effect of reducing the reported unemployment rate—just what the government wanted to see.

The Question of Social Statistics

National economic statistics are well established with the government, the media, and the public. The government also produces a lot of data on social issues such as education, health, housing, and crime. Social statistics are less complete than economic statistics. We have good data about how much money is spent on food but less information about how many people are poorly nourished. Social data are also less carefully produced than economic data. Economic statistics are generally based on larger samples, are compiled more often, and are published with a shorter time lag. The reason is clear: economic data are used by the government to guide economic policy month by month. Social data help us understand our society and address its problems but are not needed for short-term management.

There are other reasons the government is reluctant to produce social data. Many people don't want the government to ask about their sexual behavior or religion. Many people feel that the government should avoid asking about our opinions—apparently it's OK to ask, "When did you last visit a doctor?" but not "How satisfied are you with the quality of your health care?" These hesitations reflect the American suspicion of government intrusion. Yet issues such as sexual behaviors that contribute to the spread of HIV and satisfaction with health care are clearly important to citizens. Both facts and opinions on these issues can sway elections and influence policy. How can we get accurate information about social issues, collected consistently over time, and yet not entangle the government with sex, religion, and other touchy subjects?

The solution in the United States has been government funding of university sample surveys. After first deciding to undertake a sample survey asking people about their sexual behavior, in part to guide AIDS policy, the government backed away. Instead, it funded a much smaller survey of 3452 adults by the University of Chicago's National Opinion Research Center (NORC). NORC's General Social Survey (GSS), funded by the government's National Science Foundation, belongs with the Current Population Survey and the samples that undergird the CPI on any list of the most important sample surveys in the United States. The GSS includes both "fact" and "opinion" items. Respondents answer questions about their job security, their job satisfaction, and their satisfaction with their city, their friends, and their families. They talk about race, religion, and sex. Many Americans would object if the government asked whether they had seen an X-rated movie in the past year, but they reply when the GSS asks this question. The website for the GSS is www.norc.org/GSS+Website//.

This indirect system of government funding of a university-based sample survey fits the American feeling that the government itself should not be unduly invasive. It also insulates the survey from most political pressure. Alas, the government's budget cutting extends to the GSS, which now describes itself as an "almost annual" survey because lack of funds has prevented taking samples in some years. The GSS is, we think, a bargain.

This might be a good place to read the "What's the verdict?" story on page 390, and to answer the questions. These questions ask you to consider the issue of how much control government should exercise over what research can and cannot be funded by government agencies.

Chapter 16: Statistics in Summary

- An **index number** describes the value of a variable relative to its value at some **base period.**

- A **fixed market basket price index** is an index number that describes the total cost of a collection of goods and services.

- Think of the government's **Consumer Price Index** (CPI) as a fixed market basket price index for the collection of all the goods and services that consumers buy.

- Because the CPI shows how consumer prices change over time, we can use it to change a dollar amount at one time into the amount at another time that has the same buying power. This is needed to compare dollar values from different times in **real terms.**

- The details of the CPI are complex. It uses data from several large sample surveys. It is not a true fixed market basket price index because of adjustments for changing buying habits, new products, and improved quality.

- **Government statistical offices** produce data needed for government policy and decisions by businesses and individuals. The data should be accurate, timely, and free from political interference. Citizens therefore have a stake in the competence and independence of government statistical offices.

This chapter summary will help you evaluate the Case Study.

Link It

In Chapters 10 to 15, we studied methods for summarizing large amounts of data to help us see what the data are telling us. In this chapter, we discussed numbers that summarize large amounts of data on consumers to help us see what these data are telling us about the costs of goods and services. Because these index numbers, in particular the CPI, are used by the government and media to describe the cost of living, understanding how they are computed and what they represent will help us be better informed citizens.

Case Study Evaluated

Use what you have learned in this chapter to evaluate the Case Study that opened the chapter. Start by reviewing the Chapter Summary. Then answer each of the following questions in complete sentences. Be sure to communicate clearly enough for any of your classmates to understand what you are saying.

The average CPI for 1997 and 2018 can be found in Table 16.1. Use what you have learned in this chapter to convert Michael Jordan's 1997 salary, given in the Case Study at the beginning of this chapter, to 2018 dollars. Who had the largest salary in terms of 2018 dollars: Michael Jordan or LeBron James? Does this convince you which basketball player is the greatest of all time?

In this chapter you:

- Learned about the Consumer Price Index and other index numbers.
- Studied methods to compare the buying power of the dollar across different years.
- Understood the collection and use of statistics by the government.

macmillan learning Online Resources

■ LearningCurve has good questions to check your understanding of the concepts.

Check the Basics

For Exercise 16.1, see page 376. For Exercise 16.2, see page 377.

16.3 Index numbers. The value of a variable relative to its value at a base period is measured by

(a) an index number.

(b) a fixed market basket of goods.

(c) the ratio of the average value and the standard deviation.

(d) the ratio of the median and the interquartile range.

16.4 Index numbers. To find the index number for a value of a variable relative to its value at a base period, we use the formula

(a) index number $= \dfrac{\text{value}}{\text{base value}}$.

(b) index number $= \dfrac{\text{base value}}{\text{value}}$.

(c) index number $= \dfrac{\text{value}}{\text{base value}} \times 100$.

(d) index number $= \dfrac{\text{base value}}{\text{value}} \times 100$.

16.5 Fixed market baskets. An index number that describes the total cost of a collection of goods and services

(a) is called an inflation factor.

(b) is called a fixed market basket price index.

(c) must be larger than 100.

(d) all of the above.

16.6 The Consumer Price Index (CPI). The CPI

(a) can be thought of as a fixed market basket price index for the collection of all the goods and services that consumers buy.

However, it is not a true fixed market basket price index because of adjustments for changing buying habits, new products, and improved quality.

(b) can be used to change dollar amounts at one time into the amount at another time that has the same buying power.

(c) is determined from data from several large sample surveys.

(d) all of the above.

16.7 Government statistics. Government statistical offices

(a) are always a single, centralized agency.

(b) provide complete social and economic data for a country.

(c) produce data needed for government policy and decisions by businesses and individuals.

(d) all of the above.

Chapter 16 Exercises

When you need the CPI for a year that does not appear in Table 16.1, use the table entry for the year that most closely follows the year you want.

16.8 The price of gasoline. The yearly average price of unleaded regular gasoline has fluctuated as follows:

1998: $1.09 per gallon
2008: $3.09 per gallon
2018: $2.44 per gallon

Give the gasoline price index numbers (2008 = 100) for 1998, 2008, and 2018. Relatively speaking, which year had the highest gas price?

16.9 The cost of college. The part of the CPI that measures the cost of college tuition (1982–84 = 100) was 748.4 in October 2018. The overall CPI was 252.9 that month.

(a) Explain exactly what the index number 748.4 tells us about the rise in

college tuition between the base period and October 2018.

(b) College tuition has risen much faster than consumer prices in general. How do you know this?

16.10 The price of gasoline. Use your results from Exercise 16.8 to answer these questions.

(a) By how many points did the gasoline price index number change between 1998 and 2018? What percentage change was this?

(b) By how many points did the gasoline price index number change between 2008 and 2018? What percentage change was this?

You see that the point change and the percentage change in an index number are the same if we start in the base period, but not otherwise.

16.11 Toxic releases. The Environmental Protection Agency requires industry to report releases of any of a list of toxic chemicals. The total amounts released (in thousands of pounds, total on- and off-site disposal or other releases, all chemicals, all industries) were 6,939,299 in 1988, 6,655,831 in 2000, and 4,137,328 in 2013. Give an index number for toxic chemical releases in each of these years, with 1988 as the base period. By what percentage did releases increase or decrease between 1988 and 2000? Between 1988 and 2013?

16.12 How much can a dollar buy? The buying power of a dollar changes over time. The Bureau of Labor Statistics measures the cost of a "market basket" of goods and services to compile its Consumer Price Index (CPI). If the CPI is 120, goods and services that cost $100 in the base period now cost $120. Here are the average values of the CPI for the years between 1970 and 2018. The base period is the years 1982 to 1984.

Year	CPI	Year	CPI	Year	CPI
1970	38.8	1990	130.7	2010	218.1
1972	41.8	1992	140.3	2011	224.9
1974	49.3	1994	148.2	2012	229.6
1976	56.9	1996	156.9	2013	233.0
1978	65.2	1998	163.0	2014	236.7
1980	82.4	2000	172.2	2016	240.0
1982	96.5	2002	179.9	2018	251.0
1984	103.9	2004	188.9		
1986	109.6	2006	201.6		
1988	118.3	2008	215.3		

(a) Make a graph that shows how the CPI has changed over time.

(b) What was the overall trend in prices during this period? Were there any years in which this trend was reversed?

(c) In which years were prices rising fastest, in terms of percentage increase? In what period were they rising slowest?

16.13 Los Angeles and New York. The Bureau of Labor Statistics publishes separate consumer price indexes for major metropolitan areas in addition to the national CPI. The CPI (1982–84 = 100) in October 2018 was 269.5 in Los Angeles and 275.1 in New York.

(a) These numbers tell us that prices rose faster in New York than in Los Angeles between the base period and October 2018. Explain how we know this.

(b) These numbers do *not* tell us that prices in October 2018 were higher in New York than in Los Angeles. Explain why.

16.14 The Food Faddist Price Index. A food faddist eats only steak, rice, and ice cream. In 1998, he bought:

Item	1998 quantity	1998 price
Steak	200 pounds	$4.15/pound
Rice	300 pounds	0.55/pound
Ice cream	50 gallons	6.08/gallon

After a visit from his mother, he adds oranges to his diet. Oranges cost $0.53/pound in 1998. Here are the food faddist's food purchases in 2018:

Item	2018 quantity	2018 price
Steak	100 pounds	$8.21/pound
Rice	350 pounds	0.68/pound
Ice cream	75 gallons	9.57/gallon
Oranges	100 pounds	1.33/pound

Find the fixed market basket Food Faddist Price Index (1998 = 100) for the year 2018.

16.15 The Guru Price Index. A guru purchases only olive oil, loincloths, and copies of the *Atharva Veda,* from which he selects mantras for his disciples. Here are the quantities and prices of his purchases in 1985 and 2015:

Item	1985 quantity	1985 price	2015 quantity	2015 price
Olive oil	20 pints	$2.50/pint	18 pints	$7.50/pint
Loincloth	2	2.75 each	3	4.00 each
Atharva Veda	1	10.95	1	18.95

From these data, find the fixed market basket Guru Price Index (1985 = 100) for 2015.

16.16 The curse of the Bambino. In 1920 the Boston Red Sox sold Babe Ruth to the New York

Yankees for $125,000. Between 1920 and 2004, the Yankees won 26 World Series and the Red Sox won 1 (and that occurred in 2004). The Red Sox victory in 2004 supposedly broke the curse. In fact, the Red Sox won the World Series again in 2007, 2013, and 2018.

(a) How much is Ruth's 1920 salary of $125,000 worth in 2004 dollars?

(b) How much is Ruth's 1920 salary of $125,000 worth in 2018 dollars?

16.17 The cost of reading. A 2002 *Salon* article attempted to tackle the question "Why do books cost so much?" The article posited that the move from mass-market paperback to the larger and higher quality trade paperback may have contributed to increasing costs of books. *Salon* used John Updike's *Rabbit, Run* as an example. A mass-market paperback cost only 65 cents in the 1960s, but a 1991 mass-market paperback cost $5.99. In 2002, *Rabbit, Run* was only available in trade paperback and cost $14. In 2018 the same trade paperback cost $17. With discounts from online venues, a reader could purchase the 2002 trade paperback for $11.20 and can currently purchase the 2018 trade paperback for $14.50. Compare the discounted prices from 2002 and 2018. Relatively speaking, which was a better deal for the trade paperback of *Rabbit, Run*?

16.18 Blockbuster movies. According to Box Office Mojo, *Star Wars: Episode VII–The Force Awakens*, released in 2015, is the highest domestic grossing movie with sales amounting to $936,662,225. *Gone with the Wind*, released in 1939, grossed a mere $198,676,459. Due to inflation, it seems unfair to judge the two movies in original dollar amounts. Convert both movies' gross earnings to 2018 dollars and determine which movie earned more, relatively speaking.

16.19 Dream on. When Sonia started college in 2014, she set a goal of making $50,000 when she graduated. Sonia graduated in 2018. What must Sonia earn in 2018 in order to have the same buying power that $50,000 had in 2014?

16.20 Living too long? If both husband and wife are alive at age 65, in half the cases at least one will still be alive at age 93, 28 years later. Tom and Phyllis retired in 1990 with an income of $35,000 per year. They were quite comfortable—that was about the median family income in 1990. How much income did they need 28 years later, in 2018, to have the same buying power?

16.21 Microwaves on sale. The prices of new gadgets often start high and then fall rapidly. The first home microwave oven cost $1300 in 1955. You can now buy a better microwave oven for $100. Find the latest value of the CPI (it's on the website, www.bls.gov/cpi/home.htm) and use it to restate $100 in present-day dollars in 1955 dollars. Compare this with $1300 to see how much microwave oven real prices have come down.

16.22 Good golfers. In 2017, Justin Thomas won $9,921,560 on the Professional Golfers Association tour. The leading money winner in 1946 was Ben Hogan, at $42,566. Arnold Palmer, the leader in 1963, won $128,230, and Jack Nicklaus, the leader in 1972, won $320,542 that year. How do these amounts compare in real terms?

16.23 Joe DiMaggio. Yankee center fielder Joe DiMaggio was paid $32,000 in 1940 and $100,000 in 1950. Express his 1940 salary in 1950 dollars. By what percentage did DiMaggio's real income change in the decade?

16.24 New car prices. In 1980, a new Honda Accord Sedan cost $6515. The same car model and trim cost $22,750 in 1998 and $31,970 in 2018. Compare the real costs of these cars By what percentage did the real cost change between 1980 and 1998? Between 1980 and October 2018? Between 1998 and October 2018? The CPI in October 2018 was 252.9.

16.25 Paying for college. Todd thought tuition was expensive when his tuition at Washington University in St. Louis (WashU) was $10,500 in 1986. Now Todd is planning to send his daughter to college. The tuition at WashU was $48,859 in 2018. Express WashU's 1986 tuition in 2018 dollars. Did the cost of going to WashU go up faster or slower than consumer prices in general? How do you know?

16.26 Paying for college. Nineveh is looking at colleges and knows that in-state tuition at UCLA will cost her $13,749 in 2018. During her search, Nineveh stumbles across a brochure that lists the in-state tuition at UCLA as $3863 in 1998 and as $7551 in 2008. She wonders which of the three years had the highest tuition, relatively speaking, and calculates all tuitions in 2018 dollars. Perform the same calculations that Nineveh did and state which tuition year was the highest, relatively speaking.

16.27 Paying for college. Arjun is considering going to college at University of the Pacific in Stockton, CA. The tuition is $46,346 in 2018, and the cost of room and board is $13,356 that same year. Arjun's family started saving money for college in 2008, when tuition at University of the Pacific was $30,880 and the cost of room and board was $10,118. Which of the two costs, tuition or room and board, increased more between 2008 and 2018? How do you know?

16.28 The minimum wage. The federal government sets the minimum hourly wage that employers can pay a worker. Labor wants a high minimum wage, but many economists argue that too high a minimum wage makes employers reluctant to hire workers with low skills and it therefore increases unemployment. Here is information on changes in the federal minimum wage, in dollars per hour:

Year:	1960	1965	1970	1975	1980	1985
Min. wage ($):	1.00	1.25	1.60	2.10	3.10	3.35

Year:	1990	1995	2005	2007	2008	2009
Min. wage ($):	3.80	4.25	5.15	5.85	6.55	7.25

Use annual average CPIs from Table 16.1 to restate the minimum wage in constant 1960 dollars. Make two line graphs on the same axes, one showing the actual minimum wage during these years and the other showing the minimum wage in constant dollars. Explain carefully to someone who knows no statistics what your graphs show about the history of the minimum wage.

16.29 College tuition. Tuition for Michigan residents at the University of Michigan has increased as follows (use tuition at your own college if you have those data):

Year:	1998	2000	2002	2004	2006	2008
Tuition ($):	6098	6513	7411	8202	9723	11,037

Year:	2010	2012	2014	2016	2018
Tuition ($):	11,837	12,994	13,486	14,402	15,262

Use annual average CPIs from Table 16.1 to restate each year's tuition in constant 1998 dollars. Make two line graphs on the same axes, one showing actual dollar tuition for these years and the other showing constant dollar tuition. Then explain to someone who knows no statistics what your graphs show.

16.30 Rising incomes? In Example 4, we saw that the median real income (in 2017 dollars) of all households rose from $52,876 in 1984 to $61,372 in 2017. The real income that marks off the top 5% of households rose from $161,591 to $237,034 in the same period. Verify the claim in Example 4 that median income rose 16.1% and that the real income of top earners rose 46.7%.

16.31 Cable TV. Suppose that cable television systems across the country add channels to their lineup and raise the monthly fee they charge subscribers. The part of the CPI that tracks cable TV prices might not go up at all, even though consumers must pay more. Explain why.

16.32 Item weights in the CPI. The cost of buying a house (with the investment component removed) makes up about 24% of the CPI. The cost of renting a place to live makes up about 6%. Where do the weights 24% and 6% come from? Why does the cost of buying get more weight in the index?

16.33 The CPI doesn't fit me. The CPI may not measure your personal experience with changing prices. According to the Bureau of Labor Statistics, the CPI reflects the spending patterns of all urban consumers and urban wage earners and clerical workers. Those who live in rural nonmetropolitan areas or in farm households, those who are in the military, and those living in institutions (e.g., prisons, residential psychiatric facilities) are not reflected in the CPI. Explain why the CPI will not fit each of these people:

(a) Aisha lives on a cattle ranch in Montana.

(b) Mohammed heats his home with a wood stove and does not have air-conditioning.

(c) Luis and Maria were in a serious auto accident and spent much of last year in a rehabilitation center.

(d) Gabriel was homeless and unemployed last year.

(e) Aliyah is in the U.S. Marine Corps and lived on base at Okinawa, Japan.

16.34 Seasonal adjustment. Like many government data series, the CPI is published in both unadjusted and seasonally adjusted forms. The BLS says that it "strongly recommends using indexes unadjusted for seasonal variation (i.e., not seasonally adjusted indexes) for escalation." "Escalation" here means adjusting wage or other payments to keep up with changes in the CPI. Why is the unadjusted CPI preferred for this purpose?

16.35 More CPIs. In addition to the national CPI, the BLS publishes CPIs for 4 regions and for 37 local areas. Each regional and local CPI is based on just the part of the national sample of prices that applies to the region or local area. The BLS says that the local CPIs should be used with caution because they are much more variable than national or regional CPIs. Why is this?

16.36 The poverty line. The federal government announces "poverty lines" each year for households of different sizes. Households with income below the announced levels are considered to be living in poverty. An economist looked at the poverty lines over time and said that they "show a pattern of getting higher in real terms as the real income of the general population rises." What does "getting higher in real terms" say about the official poverty level?

16.37 Real wages (optional). In one of the many reports on stagnant incomes in the United States, we read that in the 1980s all income groups realized a decline in real wages. The decline was not the same for all income groups. Workers at the 33rd percentile saw a

14% decrease in real wage, while those at the 66th percentile saw just a 6% decrease. And those who earned wages at the highest rates actually saw a 1% increase in real wages.

(a) What is meant by "the 33rd percentile" of the income distribution?

(b) What does "real wages" mean?

16.38 Saving money? One way to cut the cost of government statistics is to reduce the sizes of the samples. We might, for example, cut the Current Population Survey from 50,000 households to 20,000. Explain clearly, to someone who knows no statistics, why such cuts reduce the accuracy of the resulting data.

16.39 The General Social Survey. The General Social Survey places much emphasis on asking many of the same questions year after year. Why do you think it does this?

16.40 Measuring the effects of crime. We wish to include, as part of a set of social statistics, measures of the amount of crime and of the impact of crime on people's attitudes and activities. Suggest some possible measures in each of the following categories:

(a) Statistics to be compiled from official sources such as police records.

(b) Factual information to be collected using a sample survey of citizens.

(c) Information on opinions and attitudes to be collected using a sample survey.

16.41 Statistical agencies. Write a short description of the work of one of these government statistical agencies. You can find information by starting at the FedStats website (www.fedstats.gov) and going to *Agencies*.

(a) Bureau of Economic Analysis (Department of Commerce)

(b) National Center for Education Statistics

(c) National Center for Health Statistics

Exploring the Web

Access these exercises on the text website: macmillanlearning.com/scc10e.

What's the Verdict?

The following "What's the verdict?" story asks you to think about what research should and should not be funded by the government. Although answers may not be clear-cut, you will use what you have learned in this chapter to help you think about this issue.

In 1996, the U.S. Congress passed the Dickey Amendment, which was lobbied for by the National Rifle Association and stated that the Centers for Disease Control and Prevention (CDC) could not use any federal funding for injury prevention to advocate for gun control. This amendment basically prevented the CDC from performing research on gun violence, treatment, and prevention so that the CDC would not lose important funding. In 2011, the restrictions were also extended to other agencies, such as the National Institutes of Health. As a result, the United States is not as well equipped as it could be to deal with mass shootings. For example, hospital officials from Aurora, Colorado (the location of the *Batman* movie theater shooting), and Boston, Massachusetts (the Boston Marathon bombing), could share their stories anecdotally, but very little scientific study has been done on the best methods to transport, triage, and treat gun violence victims in mass casualty events. Public health officials, emergency services workers, and medical practitioners cannot adequately prepare for these devastating situations without high-quality studies on best practices in prevention and treatment. President Obama signed an Executive Order after the mass shooting at an elementary school in Newton, Connecticut, to allow some gun violence research by the CDC, but not very much has happened since then because little funding was given to the CDC for this research. In 2018, the amendment still exists, but much of its power is reduced due to newer legislation.

Questions

WTV16.1. What do you think about the power government has to decide which research can and cannot be done?

WTV16.2. Can you think of any research topics that you do not think would be acceptable to receive government funding?

WTV16.3. Can you think of other examples of how the government can regulate what data are collected or how they are analyzed?

What's the verdict? Government policies concerning research can be controversial, and people will disagree about what should be regulated and how much control the government should exercise.

PART II Review

Data analysis is the art of describing data using graphs and numerical summaries. The purpose of data analysis is to help us see and understand the most important features of a set of data. Chapter 10 commented on basic graphs, especially pie charts, bar graphs, and line graphs. Chapters 11, 12, and 13 showed how data analysis works by presenting statistical ideas and tools for describing the distribution of one variable. Figure II.1 organizes the big ideas. We plot our data, then describe their center and spread using either the mean and standard deviation or the five-number summary. The last step, which makes sense only for some data, is to summarize the data in compact form by using a Normal curve as a model for the overall pattern. The question marks at the last two stages remind us that the usefulness of numerical summaries and Normal distributions depends on what we find when we examine graphs of our data. No short summary does justice to irregular shapes or to data with several distinct clusters.

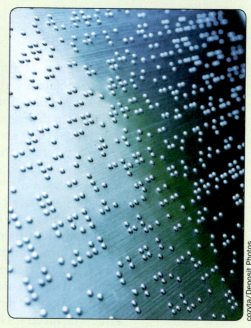

cozyta/Deposit Photos

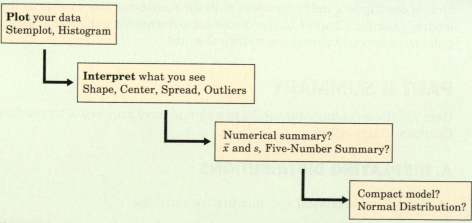

Plot your data
Stemplot, Histogram

Interpret what you see
Shape, Center, Spread, Outliers

Numerical summary?
$\bar{x}$ and s, Five-Number Summary?

Compact model?
Normal Distribution?

Figure II.1

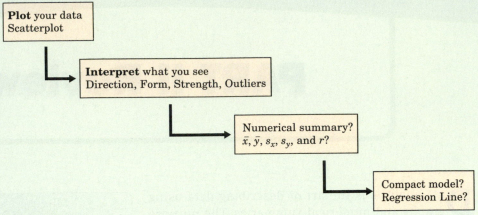

Figure II.2

Chapters 14 and 15 applied the same ideas to relationships between two quantitative variables. Figure II.2 retraces the big ideas from Figure II.1, with details that fit the new setting. We always begin by making graphs of our data. In the case of a scatterplot, we have learned a numerical summary only for data that show a roughly straight-line pattern on the scatterplot. The summary is then the means and standard deviations of the two variables and their correlation. A regression line drawn on the plot gives us a compact model of the overall pattern that we can use for prediction. Once again, there are question marks at the last two stages to remind us that correlation and regression describe only straight-line relationships.

Relationships often raise the question of causation. We know that evidence from randomized comparative experiments is the "gold standard" for deciding that one variable causes changes in another variable. Chapter 15 reminded us in more detail that strong associations can appear in data even when there is no direct causation. We must always think about the possible effects of variables lurking in the background. In Chapter 16, we met a new kind of description, index numbers, with the Consumer Price Index as the leading example. Chapter 16 also discussed government statistical offices, a quiet but important part of the statistical world.

PART II SUMMARY

Here are the most important skills you should have acquired after reading Chapters 10 through 16.

A. DISPLAYING DISTRIBUTIONS

1. Recognize categorical and quantitative variables.

2. Recognize when a pie chart can and cannot be used.

3. Make a bar graph of the distribution of a categorical variable or in general to compare related quantities.

4. Interpret pie charts and bar graphs.

5. Make a line graph of a quantitative variable over time.

6. Recognize patterns such as trends and seasonal variation in line graphs.

7. Be aware of graphical abuses, especially pictograms and distorted scales in line graphs.

8. Make a histogram of the distribution of a quantitative variable.

9. Make a stemplot of the distribution of a small set of observations. Round data as needed to make an effective stemplot.

B. DESCRIBING DISTRIBUTIONS (QUANTITATIVE VARIABLE)

1. Look for the overall pattern of a histogram or stemplot and for major deviations from the pattern.

2. Assess from a histogram or stemplot whether the shape of a distribution is roughly symmetric, distinctly skewed, or neither. Assess whether the distribution has one or more major peaks.

3. Describe the overall pattern by giving numerical measures of center and spread in addition to a verbal description of shape.

4. Decide which measures of center and spread are more appropriate: the mean and standard deviation (especially for symmetric distributions) or the five-number summary (especially for skewed distributions).

5. Recognize outliers and give plausible explanations for them.

C. NUMERICAL SUMMARIES OF DISTRIBUTIONS

1. Find the median M and the quartiles Q_1 and Q_3 for a set of observations.

2. Give the five-number summary and draw a boxplot; assess center, spread, symmetry, and skewness from a boxplot.

3. Find the mean $\bar{x}$ and (using a calculator) the standard deviation s for a small set of observations.

4. Understand that the median is less affected by extreme observations than the mean. Recognize that skewness in a distribution moves the mean away from the median toward the long tail.

5. Know the basic properties of the standard deviation: $s \geq 0$ always; $s = 0$ only when all observations are identical and increases as the spread increases; s has the same units as the original measurements; s is greatly increased by outliers or skewness.

D. NORMAL DISTRIBUTIONS

1. Interpret a density curve as a description of the distribution of a quantitative variable.

2. Recognize the shape of Normal curves, and estimate by eye both the mean and the standard deviation from such a curve.

3. Use the 68–95–99.7 rule and symmetry to state what percentage of the observations from a Normal distribution fall between two points when the points lie at the mean or one, two, or three standard deviations on either side of the mean.

4. Find and interpret the standard score of an observation.

5. (Optional) Use Table B to find the percentile of a value from any Normal distribution and the value that corresponds to a given percentile.

E. SCATTERPLOTS AND CORRELATION

1. Make a scatterplot to display the relationship between two quantitative variables measured on the same subjects. Place the explanatory variable (if any) on the horizontal scale of the plot.

2. Describe the direction, form, and strength of the overall pattern of a scatterplot. In particular, recognize positive or negative association and straight-line patterns. Recognize outliers in a scatterplot.

3. Judge whether it is appropriate to use correlation to describe the relationship between two quantitative variables. Use a calculator to find the correlation r.

4. Know the basic properties of correlation: r measures the strength and direction of only straight-line relationships; r is always a number between -1 and 1; $r = \pm 1$ only for perfect straight-line relations; r moves away from 0 toward ± 1 as the straight-line relation gets stronger.

F. REGRESSION LINES

1. Explain what the slope b and the intercept a mean in the equation $y = a + bx$ of a straight line.

2. Draw a graph of the straight line when you are given its equation.

3. Use a regression line, given on a graph or as an equation, to predict y for a given x. Recognize the danger of prediction outside the range of the available data.

4. Use r^2, the square of the correlation, to describe how much of the variation in one variable can be accounted for by a straight-line relationship with another variable.

G. STATISTICS AND CAUSATION

1. Understand that an observed association can be due to direct causation, common response, or confounding.

2. Give plausible explanations for an observed association between two variables: direct cause and effect, the influence of lurking variables, or both.

3. Assess the strength of statistical evidence for a claim of causation, especially when experiments are not possible.

H. THE CONSUMER PRICE INDEX AND RELATED TOPICS

1. Calculate and interpret index numbers.

2. Calculate a fixed market basket price index for a small market basket.

3. Use the CPI to compare the buying power of dollar amounts from different years. Explain phrases such as "real income."

PART II REVIEW EXERCISES

Review exercises are short and straightforward exercises that help you solidify the basic ideas and skills in each part of this book. We have provided "hints" that indicate where you can find the relevant material for the odd-numbered problems.

II.1 Poverty in the states. Table II.1 gives the percentages of people living below the poverty line in the 26 states east of the Mississippi River. Make a stemplot of these data. Is the distribution roughly symmetric, skewed to the right, or skewed to the left? Which states (if any) are outliers? (*Hint*: See page 253.)

II.2 Quarterbacks. Table II.2 gives the total passing yards for National Football League starting quarterbacks during the 2017 season. (These are the quarterbacks with the most games started on each team.) Make a histogram of these data. Does the distribution have a clear shape: roughly symmetric, clearly skewed to the left, clearly skewed to the right, or none of these? Which quarterbacks (if any) are outliers?

Table II.1 Percentages of state residents east of the Mississippi River living in poverty, 2016–2017 two-year average

State	Percent	State	Percent	State	Percent
Alabama	15.6	Connecticut	10.4	Delaware	10.4
Florida	13.3	Georgia	14.3	Illinois	11.5
Indiana	11.6	Kentucky	14.8	Maine	12.3
Maryland	7.5	Massachusetts	10.1	Michigan	11.9
Mississippi	19.7	New Hampshire	6.5	New Jersey	9.0
New York	12.6	North Carolina	14.0	Ohio	13.2
Pennsylvania	11.2	Rhode Island	11.8	South Carolina	14.8
Tennessee	13.2	Vermont	9.9	Virginia	10.8
West Virginia	17.7	Wisconsin	10.1		

Data from U.S. Census Bureau, *Income and Poverty in the United States: 2017*.

Table II.2 Passing yards for NFL quarterbacks in 2017

Quarterback	Yards	Quarterback	Yards
Blake Bortles	3687	Josh McCown	2926
Tom Brady	4577	Cam Newton	3302
Drew Brees	4334	Carson Palmer	1978
Jacoby Brissett	3098	Dak Prescott	3324
Derek Carr	3496	Philip Rivers	4515
Kirk Cousins	4093	Ben Roethlisberger	4251
Jay Cutler	2666	Matt Ryan	4095
Andy Dalton	3320	Tom Savage	1412
Joe Flacco	3141	Trevor Siemian	2285
Jimmy Garoppolo	1560	Alex Smith	4042
Jared Goff	3804	Matthew Stafford	4446
Brett Hundley	1836	Tyrod Taylor	2799
Case Keenum	3547	Mitchell Trubisky	2193
DeShone Kizer	2894	Carson Wentz	3296
Eli Manning	3468	Russell Wilson	3983
Marcus Mariota	3232	Jameis Winston	3504

Data from pro-football-reference.com.

II.3 Poverty in the states. Give the five-number summary for the data on poverty from Table II.1. (*Hint*: See page 272.)

II.4 Quarterbacks. Give the five-number summary for the data on passing yards for NFL quarterbacks from Table II.2.

II.5 Poverty in the states. Find the mean percentage of state residents living in poverty from the data in Table II.1. If we removed Mississippi from the data, would the mean increase or decrease? Why? Find the mean for the 25 remaining states to verify your answer. (*Hint*: See page 277.)

II.6 Big heads? The army reports that the distribution of head circumference among male soldiers is approximately Normal with mean 22.8 inches and standard deviation 1.1 inches. Use the 68–95–99.7 rule to answer these questions.
(a) Between what values do the middle 95% of head circumferences fall?
(b) What percentage of soldiers have head circumferences greater than 23.9 inches?

II.7 SAT scores. The scale for SAT exam scores is set so that the distribution of scores is approximately Normal with mean 500 and standard deviation 100. Answer these questions without using a table.
(a) What is the median SAT score? (*Hint*: See page 299.)
(b) You run a tutoring service for students who score between 400 and 600 and hope to attract many students. What percentage of SAT scores are between 400 and 600? (*Hint*: See page 300.)

II.8 Explaining correlation. You have data on the yearly wine consumption (liters of alcohol from drinking wine per person) and yearly deaths from cirrhosis of the liver for several developed countries. What does the correlation r between yearly wine consumption and yearly deaths from cirrhosis of the liver measure? Be specific, but brief.

II.9 Data on alligators. A wildlife researcher measures the length (inches) and weight (pounds) of 25 young alligators in Florida. What units of measurement do each of the following have?
(a) The mean length of the alligators. (*Hint*: See page 277.)
(b) The first quartile of the alligator lengths. (*Hint*: See page 270.)
(c) The standard deviation of the alligator lengths. (*Hint*: See page 277.)
(d) The correlation between length and alligator weight. (*Hint*: See page 325.)

II.10 More data on alligators. A wildlife researcher measures the length (inches) and weight (pounds) of 25 young alligators in Florida.

(a) Explain why you expect the correlation between length and weight to be positive.
(b) The mean length turns out to be 84.96 inches. What is the mean length in centimeters? (There are 2.54 centimeters in an inch.)
(c) The correlation between length and weight turns out to be $r = 0.9$. If you measured length in centimeters instead of inches, what would be the new value of r?

Figure II.3 plots the average brain weight in grams versus average body weight in kilograms for many species of mammals. There are many small mammals whose points at the lower left overlap. Exercises II.11 through II.16 are based on this scatterplot.

II.11 Dolphins and hippos. The points for the dolphin and hippopotamus are labeled in Figure II.3. Read from the graph the approximate body weight and brain weight for these two species. (*Hint*: See page 318.)

II.12 Dolphins and hippos. One reaction to this scatterplot is "Dolphins are smart; hippos are dumb." What feature of the plot lies behind this reaction?

II.13 Outliers. African elephants are much larger than any other mammal in the data set but lie roughly in the overall straight-line pattern. Dolphins, humans, and hippos lie outside the overall pattern. The correlation between body weight and brain weight for the entire data set is $r = 0.86$.
(a) If we removed elephants, would this correlation increase or decrease or not change much? Explain your answer. (*Hint*: See page 326.)
(b) If we removed dolphins, hippos, and humans, would this correlation increase or decrease or not change much? Explain your answer. (*Hint*: See page 326.)

II.14 Brain and body. The correlation between body weight and brain weight is $r = 0.86$. How well does body weight explain brain weight for mammals? Compute r^2 to answer this question, and briefly explain what r^2 tells us.

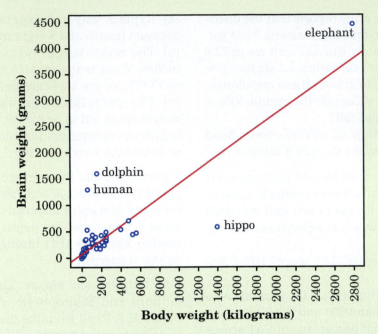

Figure II.3 Scatterplot of the average brain weight (grams) against the average body weight (kilograms) for 96 species of mammals, Exercises II.11 through II.16.

II.15 Prediction. The line on the scatterplot in Figure II.3 is the least-squares regression line for predicting brain weight from body weight. Suppose that a new mammal species with body weight 600 kilograms is discovered hidden in the rain forest. Predict the brain weight for this species. (*Hint*: See page 340.)

II.16 Slope. The line on the scatterplot in Figure II.3 is the least-squares regression line for predicting brain weight from body weight. The slope of this line is one of the numbers below. Which number is the slope? Why?

(a) $b = 0.5$

(b) $b = 1.3$

(c) $b = 3.2$

From Rex Boggs in Australia comes an unusual data set: before showering in the morning, he weighed the bar of soap in his shower stall. The weight goes down as the soap is used. The data appear in Table II.3 (weights in grams).

Table II.3 Weight (grams) of a bar of soap used to shower

Day	Weight	Day	Weight	Day	Weight
1	124	8	84	16	27
2	121	9	78	18	16
5	103	10	71	19	12
6	96	12	58	20	8
7	90	13	50	21	6

Data from Rex Boggs.

Notice that Mr. Boggs forgot to weigh the soap on some days. Exercises II.17, II.18, and II.19 are based on the soap data set.

II.17 Scatterplot. Plot the weight of the bar of soap against the day. Is the overall pattern roughly straight-line? Based on your scatterplot, is the correlation between day and weight close to 1, positive but not close to 1, close to 0, negative but not close to −1, or close to −1? Explain your answer. (*Hint*: See page 325.)

II.18 Regression. The equation for the least-squares regression line for the data in Table II.3 is

$$weight = 133.2 - 6.31 \times day$$

(a) Explain carefully what the slope $b = -6.31$ tells us about how fast the soap lost weight.
(b) Mr. Boggs did not measure the weight of the soap on Day 4. Use the regression equation to predict that weight.
(c) Draw the regression line on your scatterplot from the previous exercise.

II.19 Prediction? Use the regression equation in the previous exercise to predict the weight of the soap after 30 days. Why is it clear that your answer makes no sense? What's wrong with using the regression line to predict weight after 30 days? (*Hint*: See page 345.)

II.20 Keeping up with the Joneses. The Jones family had a household income of $60,000 in 2000, when the average CPI (1982–84 = 100) was 172.2. The average CPI for 2018 was 251.0. How much must the Joneses earn in 2018 to have the same buying power they had in 2000?

II.21 Affording a Subaru. A Subaru Outback 2.5i Limited cost $27,245 in 2005, when the average CPI (1982–84 = 100) was 195.3. The average CPI for 2018 was 251.0. How many 2018 dollars must you earn to have the same buying power as $27,245 had in 2005? (*Hint*: See page 374.)

II.22 Affording a Steinway. A Steinway concert grand piano cost $13,500 in 1976. A similar Steinway cost $156,200 in October 2018. Has the cost of the piano gone up or down in real terms? Using Table 16.1 and the fact that the October 2018 CPI was 252.9, give a calculation to justify your answer.

II.23 The price of gold. Some people recommend that investors buy gold "to protect against inflation." Here are the prices of an ounce of gold at the end of the year for the years between 1985 and 2018. Using Table 16.1, make a graph that shows how the price of gold

changed in real terms over this period. Would an investment in gold have protected against inflation by holding its value in real terms?

Year:	1990	1992	1994	1996	1998
Gold price:	$384	$344	$384	$388	$294

Year:	2000	2002	2004	2006	2008
Gold price:	$279	$310	$410	$603	$872

Year:	2010	2012	2014	2016	2018
Gold price:	$1225	$1669	$1266	$1251	$1269

(*Hint*: See page 372.)

II.24 Friday the 13th. When planning a wedding, are certain dates more or less popular than others? The bar graphs in Figure II.4 are based on data from the Office for National Statistics, National Records of Scotland. They show the average number of marriages in Great Britain that occurred either on any date of the month other than the 13th or on the 13th of the month, for each day of the week, during the period from 1995 to 2015.
(a) Explain why it is not unusual for the average number of weddings to be highest on Fridays and Saturdays.
(b) When a Friday occurs on the 13th day of a month, the average number of weddings is half as much as when a Friday occurs on any other date. What might explain this?

II.25 Drive time. Professor Moore, who lives a few miles outside a college town, records the time he takes to drive to the college each morning. Here are the times (in minutes) for 42 consecutive weekdays, with the dates in order along the rows:

8.25	7.83	8.30	8.42	8.50	8.67	8.17
9.00	9.00	8.17	7.92	9.00	8.50	9.00
7.75	7.92	8.00	8.08	8.42	8.75	8.08
9.75	8.33	7.83	7.92	8.58	7.83	8.42
7.75	7.42	6.75	7.42	8.50	8.67	10.17
8.75	8.58	8.67	9.17	9.08	8.83	8.67

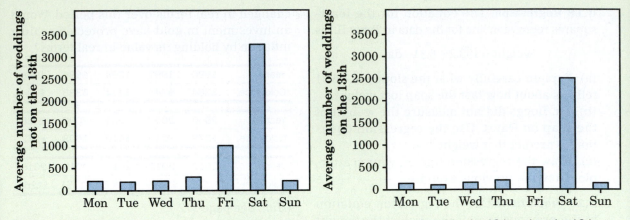

Figure II.4 Bar graphs of the average number of weddings that occur on dates other than the 13th and on the 13th, for each day of the week, from 1995 to 2015, Exercise II.24.

(a) Make a histogram of these drive times. Is the distribution roughly symmetric, clearly skewed, or neither? Are there any clear outliers? (*Hint*: See page 247.)

(b) Make a line graph of the drive times. (Label the horizontal axis in days, 1 to 42.) The plot shows no clear trend, but it does show one unusually low drive time and two unusually high drive times. Circle these observations on your plot. (*Hint*: See page 223.)

II.26 Drive time outliers. In the previous exercise, there are three outliers in Professor Moore's drive times to work. All three can be explained. The low time is the day after Thanksgiving (no traffic on campus). The two high times reflect delays due to an accident and icy roads. Remove these three observations. To summarize normal drive times, use a calculator to find the mean $\bar{x}$ and standard deviation s of the remaining 39 times.

II.27 House prices. A May 23, 2018, article in the *Los Angeles Times* reported that the median housing price in Southern California was about $520,000. Would the mean housing price be higher, about the same, or lower? Why? (*Hint*: See page 281.)

II.28 The 2016 election. Donald Trump was elected president in 2016 with 46.4% of the popular vote. His Democrat opponent, Hilary Clinton, received 48.5% of the vote, with minor candidates taking the remaining votes. Table II.4 gives the percentage of the popular vote won by President Trump in each state. Describe these data with a graph, a numerical summary, and a brief verbal description.

II.29 Statistics for investing. Joe's retirement plan invests in stocks through an "index fund" that follows the behavior of the stock market as a whole, as measured by the Standard & Poor's 500 index. Joe wants to buy a mutual fund that does not track the index closely. He reads that monthly returns from Fidelity Technology Fund have correlation $r = 0.77$ with the S&P 500 index and that Fidelity Real Estate Fund has correlation $r = 0.37$ with the index.

(a) Which of these funds has the closer relationship to returns from the stock market as a whole? How do you know? (*Hint*: See page 325.)

(b) Does the information given tell Joe anything about which fund has had higher returns? (*Hint*: See page 328.)

Table II.4 Percentage of votes for President Trump in 2016

State	Percent	State	Percent	State	Percent
Alabama	62.1	Louisiana	58.1	Ohio	51.7
Alaska	51.3	Maine	44.9	Oklahoma	65.3
Arizona	48.7	Maryland	33.9	Oregon	39.1
Arkansas	60.6	Massachusetts	32.8	Pennsylvania	48.2
California	31.6	Michigan	47.5	Rhode Island	38.9
Colorado	43.3	Minnesota	44.9	South Carolina	54.9
Connecticut	40.9	Mississippi	57.9	South Dakota	61.5
Delaware	41.7	Missouri	56.8	Tennessee	60.7
Florida	49.0	Montana	56.2	Texas	52.2
Georgia	50.8	Nebraska	58.8	Utah	45.5
Hawaii	30.0	Nevada	45.5	Vermont	30.3
Idaho	59.3	New Hampshire	46.6	Virginia	44.4
Illinois	38.8	New Jersey	41.0	Washington	36.8
Indiana	56.8	New Mexico	40.0	West Virginia	68.5
Iowa	51.2	New York	36.5	Wisconsin	47.2
Kansas	56.7	North Carolina	49.8	Wyoming	67.4
Kentucky	62.5	North Dakota	63.0		

Data from https://www.nytimes.com/elections/2016/results/president.

PART PROJECTS

Projects are longer exercises that require gathering information or producing data and that emphasize writing a short essay to describe your work. Many are suitable for teams of students.

Access these projects on the text website: **macmillanlearning.com/scc10e**.

PART III

Chance

"If chance will have me king, why, chance will crown me." So said Macbeth in Shakespeare's great play. Chance does indeed play with us all, and we can do little to understand or manage it. Sometimes, however, chance is tamed. A roll of dice, a simple random sample, even the inheritance of eye color or blood type, represent chance tied down so that we can understand and manage it. Unlike Macbeth's life or ours, we can roll the dice again. And again, and again. The outcomes are governed by chance, but in many repetitions a pattern emerges. Chance is no longer mysterious because we can describe its pattern.

We use mathematics to describe regular patterns, whether the circles and triangles of geometry or the movements of the planets. We use mathematics to understand the regular patterns of chance behavior when chance is tamed in a setting where we can repeat the same chance phenomenon again and again. The mathematics of chance is called probability. Probability is the topic of this part of the book, though we will go light on the math in favor of experimenting and thinking.

Tom Schoumakers/Shutterstock

Thinking about Chance

17

Case Study: Shared Birthdays

> **In this chapter you will:**
>
> - Learn some fundamental concepts about chance behavior.
> - Interpret probabilities like 1 in 133,000.
> - Identify some common "myths" concerning chance behavior, such as misinterpretations of the law of averages and the notion of the "hot hand" in sports.

On January 14, 2017, CBS News reported that in the Gardner family, mom, dad, and son all share the same birthday. The news article went on to say the odds of that happening are about 1 in 133,000. That's a lot less likely than getting hit by lightning sometime in your lifetime, which some put at roughly 1 in 12,000. The rarity of the event is what made the story newsworthy.

Just how amazing is this event? By the end of this chapter, you will be able to assess coincidences such as a mother, father, and son all having the same birthday. Are these events as surprising as they seem?

The Idea of Probability

Chance is a slippery subject. We start by thinking about "what would happen if we perform the same action over and over many times?" We will consider examples like the 1-in-2 chance of a head in tossing a coin before we try to think about more-complicated situations.

Even the rules of football agree that tossing a coin avoids favoritism. Favoritism in choosing subjects for a sample survey or allotting patients to treatment and placebo groups in a medical experiment is as undesirable as it is in awarding first possession of the ball in football. That's why statisticians recommend random samples and randomized experiments, which are fancy versions of tossing a coin. A big fact emerges when we watch coin tosses or the results of random samples closely: **chance behavior is unpredictable in the short run but has a regular and predictable distribution of outcomes in the long run.**

Toss a coin or choose a simple random sample. The result can't be predicted in advance because the result will vary when you toss the coin or choose the sample repeatedly. But there is still a regular pattern in the results, a pattern that emerges clearly only after many repetitions. This remarkable fact is the basis for the idea of probability.

Example 1   *Coin tossing*

When you toss a coin, there are only two possible outcomes, heads or tails. Figure 17.1 shows the results of tossing a coin 1000 times. For each number of tosses from 1 to 1000, we have plotted the proportion of those tosses that gave a head. The first toss was a head, so the proportion of heads starts at 1. The second toss was a tail, reducing the proportion of heads to 0.5 after two tosses. The next four tosses were tails followed by a head, so the proportion of heads after seven tosses is 2/7, or 0.286.

The proportion of tosses that produce heads is quite variable at first, but it settles down as we make more and more tosses. Eventually, this proportion gets close to 0.5 and stays there. We say that 0.5 is the *probability* of a head. The probability 0.5 appears as a horizontal line on the graph.

jamesgroup/
Deposit Photos

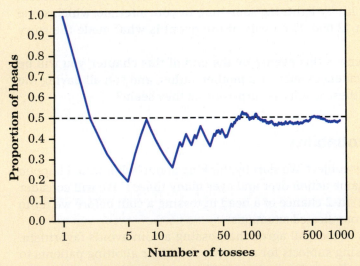

Figure 17.1 Toss a coin many times. The proportion of heads changes as we make more tosses but eventually gets very close to 0.5. This is what we mean when we say, "The probability of a head is one-half."

"Random" in statistics is a description of events that are unpredictable in the short run but that exhibit a kind of order that emerges only in the long run. It is not a synonym for "haphazard," which is defined as lacking any principle of organization. We encounter the unpredictable side of randomness in our everyday experience, but we rarely see enough repetitions of the same random phenomenon to observe the long-term regularity that probability describes. You can see that regularity emerging in Figure 17.1. In the very long run, the proportion of tosses that gives a head is 0.5. This is the intuitive idea of probability. A probability of 0.5 means that the outcome "occurs half the time in a very large number of repetitions, or trials, of a random phenomenon."

We might suspect that a coin has probability 0.5 of coming up heads just because the coin has two sides. We might be tempted to theorize that for events with two seemingly equally likely outcomes, each outcome should have probability 0.5 of occurring. But babies must have one of the two sexes, and the probabilities aren't equal—the probability of a boy is about 0.51, not 0.50. The idea of probability is empirical. That is, it is based on data rather than theorizing alone. Probability describes what happens in very many trials, and we must actually observe many coin tosses or many babies to pin down a probability. In the case of tossing a coin, some diligent people have in fact made thousands of tosses.

> ## Key Terms
>
> We call a phenomenon **random** if individual outcomes are uncertain but there is, nonetheless, a regular distribution of outcomes in a large number of repetitions.
>
> The **probability** of any outcome of a random phenomenon is a number between 0 and 1 that describes the proportion of times the outcome would occur in a very long series of repetitions.

Example 2   *Some coin tossers*

The French naturalist Count Buffon (1707–1788) tossed a coin 4040 times. Result: 2048 heads, or proportion 2048/4040 = 0.5069 for heads.

Around 1900, the English statistician Karl Pearson heroically tossed a coin 24,000 times. Result: 12,012 heads, a proportion of 0.5005.

While imprisoned by the Germans during World War II, the South African mathematician John Kerrich tossed a coin 10,000 times. Result: 5067 heads, a proportion of 0.5067.

An outcome with probability 0 never occurs. An outcome with probability 1 happens on every repetition. An outcome with probability one-half, or 1-in-2, happens half the time in a very long series of trials. Of course, we can never observe a probability exactly. We could always continue tossing the coin, for example. Mathematical probability is an

Does God play dice? Few things in the world are truly random in the sense that no amount of information will allow us to predict the outcome. We could, in principle, apply the laws of physics to a tossed coin, for example, and calculate whether it will land heads or tails. But randomness does rule events inside individual atoms. Albert Einstein didn't like this feature of the new quantum theory. "God does not play dice with the universe," said the great scientist. Many years later, it appears that Einstein was wrong.

idealization based on imagining what would happen in an infinitely long series of trials.

We aren't thinking deeply here. That some things are random is simply an observed fact about the world. Probability just gives us a language to describe the long-term regularity of random behavior. The outcome of a coin toss, the time between emissions of particles by a radioactive source, and the sexes of the next litter of lab rats are all random. So is the outcome of a random sample or a randomized experiment. The behavior of large groups of individuals is often as random as the behavior of many coin tosses or many random samples. Life insurance, for example, is based on the fact that deaths occur at random among many individuals.

Example 3

The probability of dying

We can't predict whether a particular person will die in the next year. But if we observe millions of people, deaths are random. In 2016, the National Center for Health Statistics reported that the proportion of men aged 20 to 24 years who die in any one year is 0.0014. This is the *probability* that a young man will die next year. For women that age, the probability of death is about 0.0005.

If an insurance company sells many policies to people aged 20 to 24, it knows that it will have to pay off next year on about 0.14% of the policies sold on men's lives and on about 0.05% of the policies sold on women's lives. It will charge more to insure a man because the probability of having to pay is higher.

The Ancient History of Chance

Randomness is most easily noticed in many repetitions of games of chance: rolling dice, dealing shuffled cards, spinning a roulette wheel. Chance devices similar to these have been used from remote antiquity to discover the will of the gods. The most common method of randomization in ancient times was "rolling the bones"—that is, tossing several astragali. The astragalus (Figure 17.2) is a six-sided animal heel bone that, when thrown, will come to rest on one of four sides (the other two sides are rounded). Cubical dice, made of pottery or bone, came later, but even dice existed before 2000 B.C. Gambling on the throw of astragali or dice is, compared with divination, almost a modern

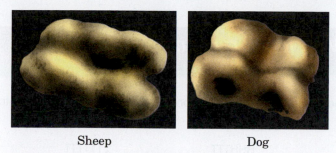

Sheep   Dog

Figure 17.2 Animal heel bones (astragali).

development. There is no clear record of this vice before about 300 B.C. Gambling reached flood tide in Roman times, then temporarily receded (along with divination) in the face of Christian displeasure.

Chance devices such as astragali have been used from the beginning of recorded history. Yet none of the great mathematicians of antiquity studied the regular pattern of many throws of bones or dice. Perhaps this is because astragali and most ancient dice were so irregular that each had a different pattern of outcomes. Or perhaps the reasons lie deeper, in the classical reluctance to engage in systematic experimentation.

Professional gamblers, who are not as inhibited as philosophers and mathematicians, did notice the regular pattern of outcomes of dice or cards and tried to adjust their bets to the odds of success. "How should I bet?" is the question that launched mathematical probability. The systematic study of randomness began (we oversimplify, but not too much) when seventeenth-century French gamblers asked French mathematicians for help in figuring out the "fair value" of bets on games of chance. *Probability theory*, the mathematical study of randomness, originated with Pierre de Fermat and Blaise Pascal in the seventeenth century and was well developed by the time statisticians took it over in the twentieth century.

Myths about Chance Behavior

The idea of probability seems straightforward. It answers the question, "What would happen if we did this many times?" In fact, both the behavior of random phenomena and the idea of probability are a bit subtle. We meet chance behavior constantly, and psychologists tell us that we deal with it poorly.

The myth of short-run regularity The idea of probability is that randomness is regular *in the long run*. Unfortunately, our intuition about randomness tries to tell us that random phenomena should also be regular in the short run. When they aren't, we look for some explanation other than chance variation.

Example 4

What looks random?

Toss a fair coin six times and record heads (H) or tails (T) on each toss. Which of these outcomes is most probable?

HTHTTH   HHHTTT   TTTTTT

Almost everyone says that HTHTTH is more probable because TTTTTT and HHHTTT do not "look random." In fact, all three are equally probable. That heads and tails are equally probable says all specific outcomes of heads and tails in six tosses are equally likely. That heads and tails are equally probable says only that about half of a very long sequence of tosses will be heads. This is because in very long sequences of tosses, the number of outcomes for which the proportion of heads is approximately one-half is much larger than the number of outcomes for which the proportion is not near one-half. That heads and tails are equally probable doesn't say that heads and tails must come close to alternating in the short run. It doesn't say that specific outcomes that balance the number of heads and tails are more likely than specific outcomes that don't. The coin has no memory. It doesn't know what past outcomes were, and it can't try to create a balanced sequence.

The outcome TTTTTT in tossing six coins looks unusual because of the run of six straight tails. The outcome HHHTTT also looks unusual because of the pattern of a run of three straight heads followed by a run of three straight tails. Runs seem "not random" to our intuition but are not necessarily unusual. Here's an example more striking than tossing coins.

Example 5

The hot hand in basketball

On January 23, 2015, against the Sacramento Kings, Klay Thompson was ice cold. He had missed his previous five shots before the end of the first half, including an uncontested layup. But after half-time, Thompson erupted. He made his next 13 shots in a row and finished the quarter with a record-smashing 37 points. After the game, Green was asked whether someone could possibly do that in the "NBA 2K" video game. His response? "Nah, you don't get that hot in 2K"

Belief that runs must result from something other than "just chance" influences behavior. If a basketball player makes several consecutive shots, both the fans and his teammates believe that he has a "hot hand" and is more likely to make the next shot. This is not supported by data. Careful study has shown that runs of baskets made or missed are no more frequent in basketball than would be expected if

OSTILL is Franck Camhi/
Shutterstock

each shot is independent of the player's previous shots. Players perform consistently, not in streaks. If a player makes half her shots in the long run, her hits and misses behave just like tosses of a coin—and that means that runs of hits and misses are more common than our intuition expects.

Now it's your turn

17.1 Coin tossing and randomness. Toss a coin 10 times and record heads (H) or tails (T) on each toss. Which of these outcomes is most probable? Least probable?

HTHTTHHTHT   TTTTTHHHHH   HHHHHHHHHH

The myth of the surprising coincidence On November 18, 2006, Ohio State beat Michigan in football by a score of 42 to 39. Later that day, the winning numbers in the Pick 4 Ohio lottery were 4239. What an amazing coincidence!

Well, maybe not. It is certainly unlikely that the Pick 4 lottery would match the Ohio State versus Michigan score that day, but it is not so unlikely that sometime during the 2006 season, the winning number of some state lottery would match the recent score of some professional, college, or high school football game involving a team in the state. There are 32 NFL teams, 235 NCAA Division I teams, 150 NCAA Division II teams, and 231 NCAA Division III teams. There are also more than 25,000 high school football teams. All play a number of games during the season. Almost all states have a Pick 3 or Pick 4 lottery game, with winning numbers often drawn multiple times per week. That's a lot of opportunities to match a Pick 3 or Pick 4 lottery number that has digits that could conceivably be a football score like 217 or 4239.

When something unusual happens, we look back and say, "Wasn't that unlikely?" We would have said the same if any of thousands of other unlikely things had happened. Here's an example where it was possible to actually calculate the probabilities.

Example 6
Winning the lottery twice

In 1986, Evelyn Marie Adams won the New Jersey State lottery for the second time, adding $1.5 million to her previous $3.9 million jackpot. The *New York Times* (February 14, 1986) claimed that the odds of one person winning the big prize twice were about 1 in 17 trillion. Nonsense, said two statistics professors in a letter that appeared in the *Times* two weeks later. The chance that Evelyn

Marie Adams would win twice in her lifetime is indeed tiny, but it is almost certain that *someone* among the millions of regular lottery players in the United States would win two jackpot prizes. The statisticians estimated even odds (a probability of one-half) of another double winner within seven years. Sure enough, Robert Humphries won his second Pennsylvania lottery jackpot ($6.8 million total) in May 1988. You might find it interesting to do an Internet search of "man wins state lottery two times" or "woman wins state lottery two times." A recent double winner was a man who won million-dollar prizes in the Canadian lottery in April 2018 and again in August 2018.

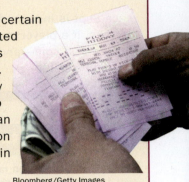
Bloomberg/Getty Images

Unusual events—especially distressing events—bring out the human desire to pinpoint a reason, a *cause*. Here's a sequel to our earlier discussion of causation: sometimes it's just the play of chance.

Example 7   *Cancer clusters*

Between 1996 and 2013, 37 children in Clyde, Ohio, a town of 6000 halfway between Toledo and Cleveland, were diagnosed with cancer. Four of the children had died. With many of the diagnoses coming between 2002 and 2006, state health authorities declared it a cancer cluster, saying the number and type of diagnoses exceed what would be expected statistically for so small a population over that time. In the fall of 2012, the EPA found high levels of toxic, possibly cancer-causing chemical compounds in soil samples from Whirlpool Park, formerly a residential area owned by home appliance manufacturer Whirlpool Corporation from the 1950s until 2008. Locals told news reporters that "black sludge" had been dumped in the area during that time. However, a February 2015 article in the *Akron Beacon Journal* reported that a suit against Whirlpool was dropped and that "Ohio health and environmental investigators have spent years testing the air and water around Clyde and talking with the children and their families about where they live and work and what they might have been exposed to. But they've never come up with the answer."

Cancer is a common disease, accounting for more than 23% of all deaths in the United States. That cancer cases sometimes occur in clusters in the same neighborhood is not surprising; there are bound to be clusters *somewhere* simply by chance. But when a cancer cluster occurs in *our* neighborhood, we tend to suspect the worst and look for someone to blame. State authorities get several thousand calls a year from people worried about "too much cancer" in their area. But the National Cancer Institute finds that the majority of reported cancer clusters are simply the result of chance.

Two Massachusetts cancer clusters, one in Randolph, Massachusetts, and one in Woburn, Massachusetts, were investigated by statisticians from the Harvard School of Public Health in the 1980s. The investigators tried to obtain complete data on everyone who had lived in the neighborhoods in the periods in question and to estimate their exposure to the suspect drinking water. They also tried to obtain data on other factors that might explain cancer, such as smoking and occupational exposure to toxic substances. The verdict: chance is the likely explanation of the Randolph cluster, but there is evidence of an association between drinking water from the two Woburn wells and developing childhood leukemia.

The myth of the law of averages Roaming the gambling floors in Las Vegas, watching money disappear into the drop boxes under the tables, is revealing. You can see some interesting human behavior in a casino. When the shooter in the dice game craps rolls several winners in a row, some gamblers think she has a "hot hand" and bet that she will keep on winning. Others say that "the law of averages" means that she must now lose so that wins and losses will balance out. Believers in the law of averages think that if you toss a coin six times and get TTTTTT, the next toss must be more likely to give a head. It's true that in the long run heads must appear half the time. What is myth is that future outcomes must make up for an imbalance like six straight tails.

Coins and dice have no memories. A coin doesn't know that the first six outcomes were tails, and it can't try to get a head on the next toss to even things out. Of course, things do even out *in the long run*. After 10,000 tosses, the results of the first six tosses don't matter. They are overwhelmed by the results of the next 9994 tosses, not compensated for.

Statistics in Your World

The probability of rain is . . .
You work all week. Then it rains on the weekend. Can there really be a statistical truth behind our perception that the weather is against us? At least on the East Coast of the United States, the answer is Yes. Going back to 1946, it seems that Sundays receive 22% more precipitation than Mondays. The likely explanation is that the pollution from all those workday cars and trucks forms the seeds for raindrops—with just enough delay to cause rain on the weekend.

Example 8
We want a boy

Belief in this phony "law of averages" can lead to consequences close to disastrous. A few years ago, "Dear Abby" published in her advice column a letter from a distraught mother of eight girls. It seems that she and her husband had planned to limit their family to four children. When all four were girls, they tried again—and again, and again. After seven straight girls, even her doctor had assured her that "the law of averages was in our favor 100 to 1." Unfortunately for this couple, having children is like tossing coins. Eight girls in a row is highly unlikely, but once seven girls have been born, it is not at all unlikely that the next child will be a girl—and it was.

digifun/Deposit Photos

What is the law of averages? Is there a "law of averages"? There is, although it is sometimes referred to as the "law of large numbers." It states that in a large number of "independent" repetitions of a random phenomenon (such as coin tossing), averages or proportions are likely to become more stable as the number of trials increases, whereas sums or counts are likely to become more variable. This does not happen by compensation for a run of bad luck because, by "independent," we mean that knowing the outcome of one trial does not change the probabilities for the outcomes of any other trials. The trials have no memory.

Figures 17.1 and 17.3 show what happens when we toss a coin repeatedly many times. In Figure 17.1 (page 406), we see that the *proportion* of heads gradually becomes closer and closer to 0.5 as the number of tosses increases. This illustrates the law of large numbers. However, Figure 17.3 shows, for these same tosses, how far the *total* number of heads differs from exactly half of the tosses being heads. We see that how this difference

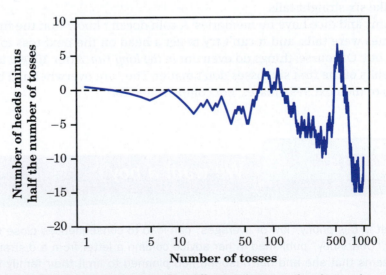

Figure 17.3 Toss a coin many times. The difference between the observed number of heads and exactly one-half the number of tosses becomes more variable as the number of tosses increases.

varies more and more as the number of tosses increases. The law of large numbers does not apply to sums or counts.

It is not uncommon to see the law of averages misstated in terms of the sums or counts rather than means or proportions. For example, assuming that the birth rates for boys and girls in the United States are equal, you may hear someone state that the *total* number of males and females in the United States should be nearly equal rather than stating that the *proportion* of males and females in the United States should be nearly equal.

Now it's your turn

17.2 Coin tossing and the law of averages. The author C. S. Lewis once wrote the following, referring to the law of averages: "If you tossed a coin a billion times, you could predict a nearly equal number of heads and tails." Is this a correct statement of the law of averages? If not, how would you rewrite the statement so that it is correct?

Personal Probabilities

Joe sits staring into his beer as his favorite baseball team, the Chicago Cubs, loses another game. The Cubbies have made some good off-season acquisitions, so let's ask Joe, "What's the chance that the Cubs will go to the World Series next year?" Joe brightens up. "Oh, about 10%," he says.

Does Joe assign probability 0.10 to the Cubs' appearing in the World Series? The outcome of next year's pennant race is certainly unpredictable, but we can't reasonably ask what would happen in many repetitions. Next year's baseball season will happen only once and will differ from all other seasons in players, weather, and many other ways. The answer to our question seems clear: if probability measures "what would happen if we did this many times," Joe's 0.10 is not a probability. Probability is based on data about many repetitions of the same random phenomenon. Joe is giving us something else, his personal judgment.

Yet we often use the term "probability" in a way that includes personal judgments of how likely it is that some event will happen. We make decisions based on these judgments—we take the bus downtown because we think the probability of finding a parking spot is low. More serious decisions also take judgments about likelihood into account. A company deciding whether to build a new plant must judge how likely it is that there will be high demand for its products three years from now when the plant is ready. Many companies express "How likely is it?" judgments as numbers— probabilities—and use these numbers in their calculations. High demand in three years, like the Cubs' winning next year's pennant, is a one-time event that doesn't fit the "do it many times" way of thinking. What is more, several company officers may give several different probabilities, reflecting differences in their individual judgment. We need another kind of probability, *personal probability*.

Key Terms

A **personal probability** of an outcome is a number between 0 and 1 that expresses an individual's judgment of how likely the outcome is.

Personal probabilities have the great advantage that they aren't limited to repeatable settings. They are useful because we base decisions on them: "I think the probability that the Patriots will win the Super Bowl is 0.75, so I'm going to bet on the game." Just remember that personal probabilities are different in kind from probabilities as "proportions in many repetitions." Because they express individual opinion, they can't be said to be right or wrong.

This is true even in a "many repetitions" setting. If Craig has a gut feeling that the probability of a head on the next toss of this coin is 0.7, that's what Craig thinks and that's all there is to it. Tossing the coin many times may show that the proportion of heads is very close to 0.5, but that's another matter. **There is no reason a person's degree of confidence in the outcome of one try must agree with the results of many tries.** We stress this because it is common to say that "personal probability" and "what happens in many trials" are somehow two interpretations of the same idea. In fact, they are quite different ideas.

Why do we even use the word "probability" for personal opinions? There are two good reasons. First, we usually do base our personal opinions on data from many trials when we have such data. Data from Buffon, Pearson, and Kerrich (Example 2) and perhaps from our own experience convince us that coins come up heads very close to half the time in many tosses. When we say that a coin has probability one-half of coming up heads *on this toss,* we are applying to a single toss a measure of the chance of a head based on what would happen in a long series of tosses. Second, personal probability and probability as long-term proportion both obey the same mathematical rules. Both kinds of probabilities are numbers between 0 and 1. We will look at more of the rules of probability in the next chapter. These rules apply to both kinds of probability.

Although "personal probability" and "what happens in many trials" are different ideas, what happens in many trials often causes us to revise our personal probability of an event. If Craig has a gut feeling that the probability of a head when he tosses a particular coin is 0.7, that's what Craig thinks. If he tosses it 20 times and gets nine heads, he may continue to believe that the probability of heads is 0.7—because personal probabilities need not agree with the results of many trials. But he may also decide to revise his personal probability downward based on what he has observed. Is there a sensible way to do this, or is this also just a matter of personal opinion?

In statistics, there are formal methods for using data to adjust personal probabilities. These are called *Bayes' procedures.* The basic rule, called *Bayes' theorem,* is attributed to the Reverend Thomas Bayes, who discussed the rule in "An Essay towards Solving a Problem in the Doctrine of Chances" published in 1764. The mathematics is somewhat complicated, and we will not discuss the details. However, the use of Bayes' procedures is becoming increasingly common among practitioners.

Probability and Risk

Once we understand that "personal judgment of how likely" and "what happens in many repetitions" are different ideas, we have a good start toward understanding why the public and the experts disagree so strongly about what is risky and what isn't. The experts use probabilities from data to describe the risk of an unpleasant event. Individuals and society, however, seem to ignore data. We worry about some risks that almost never occur while ignoring others that are much more probable.

Example 9
Asbestos in the schools

High exposure to asbestos is dangerous. Low exposure, such as that experienced by teachers and students in schools where asbestos is present in the insulation around pipes, is not very risky. The probability that a teacher who works for 30 years in a school with typical asbestos levels will get cancer from the asbestos is around 15/1,000,000. The risk of dying in a car accident during a lifetime of driving is about 15,000/1,000,000. That is, driving regularly is about 1000 times more risky than teaching in a school where asbestos is present.

Risk does not stop us from driving. Yet the much smaller risk from asbestos launched massive cleanup campaigns and a federal requirement that every school inspect for asbestos and make the findings public.

Why do we take asbestos so much more seriously than driving? Why do we worry about very unlikely threats such as tornadoes and terrorists more than we worry about heart attacks?

- We feel safer when a risk seems under our control than when we cannot control it. We are in control (or so we imagine) when we are driving, but we can't control the risk from asbestos or tornadoes or terrorists.

- It is hard to comprehend very small probabilities. Probabilities of 15 per million and 15,000 per million are both so small that our intuition cannot distinguish between them. Psychologists have shown that we generally overestimate very small risks and underestimate higher risks. Perhaps this is part of the general weakness of our intuition about how probability operates.

Statistics in Your World

What are the odds? Gamblers often express chance in terms of *odds* rather than probability. Odds of *A* to *B* in favor of an outcome means that the probability of that outcome is $A/(A+B)$. So "odds of 5 to 1 in favor of an outcome" is another way of saying "probability 5/6." A probability is always between 0 and 1, but odds range from 0 to infinity. Although odds are mainly used in gambling, they give us a way to make very small probabilities clearer. "Odds of 999 to 1" may be easier to understand than "probability 0.999."

- The probabilities for risks like asbestos in the schools are not as certain as probabilities for tossing coins. They must be estimated by experts from complicated statistical studies. Perhaps it is safest to suspect that the experts may have underestimated the level of risk.

Our reactions to risk depend on more than probability, even if our personal probabilities are higher than the experts' data-based probabilities. We are influenced by our psychological makeup and by social standards. As one writer noted, "Few of us would leave a baby sleeping alone in a house while we drove off on a 10-minute errand, even though car-crash risks are much greater than home risks."

Chapter 17: Statistics in Summary

- Some things in the world, both natural and of human design, are **random.** That is, their outcomes have a clear pattern in very many repetitions even though the outcome of any one trial is unpredictable.

- **Probability** describes the long-term regularity of random phenomena. The probability of an outcome is the proportion of very many repetitions on which that outcome occurs. A probability is a number between 0 (the outcome never occurs) and 1 (always occurs). We emphasize this kind of probability because it is based on data.

- Probabilities describe only what happens in the long run. Short runs of random phenomena like tossing coins or shooting a basketball often don't look random to us because they do not show the regularity that in fact emerges only in very many repetitions.

- **Personal probabilities** express an individual's personal judgment of how likely outcomes are. Personal probabilities are also numbers between 0 and 1. Different people can have different personal probabilities, and a personal probability does not need to agree with a proportion based on data about similar cases.

This chapter summary will help you evaluate the Case Study.

Link It

This chapter begins our study of the mathematics of chance or "probability." The important fact is that random phenomena are unpredictable in the short run but have a regular and predictable behavior in the long run.

The long-run behavior of random phenomena will help us understand both why and in what way we can trust random samples and randomized comparative experiments, the subjects of Chapters 2 through 6. It is the key to generalizing what we learn from data produced by random samples and randomized comparative experiments to some wider universe or population. We will study how this is done in Part IV. As a first step in this direction, we will look more carefully at the basic rules of probability in the next chapter.

Case Study Evaluated

Use what you have learned in this chapter to evaluate the Case Study. Start by reviewing the Chapter Summary. Then answer each of the following questions in complete sentences. Be sure to communicate clearly enough for any of your classmates to understand what you are saying.

In the Case Study described at the beginning of this chapter, you were told that the odds of all three members of the Gardner family having the same birthday is about one in 133,000. And a statistician can show that if birth dates are random and independent, the chance that three people, selected at random, all have the same birthday is, in fact, about one in 133,000.

1. Go to the most recent Census Bureau data on families (online at www.census.gov /data/tables/2016/demo/families/cps-2016.html) and look under *Table H1. Households By Type and Tenure of Householder for Selected Characteristics: 2016.* How many family households in the United States consist of a married couple with three members?

2. Assume that the families you found in the previous question all consist of a married couple with one child. Explain why the probability that there is *some 3-person family* in the United States in which the parents and child all have the same birthday is much larger than the 1 in 133,000 probability that in one randomly selected 3-person family the parents and child all have the same birthday.

3. Write a paragraph discussing whether the "surprising" coincidence described in the Case Study that began this chapter is as surprising as it might first appear.

In this chapter you:

- Learned some fundamental concepts about chance behavior such as the notion of randomness.
- Interpreted probabilities like 1 in 133,000 as the proportion of times an outcome would occur in a very long series of trials.
- Identified some common "myths" concerning chance behavior such as misinterpretations of the law of averages and the notion of the "hot hand" in sports.

| macmillan learning **Online Resources**

■ The first half of the Snapshots video, *Probability*, introduces the concepts of randomness and probability in the context of weather forecasts.

Check the Basics

For Exercise 17.1, see page 411; for Exercise 17.2, see page 415.

17.3 Randomness. Random phenomena have which of the following characteristics?

(a) They must be natural events. Man-made events cannot be random.

(b) They exhibit a clear pattern in very many repetitions, although any one trial of the phenomenon is unpredictable.

(c) Future outcomes must compensate for an imbalance that may occur in the short run, thus preserving the law of averages.

(d) They are completely unpredictable; that is, they display no clear pattern no matter how often the phenomenon is repeated.

17.4 Probability. The probability of a specific outcome of a random phenomenon is

(a) the number of times it occurs in very many repetitions of the phenomenon.

(b) the number repetitions of the phenomenon it takes for the outcome to first occur.

(c) the proportion of times it occurs in very many repetitions of the phenomenon.

(d) the ratio of the number of times it occurs to the number of times it does not occur in very many repetitions of the phenomenon.

17.5 Probability. Which of the following is true of probability?

(a) It is a number between 0 and 1.

(b) A probability of 0 means the outcome never occurs.

(c) A probability of 1 means the outcome always occurs.

(d) All of the above are true.

17.6 Probability. I toss a coin 1000 times and observe the outcome "heads" 481 times. Which of the following can be concluded from this result?

(a) This is suspicious because we should observe exactly 500 heads if the coin is tossed 1000 times.

(b) The probability of heads is approximately 481.

(c) Our best estimate of the probability of heads is 0.481, but a longer sequence of flips should yield a better estimate of this probability.

(d) If we flip the coin several more times, we will get more heads than tails in order to balance out the fact that the first thousand flips had too few heads.

17.7 Personal probabilities. Which of the following is true of a personal probability about the outcome of a phenomenon?

(a) It expresses an individual's judgment of how likely an outcome is.

(b) It can be any number because personal probabilities need not be restricted to values between 0 and 1.

(c) It must closely agree with the proportion of times the outcome would occur if the phenomenon was repeated a large number of times.

(d) Very small values indicate *strong* disagreement with the probability most people would assign to the outcome, and very large values indicate *strong* agreement with the probability most people would assign to the outcome.

Chapter 17 Exercises

17.8 Nickels spinning. Hold a nickel upright on its edge under your forefinger on a hard surface, then snap it with your other forefinger so that it spins for some time before falling. Based on 50 spins, estimate the probability of heads.

17.9 Nickels falling over. You may feel that it is obvious that the probability of a head in tossing a coin is about 1-in-2 because the coin has two faces. Such opinions are not always correct. The previous exercise asked you to spin a nickel rather than toss it—that changes the probability of a head. Now try another variation. Stand a nickel on edge on a hard, flat surface. Pound the surface with your hand so that the nickel falls over. What is the probability that it falls with heads upward? Make at least 50 trials to estimate the probability of a head.

17.10 Random digits. The table of random digits (Table A) was produced by a random mechanism that gives each digit probability 0.1 of being a 0. What proportion of the 400 digits in lines 120 to 129 in the table are 0s? This proportion is an estimate, based on 400 repetitions, of the true probability, which in this case is known to be 0.1.

17.11 How many tosses to get a head? When we toss a penny, experience shows that the probability (long-term proportion) of a head is close to 1-in-2. Suppose now that we toss the penny repeatedly until we get a head. What is the probability that the first head comes up in an even number of tosses (two, four, six, and so on)? To find out, repeat this experiment 50 times, and keep a record of the number of tosses needed to get the first head on each of your 50 trials.

(a) From your experiment, estimate the probability of the first head on the second toss. What value should we expect this probability to have?

(b) Use your results to estimate the probability that the first head appears on an even-numbered toss.

17.12 Tossing a thumbtack. Toss a thumbtack on a hard surface 50 times. How many times did

it land with the point up? What is the approximate probability of landing point up?

17.13 Rolling dice. Roll a pair of dice 100 times. How many times did you roll a 5? What is the approximate probability of rolling a 5?

17.14 Four-of-a-kind. You read in a book on poker that the probability of being dealt four-of-a-kind (a hand containing four cards of the same value, such as four kings) in a five-card poker hand is about 0.00024. Explain in simple language what this means.

17.15 From words to probabilities. Probability is a measure of how likely an event is to occur. Match one of the probabilities that follow with each statement of likelihood given. (The probability is usually a more exact measure of likelihood than is the verbal statement.)

$$0 \quad 0.01 \quad 0.4 \quad 0.6 \quad 0.99 \quad 1$$

(a) This event is impossible. It can never occur.

(b) This event is certain. It will occur on every trial.

(c) This event is very unlikely, but it will occur once in a while in a long sequence of trials.

(d) This event will occur more often than not.

17.16 Winning a baseball game. Over the period from 1965 to 2018, the champions of baseball's two major leagues won 63% of their home games during the regular season. At the end of each season, the two league champions meet in the baseball World Series. Would you use the results from the regular season to assign probability 0.63 to the event that the home team wins a World Series game? Explain your answer.

17.17 Will you have an accident? The probability that a randomly chosen driver will be involved in an accident in the next year is about 0.3. This is based on the proportion of millions of drivers who have accidents. "Accident" includes things like crumpling a fender in your own driveway, not just highway accidents.

(a) What do you think is your own probability of being in an accident in the next year? This is a personal probability.

(b) Give some reasons your personal probability might be a more accurate prediction of your "true chance" of having an accident than the probability for a random driver.

(c) Almost everyone says that their personal probability is lower than the random driver probability. Why do you think this is true?

17.18 Marital status. Based on 2018 data, the probability that a randomly chosen woman over 64 years of age is divorced is about 0.14. This probability is a long-run proportion based on all the millions of women over 64. Let's suppose that the proportion stays at 0.14 for the next 45 years. Bridget is now 20 years old and is not married.

(a) Bridget thinks her own chances of being divorced after age 64 are about 5%. Explain why this is a personal probability.

(b) Give some good reasons Bridget's personal probability might differ from the proportion of all women over 64 who are divorced.

(c) You are a government official charged with looking into the impact of the Social Security system on retirement-aged divorced women. You care only about the probability 0.14, not about anyone's personal probability. Why?

17.19 Personal probability versus data. Give an example in which you would rely on a probability found as a long-term proportion from data on many trials. Give an example in which you would rely on your own personal probability.

17.20 Personal probability? When there are few data, we often fall back on personal probability. There had been just 24 space shuttle launches, all successful, before the *Challenger* disaster in January 1986. The shuttle program management thought the chances of such a failure were only 1 in 100,000.

(a) Suppose 1 in 100,000 is a correct estimate of the chance of such a failure. If a shuttle was launched every day, about how many failures would one expect in 300 years?

(b) Give some reasons such an estimate is likely to be too optimistic.

17.21 Personal random numbers? Ask several of your friends (at least 10 people) to choose a four-digit number "at random." How many of the numbers chosen start with 1 or 2? How many start with 8 or 9? (There is strong evidence that people in general tend to choose numbers starting with low digits.)

17.22 Playing Pick 4. The Pick 4 games in many state lotteries announce a four-digit winning number each day. The winning number is essentially a four-digit group from a table of random digits. You win if your choice matches the winning digits, in exact order. The winnings are divided among all players who matched the winning digits. That suggests a way to get an edge.

(a) The winning number might be, for example, either 2873 or 9999. Explain why these two outcomes have exactly the same probability. (It is 1 in 10,000.)

(b) If you asked many people which outcome is more likely to be the randomly chosen winning number, most would favor one of them. Use the information in this chapter to say which one and to explain why. If you choose a number that people think is unlikely, you have the same chance to win, but you will win a larger amount because few other people will choose your number.

17.23 Surprising? During the Michigan versus Ohio State football game in 2018, news media reported that Jim Harbaugh and Urban Meyer, the head coaches of Michigan and Ohio State, respectively, were born in the same hospital in Toledo, Ohio. That a pair of coaches from two arch rivals were also born in the same hospital was reported as an extraordinarily improbable event. Should this fact (that they are the head coaches of two famous arch rivals and both were born in the same hospital) surprise you? Explain your answer.

17.24 An eerie coincidence? A September 13, 2011, *New York Post* article reported that the first three winners at Belmont Park were horses wearing the numbers 9, 1, 1 on the tenth-year anniversary of the 9/11 attacks on America. Should this fact surprise you? Explain your answer.

17.25 Curry's free throws. The basketball player Stephen Curry is the all-time career free-throw shooter among active players. He makes 90.4% of his free throws. In today's game, Curry misses his first two free throws. The TV commentator says, "Curry's technique looks out of rhythm today." Explain why the claim that Curry's technique has deteriorated is not justified.

17.26 In the long run. Probability works not by compensating for imbalances but by overwhelming them. Suppose that the first 10 tosses of a coin give 10 tails and that tosses after that are exactly half heads and half tails. (Exact balance is unlikely, but the example illustrates how the first 10 outcomes are swamped by later outcomes.) What is the proportion of heads after the first 10 tosses? What is the proportion of heads after 100 tosses if half of the last 90 produce heads (45 heads)? What is the proportion of heads after 1000 tosses if half of the last 990 produce heads? What is the proportion of heads after 10,000 tosses if half of the last 9990 produce heads?

17.27 The "law of averages." The baseball player Jose Altuve gets a hit about 31.6% of the time over an entire season. After he has failed to hit safely in nine straight at-bats, the TV commentator says, "Jose is due for a hit by the law of averages." Is that right? Why?

17.28 Snow coming. A meteorologist, predicting below-average snowfall this winter, says, "First, in looking at the past few winters, there has been above-average snowfall. Even though we are not supposed to use the law of averages, we are due." Do you think that "due by the law of averages" makes sense in talking about the weather? Explain.

17.29 An unenlightened gambler.

(a) A gambler knows that red and black are equally likely to occur on each spin of a roulette wheel. He observes five consecutive reds occur and bets heavily on black on the next spin. Asked why, he explains that black is "due by the law of averages." Explain to the gambler what is wrong with this reasoning.

(b) After listening to you explain why red and black are still equally likely after five reds on the roulette wheel, the gambler moves to a poker game. He is dealt five straight red cards. He remembers what you said and assumes that the next card dealt in the same hand is equally likely to be red or black. Is the gambler right or wrong, and why?

17.30 Reacting to risks. The probability of dying if you play high school football is about 10 per million each year you play. The risk of getting cancer from asbestos if you attend a school in which asbestos is present for 10 years is about 5 per million. If we ban asbestos from schools, should we also ban high school football? Briefly explain your position.

17.31 Reacting to risks. National newspapers such as *USA Today* and the *New York Times* carry many more stories about deaths from airplane crashes than about deaths from motor vehicle crashes. Motor vehicle accidents killed about 32,700 people in the United States in 2013. Crashes of all scheduled air carriers worldwide, including commuter carriers, killed 325 people in 2016, and nobody died in a crash on a U.S.–certificated scheduled airline operating anywhere in the world.

(a) Why do the news media give more attention to airplane crashes?

(b) How does news coverage help explain why many people consider flying more dangerous than driving?

17.32 What probability doesn't say. The probability of a head in tossing a coin is 1-in-2. This means that as we make more tosses, the *proportion* of heads will eventually get close to 0.5. It does not mean that the *count* of heads will get close to one-half the number of tosses. To see why, imagine that the proportion of heads is 0.49 in 100 tosses, 0.493 in 1000 tosses, 0.4969 in 10,000 tosses, and 0.49926 in 100,000 tosses of a coin. How many heads came up in each set of tosses? How close is the number of heads to half the number of tosses?

Exploring the Web

Access these exercises on the text website: macmillanlearning.com/scc10e.

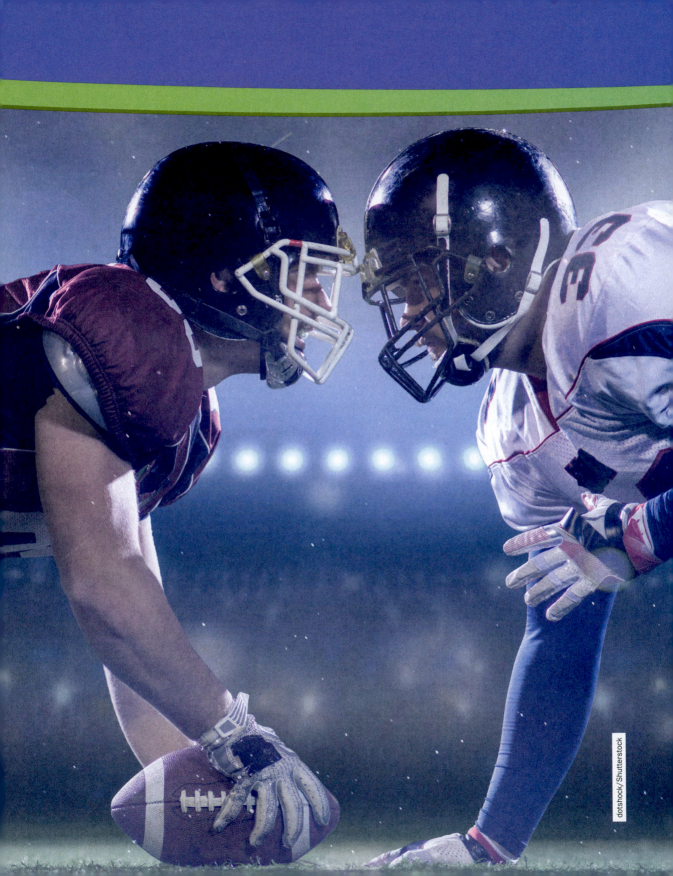

Probability Models

In this chapter you will:

- Describe random phenomena by listing the possible outcomes and their associated probabilities.

- Learn the basic rules that probabilities must obey.

- Understand the relation between the odds of an event and the probability of an event.

Shortly after the Philadelphia Eagles won Super Bowl 52, websites were already posting the probabilities of winning Super Bowl 53 for the various National Football League teams. These probabilities were updated regularly. Table 18.1 lists the probabilities posted on one website prior to the start of the 2018 season. These are perhaps best interpreted as personal probabilities. They are likely to change as the 2018 season progresses because the players on teams will change due to trades and injuries.

Because these are personal probabilities, are there any restrictions on the choices for the probabilities? In this chapter, we will answer this question. We will learn that probabilities must obey certain rules in order to make sense. By the end of this chapter, you will be able to assess whether the probabilities in Table 18.1 make sense.

Probability Models

Choose a woman aged 25 to 29 years old at random and record her marital status. "At random" means that we give every such woman the same chance to be the one we choose. That is, we choose a random sample of size 1. The probability of any marital status is just the proportion of all women aged 25

Table 18.1 Probabilities of winning Super Bowl 53

Team	Probability	Team	Probability
New England Patriots	1/7	Los Angeles Chargers	1/34
Los Angeles Rams	1/11	Baltimore Ravens	1/34
Pittsburgh Steelers	1/11	Indianapolis Colts	1/34
Philadelphia Eagles	1/11	Jacksonville Jaguars	1/34
Minnesota Vikings	1/13	Tennessee Titans	1/41
Green Bay Packers	1/13	Detroit Lions	1/41
San Francisco 49ers	1/15	New York Giants	1/51
Houston Texans	1/17	Tampa Bay Buccaneers	1/51
New Orleans Saints	1/21	Arizona Cardinals	1/67
Dallas Cowboys	1/21	Chicago Bears	1/81
Denver Broncos	1/21	Washington Redskins	1/81
Atlanta Falcons	1/21	Cincinnati Bengals	1/81
Oakland Raiders	1/26	Buffalo Bills	1/81
Seattle Seahawks	1/29	Miami Dolphins	1/101
Carolina Panthers	1/29	Cleveland Browns	1/101
Kansas City Chiefs	1/34	New York Jets	1/101

Data from www.betvega.com/super-bowl-odds/ (opening probabilities for the 2018 season).

to 29 who have that status—if we chose many women, this is the proportion we would get. Here is the set of probabilities:

Marital status:	Never married	Married	Widowed	Divorced
Probability:	0.580	0.388	0.002	0.030

Statistics in Your World

Politically correct In 1950, the Russian mathematician B. V. Gnedenko (1912–1995) wrote a book, *The Theory of Probability,* that was popular around the world. The introduction contains a mystifying paragraph that begins, "We note that the entire development of probability theory shows evidence of how its concepts and ideas were crystallized in a severe struggle between materialistic and idealistic conceptions." It turns out that "materialistic" is jargon for "Marxist-Leninist." It was good for the health of Russian scientists in the Stalin era to add such statements to their books.

This table gives a *probability model* for drawing a young woman at random and finding out her marital status. It tells us what are the possible outcomes (there are only four) and it assigns probabilities to these outcomes. The probabilities here are the proportions of all women in each marital class. That makes it clear that the probability that a woman is not married is just the sum of the probabilities of the three classes of unmarried women:

$$P(\text{not married}) = P(\text{never married}) + P(\text{widowed})$$
$$+ P(\text{divorced})$$
$$= 0.580 + 0.002 + 0.030 = 0.612$$

As a shorthand, we often write $P(\text{not married})$ for "the probability that the woman we choose is not married." You see that our model does more than assign a probability to each individual outcome—we can find the probability of any collection of outcomes by adding up individual outcome probabilities.

Probability Rules

Because the probabilities in this example are just the proportions of all women who have each marital status, they follow rules that say how proportions behave. Here are some basic rules that any probability model must obey:

A. **Any probability is a number between 0 and 1.** Any proportion is a number between 0 and 1, so any probability is also a number between 0 and 1. An event with probability 0 never occurs, and an event with probability 1 occurs on every trial. An event with probability 0.5 occurs in half the trials in the long run.

B. **The sum of probabilities of all possible outcomes must have probability 1.** Because some outcome must occur on every trial, the sum of the probabilities for all possible outcomes must be exactly 1.

C. **The probability that an event does not occur is 1 minus the probability that the event does occur.** If an event occurs in (say) 70% of all trials, it fails to occur in the other 30%. The probability that an event occurs and the probability that it does not occur always add to 100%, or 1.

D. **If two events have no outcomes in common, the probability that one or the other occurs is the sum of their individual probabilities.** If one event occurs in 40% of all trials, a different event occurs in 25% of all trials, and the two can never occur together, then one or the other occurs on 65% of all trials because $40\% + 25\% = 65\%$.

Example 1   *Marital status of young women*

Look again at the probabilities for the marital status of young women. Each of the four probabilities is a number between 0 and 1. Their sum is

$$0.5808 + 0.388 + 0.002 + 0.030 = 1$$

Rob Marmion/Shutterstock

This assignment of probabilities satisfies Rules A and B. **Any assignment of probabilities to all individual outcomes that satisfies Rules A and B is legitimate.** That is, it makes sense as a set of probabilities. Rules C and D are then automatically true. Here is an example of the use of Rule C.

The probability that the woman we draw at random is not married is, by Rule C,

$$P(\text{not married}) = 1 - P(\text{married})$$
$$= 1 - 0.388 = 0.612$$

That is, if 38.8% are married, then the remaining 61.2% are not married. Rule D says that you can also find the probability that a randomly selected woman is not married by adding the probabilities of the three distinct ways of being not married, as we did earlier. This gives the same result.

Example 2

Rolling two dice

Rolling two dice is a common way to lose money in casinos. There are 36 possible outcomes when we roll two dice and record the up-faces in order (first die, second die). Figure 18.1 displays these outcomes. What probabilities should we assign?

Casino dice are carefully made. Their spots are not hollowed out, which would give the faces different weights, but are filled with white plastic of the same density as the red plastic of the body. For casino dice, it is reasonable to assign the same probability to each of the 36 outcomes in Figure 18.1.

Arthur-studio10/Shutterstock

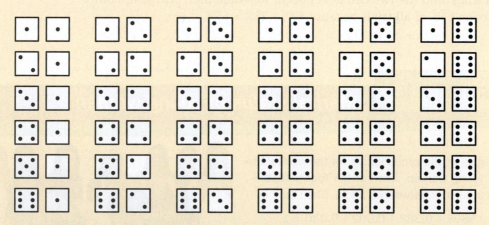

Figure 18.1 The 36 possible outcomes from rolling two dice, Example 2.

Because these 36 probabilities must have sum 1 (Rule B), each outcome must have probability 1/36, or 1-in-36.

We are interested in the sum of the spots on the up-faces of the dice. What is the probability that this sum is 5? The event "roll a 5" contains four outcomes, and its probability is the sum of the probabilities of these outcomes:

$$P(\text{roll a 5}) = P\left(\boxed{\,\cdot\,}\ \boxed{\vdots\vdots}\right) + P\left(\boxed{\therefore}\ \boxed{\because}\right) + P\left(\boxed{\because}\ \boxed{\therefore}\right) + P\left(\boxed{\vdots\vdots}\ \boxed{\,\cdot\,}\right)$$

$$= \frac{1}{36} + \frac{1}{36} + \frac{1}{36} + \frac{1}{36}$$

$$= \frac{4}{36} = 0.111$$

Now it's your turn

18.1 Rolling dice. Suppose you roll two casino dice, as in Example 2. What is the probability that the sum of the spots on the up-faces is 7? 11? 7 or 11?

The rules tell us only what probability models *make sense*. They don't tell us whether the probabilities are *correct*, that is, whether they describe what actually happens in the long run. The probabilities in Example 2 are correct for casino dice. Inexpensive dice with hollowed-out spots are not balanced, and this probability model does not describe their behavior.

What about personal probabilities? Because they are personal, can't they be anything you want them to be? If your personal probabilities don't obey Rules A and B, you are entitled to your opinion, but we can say that your personal probabilities are **incoherent.** That is, they don't go together in a way that makes sense. So we usually insist that personal probabilities for all the outcomes of a random phenomenon obey Rules A and B. That is, the same rules govern both kinds of probability. For example, if you believe that the Philadelphia Eagles, the Los Angeles Rams, and the New England Patriots each have probability of 0.4 of winning Super Bowl LIII, your personal probabilities would not obey Rule B.

You now have enough information to read the "What's the verdict?" story on page 441, and to answer the two questions.

Probability and Odds

Speaking of sports, many newspapers, magazines, and websites often give probabilities as betting odds. For example, at the end of February 2019, one website gave the betting odds of 4 to 9 that the Golden State Warriors would win the 2018–19 National Basketball Association finals. This means that a bet of $9 will pay you $4 if the team wins and cost you the $9 you bet if the team loses. If this is a fair bet, you expect that in the long run you should break even, winning as much money as you lose. In particular, if you bet $4 + 9 = 13$ times, on average you should win $4 to 9 times and lose $9 the other 4 times. Thus, on average, if you bet $4 + 9 = 13$ times, you win 9 of those bets. Odds of 4 to 9 therefore represent a probability of $9/(4 + 9) = 9/13$ of winning.

More generally, betting odds take the form "Y to Z." Odds of Y to Z represent a probability of $Z/(Y + Z)$ of winning.

As another example, on the web one can find the current odds of winning the next Super Bowl for each NFL team. We found a list of such odds at www.betvega.com/super-bowl-odds/. When we checked this website on December 5, 2018, the odds that the Los Angeles Rams would win Super Bowl 53 were 3-to-1. This corresponds to a probability of winning of $1/(3 + 1) = 1/4$. The odds that the New York Jets would win Super Bowl 53 were 5000 to 1. This corresponds to a probability of $1/(5000 + 1) = 1/5001$.

Key Terms

The **sampling distribution** of a statistic tells us what values the statistic takes in repeated samples from the same population and how often it takes those values.

We think of a sampling distribution as assigning probabilities to the values the statistic can take. Because there are usually many possible values, sampling distributions are often described by a **density curve** such as a Normal curve.

Probability Models for Sampling

Choosing a random sample from a population and calculating a statistic such as the sample proportion is certainly a random phenomenon. The *distribution* of the statistic tells us what values it can take and how often it takes those values. That sounds a lot like a probability model.

Example 3   *A sampling distribution*

Take a simple random sample of 1004 adults. Ask each whether they feel childhood vaccinations are very important. The proportion who say Yes

$$\hat{p} = \frac{\text{number who say Yes}}{1004}$$

is the sample proportion $\hat{p}$. Do this 1000 times and collect the 1000 sample proportions $\hat{p}$ from the 1000 samples. The histogram in Figure 18.2 shows the distribution of 1000 sample proportions when the truth about the population is that 71% would agree that childhood vaccinations are important. The results of random sampling are of course random: we can't predict the outcome of one sample, but the figure shows that the outcomes of many samples have a regular pattern.

This repetition reminds us that the regular pattern of repeated random samples is one of the big ideas of statistics. The Normal curve in the figure is a good approximation to the histogram. The histogram is the result of these particular 1000 simple random samples (SRSs). Think of the Normal curve as the idealized pattern we would get if we kept on taking SRSs from this population forever. That's exactly the idea of probability—the pattern we would see in the very long run. *The Normal curve assigns probabilities to sample proportions computed from random samples.*

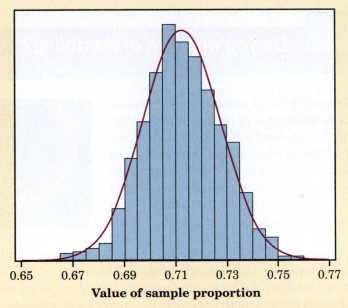

Figure 18.2 The sampling distribution of a sample proportion $\hat{p}$ from simple random samples of size 1004 drawn from a population in which 71% of the members would give positive answers, Example 3. The histogram shows the distribution from 1000 samples. The Normal curve is the ideal pattern that describes the results of a very large number of samples. (This figure was created using the Minitab software package.)

This Normal curve has mean 0.710 and standard deviation about 0.014. The "95" part of the 68–95–99.7 rule says that 95% of all samples will give a $\hat{p}$ falling within 2 standard deviations of the mean. That's within 0.028 of 0.710, or between 0.682 and 0.738. We now have more concise language for this fact: the *probability* is 0.95 that between 68.2% and 73.8% of the people in a sample will say Yes. The word "probability" says we are talking about what would happen in the long run, in very many samples.

We note that of the 1000 SRSs, 94.3% of the sample proportions were between 0.682 and 0.738, which agrees quite well with the calculations based on the Normal curve. This confirms our assertion that the Normal curve is a good approximation to the histogram in Figure 18.2.

A statistic from a large sample has a great many possible values. Assigning a probability to each individual outcome worked well for four marital classes or 36 outcomes of rolling two dice but is awkward when there are thousands of possible outcomes. Example 3 uses a different approach: assign probabilities to intervals of outcomes by using areas under a Normal density curve. Density curves have area of 1 underneath them, which lines up nicely with total probability 1. The total area under the Normal curve in Figure 18.2 is 1, and the area between 0.682 and 0.738 is 0.95, which is the probability that a sample gives a result in that interval. When a Normal curve assigns probabilities, you can calculate probabilities from the 68–95–99.7 rule or from Table B of percentiles of Normal distributions. These probabilities satisfy Rules A through D.

Example 4   *Do you approve of gambling?*

An opinion poll asks an SRS of 501 teens, "Generally speaking, do you approve or disapprove of legal gambling or betting?" Suppose that, in fact, exactly 50% of all teens would say Approve if asked. (This is close to what polls show to be true.) The poll's statisticians tell us that the sample proportion who say Approve will vary in repeated samples according to a Normal distribution with mean 0.5 and standard deviation about 0.022. This is the *sampling distribution* of the sample proportion $\hat{p}$.

logoboom/Deposit Photos

The 68–95–99.7 rule says that the probability is 0.16 that the sample proportion is more than one standard deviation below the mean. The standard deviation in this case is 0.022 and the mean is 0.50, so this implies that the probability is 0.16 that the poll gets a sample in which fewer than 47.8% say Approve. Figure 18.3 shows how to get this result from the Normal curve of the sampling distribution.

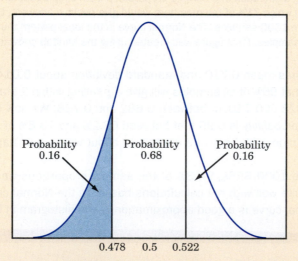

Probability 0.16

Probability 0.68

Probability 0.16

0.478   0.5   0.522

Figure 18.3 The Normal sampling distribution, Example 4. Because 0.478 is one standard deviation below the mean, the area under the curve to the left of 0.478 is 0.16.

Now it's your turn

18.2 Teen opinion poll. Refer to Example 4. Using the 68–95–99.7 rule, what is the probability that fewer than 54.4% say Yes?

Example 5

*Using Normal percentiles**

What is the probability that the opinion poll in Example 4 will get a sample in which 52% or more say Yes? Because 0.52 is not 1, 2, or 3 standard deviations away from the mean, we can't use the 68–95–99.7 rule. We will use Table B of percentiles of Normal distributions.

To use Table B, first turn the outcome $\hat{p} = 0.52$ into a standard score by subtracting the mean of the distribution and dividing by its standard deviation:

$$\frac{0.52 - 0.50}{0.022} = 0.91$$

which we will round off to 0.9 in order to use Table B. Now look in Table B. A standard score of 0.9 is the 81.59 percentile of a Normal distribution. This means that the probability is 0.8159 that the poll gets a smaller result. By Rule C (or just the fact that the total area under the curve is 1), this leaves probability 0.1841 for outcomes with 52% or more answering Yes. Figure 18.4 shows the probabilities as areas under the Normal curve.

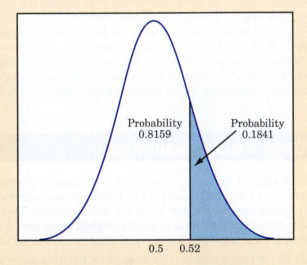

Figure 18.4 The Normal sampling distribution, Example 5. The outcome 0.52 has standard score 0.9, so Table B tells us that the area under the curve to the left of 0.52 is 0.8159.

*Example 5 is optional.

Chapter 18: Statistics in Summary

- A **probability model** describes a random phenomenon by telling what outcomes are possible and how to assign probabilities to them.

- There are two simple ways to give a probability model. The first assigns a probability to each individual outcome. These probabilities must be numbers between 0 and 1 (Rule A), and they must add to exactly 1 (Rule B). To find the probability of any **event,** add the probabilities of the outcomes that make up the event.

- The second kind of probability model assigns probabilities as areas under a **density curve,** such as a Normal curve. The total probability is 1 because the total area under the curve is 1. This kind of probability model is often used to describe the **sampling distribution** of a statistic. This is the pattern of values of the statistic in many samples from the same population.

- All legitimate assignments of probability, whether data based or personal, obey the same **probability rules.** So the mathematics of probability is always the same.

- **Odds** of Y to Z that an event occurs correspond to a probability of $Z/(Y + Z)$.

This chapter summary will help you evaluate the Case Study.

Link It

This chapter continues the discussion of probability that we began in Chapter 17. Here, we examine the formal mathematics of probability, embodied in probability models and probability rules. Probability models and rules provide the tools for describing and predicting the long-run behavior of random phenomena.

In this chapter, we also begin to formalize this process, first mentioned in Chapter 3, of using a statistic to estimate an unknown parameter. In particular, the sampling distribution will be the "probabilistic" tool we use to generalize from data produced by random samples and randomized comparative experiments to some wider population. Exactly how we do this will be the subject of Part IV.

Case Study Evaluated

Look again at Table 18.1, discussed in the Case Study that opened the chapter. Use what you have learned in this chapter to evaluate the Case Study. Start by reviewing the information in the Chapter Summary. Then answer the following questions about the Case Study in complete sentences. Be sure to communicate clearly enough for any of your classmates to understand what you are saying.

1. Do the probabilities in this table follow the rules given on page 427?

2. On some websites, the probabilities are given as betting odds of winning. If you are a bookie setting betting odds, should you set the odds so the corresponding probabilities, after converting odds to probabilities, sum to greater than 1, exactly 1, or less than 1? Discuss. (*Hint*: If you are a bookie, will you make more money if you set the betting odds so that the probabilities of winning are too high or are too low?)

In this chapter you have:

- Described random phenomena using a probability model that lists the possible outcomes and their associated probabilities.
- Learned the basic rules that probabilities must obey.
- Understood the relation between the odds of an event and the probability of an event.

macmillan learning **Online Resources**

- The Snapshots video, *Probability*, introduces the concepts of randomness, probability, and probability models in the context of some examples.
- The StatBoards video, *The Four Basic Probability Rules*, discusses the basic probability rules in the context of some simple examples.

Check the Basics

For Exercise 18.1, see page 429; for Exercise 18.2, see page 433.

18.3 **Probability models.** A probability model describes a random phenomenon by telling us which of the following?

(a) Whether we are using data-based or personal probabilities.

(b) What outcomes are possible and how to assign probabilities to these outcomes.

(c) Whether the probabilities of all outcomes sum to 1 or sum to a number different than 1.

(d) All of the above.

18.4 **Probability models.** Which of the following is true of any legitimate probability model?

(a) The probabilities of the individual outcomes must be numbers between 0 and 1, and they must sum to no more than 1.

(b) The probabilities of the individual outcomes must be numbers between 0 and 1, and they must sum to at least 1.

(c) The probabilities of the individual outcomes must be numbers between 0 and 1, and they must sum to exactly 1.

(d) Probabilities can be computed using the Normal curve.

18.5 **Density curves.** Which of the following is true of density curves?

(a) Areas under a density curve determine probabilities of outcomes.

(b) The total area under a density curve is 1.

(c) The Normal curve is a density curve.

(d) All of the above are true.

18.6 **Probability rules.** To find the probability of any event,

(a) add up the probabilities of the outcomes that make up the event.

(b) use the probability of the outcome that best approximates the event.

(c) assign it a random, but plausible, value between 0 and 1.

(d) average together the personal probabilities of several experts.

18.7 Sampling distributions. The sampling distribution of a statistic is

(a) the method of sampling used to obtain the data from which the statistic is computed.

(b) the possible methods of computing a statistic from the data.

(c) the pattern of the data from which the statistic is computed.

(d) the pattern of values of the statistic in many samples from the same population.

Chapter 18 Exercises

18.8 Moving up. Sociologists studying social mobility in the United States find that the probability that someone who began their career in the bottom 10% of earnings remains in the bottom 10% 15 years later is 0.59. What is the probability that such a person moves to one of the higher income classes 15 years later?

18.9 Causes of death. Government data assign a single cause for each death that occurs in the United States. Data from 2016 show that the probability is 0.23 that a randomly chosen death was due to heart disease and 0.22 that it was due to cancer. What is the probability that a death was due either to heart disease or to cancer? What is the probability that the death was due to some other cause?

18.10 Land in Canada. Choose an acre of land in Canada at random. The probability is 0.38 that it is forest and 0.07 that it is agricultural.

(a) What is the probability that the acre chosen is not forested?

(b) What is the probability that it is either forest or agricultural?

(c) What is the probability that a randomly chosen acre in Canada is something other than forest or agricultural?

18.11 Our favorite president. A 2018 poll by the Pew Research Survey Center interviewed a random sample of 2002 adult Americans. Those in the sample were asked which president has done the best job in their lifetime. Here are the results:

Outcome	Probability
Barack Obama	0.31
Ronald Reagan	0.21
Bill Clinton	0.13
Donald Trump	0.10
Someone else	?

These proportions are probabilities for the random phenomenon of choosing an adult American at random and asking her or his opinion.

(a) What must be the probability that the person chosen selects someone other than Barack Obama, Ronald Reagan, Bill Clinton, or Donald Trump? Why?

(b) The event "I would select either Barack Obama or Ronald Reagan" contains the first two outcomes. What is its probability?

18.12 Rolling a die. Figure 18.5 displays several assignments of probabilities to the six faces of a die. We can learn which assignment is actually *correct* for a particular die only by rolling the die many times. However, some of the assignments are not *legitimate* assignments of probability. That is, they do not obey the rules. Which are legitimate and which are not? In the

Probability

Outcome	Model 1	Model 2	Model 3	Model 4
⚀	1/7	1/3	1/3	1
⚁	1/7	1/6	1/6	1
⚂	1/7	1/6	1/6	2
⚃	1/7	0	1/6	1
⚄	1/7	1/6	1/6	1
⚅	1/7	1/6	1/6	2

Figure 18.5 Four probability models for rolling a die, Exercise 18.12.

case of the illegitimate models, explain what is wrong.

18.13 Political views of college students. Select a first-year college student at random and ask how they would characterize their political views. Here are the probabilities, based on proportions from a large sample survey in 2016 of first-year students:

Political view:	Far left	Liberal	Middle of the road	Conservative	Far right
Probability:	0.04	0.31	0.43	0.20	0.02

(a) What is the sum of these probabilities? Why do you expect the sum to have this value?

(b) What is the probability that a randomly chosen first-year college student had political views that are not far left?

(c) What is the probability that a first-year student had political views that are either conservative or far right?

18.14 Tetrahedral dice. Psychologists sometimes use tetrahedral dice to study our intuition about chance behavior. A tetrahedron (Figure 18.6) is a pyramid with four faces, each a triangle with all sides equal in length. Label the four faces of a tetrahedral die with one, two, three, and four spots. Give a probability model for rolling such a die and recording the number of spots on the down-face. Explain why you think your model is at least close to correct.

18.15 Birth order. A couple plan to have three children. There are eight possible arrangements of girls and boys. For example, GGB means the first two children are girls and the third

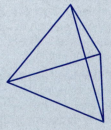

Figure 18.6 A tetrahedron. Exercises 18.14 and 18.16 concern dice with this shape.

child is a boy. All eight arrangements are (approximately) equally likely.

(a) Write down all eight arrangements of the sexes of three children. What is the probability of any one of these arrangements?

(b) What is the probability that the couple's children are two girls and one boy?

18.16 More tetrahedral dice. Tetrahedral dice are described in Exercise 18.14. Give a probability model for rolling two such dice. That is, write down all possible outcomes and give a probability to each. (Example 2 and Figure 18.1 may help you.) What is the probability that the sum of the down-faces is 5?

18.17 Roulette. A roulette wheel has 38 slots, numbered 0, 00, and 1 to 36. The slots 0 and 00 are colored green, 18 of the others are red, and 18 are black. The dealer spins the wheel and, at the same time, rolls a small ball along the wheel in the opposite direction. The wheel is carefully balanced so that the ball is equally likely to land in any slot when the wheel slows. Gamblers can bet on various combinations of numbers and colors.

(a) What is the probability of any one of the 38 possible outcomes? Explain your answer.

(b) If you bet on "red," you win if the ball lands in a red slot. What is the probability of winning?

(c) A friend tells you that the odds that a bet on "red" will win are 10 to 9. Is your friend correct? If not, what are the correct odds?

(d) The slot numbers are laid out on a board on which gamblers place their bets. One column of numbers on the board contains all multiples of 3, that is, 3, 6, 9, . . . , 36. You place a "column bet" that wins if any of these numbers comes up. What is your probability of winning?

18.18 Colors of M&Ms. If you draw an M&M candy at random from a bag of the candies, the candy you draw will have one of six colors. The probability of drawing each color depends on the proportion of each color among all candies made.

(a) Here are the probabilities of each color for a randomly chosen plain M&M produced at the New Jersey factory:

Color:	Blue	Orange	Green	Yellow	Red	Brown
Probability:	0.250	0.250	0.125	0.125	0.125	?

What must be the probability of drawing a brown candy?

(b) The probabilities for plain M&Ms produced at the Tennessee factory are a bit different. Here they are:

Color:	Blue	Orange	Green	Yellow	Red	Brown
Probability:	0.207	0.205	0.198	0.135	0.131	?

What is the probability that an M&M chosen at random from the Tennessee factory is brown?

(c) What is the probability that a plain M&M from the New Jersey factory is any of blue, yellow, or orange? What is the probability that a plain M&M from the Tennessee factory has one of these colors?

18.19 Legitimate probabilities? In each of the following situations, state whether or not the given assignment of probabilities to individual outcomes is legitimate, that is, whether it satisfies the rules of probability. If not, give specific reasons for your answer.

(a) When a coin is spun, $P(H) = 0.55$ and $P(T) = 0.45$.

(b) When two coins are tossed, $P(HH) = 0.4$, $P(HT) = 0.4$, $P(TH) = 0.4$, and $P(TT) = 0.4$.

(c) Plain M&Ms have not always had the mixture of colors given in Exercise 18.18. In 1997, brown was the most popular color and had probability 0.30. Yellow and red each had probability 0.20. Orange, green, and blue each had probability 0.10.

18.20 Immigration. Suppose that 29% of all adult Americans think that the level of immigration to the United States should be decreased. An opinion poll interviews 1520 randomly chosen

Americans and records the sample proportion who feel that the level of immigration to this country should be decreased. This statistic will vary from sample to sample if the poll is repeated. The sampling distribution is approximately Normal with mean 0.29 and standard deviation about 0.012. Sketch this Normal curve and use it to answer the following questions.

(a) The mean of the population is 0.29. In what range will the middle 95% of all sample results fall?

(b) What is the probability that the poll gets a sample in which fewer than 27.8% say the level of immigration to this country should be decreased?

18.21 Airplane safety. Suppose that 44% of all adults think that airline travel is safer than driving. An opinion poll plans to ask an SRS of 1021 adults about airplane safety. The proportion of the sample who think that airline travel is safer than driving will vary if we take many samples from this same population. The sampling distribution of the sample proportion is approximately Normal with mean 0.44 and standard deviation about 0.016. Sketch this Normal curve and use it to answer the following questions.

(a) What is the probability that the poll gets a sample in which more than 40.8% of the people think that airline travel is safer than driving?

(b) What is the probability of getting a sample that misses the truth (44%) by 3.2% or more?

18.22 Immigration (optional). In the setting of Exercise 18.20, what is the probability of getting a sample in which more than 30% of those sampled think that the level of immigration to this country should be decreased? (Use Table B.)

18.23 Airplane safety (optional). In the setting of Exercise 18.21, what is the probability of getting a sample in which more than 45% think that airline travel is safer than driving? (Use Table B.)

18.24 Do you jog? An opinion poll asks an SRS of 1500 adults, "Do you jog?" Suppose (as is approximately correct) that the population proportion who jog is $p = 0.20$. In a large number of samples, the proportion $\hat{p}$ who answer Yes will be approximately Normally distributed with mean 0.20 and standard deviation 0.01. Sketch this Normal curve and use it to answer the following questions.

(a) What percentage of many samples will have a sample proportion who jog that is 0.20 or less? Explain clearly why this percentage is the probability that $\hat{p}$ is 0.20 or less.

(b) What is the probability that $\hat{p}$ will take a value between 0.18 and 0.22? (Use the 68–95–99.7 rule.)

(c) Now use Rule C to determine the following probability: what is the probability that $\hat{p}$ does not lie between 0.18 and 0.22?

18.25 Applying to college. You ask an SRS of 1500 college students whether they applied for admission to any other college. Suppose that, in fact, 35% of all college students applied to colleges besides the one they are attending. (That's close to the truth.) The sampling distribution of the proportion $\hat{p}$ of your sample who say Yes is approximately Normal with mean 0.35 and standard deviation 0.01. Sketch this Normal curve and use it to answer the following questions.

(a) Explain in simple language what the sampling distribution tells us about the results of our sample.

(b) What percentage of many samples would have a $\hat{p}$ larger than 0.37? (Use the 68–95–99.7 rule.) Explain in simple language why this percentage is the probability of an outcome larger than 0.37.

(c) What is the probability that your sample will have a $\hat{p}$ less than 0.33?

(d) Use Rule D to answer the following question: what is the probability that your sample result will be either less than 0.33 or greater than 0.35?

18.26 **Generating a sampling distribution.** Let us illustrate the idea of a sampling distribution in the case of a very small sample from a very small population. The population is the scores of 10 students on an exam:

Student:	0	1	2	3	4	5	6	7	8	9
Score:	82	62	80	58	72	73	65	66	74	62

The parameter of interest is the mean score in this population. The sample is an SRS of size $n = 4$ drawn from the population. Because the students are labeled 0 to 9, a single random digit from Table A chooses one student for the sample.

(a) Find the mean of the 10 scores in the population. This is the population mean.

(b) Use Table A, or software (see Example 3 in Chapter 2, page 25) to draw an SRS of size 4 from this population. Write the four scores in your sample and calculate the mean $\bar{x}$ of the sample scores. This statistic is an estimate of the population mean.

(c) Repeat this process 10 times using different parts of Table A or using software. Make a histogram of the 10 values of $\bar{x}$. You are constructing the sampling distribution of $\bar{x}$. Is the center of your histogram close to the population mean you found in part (a)?

18.27 **Odds and personal probability.** One way to determine your personal probability about an event is to ask what you would consider a fair bet that the event will occur. Suppose in August 2018 you believed it fair that a bet of $2 should win $10 if the Philadelphia Eagles win Super Bowl 53.

(a) What does this bet say you believe the odds to be that the Philadelphia Eagles will win Super Bowl 53?

(b) Convert the odds in part (a) to a probability. This would be your personal probability that the Philadelphia Eagles will win Super Bowl 53.

18.28 **The addition rule (optional).** Probability rule D states: *If two events have no outcomes in common, the probability that one or the other occurs is the sum of their individual probabilities.* This is sometimes called the addition rule for disjoint events. A more general form of the addition rule

is: *the probability that one or the other of two events occurs is the sum of their individual probabilities minus the probability that both occur.* To verify this rule, suppose you roll two casino dice as in Example 2. Refer to the outcomes in Figure 18.1 to answer the following.

(a) How many of the outcomes in Figure 18.1 have the sum of the spots on the up-faces equal to 6? What is the probability that the sum of the spots on the up-faces is 6?

(b) How many of the outcomes in Figure 18.1 have at least one of the up-faces showing a single spot? What is the probability that at least one of the up-faces has only a single spot?

(c) How many of the outcomes in Figure 18.1 have the sum of the spots on the up-faces equal to 6 and at least one of the up-faces showing a single spot? What is the probability that the sum of the spots on the up-faces is 6 and at least one of the up-faces has only a single spot?

(d) How many of the outcomes in Figure 18.1 have either the sum of the spots on the up-faces equal to 6 or have at least one of the up-faces showing a single spot? What is the probability that the sum of the spots on the up-faces is either a 6 or at least one of the up-faces has a single spot?

(e) The addition rule says that your answer to part (d) should be equal to the sum of your answers to parts (a) and (b) minus your answer to part (c). Verify that this is the case.

Note: The outcomes you were asked to count in part (c) are among those counted in parts (a) and (b). When we combine the outcomes in parts (a) and (b), we "double count" the outcomes in part (c). One of these "double counts" must be eliminated so that the combination corresponds to the outcomes in part (d). This is the reason that, in the addition rule, you subtract the probability that both occur from the sum of their individual probabilities.

Exploring the Web

Access these exercises on the text website: macmillanlearning.com/scc10e.

What's the Verdict?

On Tuesday, October 23, 2018, the Mega Millions lottery had a grand prize jackpot of $1.6 billion. This unusually large jackpot excited many people who would not normally buy tickets to play the game for the first time.

For the Mega Millions game, each player selects 6 numbers that they hope will match with those randomly selected by the company on the big night.

According to the Mega Millions website: **http://www.megamillions.com/how-to -play**, the odds of winning the jackpot and some lesser prizes are listed below.

Figure 18.7 The odds of winning as posted on the Mega Millions website. Exercises WTV18.1 and WTV18.2.

Questions

WTV18.1. Do you have a better chance of winning if you play the birthdates of everybody in your family instead of letting a machine at the convenience store where you purchase your ticket pick 6 random numbers for you? Why or why not?

WTV18.2. Make a valid probability model for all the possible outcomes and their probabilities (leave the probabilities as fractions). Don't forget to include the possibility of not winning anything. Explain why this is a valid probability model.

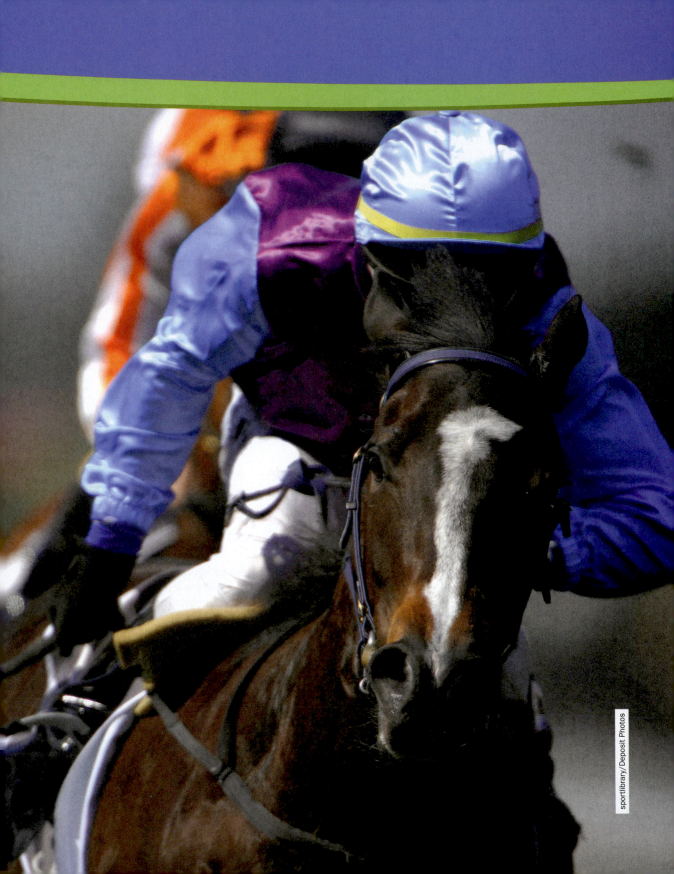

Simulation

Case Study: Luck or Cheating?

In this chapter you will:

- Learn a method for estimating the probability of a complicated event.

- Express a probability model in graphical form.

In a horse race, the starting position can affect the outcome. It is advantageous to have a starting position that is near the inside of the track. To ensure fairness, starting position is determined by a random draw before the race. All positions are equally likely, so no horse has an advantage.

During the summer and autumn of 2007, the members of the Ohio Racing Commission noticed that one trainer appeared to have an unusually good run of luck. In 35 races, the horses of this trainer received one of the three inner positions 30 times. The number of horses in a race ranges from 6 to 10, but most of the time the number of entries is 9. Thus, for most races the chance of getting one of the three inside positions is 1-in-3.

The Ohio Racing Commission believed that the trainer's run of luck was too good to be true. A mathematician can show that the chance of receiving one of the three inside positions at least 30 times in 35 races is very small. The Ohio Racing Commission, therefore, suspected that cheating had occurred. But the trainer had entered horses in nearly 1000 races over several years. Perhaps it was inevitable that at some time over the course of these nearly 1000 races, the trainer would have a string of 35 races in which he received one of the three inside positions at least 30 times. It came to the attention of the Ohio Racing Commission only because it was one of those seemingly surprising coincidences discussed in Chapter 17. Perhaps the accusation of cheating was unfounded.

Calculating the probability that in a sequence of 1000 races with varying numbers of horses, there would occur a string of 35 consecutive races in which one would receive one of the three inside positions at least 30 times is very difficult. By the end of the chapter, you will be able to describe how to find the probabilities of more complicated events.

443

Where Do Probabilities Come From?

The probabilities of heads and tails in tossing a coin are very close to 1-in-2. In principle, these probabilities come from *data* on many coin tosses. Joe's personal probabilities for the winner of next year's Super Bowl come from Joe's own *individual judgment*. What about the probability that we get a run of three straight heads somewhere in 10 tosses of a coin? We can find this probability by *calculation from a model* that describes tossing coins. That is, once we have used data to give a probability model for the random fall of coins, we don't have to go back to the beginning every time we want the probability of a new event.

The big advantage of probability models is that they allow us to calculate the probabilities of complicated events starting from an assignment of probabilities to simple events like "heads on one toss." This is true whether the model reflects probabilities from data or personal probabilities. Unfortunately, the math needed to do probability calculations is often tough. Technology rides to the rescue: once we have a probability model, we can use a computer to *simulate* many repetitions. This is easier than math and much faster than actually running many repetitions in the real world. You might compare finding probabilities by simulation to practicing flying in a computer-controlled flight simulator. Both kinds of simulation are in wide use. Both have similar drawbacks: they are only as good as the model you start with. Flight simulators use a software model of how an airplane reacts. Simulations of probabilities use a probability model. We set the model in motion by using our old friends the random digits from Table A.

> ### Key Terms
> Using random digits from a table or from computer software to imitate chance behavior is called **simulation**.

We look at simulation partly because it is how scientists and engineers really do find probabilities in complex situations. Simulations are used to develop strategies for reducing waiting times in lines to speak to a teller at banks, in lines to check in at airports, and in lines to vote during elections. Simulations are used to study the effects of changes in greenhouse gases on the climate. Simulations are used to study the effects of catastrophic events, such as the failure of a nuclear power plant, the effects of the explosion of a nuclear device on a structure, or the progression of a deadly, infectious disease in a densely populated city.

We also look at simulation because simulation forces us to think clearly about probability models. We'll do the hard part (setting up the model) and leave the easy part (telling a computer to do 10,000 repetitions) to those who really need the right probability at the end.

Simulation Basics

Simulation is an effective tool for finding probabilities of complex events once we have a trustworthy probability model. We can use random digits to simulate many repetitions quickly. The proportion of repetitions on which an event occurs will eventually be close to its probability, so simulation can give good estimates of probabilities. The art of simulation is best learned from a series of examples.

Example 1
Doing a simulation

Toss a coin 10 times. What is the probability of a run of at least 3 consecutive heads or 3 consecutive tails?

Step 1. Give a probability model. Our model for coin tossing has two parts:

- Each toss has probabilities 0.5 for a head and 0.5 for a tail.

- Tosses are *independent* of each other. That is, knowing the outcome of one toss does not change the probabilities for the outcomes of any other toss.

Step 2. Assign digits to represent outcomes. Digits in Table A of random digits will stand for the outcomes, in a way that matches the probabilities from Step 1. We know that each digit in Table A has probability 0.1 of being any one of 0, 1, 2, 3, 4, 5, 6, 7, 8, or 9 and that successive digits in the table are independent. Here is one assignment of digits for coin tossing:

- One digit simulates one toss of the coin.

- Odd digits represent heads; even digits represent tails.

This works because the five odd digits give probability 5/10 to heads (but any other assignment where half the digits represent heads is equally good). Successive digits in the table simulate independent tosses.

Step 3. Simulate many repetitions. Ten digits simulate 10 tosses, so looking at 10 consecutive digits in Table A simulates one repetition. Read many groups of 10 digits from the table to simulate many repetitions. Be sure to keep track of whether or not the event we want (a run of three heads or three tails) occurs on each repetition.

Here are the first three repetitions, starting at line 101 in Table A. We have underlined all runs of three or more heads or tails.

	Repetition 1	Repetition 2	Repetition 3
Digits	1 9 2 2 3 9 5 0 3 4	0 5 7 5 6 2 8 7 1 3	9 6 4 0 9 1 2 5 3 1
Heads/tails	H H T T H H H T H T	T H H H T T T H H H	H T T T H H T H H H
Run of 3?	YES	YES	YES

Continuing in Table A, we did 25 repetitions; 23 of them did have a run of three or more heads or tails. So we estimate the probability of a run by the proportion

$$\text{estimated probability} = \frac{23}{25} = 0.92$$

Of course, 25 repetitions are not enough to be confident that our estimate is accurate. Now that we understand how to do the simulation, we can tell a computer to do many thousands of repetitions, and using a computer to do these many simulations is how simulations are conducted in practice. A long simulation (or hard mathematics) finds that the true probability is about 0.826. Most people think runs are somewhat unlikely, so even our short simulation challenges our intuition by showing that runs of three occur most of the time in 10 tosses.

Key Terms

Two random phenomena are **independent** if knowing the outcome of one does not change the probabilities for outcomes of the other.

Once you have gained some experience in simulation, setting up the probability model (Step 1) is usually the hardest part of the process. Although coin tossing may not fascinate you, the model in Example 1 is typical of many probability problems because it consists of *independent trials* (the tosses) all having the *same possible outcomes* with the *same probabilities*. Shooting 10 free throws and observing the sexes of 10 children have similar models and are simulated in much the same way. The new part of the model is independence, which simplifies our work because it allows us to simulate each of the 10 tosses in exactly the same way.

Independence, like all aspects of probability, can be verified only by observing many repetitions. It is plausible that repeated tosses of a coin are independent (the coin has no memory), and observation shows that they are. It seems less plausible that successive shots by a basketball player are independent, but observation shows that they are at least very close to independent.

Step 2 (assigning digits) rests on the properties of the random digit table. Here are some examples of this step.

Example 2

Assigning digits for simulation

In the United States, the Age Discrimination Employment Act (ADEA) forbids age discrimination against people who are age 40 and older. Terminating an "unusually" large proportion of employees age 40 and older can trigger legal action. Simulation can help determine what might be an "unusually" large proportion of terminations if terminations are unrelated to age. How might we simulate randomly selecting employees for termination?

(a) Choose one employee at random from a group of which 40% are age 40 and older. One digit simulates one employee:

$$0, 1, 2, 3 = \text{age 40 and older}$$
$$4, 5, 6, 7, 8, 9 = \text{under age 40}$$

(b) Choose one employee at random from a group of which 43% are age 40 and older. Now *two* digits simulate one person:

$$00, 01, 02, \ldots, 42 = \text{age 40 and older}$$
$$43, 74, 75, \ldots, 99 = \text{under age 40}$$

We assigned 43 of the 100 two-digit pairs to "age 40 and older" to get probability 0.43. Representing "age 40 and older" by $01, 02, \ldots, 43$ and "under age 40" by $44, 45, \ldots, 99, 00$ would also be correct.

(c) Choose one employee at random from a group of which 30% are age 40 and older and have no plans to retire, 10% are age 40 and older and plan to retire in the next few months, and 60% are under age 40. There are now three possible outcomes, but the principle is the same. One digit simulates one person:

$$0, 1, 2 = \text{age 40 and older and have no plans to retire}$$
$$3 = \text{age 40 and older and plan to retire in the next few months}$$
$$4, 5, 6, 7, 8, 9 = \text{under age 40}$$

Now it's your turn

19.1 Political affiliation of a randomly selected college student. Recent surveys of college students suggest that 40% are liberal, 40% are middle-of-the-road, and 20% are conservative. How would you assign digits for a simulation to determine the political affiliation (liberal, middle-of-the-road, or conservative) of a college student chosen at random from the population of all college students?

Statistics in Your World

Was he good or was he lucky?

When a baseball player hits .300, everyone applauds. A .300 hitter gets a hit in 30% of times at bat. Could a .300 year just be luck? Typical major leaguers bat about 500 times a season and hit about .260. A hitter's successive tries seem to be independent. From this model, we can calculate or simulate the probability of an "average" player hitting .300. It is about 0.025. Out of 100 run-of-the-mill major league hitters, two or three each year will bat .300 just because they were lucky.

Thinking about Independence

Before discussing more elaborate simulations, it is worth discussing the concept of independence further. We said earlier that independence can be verified only by observing many repetitions of random phenomena. It is probably more accurate to say that a *lack* of independence can be verified only by observing many repetitions of random phenomena. How does one recognize that two random phenomena are not independent? For example, how can we tell if tosses of a fair coin (that is, one for which the probability of a head is 0.5 and the probability of a tail is 0.5) are not independent?

One approach might be to apply the definition of "independence." For a sequence of tosses of a fair coin, one could compute the proportion of times in the sequence that a toss is followed by the same outcome—in other words, the frequency with which a head is followed by a head or a tail is followed by a tail. This proportion should be close to 0.5 if tosses are independent (knowing the outcome of one toss does not change the probabilities for outcomes of the next) and if many tosses have been observed.

Example 3 — *Investigating independence*

Suppose we tossed a fair coin 15 times and obtained the following sequence of outcomes:

Toss:	1	2	3	4	5	6	7	8	9	10	11	12	13	14	15
Outcome:	H	H	H	T	H	H	T	T	T	T	T	T	H	H	T	T

For the first 14 tosses, the following toss is the same nine times. The proportion of times a toss is followed by the same outcome is

$$\text{proportion} = \frac{9}{14} = 0.64$$

For so few tosses, this would not be considered a large departure from 0.5.

Unfortunately, if the proportion of times a head is followed by a head or a tail is followed by a tail is close to 0.5, this does not necessarily imply that the tosses are independent. For example, suppose that instead of tossing the coin, we simply placed the coin heads up or tails up according to the following pattern:

Trial:	1	2	3	4	5	6	7	8	9	10	11	12	13	14	15
Outcome:	H	H	T	T	H	H	T	T	H	H	T	T	H	H	T

We begin with two heads, followed by two tails, followed by two heads, etc. If we know the previous outcomes, we know exactly what the next outcome will be. Successive outcomes are not independent. However, looking at the first 14 outcomes, the proportion of times in this sequence that a head is followed by a head or a tail is followed by a tail is

$$\text{proportion} = \frac{7}{14} = 0.5$$

Thus, our approach can help us recognize when independence is lacking, but not when independence is present.

Another method for assessing independence is based on the concept of correlation, which we discussed in Chapter 14. If two random phenomena have numerical outcomes, and we observe both phenomena in a sequence of n trials, we can compute the correlation for the resulting data. If the random phenomena are independent, there will be no straight-line relationship between them, and the correlation should be close to 0.

It is not necessarily true that two random phenomena are independent if their correlation is 0. In Exercise 14.24 (page 337), there is a clear curved relationship between speed and mileage, but the correlation is 0. Independence implies no relationship at all, but correlation measures the strength of only a straight-line relationship.

Because independence implies no relationship, we would expect to see no overall pattern in a scatterplot of the data if the variables are independent. Looking at scatterplots is another method for determining if independence is lacking.

Many methods for assessing independence exist. For example, if trials are not independent and, say, tossing a head increases the probability that the next toss is also a head, then in a sequence of tosses, we might expect to see unusually long runs of heads. We mentioned this idea of unusually long runs in Example 5 (page 410) of Chapter 17. Unusually long runs of made free throws would be expected if a basketball player has a "hot hand." However, careful study has shown that runs of baskets made or missed are no more frequent in basketball than would be expected if each shot is independent of the player's previous shots.

Statistics in Your World

Weather forecasting Modern weather forecasting uses complicated simulations to predict tomorrow's weather. These simulations must be run on computers and are based on complex equations relating meteorological variables such as air pressure, temperature, moisture, and wind. The mathematical instructions that produce such a virtual weather forecast are often referred to as a computer model. By varying the initial conditions according to some probability model, the computer model can produce many simulations, and from these, estimates of the probability of rain, for example, can be made.

More Elaborate Simulations

The building and simulation of random models constitute a powerful tool of contemporary science, but a tool that can be understood without advanced mathematics. What is more, doing a few simulations will increase your understanding of probability more than many pages of our prose. Having in mind these two goals of understanding simulation for itself and understanding simulation to understand probability, let us look at two more elaborate examples. The first still has independent trials, but there is no longer a fixed number of trials as there was when we tossed a coin 10 times.

Example 4

We want a girl

A couple plan to have children until they have a girl or until they have three children, whichever comes first. What is the probability that they will have a girl among their children?

Step 1. The probability model is like that for coin tossing:

- Each child has probability 0.49 of being a girl and 0.51 of being a boy. (Yes, more boys than girls are born. Boys have higher infant mortality, so the sexes even out soon.)

- The sexes of successive children are independent.

Step 2. Assigning digits is also easy. Two digits simulate the sex of one child. We assign 49 of the 100 pairs to "girl" and the remaining 51 to "boy":

$$00, 01, 02, \ldots, 48 = \text{girl}$$
$$49, 50, 51, \ldots, 99 = \text{boy}$$

digifun/Deposit Photos

Step 3. To simulate one repetition of this childbearing strategy, read pairs of digits from Table A until the couple have either a girl or three children. The number of pairs needed to simulate one repetition depends on how quickly the couple get a girl. Here are 10 repetitions, simulated using line 130 of Table A. To interpret the pairs of digits, we have written G for girl and B for boy under them, have added space to separate repetitions, and under each repetition have written "+" if a girl was born and "−" if not.

6905	16	48	17	8717	40	9517	845340	648987	20
B G	G	G	G	B G	G	B G	B B G	B B B	G
+	+	+	+	+	+	+	+	−	+

In these 10 repetitions, a girl was born nine times. Our estimate of the probability that this strategy will produce a girl is, therefore,

$$\text{estimated probability} = \frac{9}{10} = 0.9$$

Some mathematics shows that, if our probability model is correct, the true probability of having a girl is 0.867. Our simulated answer came quite close. Unless the couple are unlucky, they will succeed in having a girl.

Our final example has stages that are *not independent*. That is, the probabilities at one stage depend on the outcome of the preceding stage.

Example 5

A kidney transplant

Morris's kidneys have failed and he is awaiting a kidney transplant. His doctor gives him this information for patients in his condition: 90% survive the transplant operation, and 10% die. The transplant succeeds in 60% of those who survive, and the other 40% must return to kidney dialysis. The proportions who survive for at least five years are 70% for those with a new kidney and 50% for those who return to dialysis. Morris wants to know the probability that he will survive for at least five years.

dlpn/Deposit Photos

Step 1. The **tree diagram** in Figure 19.1 organizes this information to give a probability model in graphical form. The tree shows the three stages and the possible outcomes and probabilities at each stage. Each path through the tree leads to either survival for five years or to death in less than five years. To simulate Morris's fate, we must simulate each of the three stages. The probabilities at Stage 3 depend on the outcome of Stage 2.

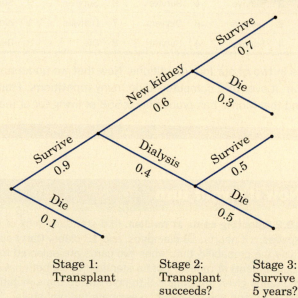

Stage 1:
Transplant

Stage 2:
Transplant
succeeds?

Stage 3:
Survive
5 years?

Figure 19.1 A tree diagram for the probability model, Example 5. Each branch point starts a new stage, with outcomes and their probabilities written on the branches. One repetition of the model follows the tree to one of its endpoints.

Step 2. Here is our assignment of digits to outcomes:

Stage 1:

$$0 = \text{die}$$
$$1, 2, 3, 4, 5, 6, 7, 8, 9 = \text{survive}$$

Stage 2:

$$0, 1, 2, 3, 4, 5 = \text{transplant succeeds}$$
$$6, 7, 8, 9 = \text{return to dialysis}$$

Stage 3 with kidney:

$$0, 1, 2, 3, 4, 5, 6 = \text{survive five years}$$
$$7, 8, 9 = \text{die}$$

Stage 3 with dialysis:

$$0, 1, 2, 3, 4 = \text{survive five years}$$
$$5, 6, 7, 8, 9 = \text{die}$$

The assignment of digits at Stage 3 depends on the outcome of Stage 2. That's lack of independence.

Step 3. Here are simulations of several repetitions, each arranged vertically. We used random digits from line 110 of Table A.

	Repetition 1		Repetition 2		Repetition 3		Repetition 4	
Stage 1	3	Survive	4	Survive	8	Survive	9	Survive
Stage 2	8	Dialysis	8	Dialysis	7	Dialysis	1	Kidney
Stage 3	4	Survive	4	Survive	8	Die	8	Die

Morris survives five years in two of our four repetitions. Now that we understand how to arrange the simulation, we should turn it over to a computer to do many repetitions. From a long simulation or from mathematics, we find that Morris has probability 0.558 of living for at least five years.

Now it's your turn

19.2 Selecting cards at random. In a standard deck of 52 cards, there are 13 spades, 13 hearts, 13 diamonds, and 13 clubs. Carry out a simulation to determine the probability that, when two cards are selected together (that is, without replacement) at random from a standard deck, both have the same suit. Follow the steps given in Example 5 to help set up the simulation. Do your simulation 10 times and use the result to estimate the probability.

Chapter 19: Statistics in Summary

■ We can use random digits to **simulate** random outcomes if we know the probabilities of the outcomes. Use the fact that each random digit has probability 0.1 of taking any one of the 10 possible digits and that all digits in the random number table are **independent** of each other.

■ To simulate more complicated random phenomena, string together simulations of each stage. A common situation is several independent trials with the same possible outcomes and probabilities on each trial. Other simulations may require varying numbers of trials or different probabilities at each stage or may have stages that are not independent so that the probabilities at some stage depend on the outcome of earlier stages.

■ The key to successful simulation is thinking carefully about the probability model. A **tree diagram** can be helpful by giving the probability model in graphical form.

This chapter summary will help you evaluate the Case Study.

Link It

In Chapter 18, we discussed probability models and the basic rules of probability. These allow us to compute the probabilities of simple events, but the math needed to find the probabilities of complicated events is often tough. In this chapter, we learn how we can use simulations to determine the probabilities of complicated events. The underlying idea, introduced in Chapter 17, is that probability is the long-run proportion of times an event occurs. Simulating such an event many, many times using technology allows us to estimate this long-run proportion.

Case Study Evaluated

Use what you have learned in this chapter to evaluate the Case Study that opened the chapter. Start by reviewing the chapter summary. Then answer each of the following questions in complete sentences. Be sure to communicate clearly enough for any of your classmates to understand what you are saying.

For simplicity, assume that there are always nine horses in a race so that the probability of receiving one of the three inside positions is 1-in-3. Describe how you would do a simulation to estimate the probability that, somewhere in a sequence of 1000 races, a string of 35 consecutive races would occur in which at least 30 times one of the three inside positions was assigned. Begin by describing how you would simulate one case of 1000 races and, for these races, how you would determine whether somewhere in the 1000 races there was a string of 35 consecutive races in which at least 30 times one of the three inside positions was assigned. Do not try to carry out the simulation. This would be very time-consuming and is best left to a computer. (In fact, the Ohio Racing Commission hired a statistician to compute the probability that in a sequence of 1000 races with varying numbers of horses, there would occur a string of 35 consecutive races in which one would receive one of the three inside positions at least 30 times. The statistician used simulation to estimate this probability.)

In this chapter you:

- Estimated the probability of a complicated event using simulation.
- Expressed a probability model in graphical form using a tree diagram.

macmillan learning **Online Resources**

- LearningCurve has good questions to check your understanding of the concepts.

Check the Basics

For Exercise 19.1, see page 447; for Exercise 19.2, see page 452.

19.3 Simulations. To simulate random outcomes, we need to know

(a) the probabilities of the outcomes.

(b) whether the probabilities are personal probabilities.

(c) both the mean and standard deviation so we can use the appropriate Normal curve.

(d) that the random outcome has a probability of 0.1 so we can use a table of random digits.

19.4 A simulation. To simulate the toss of a fair coin (the probability of heads and tails are both 0.5) using a table of random digits,

(a) assign the digits 0, 1, 2, 3, and 4 to represent heads and the digits 5, 6, 7, 8, and 9 to represent tails.

(b) assign the digits 0, 2, 4, 6, and 8 to represent heads and the digits 1, 3, 5, 7, and 9 to represent tails.

(c) assign the digits 0, 1, 5, 8, and 9 to represent heads and the digits 2, 3, 5, 6, and 7 to represent tails.

(d) use any of the above. All are correct.

19.5 Independence. Two random phenomena are independent if

(a) knowing that one of the outcomes has occurred means the other cannot occur.

(b) knowing the outcomes of one does not change the probabilities for outcomes of the other.

(c) both have the same probability of occurring.

(d) both have different and unrelated probabilities of occurring.

19.6 Elaborate simulations. A tree diagram

(a) was originally used by biologists for simulations involving trees.

(b) is used to determine if two random phenomena are independent.

(c) is used when two random phenomena are independent.

(d) specifies a probability model in graphical form.

19.7 Elaborate simulations. The key to successful simulation is

(a) keeping the tree diagram as simple as possible.

(b) thinking carefully about the probability model for the simulation.

(c) using as few trials as possible so that the chance of an incorrect trial is kept small.

(d) using *all* the digits in a random digits table.

Chapter 19 Exercises

19.8 Approval ratings of Democrats in Congress. An opinion poll selects adult Americans at random and asks them, "Do you approve or disapprove of the way Democrats in Congress are handling their job?" Explain carefully how you would assign digits from Table A to simulate the response of one person in each of the following situations.

(a) Of all adult Americans, 50% would say approve and 50% disapprove.

(b) Of all adult Americans, 40% would say approve and 60% disapprove.

(c) Of all adult Americans, 40% would say approve, 50% disapprove, and 10% undecided.

(d) Of all adult Americans, 31% would say approve, 61% disapprove, and 8% undecided. (These were the percentages in a September 2017 Gallup Poll.)

19.9 On time flights. Suppose that 80% of American Airlines flights are on time (this is approximately the percentage of American Airline flights on time in November 2018). You check 10 American Airline flights chosen at random. What is the probability that all 10 are on time?

(a) Give a probability model for checking 10 flights chosen independently of each other.

(b) Assign digits to represent on time and not on time.

(c) Simulate 25 repetitions, starting at line 129 of Table A. What is your estimate of the probability?

19.10 Basic simulation. Use Table A to simulate the responses of 10 independently chosen adults in each of the four situations of Exercise 19.8.

(a) For situation (a), use line 110.

(b) For situation (b), use line 111.

(c) For situation (c), use line 112.

(d) For situation (d), use line 113.

19.11 Simulating an opinion poll. A 2018 poll by the Pew Research Survey Center interviewed a random sample of 2002 adult Americans. Those in the sample were asked which president has done the best job in their lifetime. The poll showed that about 30% of adult Americans regarded Barack Obama as having done the best job in their lifetime. Assume the poll accurately reflects the opinions of all adult Americans. Choosing an adult American at random then has probability 0.3 of getting one who would identify Barack Obama as having done the best job in their lifetime. If we interview adult Americans separately, we can assume that their responses are independent. We want to know the probability that a simple random sample of 100 adult Americans will contain at least 35 who say that Barack Obama has done the best job in their lifetime. Explain carefully how to do this simulation and simulate *one* repetition of the poll starting at line 112 of Table A. How many of the 100 adult Americans said that

Barack Obama has done the best job in their lifetime? Explain how you would estimate the probability by simulating many repetitions.

19.12 An easy A? Choose a student at random from all who took the large accelerated introductory statistics courses at Hudson River College in the last 10 years. The probabilities for the student's grade are

Grade:	A	B	C	D or F
Probability:	0.4	0.3	0.2	?

(a) What must be the probability of getting a D or an F?

(b) To simulate the grades of randomly chosen students, how would you assign digits to represent the four possible outcomes listed?

19.13 First-year college students. Select a first-year college student at random and ask how many hours during a typical week did they spend studying or doing homework during their last year in high school? Probabilities for the outcomes are

Time:	Less than one hour	1 to 5 hours	6 to 10 hours	More than 10 hours
Probability:	0.1	0.5	0.2	?

(a) What must be the probability that a randomly chosen first-year college student says they spent more than 10 hours per week studying or doing homework during their last year in high school?

(b) To simulate the responses of randomly chosen first-year college students, how would you assign digits to represent the four possible outcomes listed?

19.14 More on an easy A. In Exercise 19.12, you explained how to simulate the grade of a randomly chosen student who took the accelerated statistics course in the last 10 years. Suppose you select five students at random who took the course in the last 10 years. Use simulation to estimate the probability that all five got a C or better in the course. (Simulate 20 repetitions and assume the student grades are independent of each other.)

19.15 More on first-year college students. In Exercise 19.13, you explained how to simulate the response of a randomly chosen first-year college student to the question of how many hours during a typical week they spent studying or doing homework during their last year in high school. The Tutoring Center decides to choose eight students at random and provide them with training on study habits. What is the probability that at least five of the eight students chosen were among those who reported that they spent either less than 1 hour or only 1 to 5 hours per week on studying or doing homework? Simulate 10 repetitions of the center's choices to estimate this probability.

19.16 LeBron's three-point shooting. The basketball player LeBron James makes about 34% of his three-point shots over an entire season. Take his probability of a success to be 0.34 on each shot. Using line 122 of Table A, simulate 25 repetitions of his performance in a game in which he shoots 10 three-point shots.

(a) Estimate the probability that LeBron makes at least half of his three-point shots.

(b) Examine the sequence of hits and misses in your 25 repetitions. How long was the longest run of shots made?

19.17 Sue's three-point shooting. Sue Bird of the Women's National Basketball Association team Seattle Storm makes 39% of her three-point shots. In an important game, she shoots four three-point shots late in the game and misses all of them. The fans think she was nervous, but the misses may simply be chance. Let's shed some light by estimating a probability.

(a) Describe how to simulate a single three-point shot if the probability of making each shot is 0.39. Then describe how to simulate four independent three-point shots.

(b) Simulate 50 repetitions of the four three-point shots and record the number missed on each repetition. Use Table A, starting at line 125. What is the approximate probability that Sue will miss all four shots?

19.18 Repeating an exam. Elaine is enrolled in a self-paced course that allows three attempts to pass an examination on the material. She skims the online reading and then takes the exam. Assume that after only skimming the online material, Elaine has probability 0.4 of passing on any one attempt. What is Elaine's probability of passing in three attempts? (Assume the attempts are independent because she takes a different examination on each attempt.)

(a) Explain how you would use random digits to simulate one attempt at the exam.

(b) Elaine will stop taking the exam as soon as she passes. (This is much like Example 4.) Simulate 50 repetitions, starting at line 120 of Table A. What is your estimate of Elaine's probability of passing the exam?

(c) Do you think the assumption that Elaine's probability of passing the exam is the same on each trial is realistic? Why?

19.19 A better model for repeating an exam. A more realistic probability model for Elaine's attempts to pass an exam in the previous exercise is as follows. On the first try, she has probability 0.4 of passing. If she fails on the first try, her probability on the second try increases to 0.5 because she learned something from her first attempt. If she fails on two attempts, the probability of passing on a third attempt is 0.6. She will stop as soon as she passes. The course rules force her to stop after three attempts, in any case.

(a) Make a tree diagram of Elaine's progress. Notice that she has different probabilities of passing on each successive try.

(b) Explain how to simulate one repetition of Elaine's tries at the exam.

(c) Simulate 50 repetitions and estimate the probability that Elaine eventually passes the exam. Use Table A, starting at line 130.

19.20 Gambling in ancient Rome. Tossing four astragali was the most popular game of chance in Roman times. Many throws of a present-day sheep's astragalus show that the approximate probability distribution for the four sides of the bone that can land uppermost are

Outcome	Probability
Narrow flat side of bone	1/10
Broad concave side of bone	4/10
Broad convex side of bone	4/10
Narrow hollow side of bone	1/10

The best throw of four astragali was the "Venus," when all four uppermost sides were different.

(a) Explain how to simulate the throw of a single astragalus. Then explain how to simulate throwing four astragali independently of each other.

(b) Simulate 25 throws of four astragali. Estimate the probability of throwing a Venus. Be sure to say what line of Table A you used.

19.21 The Asian stochastic beetle. We can use simulation to examine the fate of populations of living creatures. Consider the Asian stochastic beetle. Females of this insect have the following pattern of reproduction:

• 20% of females die without female offspring, 30% have one female offspring, and 50% have two female offspring.

• Different females reproduce independently.

What will happen to the population of Asian stochastic beetles: will they increase rapidly, barely hold their own, or die out? It's enough to look at the female beetles, as long as there are some males around.

(a) Assign digits to simulate the offspring of one female beetle.

(b) Make a tree diagram for the female descendants of one beetle through three generations. The second generation, for example, can have 0, 1, or 2 females. If it has 0, we stop. Otherwise, we simulate the offspring of each second-generation female. What are the possible numbers of beetles after three generations?

(c) Use line 105 of Table A to simulate the offspring of five beetles to the third generation. How many descendants does each have after three generations? Does it appear that the beetle population will grow?

19.22 Two warning systems. An airliner has two independent automatic systems that sound a warning if there is terrain ahead (that means the airplane is about to fly into a mountain). Neither system is perfect. System A signals in time with probability 0.9. System B does so with probability 0.8. The pilots are alerted if either system works.

(a) Explain how to simulate the response of System A to terrain.

(b) Explain how to simulate the response of System B.

(c) Both systems are in operation simultaneously. Draw a tree diagram with System A as the first stage and System B as the second stage. Simulate 100 trials of the reaction to terrain ahead. Estimate the probability that a warning will sound. The probability for the combined system is higher than the probability for either A or B alone.

19.23 Playing craps. The game of craps is played with two dice. The player rolls both dice and wins immediately if the outcome (the sum of the faces) is 7 or 11. If the outcome is 2, 3, or 12, the player loses immediately. If he rolls any other outcome, he continues to throw the dice until he either wins by repeating the first outcome or loses by rolling a 7.

(a) Explain how to simulate the roll of a single fair die. (*Hint*: Just use digits 1 to 6 and ignore the others.) Then explain how to simulate a roll of two fair dice.

(b) Draw a tree diagram for one play of craps. In principle, a player could continue forever, but stop your diagram after four rolls of the dice. Use Table A, beginning at line 114, to simulate plays and estimate the probability that the player wins.

19.24 Overbooking. Your company operates small commuter planes. Each plane carries eight passengers. Some passengers who reserve seats don't show up—in fact, the probability is 0.1 that a randomly chosen passenger will fail to appear. Passengers' behaviors are independent. If you allow nine reservations for each plane, what is the probability that more than eight passengers will appear? Do a simulation to estimate this probability.

19.25 A multiple-choice exam. Matt has lots of experience taking multiple-choice exams without doing much studying. He is about to take a quiz that has 10 multiple-choice questions, each with four possible answers. Here is Matt's personal probability model. He thinks that in 75% of questions, he can eliminate one answer as obviously wrong; then he guesses from the remaining three. He then has probability 1-in-3 of guessing the right answer. For the other 25% of questions, he must guess from all four answers, with probability 1-in-4 of guessing correctly.

(a) Make a tree diagram for the outcome of a single question. Explain how to simulate Matt's success or failure on one question.

(b) Questions are independent. To simulate the quiz, just simulate 10 questions. Matt needs to get at least 6 questions right to pass the quiz. You could find his probability of passing by simulating many tries at the quiz, but we ask you to simulate just one try. Did Matt pass this quiz?

19.26 More on overbooking. Let's continue the simulation of Exercise 19.24. You offer special vouchers to people who willingly give up their seats when the plane is overbooked. The probability that a passenger will volunteer to accept a special voucher when a plane is overbooked is 0.2. You want to know the probability that some passengers with reservations will be left stranded unwillingly if the plane is overbooked and not enough vouchers are voluntarily taken. Draw a tree diagram with the plane (full or not) as the first stage and the offer of a voucher (at least one passenger will volunteer to accept the voucher offer or nobody will volunteer to accept the voucher offer) is the second stage. In Exercise 19.24, you simulated a number of repetitions of the first stage. Add simulations of the second stage whenever the plane is overbooked. What is your estimate of the probability that a passenger will be stranded unwillingly?

19.27 The birthday problem. A famous example in probability theory shows that the probability that at least two people in a room have the same birthday is already greater than 1-in-2 when 23 people are in the room. The probability model is

- The birth date of a randomly chosen person is equally likely to be any of the 365 dates of the year.

- The birth dates of different people in the room are independent.

To simulate birthdays, let each three-digit group in Table A stand for one person's birth date. That is, 001 is January 1 and 365 is December 31. Ignore leap years and skip groups that don't represent birth dates. Use line 139 of Table A to simulate birthdays of randomly chosen people until you hit the same date a second time. How many people did you look at to find two with the same birthday?

With a computer, you could easily repeat this simulation many times. You could find the probability that at least 2 out of 23 people have the same birthday, or you could find the expected number of people you must question to find two with the same birthday. These problems are a bit tricky to do by math, so they show the power of simulation.

19.28 The multiplication rule. Here is another basic rule of probability: *if several events are independent, the probability that all of the events happen is the product of their individual probabilities*. We know, for example, that a child has probability 0.49 of being a girl and probability 0.51 of being a boy, and that the sexes of successive children are independent. So the probability that a couple's two children are two girls is $(0.49)(0.49) = 0.2401$. You can use this multiplication rule to calculate the probability that we simulated in Example 4.

(a) Write down all eight possible arrangements of the sexes of three children, for example, BBB and BBG. Use the multiplication rule to find the probability of each outcome. Check your work by verifying that your eight probabilities add to 1.

(b) The couple in Example 4 plan to stop when they have a girl or to stop at three children even if all are boys. Use your work from part (a) to find the probability that they get a girl.

Exploring the Web

Access these exercises on the text website: macmillanlearning.com/scc10e.

The House Edge: Expected Values

Case Study: Better Betting?

In this chapter you will:

- Compute the long-run average outcome of a random phenomenon with numerical outcomes.

- Compare games of chance that have huge jackpots but small chances of winning with games with more modest jackpots but more reasonable chances of winning.

If you gamble, you care about how often you will win. The probability of winning tells you what proportion of a large number of bets will be winners. You care even more about *how much* you will win because winning a lot is better than winning a little.

There are a lot of ways to gamble. You can play games, like some of the multistate lotteries, that have enormous jackpots but very small probabilities of winning. You can play games like roulette for which the probability of winning is much larger than for a multistate lottery, but with smaller jackpots. Which is the better gamble: an enormous jackpot with extremely small odds or a modest jackpot with more reasonable odds? By the end of this chapter, you will be able to determine whether buying a multistate lottery ticket or simply playing red in roulette is a better bet.

Expected Values

Gambling on chance outcomes goes back to ancient times and has continued throughout history. Both public and private lotteries were common in the early years of the United States. After disappearing for a century or so, government-run gambling reappeared in 1964, when New Hampshire caused a furor by introducing a lottery to raise public revenue without raising taxes. The furor subsided quickly as larger states adopted the idea. Forty-two states and all Canadian provinces now sponsor lotteries. State lotteries made gambling acceptable as entertainment. Some form of legal gambling is allowed in 48 of the 50 states. More than half of all adult Americans have gambled legally. They spend more betting than on spectator sports, video games, theme parks, and movie tickets combined. If you are going to bet, you should understand what makes a bet good or bad. As our introductory Case Study says, we care about how much we win as well as about our probability of winning.

Example 1
The DC Lottery

Here is a simple lottery wager: the "Straight" from the DC-3 game of the DC Lottery offered by Washington, DC. You pay $1.00 and choose a three-digit number. The lottery chooses a three-digit winning number at random and pays you $500 if your number is chosen. Because there are 1000 three-digit numbers, you have probability 1-in-1000 of winning. Here is the probability model for your winnings:

Outcome:	$0	$500
Probability:	0.999	0.001

What are your average winnings? The ordinary average of the two possible outcomes $0 and $500 is $250, but that makes no sense as the average winnings because $500 is much less likely than $0. In the long run, you win $500 once in every 1000 bets and $0 on the remaining 999 of 1000 bets. (Of course, if you play the game regularly, buying one ticket each time you play, after you have bought exactly 1000 DC-3 tickets, there is no guarantee that you will win exactly once. Probabilities are only *long-run* proportions.) Your long-run average winnings from a ticket are

$$\$500 \frac{1}{1000} + \$0 \frac{999}{1000} = \$0.50$$

or 50 cents. You see that in the long run the state pays out one-half of the money bet and keeps the other half.

Here is a general definition of the kind of "average outcome" we used to evaluate the bets in Example 1.

An expected value is an average of the possible outcomes, but it is not an ordinary average in which all outcomes get the same weight. Instead, each outcome is weighted by its probability so that outcomes that occur more often get higher weights.

Key Terms

The **expected value** of a random phenomenon that has numerical outcomes is found by multiplying each outcome by its probability and then adding all the products.

In symbols, if the possible outcomes are $a_1, a_2, \ldots, a_k$ and their probabilities are $p_1, p_2, \ldots, p_k$, the expected value is

$$\text{expected value} = a_1 p_1 + a_2 p_2 + \cdots + a_k p_k$$

Example 2

The DC Lottery, continued

The Straight wager in Example 1 pays off if you match the three-digit winning number exactly. You can choose instead to make a \$1 Straight/Box-6 way wager. You again choose a three-digit number, but you must choose one with all three digits different. You now have two ways to win. You win \$290 if you exactly match the winning number, and you win \$40 if your number has the same digits as the winning number, in any order. For example, if your number is 123, you win \$290 if the winning number is 123 and \$40 if the winning number is any of 132, 213, 231, 312, and 321. In the long run, you win \$290 once every 1000 bets and \$40 five times for every 1000 bets.

The probability model for the amount you win is

Outcome:	\$0	\$40	\$290
Probability:	0.994	0.005	0.001

The expected value is

$$\text{expected value} = (\$0)(0.994) + (\$40)(0.005) + (\$290)(0.001) = \$0.49$$

We see that the Straight/Box-6 bet is a slightly worse bet than the Straight bet because the state pays out slightly less than half the money bet.

The DC Lottery is an example of one type of lottery game in that it pays a fixed amount for each type of bet. Other states pay off on the "pari-mutuel" system. New Jersey's Pick 3 game is typical: the state pools the money bet and pays out half of it, equally divided among the winning tickets. You still have probability 1-in-1000 of winning a Straight bet, but the amount your number 123 wins depends both on how much was bet on Pick 3 that day and on how many other players chose the number 123. Without fixed amounts, we can't find the expected value of today's bet on 123, but one thing is constant: the state keeps half the money bet.

The idea of expected value as an average applies to random outcomes other than games of chance. It is used, for example, to describe the uncertain return from buying stocks or building a new factory. Here is a different example.

Example 3   *How many vehicles per household?*

What is the average number of motor vehicles in American households? The U.S. Energy Information Administration tells us that the distribution of vehicles per household (2017) is as follows:

Number of vehicles:	0	1	2	3	4	5	6
Proportion of households:	0.09	0.34	0.33	0.15	0.06	0.02	0.01

romakoma/Shutterstock

This is a probability model for choosing a household at random and counting its vehicles. (We ignored the very few households with more than six vehicles.) The expected value for this model is the average number of vehicles per household. This average is

$$\text{expected value} = (0)(0.09) + (1)(0.34) + (2)(0.33) + (3)(0.15) + (4)(0.06) + (5)(0.02) + (6)(0.01)$$
$$= 1.85 \text{ vehicles per household}$$

Now it's your turn

20.1 Number of children. The Census Bureau gives this distribution for the number of a family household's related children under the age of 18 in American households in 2016:

Number of children:	0	1	2	3	4
Proportion:	0.58	0.18	0.16	0.06	0.02

In this table, 4 actually represents 4 or more (the Census Bureau does not provide separate information on households with 5, 6, or more children under the age of 18). But for purposes of this exercise, assume that it means only households with exactly four children under the age of 18. This is also the probability distribution for the number of children under 18 in a randomly chosen household. The expected value of this distribution is the average number of children under 18 in a household. What is this expected value?

The Law of Large Numbers

The definition of "expected value" says that it is an average of the possible outcomes, but an average in which outcomes with higher probability count more. We argued that the expected value is also the average outcome in another sense—it represents the long-run average we will actually see if we repeat a bet many times or choose many households at random. This is more than intuition. Mathematicians can prove, starting from just the basic rules of probability, that the expected value calculated from a probability model really is the "long-run average." This famous fact is called the *law of large numbers*.

The law of large numbers is closely related to the idea of probability. In many independent repetitions, the proportion of each possible outcome will be close to its probability, and the average outcome obtained will be close to the expected value. These facts express the long-run regularity of chance events. They are the true version of the "law of averages," as we mentioned in Chapter 17.

The law of large numbers explains why gambling, which is a recreation or an addiction for individuals, is a business for a casino. The "house" in a gambling operation is not gambling at all. The average winnings of a large number of customers will be quite close to the expected value. The house has calculated the expected value ahead of time and knows what its take will be in the long run. There is no need to load the dice or stack the cards to guarantee a profit. Casinos concentrate on inexpensive entertainment and cheap bus trips to keep the customers flowing in. If enough bets are placed, the law of large numbers guarantees the house a profit. Life insurance companies operate much like casinos—they bet that the people who buy insurance will not die. Some do die, of course, but the insurance company knows the probabilities and relies on the law of large numbers to predict the average amount it will have to pay out. Then the company sets its premiums high enough to guarantee a profit.

Key Terms

According to the **law of large numbers,** if a random phenomenon with numerical outcomes is repeated many times independently, the mean of the actually observed outcomes approaches the expected value.

Statistics in Your World

Rigging the lottery We have all seen televised lottery drawings in which numbered balls bubble about and are randomly popped out by air pressure. How might we rig such a drawing? In 1980, the Pennsylvania lottery was rigged by the host and several stagehands. They injected paint into all balls bearing 8 of the 10 digits. This weighed them down and guaranteed that all 3 balls for the winning number would have the remaining two digits. The perps then bet on all combinations of these digits. When 6-6-6 popped out, they won $1.2 million. Yes, they were caught.

Thinking about Expected Values

As with probability, it is worth exploring a few fine points about expected values and the law of large numbers.

How large is a large number? The law of large numbers says that the actual average outcome of many trials gets closer to the expected value as more trials are made. It doesn't say how many trials are needed to guarantee

Statistics in Your World

High-tech gambling There are more than 700,000 slot machines in the United States. Once upon a time, you put in a coin and pulled the lever to spin three wheels, each with 20 symbols. No longer. Now the machines are video games with flashy graphics and outcomes produced by random number generators. Machines can accept many coins at once, can pay off on a bewildering variety of outcomes, and can be networked to allow common jackpots. Gamblers still search for systems, but in the long run, the random number generator guarantees the house its 5% profit.

an average outcome close to the expected value. That depends on the *variability* of the random outcomes.

The more variable the outcomes, the more trials are needed to ensure that the mean outcome is close to the expected value. Games of chance must be quite variable if they are to hold the interest of gamblers. Even a long evening in a casino has an unpredictable outcome. Gambles with extremely variable outcomes, like state lottos with their very large but very improbable jackpots, require impossibly large numbers of trials to ensure that the average outcome is close to the expected value. (The state doesn't rely on the law of large numbers—most lotto payoffs, unlike casino games, use the pari-mutuel system. In a pari-mutuel system, payoffs and payoff odds are determined by the actual amounts bet. In state lottos, for example, the payoffs are determined by the total amount bet after the state removes its share. In horse racing, payoff odds are determined by the relative amounts bet on the different horses.)

Though most forms of gambling are less variable than the lotto, the practical answer to the applicability of the law of large numbers is that the expected value of the winnings for the house is positive and the house plays often enough to rely on it. Your problem is that the expected value of your winnings is negative. As a group, gamblers play as often as the house. Because their expected value is negative, as a group they lose money over time. However, this loss is not spread evenly among the many individual gamblers. Some win big, some lose big, and some break even. Much of the psychological allure of gambling is its unpredictability for the player. The business of gambling rests on the fact that the result is not unpredictable for the house.

STATISTICAL CONTROVERSIES

The State of Legalized Gambling

Most voters think that some forms of gambling should be legal, and the voting majority has its way: lotteries and casinos are common both in the United States and in other nations. The arguments in favor of allowing gambling are straightforward. Many people find betting entertaining and are willing to lose a bit of money in exchange for some excitement. Gambling doesn't harm other people, at least not directly. A democracy should allow entertainment that a majority supports and that doesn't do harm. State lotteries raise money for good causes such as education and are a kind of voluntary tax that no one is forced to pay.

These are some of the arguments for legalized gambling. What are some of the arguments against legalized gambling? Ask yourself, from which socioeconomic class do people who tend to play the lottery come, and hence who bears the burden for this "voluntary tax"? For more information, see the sources listed in the Notes and Data Sources at the end of this chapter.

Is there a winning system? Serious gamblers often follow a system of betting in which the amount bet on each play depends on the outcome of previous plays. You might, for example, double your bet on each spin of the roulette wheel until you win—or, of course, until your fortune is exhausted. Such a system tries to take advantage of the fact that you have a memory even though the roulette wheel does not. Can you beat the odds with a system? No. Mathematicians have established a stronger version of the law of large numbers that says that if you do not have an infinite fortune to gamble with, your average winnings (the expected value) remain the same as long as successive trials of the game (such as spins of the roulette wheel) are independent. Sorry.

WTV You now have enough information to read the "What's the verdict?" story on page 475, and to answer the six questions.

Finding Expected Values by Simulation

How can we calculate expected values in practice? You know the mathematical recipe, but that requires that you start with the probability of each outcome. Expected values that are too difficult to compute in this way can be found by simulation. The procedure is as before: give a probability model, use random digits to imitate it, and simulate many repetitions. By the law of large numbers, the average outcome of these repetitions will be close to the expected value.

Example 4

We want a girl, again

A couple plan to have children until they have a girl or until they have three children, whichever comes first. We simulated 10 repetitions of this scheme in Example 4 of Chapter 19 (page 450). There, we estimated the probability that they will have a girl among their children. Now we ask a different question: how many children, on the average, will couples who follow this plan have? That is, we want the expected number of children.

The simulation is exactly as before. The probability model says that the sexes of successive children are independent and that each child has probability 0.49 of being a girl. Here are our earlier simulation results—but rather than noting whether the couple did have a girl, we now record the number of children they have. Recall that a pair of digits simulates one child, with 00 to 48 (probability 0.49) standing for a girl.

6905	16	48	17	8717	40	9517	845340	648987	20
B G	G	G	G	B G	G	B G	B B G	B B B	G
2	1	1	1	2	1	2	3	3	1

The mean number of children in these 10 repetitions is

$$\bar{x} = \frac{2+1+1+1+2+1+2+3+3+1}{10} = \frac{17}{10} = 1.7$$

We estimate that if many couples follow this plan, they will average 1.7 children each. This simulation is too short to be trustworthy. Math or a long simulation shows that the actual expected value is 1.77 children.

Now it's your turn

20.2 Stephen Curry's three-point shooting. Stephen Curry makes about 44% of the three-point shots that he attempts. On average, how many three-point shots must he take in a game before he makes his first shot? In other words, we want the expected number of shots he takes before he makes his first. Estimate this by using 10 simulations of sequences of shots, stopping when he makes his first. Use Example 4 to help you set up your simulation. What is your estimate of the expected value?

Chapter 20: Statistics in Summary

- The **expected value** is found as an average of all the possible outcomes, each weighted by its probability.

- When the outcomes are numbers, as in games of chance, we are often interested in the long-run average outcome. The **law of large numbers** says that the mean outcome in many repetitions eventually gets close to the expected value.

- If you don't know the outcome probabilities, you can estimate the expected value (along with the probabilities) by simulation.

This chapter summary will help you evaluate the Case Study.

Link It

In Chapter 17, we discussed the law of averages, both incorrect and correct interpretations. The correct interpretation is sometimes referred to as the law of large numbers. In this chapter, we formally state the law of averages and its relation to the expected value. Understanding the law of large numbers and expected values is helpful in understanding the behavior of games of chance, including state lotteries. Expected values provide a way you can compare games of chance with huge jackpots but small chances of winning with games with more modest jackpots but more reasonable chances of winning.

Case Study Evaluated

Use what you have learned in this chapter to evaluate the Case Study that opened the chapter. Start by reviewing the Chapter Summary. Then answer each of the following questions in complete sentences. Be sure to communicate clearly enough for any of your classmates to understand what you are saying.

1. An American roulette wheel has 38 slots, of which 18 are black, 18 are red, and 2 are green. When the wheel is spun, the ball is equally likely to come to rest in any of the slots. A bet of $1 on red will win $2 (and you will also get back the $1 you bet) if the ball lands in a red slot. (When gamblers bet on red or black, the two green slots belong to the house.) Give a probability model for the winnings of a $1 bet on red and find the expected value of this bet.

2. The website for the Powerball lottery game gives the following table for the various prizes and the probability of winning:

Prize:	Jackpot	$1,000,000	$50,000	$100	$100
Probability:	1 in 292,201,338	1 in 11,688,054	1 in 913,129	1 in 36,525	1 in 14,494

Prize:	$7	$7	$4	$4
Probability:	1 in 580	1 in 701	1 in 92	1 in 38

The jackpot always starts at $40 million for the cash payout for a $2 ticket and grows each time there is no winner. What is the expected value of a $2 bet when the jackpot is $40 million? If there are multiple winners, they share the jackpot, but for purposes of this problem, ignore this.

3. Do you think roulette or the Powerball is the better game to play? Discuss.

> ### In this chapter you have:
>
> - Computed the long-run average outcome of a random phenomenon with numerical outcomes by computing the expected value.
> - Compared games of chance that have huge jackpots but small chances of winning with games with more modest jackpots but more reasonable chances of winning by comparing the expected value of the winnings.

macmillan learning Online Resources

■ LearningCurve has good questions to check your understanding of the concepts.

Check the Basics

For Exercise 20.1, see page 464; for Exercise 20.2, see page 468.

20.3 Expected value. Which of the following is true of the expected value of a random phenomenon?

(a) It must be one of the possible outcomes.

(b) It cannot be one of the possible outcomes because it is an average.

(c) It can only be computed if the random phenomenon has numerical values.

(d) It is the outcome that has the highest probability of occurring.

20.4 Expected value. The expected value of a random phenomenon that has numerical outcomes is

(a) the outcome that occurs with the highest probability.

(b) the outcome that occurs more often than not in a large number of trials.

(c) the average of all possible outcomes.

(d) the average of all possible outcomes, each weighted by its probability.

20.5 Expected value. You flip a coin for which the probability of heads is 0.5 and the probability of tails is 0.5. If the coin comes up heads, you win $1. Otherwise, you lose $1. The expected value of your winnings is

(a) $0.

(b) $1.

(c) −$1.

(d) $0.5.

20.6 The law of large numbers. The law of large numbers says that the mean outcome in many repetitions of a random phenomenon having numerical outcomes

(a) gets close to the expected value as the number of repetitions increases.

(b) goes to zero as the number of repetitions increases because, eventually, positive and negative outcomes balance.

(c) increases steadily as the number of repetitions increases.

(d) must always be a number between 0 and 1.

20.7 **The law of large numbers.** I simulate a random phenomenon that has numerical outcomes many, many times. If I average together all the outcomes I observe, this average

(a) should be close to the probability of the random phenomenon.

(b) should be close to the expected value of the random phenomenon.

(c) should be close to the sampling distribution of the random phenomenon.

(d) should be close to 0.5.

Chapter 20 Exercises

20.8 **The numbers racket.** Pick 3 lotteries (Example 1) copy the numbers racket, an illegal gambling operation common in the poorer areas of large cities. States usually justify their lotteries by donating a portion of the proceeds to education. One version of a numbers racket operation works as follows. You choose any one of the 1000 three-digit numbers 000 to 999 and pay your local numbers runner $1 to enter your bet. Each day, one three-digit number is chosen at random and pays off $600. What is the expected value of a bet on the numbers? Is the numbers racket more or less favorable to gamblers than the Pick 3 game in Example 1?

20.9 **DC-4.** The "Straight" DC-4 lottery game is much like the "Straight" DC-3 game of Example 1. Winning numbers for both are reported on television and in local newspapers. You pay $1.00 and pick a four-digit number. The state chooses a four-digit number at random and pays you $5000 if your number is chosen. What are the expected winnings from a $1.00 Straight DC-4 wager?

20.10 **More DC-3.** There are other elaborate versions of DC-3 (Example 2). In the $1 Straight-Box (3-way) bet, if you choose a number with two digits the same, for example, 112, you win $330 if the randomly chosen winning number is 112, and you win $80 if the winning number has the digits 1, 1, and 2 in any other order (there are 2 such other orders, namely 121 and 211). What is the expected amount you win?

20.11 **More roulette.** An American roulette wheel has 38 slots, of which 18 are black, 18 are red, and 2 are green. When the wheel is spun, the ball is equally likely to come to rest in any of the slots. Gamblers bet on roulette by placing chips on a table that lays out the numbers and colors of the 38 slots in the roulette wheel. The red and black slots are arranged on the table in three columns of 12 slots each. A $1 column bet wins $3 if the ball lands in one of the 12 slots in that column. What is the expected amount such a bet wins? If you did the Case Study Evaluated, is a column bet more or less favorable to a gambler than a bet on red or black (see Question 1 in the Case Study Evaluated on page 469)?

20.12 **Making decisions.** The psychologist Amos Tversky did many studies of our perception of chance behavior. In its obituary of Tversky, the *New York Times* cited the following example.

(a) Tversky asked subjects to choose between two public health programs that affect 600 people. The first has probability 1/2 of saving all 600 and probability 1/2 that all 600 will die. The other is guaranteed to *save* exactly 400 of the 600 people. Find the expected number of people saved by the first program.

(b) Tversky then offered a different choice. One program has probability 1/2 of saving all 600 and probability 1/2 of losing all 600, while the other is guaranteed to *lose* exactly 200 of the 600 lives. What is the difference between this choice and that in part (a)?

(c) Given the options in part (a), most subjects choose the second program. Given the options in part (b), most subjects choose the first program. Do the subjects appear to use expected values in their choice? Why do you think the choices differ in parts (a) and (b)?

20.13 Making decisions. A six-sided die has two green and four red faces and is balanced so that each face is equally likely to come up. You must choose one of the following three sequences of colors:

RGRRR

RGRRRG

GRRRRR

Now start rolling the die. You will win $25 if the first rolls give the sequence you chose.

(a) Which sequence has the highest probability? Why? (You can see which is most probable without actually finding the probabilities.) Because the $25 payoff is fixed, the most probable sequence has the highest expected value.

(b) In a psychological experiment, 63% of 260 students who had not studied probability chose the second sequence. Based on the discussion of "myths about chance behavior" in Chapter 17, explain why most students did not choose the sequence with the best chance of winning.

20.14 Estimating sales. Gain Communications sells aircraft communications units. Next year's sales depend on market conditions that cannot be predicted exactly. Gain follows the modern practice of using probability estimates of sales. The sales manager estimates next year's sales as follows:

Units sold:	6,000	7,000	8,000	9,000	10,000
Probability:	0.1	0.2	0.4	0.2	0.1

These are personal probabilities that express the informed opinion of the sales manager. What is the sales manager's expected value of next year's sales?

20.15 Keno. Keno is a popular game in casinos. Balls numbered 1 to 80 are tumbled in a machine as the bets are placed, then 20 of the balls are chosen at random. Players select

numbers by marking a card. Here are two of the simpler Keno bets. Give the expected winnings for each.

(a) A $1 bet on "Mark 1 number" pays $3 if the single number you mark is one of the 20 chosen; otherwise, you lose your dollar.

(b) A $1 bet on "Mark 2 numbers" pays $12 if both your numbers are among the 20 chosen. The probability of this is about 0.06. Is Mark 2 a more or a less favorable bet than Mark 1?

20.16 Rolling two dice. Example 2 of Chapter 18 (page 428) gives a probability model for rolling two casino dice and recording the number of spots on each of the two up-faces. That example also shows how to find the probability that the total number of spots showing is five. Follow that method to give a probability model for the *total* number of spots. The possible outcomes are 2, 3, 4, . . . , 12. Then use the probabilities to find the expected value of the total.

20.17 The Asian stochastic beetle. We met this insect in Exercise 19.21 (page 457). Females have this probability model for their number of female offspring:

Offspring:	0	1	2
Probability:	0.2	0.3	0.5

(a) What is the expected number of female offspring?

(b) Use the law of large numbers to explain why the population should grow if the expected number of female offspring is greater than 1 and die out if this expected value is less than 1.

20.18 An expected rip-off? A "psychic" runs the following ad in a magazine:

Expecting a baby? Renowned psychic will tell you the sex of the unborn child from any photograph of the mother. Cost, $20. Money-back guarantee.

This may be a profitable con game. Suppose that the psychic simply replies "boy" to all inquiries. In the worst case, everyone who has a girl will ask for her money back. Find the expected value of the psychic's profit by filling in the table below.

Sex of child	Probability	The psychic's profit
Boy	0.51	
Girl	0.49	

20.19 The Asian stochastic beetle again. In Exercise 20.17, you found the expected number of female offspring of the Asian stochastic beetle. Simulate the offspring of 100 beetles and find the mean number of offspring for these 100 beetles. Compare this mean with the expected value from Exercise 20.17. (The law of large numbers says that the mean will be very close to the expected value if we simulate enough beetles.)

20.20 Life insurance. You might sell insurance to a 20-year-old friend. The probability that a man aged 20 will die in the next year is about 0.0007. You decide to charge $2000 for a policy that will pay $1 million if your friend dies.

(a) What is your expected profit on this policy?

(b) Although you expect to make a good profit, you would be foolish to sell a single policy only to your friend. Why?

(c) A life insurance company that sells thousands of policies, on the other hand, would do very well selling policies on exactly these same terms. Explain why.

20.21 Household size. The Census Bureau gives this distribution for the number of people in American households in 2016:

Family size:	1	2	3	4	5	6	7
Proportion:	0.28	0.35	0.15	0.13	0.06	0.02	0.01

(*Note*: In this table, 7 actually represents households of size 7 or greater. But for purposes of this exercise, assume that it means only households of size exactly 7.)

(a) This is also the probability distribution for the size of a randomly chosen household. The expected value of this distribution is the average number of people in a household. What is this expected value?

(b) Suppose you take a random sample of 1000 American households. About how many of these households will be of size 2? Sizes 3 to 7?

(c) Based on your calculations in part (b), how many people are represented in your sample of 1000 households? (*Hint*: The number of individuals in your sample who live in households of size 7 is 7 times the number of households of size 7. Repeat this reasoning to determine the number of individuals in your sample who live in households of sizes 2 to 6. Add the results to get the total number of people represented in your sample.)

(d) Calculate the probability distribution for the household size lived in by individual people. Describe the shape of this distribution. What does this shape tell you about household structure?

20.22 An easy A. The distribution of grades in an easy introductory statistics course is as follows:

Grade:	A	B	C	D	F
Probability:	0.4	0.3	0.2	0.1	0.0

To calculate student grade point averages, grades are expressed in a numerical scale with A = 4, B = 3, and so on down to F = 0.

(a) Find the expected value. This is the average grade in this course.

(b) Explain how to simulate choosing students at random and recording their grades. Simulate 50 students and find the mean of their 50 grades. Compare this estimate of the expected value with the exact expected value from part (a). (The law of large numbers says that the estimate will be very accurate if we simulate a very large number of students.)

20.23 We really want a girl. Example 4 estimates the expected number of children a couple will have if they keep going until they get a girl or until they have three children. Suppose that they set no limit on the number of children but just keep going until they get a girl. Their expected number of children must now be higher than in Example 4. How would you simulate such a couple's children? Simulate 25 repetitions. What is your estimate of the expected number of children?

20.24 Play this game, please. OK, friends, we've got a little deal for you. We have a fair coin (heads and tails each have probability 0.5). Toss it twice. If two heads come up, you win right there. If you get any result other than two heads, we'll give you another chance: toss the coin twice more, and if you get two heads, you win. (Of course, if you fail to get two heads on the second try, we win.) Pay us a dollar to play. If you win, we'll give you your dollar back plus another dollar.

(a) Make a tree diagram for this game. Use the diagram to explain how to simulate one play of this game.

(b) Your dollar bet can win one of two amounts: 0 if we win and $2 if you win. Simulate 50 plays, using Table A, starting at line 125. Use your simulation to estimate the expected value of the game.

20.25 A multiple-choice exam. Charlene takes a quiz with 10 multiple-choice questions, each with five answer choices. If she just guesses independently at each question, she has probability 0.20 of guessing right on each. Use simulation to estimate Charlene's expected number of correct answers. (Simulate 20 repetitions.)

20.26 Repeating an exam. Exercise 19.19 (page 457) gives a model for up to three attempts at an exam in a self-paced course. In that exercise, you simulated 50 repetitions to estimate Elaine's probability of passing the exam. Use those simulations (or do 50 new repetitions) to estimate the expected number of tries Elaine will make.

20.27 A common expected value. Here is a common setting that we simulated in Chapter 19: there are a fixed number of independent trials with the same two outcomes and the same probabilities on each trial. Tossing a coin, shooting basketball free throws, and observing the sex of newborn babies are all examples of this setting. Call the outcomes "hit" and "miss." We can see what the expected number of hits should be. If Stephen Curry shoots 12 three-point shots and has probability 0.44 of making each one, the expected number of hits is 44% of 12, or 5.28. By the same reasoning, if we have n trials with probability p of a hit on each trial, the expected number of hits is np. This fact can be proved mathematically. Can we verify it by simulation?

Simulate 10 tosses of a fair coin 50 times. (To do this quickly, use the first 10 digits in each of the 50 rows of Table A, with odd digits meaning a head and even digits a tail.) What is the expected number of heads by the np formula? What is the mean number of heads in your 50 repetitions?

20.28 Casino winnings. What is a secret, at least to naive gamblers, is that in the real world, a casino does much better than expected values suggest. In fact, casinos keep a bit over 20% of the money gamblers spend on roulette chips. That's because players who win keep on playing. Think of a player who gets back exactly 95% of each dollar bet. After one bet, he has 95 cents.

(a) After two bets, how much does he have of his original dollar bet? (*Hint*: This will be 95% of the 95 cents he has after the first bet.)

(b) After three bets, how much does he have of his original dollar bet?

Notice that the longer he keeps recycling his original dollar, the more of it the casino keeps. Real gamblers don't get a fixed percentage back on each bet, but even the luckiest will lose his stake if he plays long enough. The casino keeps 5.3 cents of every individual one-dollar bet but a total of 20 cents of every dollar spent on betting (due to recycling of the original bet).

Exploring the Web

Access these exercises on the text website: macmillanlearning.com/scc10e.

What's the Verdict?

On Tuesday October 23, 2018, the Mega Millions lottery had a grand prize jackpot of $1.6 billion. This unusually large prize excited many people who would not normally buy tickets to play this lottery for the first time.

For the Mega Millions game, each player selects six numbers that they hope will match with those randomly selected by the company on the big night.

According to the Mega Millions website: http://www.megamillions.com/how-to-play, the odds of winning the jackpot and some lesser prizes are listed in Figure 18.7.

Questions

WTV20.1. Find the expected value for the outcome of playing one ticket (for now, ignore the cost of the ticket and ignore the possibility of multiple jackpot winners). Use the following table based on a jackpot of $1.6 billion.

Outcome:	+$1,600,000,000	+$1,000,000
Probability:	1/302,575,350	1/12,607,306

Outcome:	+$10,000	+$500	+$200
Probability:	1/931,001	1/38,792	1/14,547

Outcome:	+$10	+$10	+$4	+$2	$0
Probability:	1/606	1/693	1/89	1/37	23/24

WTV20.2. If each lottery ticket costs $2, what is your expected value of the difference in your wallet? (*Hint:* You do not have to start over with a new table or long calculation. Use your previous expected value work and the new $2 ticket cost to do a simple calculation.)

WTV20.3. Using the expected value of the difference in your wallet, does this seem like a game that would favor you winning something? Why?

WTV20.4. Jackpots for the Mega Millions start at $40 million and grow each time there is no winner. What is the expected value in your wallet if the jackpot is $40 million? Repeat the calculations you did above for the $1.6 billion jackpot using the following table.

Outcome:	+$40,000,000	+$1,000,000	+$10,000
Probability:	1/302,575,350	1/12,607,306	1/931,001

Outcome:	+$500	+$200	+$10
Probability:	1/38,792	1/14,547	1/606

Outcome:	+$10	+$4	+$2	$0
Probability:	1/693	1/89	1/37	23/24

WTV20.5. (This is a more challenging question.) What value does the jackpot need to be for the expected value in your wallet to be 0? (Ignore the possibility of splitting the jackpot if there are multiple winners.)

WTV20.6. Do you want to play the Mega Millions? Why or why not?

PART III Review

Some phenomena are random. That is, although their individual outcomes are unpredictable, there is a regular pattern in the long run. Gambling devices (rolling dice, spinning roulette wheels) and taking a simple random sample (SRS) are examples of random phenomena. Probability and expected value give us a language to describe randomness. Random phenomena are not haphazard or chaotic any more than random sampling is haphazard. Randomness is instead a kind of order in the world, a long-run regularity as opposed to either chaos or a determinism that fixes events in advance. Chapter 17 discusses the idea of randomness, Chapter 18 presents basic facts about probability, Chapter 19 shows how to use simulation to estimate probabilities, and Chapter 20 discusses expected values.

Tom Schoumakers/Shutterstock

When randomness is present, probability answers the question, "How often in the long run?" and expected value answers the question, "How much on the average in the long run?" The two answers are connected because "expected value" is defined in terms of probabilities. Much work with probability starts with a probability model that assigns probabilities to the basic outcomes. Any such model must obey the rules of probability. Another example of a probability model uses a density curve such as a Normal curve to assign probabilities as areas under the curve. Personal probabilities express an individual's judgment of how likely some event is. Personal probabilities for several possible outcomes must also follow the rules of probability if they are to be consistent with each other.

To calculate the probability of a complicated event without using complicated math, we can use random digits to simulate many repetitions. You can also find expected values by simulation. Chapter 19 shows how to do simulations. First, give a probability model for the outcomes, then assign random digits to imitate the assignment of probabilities. The table of random digits now imitates repetitions. Keep track of the proportion of repetitions on which an event occurs to estimate its probability. Keep track of the mean outcome to estimate an expected value.

PART III SUMMARY

Here are the most important skills you should have acquired after reading Chapters 17 through 20.

A. RANDOMNESS AND PROBABILITY

1. Recognize that some phenomena are random. Understand that probability describes the long-run regularity of random phenomena.

2. Understand the idea of the probability of an event as the proportion of times the event occurs in very many repetitions of a random phenomenon. Use the idea of probability as long-run proportion to think about probability.

3. Recognize that short runs of random phenomena do not display the regularity described by probability. Understand that randomness is unpredictable in the short run. Avoid seeking causal explanations for random occurrences.

B. PROBABILITY MODELS

1. Use basic probability facts to detect illegitimate assignments of probability: any probability must be a number between 0 and 1, and the sum of the probabilities assigned to all possible outcomes must be 1.

2. Use basic probability facts to find the probabilities of events that are formed from other events: the probability that an event does not occur is 1 minus its probability. If two events cannot occur at the same time, the probability that one or the other occurs is the sum of their individual probabilities.

3. When probabilities are assigned to individual outcomes, find the probability of an event by adding the probabilities of the outcomes that make it up.

4. When probabilities are assigned by a Normal curve, find the probability of an event by finding an area under the curve.

C. SIMULATION

1. Specify simple probability models that assign probabilities to each of several stages when the stages are independent of each other.

2. Assign random digits to simulate such models.

3. Estimate either a probability or an expected value by repeating a simulation many times.

D. EXPECTED VALUE

1. Understand the idea of expected value as the average of numerical outcomes in very many repetitions of a random phenomenon.

2. Find the expected value from a probability model that lists all outcomes and their probabilities (when the outcomes are numerical).

PART III REVIEW EXERCISES

Review exercises are short and straightforward exercises that help you solidify the basic ideas and skills in each part of this book. We have provided "hints" that indicate where you can find the relevant material for the odd-numbered problems.

III.1 What's the probability? Find an online directory with at least 100 different phone numbers. Look at the last four digits of each telephone number, the digits that specify an individual number within an exchange given by the first three digits. Note the first of these four digits in each of the first 100 telephone numbers in the directory.

Three directories we found online were at education.ohio.gov/Contact/Phone-Directory, cityfone.lacity.org/department_drilldown .cfm?SECT=a, and www.somervillema.gov /staff-contact-directory. State government agencies often have directories with office phone numbers of employees. (If you happen to have access to a printed directory, you can use that instead of an online directory.)

(a) How many of the digits are 1, 2, or 3? What is the approximate probability that the first of the four "individual digits" in a telephone number is 1, 2, or 3? (*Hint*: See page 407.)
(b) If all 10 possible digits had the same probability, what would be the probability of getting a 1, 2, or 3? Based on your work in part (a), do you think the first of the four "individual digits" in telephone numbers is equally likely to be any of the 10 possible digits? (*Hint*: See page 427.)

III.2 Eye colors. Choose a person at random and record his or her eye colors. Here are the probabilities for each eye color:

Eye color:	Brown	Blue	Hazel/amber	Other
Probability:	0.7	0.1	0.1	?

(a) What must be the probability that a randomly chosen person has a color other than brown, blue, hazel, or amber?

(b) To simulate the eye colors of randomly chosen people, how would you assign digits to represent the four types?

III.3 Grades in an economics course. Indiana University posts the grade distributions for its courses online. Students in Economics 201 in the spring 2018 semester received the following: 1% A+, 7% A, 8% A−, 8% B+, 16% B, 14% B−, 10% C+, 13% C, 6% C−, 5% D+, 4% D, 3% D−. Choose an Economics 201 student at random. The probabilities for the student's grade are the following:

Grade	A+	A	A−	B+	B	B−
Probability	0.01	0.07	0.08	0.08	0.16	0.14

Grade	C+	C	C−	D+	D	D−	F
Probability	0.10	0.13	0.06	0.05	0.04	0.03	?

(a) What must be the probability of getting an F? (*Hint*: See page 427.)
(b) To simulate the grades of randomly chosen students, how would you assign digits to represent the five possible outcomes listed? (*Hint*: See page 450.)

III.4 Eye colors. Tyra has amber eyes. What is the probability that one or more of Tyra's six close friends has amber- or hazel-colored eyes? Using your work in Exercise III.2, simulate 10 repetitions and estimate this probability. (Your estimate from just 10 repetitions isn't reliable, but you have shown in principle how to find the probability.)

III.5 Grades in an economics course. If you choose four students at random from all those who have taken the course described in Exercise III.3, what is the probability that all the students chosen got a B or better? Simulate 10 repetitions of this random choosing and use your results to estimate the probability. (Your estimate from only 10 repetitions isn't reliable, but if you can do 10, you could do 10,000.) (*Hint*: See page 450.)

III.6 Grades in an economics course. Choose a student at random from the course described in Exercise III.3 and observe what grade that student earns (with A+ = 4.3, A = 4.0, A− = 3.7, B+ = 3.3, B = 3.0, B− = 2.7, C+ = 2.3, C = 2.0, C− = 1.7, D+ = 1.3, D = 1.0, D− = 0.7, and F = 0.0).
(a) What is the expected grade of a randomly chosen student?
(b) The expected grade is not one of the 12 grades possible for one student. Explain why your result nevertheless makes sense as an expected value.

III.7 Dice. What is the expected number of spots observed in rolling a carefully balanced die once? (*Hint*: See page 463.)

III.8 Profit from a risky investment. Rotter Partners is planning a major investment. The amount of profit X is uncertain, but a probabilistic estimate gives the following distribution (in millions of dollars):

Profit:	−1	0	1	2	3	5	20
Probability:	0.1	0.1	0.2	0.2	0.2	0.1	0.1

What is the expected value of the profit?

III.9 Poker. Deal a five-card poker hand from a shuffled deck. The probabilities of several types of hand are approximately as follows:

Hand:	Worthless	One pair	Two pairs	Better hands
Probability:	0.50	0.42	0.05	?

(a) What must be the probability of getting a hand better than two pairs? (*Hint*: See page 427.)
(b) What is the expected number of hands a player is dealt before the first hand better than one pair appears? Explain how you would use simulation to answer this question, then simulate just two repetitions. (*Hint*: See page 450.)

III.10 How much education? The Census Bureau gives this distribution of the highest level of education attained for a randomly chosen American over 25 years old in 2014:

Education:	Less than high school	High school graduate	College, no bachelor's
Probability:	0.104	0.288	0.163

Education:	Associate's degree	Bachelor's degree	Advanced degree
Probability:	0.103	0.213	0.129

(a) How do you know that this is a legitimate probability model?
(b) What is the probability that a randomly chosen person over age 25 has at least a high school education?
(c) What is the probability that a randomly chosen person over age 25 has at least a bachelor's degree?

III.11 Language study. Choose a student in grades K to 12 at random and ask if he or she is studying a language other than English. Here is the distribution of results:

Language:	Spanish	French	German	Chinese	All others	None
Probability:	0.136	0.024	0.006	0.004	0.026	0.804

(a) Explain why this is a legitimate probability model. (*Hint*: See page 427.)
(b) What is the probability that a randomly chosen student is studying a language other than English? (*Hint*: See page 427.)
(c) What is the probability that a randomly chosen student is studying French, German, or Spanish? (*Hint*: See page 427.)

III.12 Choosing at random. Abby, Deborah, Mei-Ling, Sam, and Roberto work in a firm's public relations office. Their employer must choose two of them to attend a conference in Paris. To avoid unfairness, the choice will be made by drawing two names from a hat. (This is an SRS of size 2.)
(a) Write down all possible choices of two of the five names. These are the possible outcomes.

(b) The random drawing makes all outcomes equally likely. What is the probability of each outcome?
(c) What is the probability that Mei-Ling is chosen?
(d) What is the probability that neither of the two men (Sam and Roberto) is chosen?

III.13 Languages in Canada. Canada has two official languages: English and French. Choose a resident of Quebec at random and ask, "What is your mother tongue?" Here is the distribution of responses, combining many separate languages from the province of Quebec:

Language:	English	French	Other
Probability:	0.083	0.789	?

(a) What is the probability that a randomly chosen resident of Quebec's mother tongue is either English or French? (*Hint*: See page 427.)
(b) What is the probability that a randomly chosen resident of Quebec's mother tongue is "Other"? (*Hint*: See page 427.)

III.14 Confidence in public schools. (From the news) A 2018 Gallup poll asked adult Americans how much confidence they had in the public schools. Assume that the results of the poll accurately reflect the opinions of all adult Americans. Here is the distribution of responses:

Response:	Great deal	Quite a lot	Some	Very little	None
Probability:	0.12	0.17	0.44	0.25	?

(a) What is the probability that a randomly chosen adult American has no confidence in the public schools?
(b) What is the probability that a randomly chosen adult American has a great deal or quite a lot of confidence in the public schools?

III.15 An IQ test. The Wechsler Adult Intelligence Scale (WAIS) is a common IQ test for adults. The distribution of WAIS scores for persons over 16 years of age is approximately Normal with mean 100 and standard deviation 15. Use the 68–95–99.7 rule to answer these questions.

(a) What is the probability that a randomly chosen individual has a WAIS score of 115 or higher? (*Hint*: See page 432.)

(b) In what range do the scores of the middle 95% of the adult population lie? (*Hint*: See page 432.)

III.16 Worrying about crime. How much do Americans worry about crime and violence? Suppose that 50% of all adults worry a great deal about crime and violence. (According to a 2018 sample survey that asked this question, 50% is about right.) A polling firm chooses an SRS of 1100 people. If they do this many times, the percentage of the sample who say they worry a great deal will vary from sample to sample following a Normal distribution with mean 50% and standard deviation 1.5%. Use the 68–95–99.7 rule to answer these questions.

(a) What is the probability that one such sample gives a result within ±1.5% of the truth about the population?

(b) What is the probability that one such sample gives a result within ±3% of the truth about the population?

III.17 An IQ test (optional). Use the information in Exercise III.15 and Table B at the end of this text to find the probability that a randomly chosen person has a WAIS score of 112 or higher. (*Hint*: See page 433.)

III.18 Worrying about crime (optional). Use the information in Exercise III.16 and Table B at the end of this text to find the probability that one sample misses the truth about the population by 2.5% or more. (This is the probability that the sample result is either less than 47.5% or greater than 52.5%.)

III.19 An IQ test (optional). How high must a person score on the WAIS test to be in the top 10% of all scores? Use the information in Exercise III.15 and Table B at the end of this text to answer this question. (*Hint*: See page 433.)

III.20 Models, legitimate and not. A bridge deck contains 52 cards, four of each of the 13 face values ace, king, queen, jack, ten, nine, . . . , two. You deal a single card from such a deck and record the face value of the card dealt. Give an assignment of probabilities to the possible outcomes that should be correct if the deck is thoroughly shuffled. Give a second assignment of probabilities that is legitimate (that is, obeys the rules of probability) but differs from your first choice. Then give a third assignment of probabilities that is *not* legitimate, and explain what is wrong with this choice.

III.21 Mendel's peas. Gregor Mendel used garden peas in some of the experiments that revealed that inheritance operates randomly. The seed color of Mendel's peas can be either green or yellow. Suppose we produce seeds by "crossing" two plants, both of which carry the G (green) and Y (yellow) genes. Each parent has probability 0.5 of passing each of its genes to a seed, independently of the other parent. A seed will be yellow unless both parents contribute the G gene. Seeds that get two G genes are green.

What is the probability that a seed from this cross will be green? Set up a simulation to answer this question, and estimate the probability from 25 repetitions. (*Hint*: See page 446.)

III.22 Predicting the winner. There are eight teams in the Eastern Division of the National Hockey League. Here's one set of personal probabilities for next year's Eastern Division champion: The Tampa Bay Lightning and the Toronto Maple Leafs have probability 0.3 of winning. The Ottawa Senators, the Detroit Red Wings, and the Florida Panthers, have no chance. That leaves three teams. The Boston Bruins, and the Montreal Canadiens both have the same probability of winning, but that probability is one-half that of the Buffalo Sabres. What probability does each of the eight teams have?

III.23 Selling cars. Bill sells new cars in a small town for a living. On a weekday afternoon, he will deal with one customer with probability 0.6, two customers with probability 0.3, and three customers with probability 0.1. Each customer has probability 0.2 of buying a car. Customers buy independently of each other.

Describe how you would simulate the number of cars Bill sells in an afternoon. You must first simulate the number of customers, then simulate the buying decisions of one, two, or three customers. Simulate one afternoon to demonstrate your procedure. (*Hint*: See page 450.)

PART PROJECTS

Projects are longer exercises that require gathering information or producing data and that emphasize writing a short essay to describe your work. Many are suitable for teams of students.

Access these projects on the text website: **macmillanlearning.com/scc10e.**

PART IV

Inference

To *infer* is to draw a conclusion from evidence. *Statistical inference* draws a conclusion about a population from evidence provided by a sample. Drawing conclusions in mathematics is a matter of starting from a hypothesis and using logical argument to prove without doubt that the conclusion follows. Statistics isn't like that. Statistical conclusions are uncertain because the sample isn't the entire population. So statistical inference has to not only state conclusions but also say how uncertain they are. We use the language of probability to express uncertainty.

Because inference must both give conclusions and say how uncertain they are, it is the most technical part of statistics. Texts and courses intended to train people to *do* statistics spend most of their time on inference. Our aim in this book is to help you *understand* statistics, which takes less technique but often more thought. We will look only at a few basic techniques of inference. The techniques are simple, but the ideas are subtle, so prepare to think. To start, think about what you already know and don't be too impressed by elaborate statistical techniques: even the fanciest inference cannot remedy basic flaws such as voluntary response samples or uncontrolled experiments.

manorial1/Shutterstock

What Is a Confidence Interval?

21

Case Study: Social Media Use

In this chapter you will:

- Learn how to construct confidence intervals for proportions and means.

- Be able to interpret what confidence intervals represent.

What is the most popular social media platform? It might depend on who you ask. In 2018, the Pew Research Center published a report on social media use. To obtain their sample, random digit dialing was used. The sample included 2002 adults, 18 years of age or older, from across the United States. A total of 1502 survey respondents were interviewed on a cell phone, and the remaining 500 survey respondents were interviewed on a landline phone. Survey respondents were 18 years of age and older.

When asked which social media platforms they used, 75% of the 2002 survey respondents said they used YouTube. Facebook was used by 68% of the 2002 respondents, and over one-third, or 35%, of the survey respondents reported using Instagram. Snapchat (27%) and Twitter (24%) were used less often by all survey respondents. When results were broken down by age group, however, a different pattern of results emerged. Of the 201 survey respondents between the ages of 18 and 24 years, 94% reported using YouTube, 80% reported using Facebook, 71% reported using Instagram, 78% reported using Snapchat, and 45% reported using Twitter. We know these numbers won't be exactly accurate for the entire population of 18- to 24-year-olds, but are they close? By the end of this chapter, you will be able to see how accurate numbers like 94%, 80%, 71%, 78%, and 45% are.

Estimating

Statistical inference draws conclusions about a population on the basis of data about a sample. One kind of conclusion answers questions like "What percentage of employed women have a college degree?" or "What is the mean survival time for patients with this type of cancer?" These questions ask about a number (a percentage, a mean) that describes a population. Numbers that describe a population are **parameters.** To estimate a population parameter, choose a sample from the population and use a **statistic,** a number calculated from the sample, as your estimate. Here's an example.

Example 1

Graduation plans

The National Survey of Student Engagement (NSSE) is administered annually to first-year and senior undergraduate students across the United States and Canada. The purpose of the NSSE is to better understand the extent to which these students engage in educational practices associated with high levels of learning and development. In 2018, the NSSE was completed by students from 511 institutions of higher learning. One question graduating seniors were asked was "After graduation, what best describes your immediate plans?" Of the 23,915 seniors

Dmytro Zinkevych/Shutterstock

who answered this question, 5038 indicated they planned to go to graduate or professional school. Based on this information, what can we say about the percentage of all college seniors who plan to go to graduate or professional school?

Our population is college seniors who reside in the United States or Canada. The parameter is the proportion who plan to go to graduate or professional school.

Call this unknown parameter p, for "proportion." The statistic that estimates the parameter p is the **sample proportion**

$$\hat{p} = \frac{\text{count in the sample}}{\text{size of the sample}} = \frac{5038}{23915} = 0.211$$

A basic move in statistical inference is to use a sample statistic to estimate a population parameter. Once we have the sample in hand, we estimate that the proportion of all college seniors in the United States and Canada who plan to go to graduate or professional school is "about 21.1%" because the proportion in the sample was exactly 21.1%. We can only estimate that the truth about the population is "about" 21.1% because we know that the sample result is unlikely to be exactly the same as the true population proportion. A confidence interval makes that "about" precise.

We will first calculate the interval for a population proportion, and then we will reflect on what we have done and generalize a bit.

Estimating with Confidence

We want to estimate the proportion p of the individuals in a population who have some characteristic—they are employed or they approve of the president's performance, for example. Let's call the characteristic we are looking for a "success." We use the proportion $\hat{p}$ of successes in a simple random sample (SRS) to estimate the proportion p of successes in the population. How good is the statistic $\hat{p}$ as an estimate of the parameter p? To find out, we ask, "What would happen if we took many samples?" Well, we know that $\hat{p}$ would vary from sample to sample. We also know that this sampling variability isn't haphazard. It has a clear pattern in the long run, a pattern that is pretty well described by a Normal curve. The facts are given in the box at the right.

These facts can be proved by mathematics, so they are a solid starting point. Figure 21.1 summarizes them in a form that also reminds us that a sampling distribution describes the results of lots of samples from the same population.

Sampling distribution of a sample proportion

The **sampling distribution** of a statistic is the distribution of values taken by the statistic in all possible samples of the same size from the same population.

Take an SRS of size n from a large population that contains proportion p of successes. Let $\hat{p}$ be the **sample proportion** of successes,

$$\hat{p} = \frac{\text{count of successes in the sample}}{n}$$

Then, if the sample size is large enough:

- The sampling distribution of $\hat{p}$ is **approximately Normal.**
- The **mean** of the sampling distribution of $\hat{p}$ is p.
- The **standard deviation** of the sampling distribution of $\hat{p}$ is

$$\sqrt{\frac{p(1-p)}{n}}$$

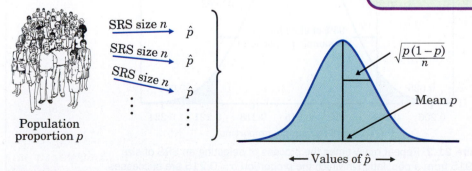

Figure 21.1 Repeat many times the process of selecting an SRS of size n from a population in which the proportion p are successes. The values of the sample proportion of successes $\hat{p}$ have this Normal sampling distribution.

Example 2

More on graduation plans

Suppose, for example, that the truth is that 21.5% of college seniors in the United States and Canada plan to go to graduate or professional school. Then, in the setting of Example 1, $p = 0.215$. The NSSE sample of size $n = 23{,}915$ would, if repeated many times, produce sample proportions $\hat{p}$ that closely follow the Normal distribution with

$$\text{mean} = p = 0.215$$

and

$$\begin{aligned}\text{standard error} &= \sqrt{\frac{p(1-p)}{n}}\\ &= \sqrt{\frac{(0.215)(0.785)}{23915}}\\ &= \sqrt{0.000007057} = 0.00266\end{aligned}$$

The center of this Normal distribution is at the truth about the population. That's in the absence of bias in random sampling once again. The standard error is small because the sample is quite large. So, almost all samples will produce a statistic $\hat{p}$ that is close to the true p. In fact, the 95 part of the 68–95–99.7 rule says that 95% of all sample outcomes will fall between

$$\text{mean} - 2 \text{ standard errors} = 0.215 - 0.0053 = 0.2097$$

and

$$\text{mean} + 2 \text{ standard errors} = 0.215 + 0.0053 = 0.2203$$

Figure 21.2 displays these facts.

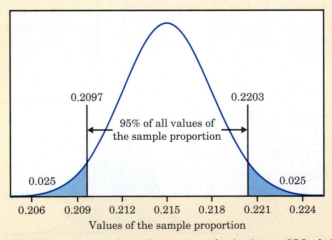

Figure 21.2 Repeat many times the process of selecting an SRS of size 23,915 from a population in which the proportion $p = 0.215$ are successes. The middle 95% of the values of the sample proportion $\hat{p}$ will lie between 0.2097 and 0.2203.

So far, we have just put numbers on what we already knew: we can trust the results of large random samples because almost all such samples give results that are close to the truth about the population. The numbers say that in 95% of all samples of size 23,915, the statistic $\hat{p}$ and the parameter p are within 0.0053 of each other. We can put this another way: 95% of all samples give an outcome $\hat{p}$ such that the population truth p is captured by the interval from $\hat{p} - 0.0053$ to $\hat{p} + 0.0053$.

The 0.0053 came from substituting $p = 0.215$ into the formula for the standard error of $\hat{p}$. For any value of p, the general fact is

> When the population proportion has the value p, 95% of all samples catch p in the interval extending 2 standard errors on either side of $\hat{p}$.

That's the interval

$$\hat{p} \pm 2\sqrt{\frac{p(1-p)}{n}}$$

Is this the 95% confidence interval we want? Not quite. The interval can't be found just from the data because the standard deviation involves the population proportion p, and in practice, we don't know p. In Example 2, we used $p = 0.215$ in the formula, but this may not be the true p.

What can we do? Well, the standard deviation of the statistic $\hat{p}$, or the standard error, does depend on the parameter p, but it doesn't change a lot when p changes. Go back to Example 2 and redo the calculation for other values of p. Here's the result:

Value of p:	0.20	0.205	0.21	0.215	0.22
Standard error:	0.00259	0.00261	0.00263	0.00266	0.00268

We see that, if we guess a value of p reasonably close to the true value, the standard error found from the guessed value will be close to the true value (or the true standard error). We know that, when we take a large random sample, the statistic $\hat{p}$ is almost always close to the parameter p. So, we will use $\hat{p}$ as the guessed value of the unknown p. Now we have an interval that we can calculate from the sample data.

Key Terms

The standard deviation of the sampling distribution of a sample statistic is commonly referred to as the **standard error.**

Statistics in Your World

Who is a smoker? When estimating a proportion p, be sure you know what counts as a "success." The news says that 20% of adolescents smoke. Shocking. It turns out that this is the percentage who smoked at least once in the past month. If we say that a smoker is someone who smoked in at least 20 of the past 30 days and smoked at least half a pack on those days, fewer than 4% of adolescents qualify.

95% confidence interval for a population proportion

Choose an SRS of size n from a large population that contains an unknown proportion p of successes. Call the proportion of successes in this sample $\hat{p}$. An **approximate 95% confidence interval** for the parameter p is

$$\hat{p} \pm 2\sqrt{\frac{\hat{p}(1-\hat{p})}{n}}$$

Example 3

A confidence interval for graduation plans

The NSSE sample of 23,915 college seniors found that 5038 reported plans to go to graduate or professional school, giving a sample proportion $\hat{p} = 0.211$. The 95% confidence interval for the proportion of all college seniors from the United States and Canada who plan to go to graduate or professional school is

$$\hat{p} \pm 2\sqrt{\frac{\hat{p}(1-\hat{p})}{n}} = 0.211 \pm 2\sqrt{\frac{(0.211)(0.789)}{23915}}$$
$$= 0.211 \pm (2)(0.00264)$$
$$= 0.211 \pm 0.0053$$
$$= 0.2057 \text{ to } 0.2163$$

Interpret this result as follows: we got this interval by using a procedure that catches the true unknown population proportion in 95% of all samples. The shorthand is that we are **95% confident** that between 20.57% and 21.63% of all college seniors in the United States and Canada plan to go to graduate or professional school.

Now it's your turn

21.1 Why do we turn to YouTube? A November 2018 Pew Research Center survey consisting of a random sample of 4594 adult Americans found that 51% reported using YouTube in order to figure out how to do things they haven't done before. Find a 95% confidence interval for the proportion of all adult Americans who use YouTube to figure out how to do things they haven't done before. How would you interpret this interval?

Understanding Confidence Intervals

Our 95% confidence interval for a population proportion has the familiar form

estimate ± margin of error

News reports of sample surveys, for example, usually give the estimate and the margin of error separately: "A new Gallup Poll shows that 65% of women favor new laws restricting guns. The margin of error is plus or minus four percentage points." News reports usually leave out the level of confidence, although it is almost always 95%.

The next time you hear a report about the result of a sample survey, consider the following. If most confidence intervals reported in the media have a 95% level of confidence, then in about 1 in 20 reported poll results, the confidence interval does not contain the true value of the population proportion.

A complete description of a confidence interval is given in the box at the right.

There are many recipes for statistical confidence intervals for use in many situations. Not all confidence intervals are expressed in the form "estimate ± margin of error." Be sure you understand how to interpret a confidence interval. The interpretation is the same for any recipe, and you can't use a calculator or a computer to do the interpretation for you.

Confidence intervals use the central idea of probability: ask what would happen if we repeated the sampling many times. The 95% in a 95% confidence interval is a probability, the probability that the method produces an interval that does capture the true parameter.

Example 4 — *How confidence intervals behave*

The NSSE sample of 23,915 college seniors found that 5038 reported plans to go to graduate or professional school, so the sample proportion was

$$\hat{p} = \frac{5038}{23915} = 0.211$$

and the 95% confidence interval was

$$\hat{p} \pm 2\sqrt{\frac{\hat{p}(1-\hat{p})}{n}} = 0.211 \pm 0.0053$$

Draw a second sample from the same population. It finds that 4976 of its 23,915 respondents plan to go to graduate or professional school. For this sample,

$$\hat{p} = \frac{4976}{23915} = 0.208$$

$$\hat{p} \pm 2\sqrt{\frac{\hat{p}(1-\hat{p})}{n}} = 0.208 \pm 0.0052$$

Draw another sample. Now the count is 5503 and the sample proportion and confidence interval are

$$\hat{p} = \frac{5503}{23915} = 0.230$$

$$\hat{p} \pm 2\sqrt{\frac{\hat{p}(1-\hat{p})}{n}} = 0.230 \pm 0.0054$$

Keep sampling. Each sample yields a new estimate $\hat{p}$ and a new confidence interval. *If we sample forever, 95% of these intervals capture the true parameter.* This is true no matter what the true value is. Figure 21.3 summarizes the behavior of the confidence interval in graphical form.

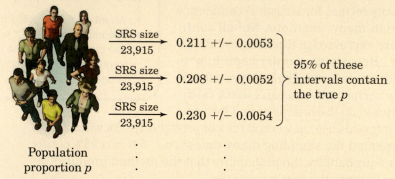

Population
proportion p

Figure 21.3 Repeated samples from the same population give different 95% confidence intervals, but 95% of these intervals capture the true population proportion p.

Example 4 and Figure 21.3 remind us that repeated samples give different results and that we are guaranteed only that 95% of the samples give a correct result. On the assumption that two pictures are better than one, Figure 21.4 gives a different view of how confidence intervals behave. Figure 21.4 goes behind the scenes. The vertical line is the true value of the population proportion p. The Normal curve at the top of the figure is the sampling distribution of the sample statistic $\hat{p}$, which is centered at the true p. We are behind the scenes because, in real-world statistics, we usually don't know p.

The 95% confidence intervals from 25 SRSs appear below the graph in Figure 21.4, one after the other. The central dots are the values of $\hat{p}$, the centers of the intervals. The arrows on either side span the confidence interval. In the long run, 95% of the intervals will capture the true p and 5% will miss. Of the 25 intervals in Figure 21.4, there are 24 hits and 1 miss. (Remember that probability describes only what happens in the long run — we don't expect exactly 95% of 25 intervals to capture the true parameter.)

Don't forget that our interval is only *approximately* a 95% confidence interval. It isn't exact for two reasons. The sampling distribution of the sample proportion $\hat{p}$ isn't exactly Normal. And we don't get the standard deviation, or the standard error, of $\hat{p}$ exactly right because we used $\hat{p}$ in place of the unknown p. We use a new estimate of the standard deviation of the sampling distribution every time, even though the true standard deviation never changes. Both of these difficulties go away as the sample size n gets

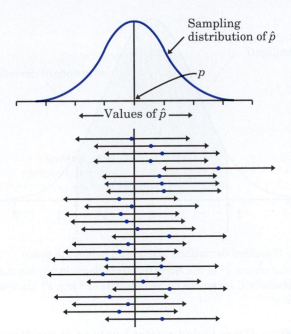

Figure 21.4 Twenty-five samples from the same population give these 95% confidence intervals. In the long run, 95% of all such intervals cover the true population proportion, marked by the vertical line.

larger. So, our recipe is good only for large samples. What is more, the recipe assumes that the population is really big—at least 10 times the size of the sample. Professional statisticians use more elaborate methods that take the size of the population into account and work even for small samples. But our method works well enough for many practical uses. More important, it shows how we get a confidence interval from the sampling distribution of a statistic. That's the reasoning behind any confidence interval.

More on Confidence Intervals for a Population Proportion*

We used the 95 part of the 68–95–99.7 rule to get a 95% confidence interval for the population proportion. Perhaps you think that a method that works 95% of the time isn't good enough. You want to be 99% confident. For that, we need to mark off the central 99% of a Normal distribution. For any probability C between 0 and 1, there is a number z^* such that any Normal distribution has probability C within z^* standard deviations of the mean. Figure 21.5 shows how the probability C and the number z^* are related.

Table 21.1 gives the numbers z^* for various choices of C. For convenience, the table gives C as a confidence level in percent. The numbers

*This section is optional.

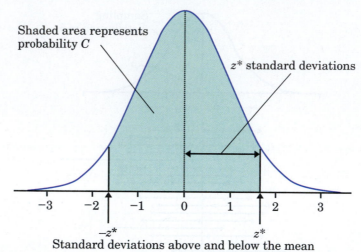

Shaded area represents probability C

z^* standard deviations

$-3 \quad -2 \quad -1 \quad 0 \quad 1 \quad 2 \quad 3$

$-z^*$ $\qquad\qquad$ z^*

Standard deviations above and below the mean

Figure 21.5 Critical values z^* of the Normal distributions. In any Normal distribution, there is area (probability) C under the curve between $-z^*$ and z^* standard deviations away from the mean.

Confidence interval for a population proportion

Choose an SRS of size n from a population of individuals of which proportion p are successes. The proportion of successes in the sample is $\hat{p}$. When n is large, an **approximate level C confidence interval for p** is

$$\hat{p} \pm z^* \sqrt{\frac{\hat{p}(1-\hat{p})}{n}}$$

where z^* is the critical value for probability C from Table 21.1.

z^* are called **critical values** of the Normal distributions. Table 21.1 shows that any Normal distribution has probability 99% within ±2.58 standard deviations of its mean. The table also shows that any Normal distribution has probability 95% within ±1.96 standard deviations of its mean. The 68–95–99.7 rule uses 2 in place of the critical value $z^* = 1.96$. That is good enough for practical purposes, but the table gives the more exact value.

From Figure 21.5 we see that, with probability C, the sample proportion $\hat{p}$ takes a value within z^* standard deviations of p. That is just to say that, with probability C, the interval extending z^* standard deviations on either side of the observed $\hat{p}$ captures the unknown p. Using the estimated standard error of $\hat{p}$ produces the formula for the confidence interval for a population proportion given in the box at the left.

Table 21.1 Critical values of the Normal distributions

Confidence level C	Critical value z^*	Confidence level C	Critical value z^*
50%	0.67	90%	1.64
60%	0.84	95%	1.96
70%	1.04	99%	2.58
80%	1.28	99.9%	3.29

Example 5 — *A 99% confidence interval*

The NSSE sample of 23,915 college seniors found that 5038 reported plans to go to graduate or professional school. We want a 99% confidence interval for the proportion p of all college seniors in the United States and Canada who plan to go to graduate or professional school. Table 21.1 says that for 99% confidence, we must go out $z^* = 2.58$ standard deviations. Here are our calculations:

$$\hat{p} = \frac{5038}{23915} = 0.211$$

$$\hat{p} \pm z^* \sqrt{\frac{\hat{p}(1-\hat{p})}{n}} = 0.211 \pm 2.58 \sqrt{\frac{(0.211)(0.789)}{23915}}$$

$$= 0.211 \pm (2.58)(0.00264)$$

$$= 0.211 \pm 0.0068$$

$$= 0.2042 \text{ to } 0.2178$$

We are 99% confident that between 20.42% and 21.78% of all college seniors in the United States and Canada plan to go to graduate or professional school. Put differently, we got this range of percentages by using a method that gives a correct answer 99% of the time.

Compare Example 5 with the calculation of the 95% confidence interval in Example 3. The only difference is the use of the critical value 2.58 for 99% confidence in place of 2 for 95% confidence. That makes the margin of error for 99% confidence larger and the confidence interval wider. Higher confidence isn't free—we pay for it with a wider interval. Figure 21.5 reminds us why this is true. To cover a higher percentage of the area under a Normal curve, we must go farther out from the center. Figure 21.6 compares the lengths of the 90%, 95%, and 99% confidence intervals.

Now it's your turn

21.2 What is America's most important problem? A July 2018 Gallup Poll consisting of a random sample of 1033 adult Americans found that 22% believe that immigration is the most important problem facing the nation. Find a 99% confidence interval for the proportion of all adult Americans who believe immigration is the most important problem facing the nation. How would you interpret this interval?

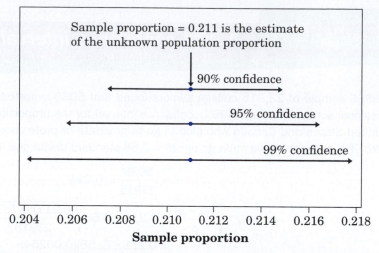

Figure 21.6 The lengths of three confidence intervals for the graduation plans example. All three are centered at the estimate $\hat{p} = 0.211$. When the data and the sample size remain the same, higher confidence results in a larger margin of error.

The Sampling Distribution of a Sample Mean*

What is the mean number of hours your college's first-year students study each week? What was their mean grade point average in high school? We often want to estimate the mean of a population. To distinguish the population mean (a parameter) from the sample mean $\bar{x}$, we write the population mean as μ (the Greek letter mu). We use the mean $\bar{x}$ of an SRS to estimate the unknown mean μ of the population.

Like the sample proportion $\hat{p}$, the sample mean $\bar{x}$ from a large SRS has a sampling distribution that is close to Normal. Because the sample mean of an SRS is an unbiased estimator of μ, the sampling distribution of $\bar{x}$ has μ as its mean. The standard deviation, or standard error, of $\bar{x}$ depends on the standard deviation of the population, which is usually written as σ (the Greek letter sigma). By mathematics, we can discover the facts given in the box at the left.

It isn't surprising that the values that $\bar{x}$ takes in many samples are centered at the true mean μ of the population. That's the lack of bias in random sampling once again. The other two facts about the sampling

Sampling distribution of a sample mean

Choose an SRS of size n from a population in which individuals have mean μ and standard deviation σ. Let $\bar{x}$ be the mean of the sample. Then:

- The sampling distribution of $\bar{x}$ is **approximately Normal** when the sample size n is large.
- The **mean** of the sampling distribution of $\bar{x}$ is equal to μ.
- The **standard deviation** or the **standard error** of the sampling distribution of $\bar{x}$ is $\sigma/\sqrt{n}$.

*This section is optional.

distribution make precise two very important properties of the sample mean $\bar{x}$:

• The mean of a number of observations is less variable than individual observations.

• The distribution of a mean of a number of observations is more Normal than the distribution of individual observations.

Figure 21.7 illustrates the first of these properties. It compares the distribution of a single observation with the distribution of the mean $\bar{x}$ of 10 observations. Both have the same center, but the distribution of $\bar{x}$ is less spread out. In Figure 21.7, the distribution of individual observations is Normal. If that is true, then the sampling distribution of $\bar{x}$ is exactly Normal for any size sample, not just approximately Normal for large samples. A remarkable statistical fact, called the **central limit theorem,** says that as we take more and more observations at random from *any* population, the distribution of the mean of these observations eventually gets close to a Normal distribution. (There are some technical qualifications to this big fact, but in practice, we can ignore them.) The central limit theorem lies behind the use of Normal sampling distributions for sample means.

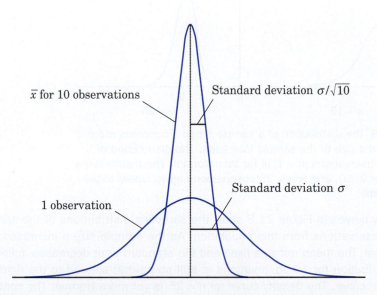

Figure 21.7 The sampling distribution of the sample mean $\bar{x}$ of 10 observations compared with the distribution of individual observations.

Example 6   *The central limit theorem in action*

Figure 21.8 shows the central limit theorem in action. The top-left density curve describes individual observations from a population. It is strongly right-skewed. Distributions like this describe the time it takes to repair a household appliance, for example. Most repairs are quickly done, but some are lengthy.

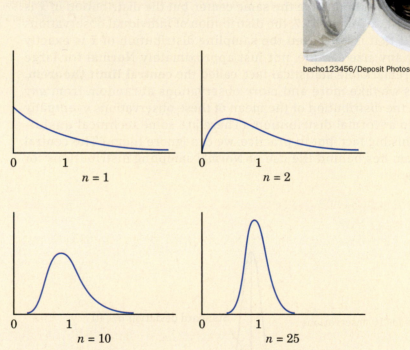

bacho123456/Deposit Photos

Figure 21.8 The distribution of a sample mean $\bar{x}$ becomes more Normal as the size of the sample increases. The distribution of individual observations ($n = 1$) is far from Normal. The distributions of means of 2, 10, and finally 25 observations move closer to the Normal shape.

The other three density curves in Figure 21.8 show the sampling distributions of the sample means of 2, 10, and 25 observations from this population. As the sample size n increases, the shape becomes more Normal. The mean remains fixed and the standard error decreases, following the pattern $\sigma/\sqrt{n}$. The distribution for 10 observations is still somewhat skewed to the right but already resembles a Normal curve. The density curve for $n = 25$ is yet more Normal. The contrast between the shapes of the population distribution and of the distribution of the mean of 10 or 25 observations is striking.

Confidence Intervals for a Population Mean*

The standard error of $\bar{x}$ depends on both the sample size n and the standard deviation σ of individuals in the population. We know n but not σ. When n is large, the sample standard deviation s is close to σ and can be used to estimate it, just as we use the sample mean $\bar{x}$ to estimate the population mean μ. The estimated standard error of $\bar{x}$ is, therefore, $s/\sqrt{n}$. Now we can find confidence intervals for μ following the same reasoning that led us to confidence intervals for a proportion p. The big idea is that to cover the central area C under a Normal curve, we must go out a distance z^* on either side of the mean. Look again at Figure 21.5 to see how C and z^* are related.

> ## Confidence interval for a population mean
>
> Choose an SRS of size n from a large population of individuals having mean μ. The mean of the sample observations is $\bar{x}$. When n is reasonably large, an **approximate level C confidence interval for μ** is
>
> $$\bar{x} \pm z^* \frac{s}{\sqrt{n}}$$
>
> where z^* is the critical value for confidence level C from Table 21.1.

The cautions we noted in estimating p apply here as well. The recipe is valid only when an SRS is drawn and the sample size n is reasonably large. How large is reasonably large? The answer depends upon the true shape of the population distribution. A sample size of $n \geq 15$ is usually adequate unless there are extreme outliers or strong skewness. For clearly skewed distributions, a sample size of $n \geq 40$ often suffices if there are no outliers.

The margin of error again decreases only at a rate proportional to $\sqrt{n}$ as the sample size n increases. And it bears repeating that $\bar{x}$ and s are strongly influenced by outliers. Inference using $\bar{x}$ and s is suspect when outliers are present. Always look at your data.

Example 7   *NAEP quantitative scores*

The National Assessment of Educational Progress (NAEP) includes a mathematics test for high school seniors. Scores on the test range from 0 to 300. Demonstrating the ability to use the Pythagorean theorem to determine the length of a hypotenuse is an example of the skills and knowledge associated with performance at the Basic level. An example of the knowledge and skills associated with the Proficient level is using trigonometric ratios to determine length.

In 2015, the NAEP randomly selected 13,200 seniors in high school who took the mathematics test. The mean mathematics score was $\bar{x} = 152$, and the standard deviation of their scores was $s = 34$. Assume that these 13,200 students were a random sample from the population of all 12th-graders.

*This section is optional.

On the basis of this sample, what can we say about the mean score μ in the population of all 12th-grade students?

The 95% confidence interval for μ uses the critical value $z* = 1.96$ from Table 21.1. The interval is

$$\bar{x} \pm z* \frac{s}{\sqrt{n}} = 152 \pm 1.96 \frac{34}{\sqrt{13,200}}$$
$$= 152 \pm (1.96)(0.296) = 152 \pm 0.58$$

We are 95% confident that the interval from 151.42 to 152.58 includes the mean score for all 12th-grade students.

Now it's your turn

21.3 Running a mile. The manager at a large fitness center keeps track of the amount of time it takes a sample of 65 clients to run a mile. She finds that the mean time to run a mile in this sample is $\bar{x} = 10.4$ minutes and the standard deviation is $s = 2.1$ minutes. Assuming the sample is a random sample of all clients at this particular fitness center, find a 95% confidence interval for μ, the unknown mean running time of all clients from the fitness center.

Chapter 21: Statistics in Summary

- **Statistical inference** draws conclusions about a population on the basis of data from a sample. Because we don't have data for the entire population, our conclusions are uncertain.

- A **confidence interval** estimates an unknown parameter in a way that tells us how uncertain the estimate is. The interval itself says how closely we can pin down the unknown parameter. The **confidence level** is a probability that says how often in many samples the method would produce an interval that does catch the parameter. We find confidence intervals starting from the **sampling distribution** of a statistic, which shows how the statistic varies in repeated sampling.

- The standard deviation of the sampling distribution of the sample statistic is commonly referred to as the **standard error.**

- We estimate a population proportion p using the sample proportion $\hat{p}$ of an SRS from the population. Confidence intervals for p are based on the **sampling distribution** of $\hat{p}$. When the sample size n is large, this distribution is approximately Normal.

- We estimate a population mean μ using the sample mean $\bar{x}$ of an SRS from the population. Confidence intervals for μ are based on the **sampling distribution** of $\bar{x}$. When the sample size n is large, the **central limit theorem** says that this distribution is approximately Normal. Although the details of the methods differ, inference about μ is quite similar to inference about a population proportion p because both are based on Normal sampling distributions.

This chapter summary will help you evaluate the Case Study.

Link It

The reason we collect data is not to learn about the individuals that we observed but to infer from the data to some wider population that the individuals represent. Chapters 1 through 6 tell us that the way we produce the data (sampling, experimental design) affects whether we have a good basis for generalizing to some wider population—in particular, whether a sample statistic provides insight into the value of the corresponding population parameter. Chapters 17 through 20 discuss probability, the formal mathematical tool that determines the nature of the inferences we make. Chapter 18 discusses sampling distributions, which tell us how statistics computed from repeated SRSs behave and hence what a statistic (in particular, a sample proportion) computed from our sample is likely to tell us about the corresponding parameter of the population (in particular, a population proportion) from which the sample was selected.

In this chapter, we discuss the basic reasoning of statistical estimation of a population parameter, with emphasis on estimating a population proportion and population mean. To an estimate of a population parameter, such as a population proportion, we attach a margin of error and a confidence level. The result is a confidence interval. The sampling distribution, first introduced in Chapter 3 and discussed more fully in Chapter 18, provides the mathematical basis for constructing confidence intervals and understanding their properties. We will provide more advice on interpreting confidence intervals in Chapter 23.

Case Study Evaluated

Use what you have learned in this chapter to evaluate the Case Study described at the beginning of the chapter. Start by reviewing the Chapter Summary. Then answer each of the following questions in complete sentences. Be sure to communicate clearly enough for any of your classmates to understand what you are saying. Based on the information provided in the Pew Research Center report, we can construct confidence intervals. For example, with 95% confidence, between 64.7% and 77.3% of 18- to 24-year-olds use Instagram and between 72.3% and 83.7% of 18- to 24-year-olds use Snapchat.

1. Interpret the above confidence intervals in plain language that someone who knows no statistics will understand.

2. Suppose we wanted to use information from the entire sample of 2002 survey respondents in order to construct a 95% confidence interval to estimate the true population percentage of Instagram users. Would you expect the margin of error for this confidence interval to be larger than, smaller than, or the same as the margin of error for the 95% confidence interval based on the sample of 201 18- to 24-year-olds? Explain your reasoning.

macmillan learning **Online Resources**

■ The Snapshots Video, *Inference for One Proportion*, discusses confidence intervals for a population proportion in the context of a sample survey conducted during the 2012 presidential elections.

■ The StatClips Examples video, *Confidence Intervals: Intervals for Proportions, Example C*, provides an example of how to calculate a 90% confidence interval for a proportion.

■ The Snapshots Video, *Confidence Intervals*, discusses confidence intervals for a population mean in the context of an example involving birds killed by wind turbines.

Check the Basics

For Exercise 21.1, see page 492; for Exercise 21.2, see page 497; and for Exercise 21.3, see page 502.

21.4 Gun ownership. In 2017, the Pew Research Center published the results of a survey on gun ownership. A random sample of 3930 American adults aged 18 years and older participated in this survey. Of these 3930 individuals, 1269 individuals reported that they owned at least one gun. Of these self-reported gun owners, 850 said the reason they owned a gun was for protection. Based on this information, we know the sample proportion, $\hat{p}$, of gun owners from this survey who own a gun for protection is

(a) 0.042.

(b) 0.216.

(c) 0.323.

(d) 0.670.

21.5 Barriers to travel. In their 2018 Travel Trends report, the American Association of Retired Persons reported the results of a survey of 374 adults between the ages of 20 and 36 years. Of the 374 individuals in this age group, 176 indicated that cost was the biggest barrier when it comes to traveling. Based on this information, the 95% confidence interval for the proportion of all adults in this age group who find cost to be the biggest barrier to travel would be

(a) 0.451 to 0.491.

(b) 0.445 to 0.497.

(c) 0.420 to 0.522.

(d) 0.397 to 0.545.

21.6 Computer use. A random sample of 197 12th-grade students from across the United States was surveyed, and it was observed that these students spent an average of 23.5 hours on the computer per week, with a standard deviation of 8.7 hours. If we plan to use these data to construct a 99% confident interval, the margin of error will be approximately

(a) 0.07.

(b) 0.62.

(c) 1.6.

(d) 8.7.

21.7 Anxiety. A particular psychological test is used to measure anxiety. The average test score for all university students nationwide is 85 points. Suppose a random sample of university students is selected and a confidence interval based on their mean anxiety score is constructed. Which of the following statements about the confidence interval are true?

(a) The resulting confidence interval will contain 85.

(b) The 95% confidence interval for a sample of size 100 will generally have a smaller margin of error than the 95% confidence interval for a sample of size 50.

(c) For a sample of size 100, the 95% confidence interval will have a smaller margin of error than the 90% confidence interval.

21.8 Pigs. A 90% confidence interval is calculated for a sample of weights of 135 randomly selected pigs, and the resulting confidence interval is from 75 to 90 pounds. Will the sample mean weight (from this particular sample of size 135) fall within the confidence interval?

(a) No

(b) Yes

(c) Maybe

Chapter 21 Exercises

21.9 A student survey. Tonya wants to estimate the proportion of the students in her dormitory like the dorm food. She interviews an SRS of 50 of the 175 students living in the dormitory. She finds that 14 think the dorm food is good.

(a) What population does Tonya want to draw conclusions about?

(b) In your own words, what is the population proportion p in this setting?

(c) What is the numerical value of the sample proportion $\hat{p}$ from Tonya's sample?

21.10 Fire the coach? A college president says, "99% of the alumni support my firing of Coach Boggs." You contact an SRS of 200 of the college's 15,000 living alumni and find that 66 of them support firing the coach.

(a) What population does the inference concern here?

(b) Explain clearly what the population proportion p is in this setting.

(c) What is the numerical value of the sample proportion $\hat{p}$?

21.11 Are teachers engaged in their work?

Results from a Gallup survey conducted in 2013 and 2014 reveal that 30% of kindergarten through grade 12 school teachers report feeling engaged in their work. The report from this random sample of 6711 teachers stated that, with 95% confidence, the margin of sampling error was ±1.0%. Explain to someone who knows no statistics what the phrase "95% confidence" means in this report.

21.12 Gun control. A March 2018 Gallup survey asked a sample of 1041 adults if they wanted stricter laws covering the sale of firearms. A total of 697 of the survey respondents said "yes." Although the samples in national polls are not SRSs, they are similar enough that our method gives approximately correct confidence intervals.

(a) Say in words what the population proportion p is for this poll.

(b) Find a 95% confidence interval for p.

(c) Gallup announced a margin of error of plus or minus 4 percentage points for this poll result. How well does your work in part (b) agree with this margin of error?

21.13 Using public libraries. In 2016, the Pew Research Center reported the results of a survey about public library usage. A nationally representative sample of 1601 people ages 16 and older who were living in the United States completed the survey. One survey question asked whether the survey respondent had visited, in person, a library or a bookmobile in the last year. A total of 768 of those surveyed answered "yes" to this question. Although the samples in national polls are not SRSs, they are similar enough that our method gives approximately correct confidence intervals.

(a) Explain in words what the parameter p is in this setting.

(b) Use the poll results to give a 95% confidence interval for p.

(c) Write a short explanation of your findings in part (b) for someone who knows no statistics.

21.14 Computer crime. Adults are spending more and more time on the Internet, and the number experiencing computer-based or Internet-based crime is rising. A 2018 Gallup Poll of 1019 adults, aged 18 and older, found that 723 of those in the sample said that they worry about having their personal, credit card, or financial information stolen by computer hackers. Although the samples in national polls are not SRSs, they are similar enough that our method gives approximately correct confidence intervals.

(a) Explain in words what the parameter p is in this setting.

(b) Use the poll results to give a 95% confidence interval for p.

21.15 Gun control. In Exercise 21.12, you constructed a 95% confidence interval based on a random sample of $n = 1041$ adults. How large a sample would be needed to get a margin of error half as large as the one in Exercise 21.12? You may find it helpful to refer to the discussion surrounding Example 5 in Chapter 3 (page 44).

21.16 The effect of sample size. A November 2018 CBS News Poll found that 35% of its sample planned to use online retailers for most holiday shopping. Give a 95% confidence interval for the proportion of all adults who plan to shop in this way, assuming that the result $\hat{p} = 0.35$ comes from a sample of size

(a) $n = 650$.

(b) $n = 1800$.

(c) $n = 5000$.

(d) Explain briefly what your results show about the effect of increasing the size of a sample on the width of the confidence interval.

21.17 Random digits. We know that the proportion of 0s among a large set of random digits is $p = 0.1$ because all 10 possible digits are equally probable. The entries in a table of random digits are a random sample from the population of all random digits. To get an SRS of 200 random digits, look at the first digit in each of the 200 five-digit groups in lines 101 to 125 of Table A in the back of the book. How many of these 200 digits are 0s? Give a 95% confidence interval for the proportion of 0s in the population from which these digits are a random sample. Does your interval cover the true parameter value, $p = 0.1$?

21.18 Tossing a thumbtack. If you toss a thumbtack on a hard surface, what is the probability that it will land point up? Estimate this probability p by tossing a thumbtack 100 times. The 100 tosses are an SRS of size 100 from the population of all tosses. The proportion of these 100 tosses that land point up is the sample proportion $\hat{p}$. Use the result of your tosses to give a 95% confidence interval for p. Write a brief explanation of your findings for someone who knows no statistics but wonders how often a thumbtack will land point up.

21.19 Don't forget the basics. Vaping involves inhaling vapors (sometimes including nicotine) from electronic devices. How prevalent is this practice among high school seniors?

In 2017, a publication entitled "Monitoring the Future: National Survey Results on Drug Use" reported that of 13,500 high school seniors who were surveyed, 35.8% indicated that they had tried vaping at some point in their lives. We can use this information to construct a confidence interval for the proportion of all high school seniors who have ever tried vaping. However, this sample survey may have bias that our confidence interval does not take into account.

(a) Why is some bias likely to be present?

(b) Does the sample proportion 35.8% probably overestimate or underestimate the true population proportion? Please explain.

(c) If 35.8% of the 13,500 high school seniors reported they had tried vaping, this means the actual number of students who reported they had tried vaping must be equal to what?

21.20 Count Buffon's coin. The eighteenth-century French naturalist Count Buffon tossed a coin 4040 times. He got 2048 heads. Give a 95% confidence interval for the proportion of all Buffon's coin tosses that land head side up. Are you confident that this proportion is not 0.5? Why?

21.21 Share the wealth. The *New York Times* conducted a nationwide poll of 1650 randomly selected American adults. Of these, 1089 felt that the distribution of money and wealth in this country should be more evenly distributed among more people. We can consider the sample to be an SRS.

(a) Give a 95% confidence interval for the proportion of all American adults who, at the time of the poll, felt that the distribution of money and wealth in this country should be more evenly distributed among more people.

(b) The news article says, "In theory, in 19 cases out of 20, the poll results will differ by no more than 3 percentage points in either direction from what would have been obtained by seeking out all American

adults." Explain how your results agree with this statement.

21.22 Harley motorcycles. In 2013, it was reported that 55% of the new motorcycles that were registered in the United States were Harley-Davidson motorcycles. You plan to interview an SRS of 600 new motorcycle owners.

(a) What is the sampling distribution of the proportion of your sample who own Harleys?

(b) How likely is your sample to contain 57% or more who own Harleys? How likely is it to contain at least 51% Harley owners? Use the 68–95–99.7 rule and your answer to part (a).

21.23 Do you jog? Suppose that 10% of all adults jog. An opinion poll asks an SRS of 400 adults if they jog.

(a) What is the sampling distribution of the proportion $\hat{p}$ in the sample who jog?

(b) According to the 68–95–99.7 rule, what is the probability that the sample proportion who jog will be 7.3% or greater?

21.24 The quick method. The quick method of Chapter 3 (pages 43–44) uses $\hat{p} \pm 1/\sqrt{n}$ as a rough recipe for a 95% confidence interval for a population proportion. The margin of error from the quick method is a bit larger than needed. It differs most from the more accurate method of this chapter when $\hat{p}$ is close to 0 or 1. An SRS of 500 motorcycle registrations finds that 68 of the motorcycles are Harley-Davidsons. Give a 95% confidence interval for the proportion of all motorcycles that are Harleys by the quick method and then by the method of this chapter. How much larger is the quick-method margin of error?

21.25 68% confidence. We used the 95 part of the 68–95–99.7 rule to give a recipe for a 95% confidence interval for a population proportion p.

(a) Use the 68 part of the rule to give a recipe for a 68% confidence interval.

(b) Explain in simple language what "68% confidence" means.

(c) Use the result of the NSSE (Example 3, page 492) to give a 68% confidence interval for the proportion of college seniors who plan to go to graduate or professional school. How does your interval compare with the 95% interval in Example 3?

21.26 Simulating confidence intervals. In Exercise 21.25, you found the recipe for a 68% confidence interval for a population proportion p. Suppose that (unknown to anyone) 20% of college seniors plan to go to graduate or professional school.

(a) How would you simulate the proportion of an SRS of 25 college seniors?

(b) Simulate choosing 10 SRSs, using a different row in Table A for each sample. What are the 10 values of the sample proportion $\hat{p}$ who plan to go to graduate or professional school?

(c) Find the 68% confidence interval for p from each of your 10 samples. How many of the intervals capture the true parameter value $p = 0.2$? (Samples of size 25 are not large enough for our recipe to be very accurate, but even a small simulation illustrates how confidence intervals behave in repeated samples.)

The following exercises concern the optional sections of this chapter.

21.27 Gun control. Exercise 21.12 reports a Gallup survey in which 697 of a sample of 1041 adults said they wanted stricter laws covering the sale of firearms. Use Table 21.1 to give a 90% confidence interval for the proportion of all adults who feel this way. How does your interval compare with the 95% confidence interval from Exercise 21.12?

21.28 Using public libraries. Exercise 21.13 reports a 2016 Pew Research Center survey on public library usage in which 768 in a sample of 1601 American adults said that they had visited, in person, a library or bookmobile in the last year. Use Table 21.1 to give a

99% confidence interval for the proportion of all American adults who have done this. How does your interval compare with the 95% confidence interval of Exercise 21.13?

21.29 Restaurant nutrition labels. A 2018 Gallup survey on food labeling found that 45% of a sample of 1033 American adults report paying a great deal or a fair amount of attention to nutritional information that is printed on restaurant menus or posted in restaurants. Use this survey result and Table 21.1 to give 70%, 80%, 90%, and 99% confidence intervals for the proportion of all adult Americans who pay a great deal or a fair amount of attention to this information. What do your results show about the effect of changing the confidence level?

21.30 Unhappy HMO patients. How likely are patients who file complaints with a health maintenance organization (HMO) to leave the HMO? In one year, 639 of the more than 400,000 members of a large New England HMO filed complaints. Fifty-four of the complainers left the HMO voluntarily. (That is, they were not forced to leave by a move or a job change.) Consider this year's complainers as an SRS of all patients who will complain in the future. Give a 90% confidence interval for the proportion of complainers who voluntarily leave the HMO.

21.31 Estimating unemployment. The Bureau of Labor Statistics (BLS) uses 90% confidence in presenting unemployment results from the monthly Current Population Survey (CPS). The November 2018 survey reported that of the 162,770 individuals surveyed in the civilian labor force, 156,795 were employed and 5975 were unemployed. The CPS is not an SRS, but for the purposes of this exercise, we will act as though the BLS took an SRS of 162,770 people. Give a 90% confidence interval for the proportion of those surveyed who were unemployed. (*Note:* Example 3 in Chapter 8 on pages 164 explains how unemployment is measured.)

21.32 Safe margin of error. The margin of error $z^*\sqrt{\hat{p}(1-\hat{p})n}$ is 0 when $\hat{p}$ is 0 or 1 and is largest when $\hat{p}$ is 1/2. To see this, calculate $\hat{p}(1-\hat{p})$ for $\hat{p} = 0, 0.1, 0.2, \ldots, 0.9,$ and 1. Plot your results vertically against the values of $\hat{p}$ horizontally. Draw a curve through the points. You have made a graph of $\hat{p}(1-\hat{p})$. Does the graph reach its highest point when $\hat{p} = 1/2$? You see that taking $\hat{p} = 1/2$ gives a margin of error that is always at least as large as needed.

21.33 The idea of a sampling distribution. Figure 21.1 (page 489) shows the idea of the sampling distribution of a sample proportion $\hat{p}$ in picture form. Draw a similar picture that shows the idea of the sampling distribution of a sample mean $\bar{x}$.

21.34 IQ test scores. Here are the IQ test scores of 31 seventh-grade students in a Midwest school district:

```
114 100  104   89 102   91 114 114 103 105
108 130  120 132 111 128 118 119   86   72
111 103   74 112 107 103   98   96 112 112  93
```

(a) We expect the distribution of IQ scores to be close to Normal. Make a histogram of the distribution of these 31 scores. Does your plot show outliers, clear skewness, or other non-Normal features? Using a calculator, find the mean and standard deviation of these scores.

(b) Treat the 31 students as an SRS of all middle-school students in the school district. Give a 95% confidence interval for the mean score in the population.

(c) In fact, the scores are those of all seventh-grade students in one of the several schools in the district. Explain carefully why we cannot trust the confidence interval from (b).

21.35 Averages versus individuals. Scores on the ACT college entrance examination vary Normally with mean $\mu = 18$ and standard deviation $\sigma = 6$. The range of reported scores is 1 to 36.

(a) What range contains the middle 95% of all individual scores?

(b) If the ACT scores of 50 randomly selected students are averaged, what range contains the middle 95% of the averages $\bar{x}$?

21.36 Blood pressure. A randomized comparative experiment studied the effect of diet on blood pressure. Researchers divided 54 healthy white males at random into two groups. One group received a calcium supplement, and the other group received a placebo. At the beginning of the study, the researchers measured many variables on the subjects. The average seated systolic blood pressure of the 27 members of the placebo group was reported to be $\bar{x} = 114.9$ with a standard deviation of $s = 9.3$.

(a) Give a 95% confidence interval for the mean blood pressure of the population from which the subjects were recruited.

(b) The recipe you used in part (a) requires an important assumption about the 27 men who provided the data. What is this assumption?

21.37 Testing a random number generator. Our statistical software has a "random number generator" that is supposed to produce numbers scattered at random from 0 to 1. If this is true, the numbers generated come from a population with $\mu = 0.5$. A command to generate 100 random numbers gives outcomes with mean $\bar{x} = 0.536$ and $s = 0.312$. Give a 90% confidence interval for the mean of all numbers produced by the software.

21.38 Will they charge more? A bank wonders whether omitting the annual credit card fee for customers who charge at least $2500 in a year will increase the amount charged on its credit cards. The bank makes this offer to an SRS of 200 of its credit card customers. It then compares how much these customers charge this year with the amount that they charged last year. The mean increase in the sample is $346, and the standard deviation is $112. Give a 99% confidence interval for the mean amount charges would have increased if this benefit had been extended to all such customers.

21.39 A sampling distribution. Exercise 21.37 concerns the mean of the random numbers generated by a computer program. The mean is supposed to be 0.5 because the numbers are supposed to be spread at random from 0 to 1. We asked the software to generate samples of 100 random numbers repeatedly. Here are the sample means $\bar{x}$ for 50 samples of size 100:

0.532 0.450 0.481 0.508 0.510 0.530 0.499 0.461 0.543 0.490

0.497 0.552 0.473 0.425 0.449 0.507 0.472 0.438 0.527 0.536

0.492 0.484 0.498 0.536 0.492 0.483 0.529 0.490 0.548 0.439

0.473 0.516 0.534 0.540 0.525 0.540 0.464 0.507 0.483 0.436

0.497 0.493 0.458 0.527 0.458 0.510 0.498 0.480 0.479 0.499

The sampling distribution of $\bar{x}$ is the distribution of the means from all possible samples. We actually have the means from 50 samples. Make a histogram of these 50 sample means. Does the distribution appear to be roughly Normal, as the central limit theorem says will happen for large enough samples?

21.40 Will they charge more? In Exercise 21.38, you carried out the calculations for a confidence interval based on a bank's experiment in changing the rules for its credit cards. You ought to ask some questions about this study.

(a) The distribution of the amount charged is skewed to the right, but outliers are prevented by the credit limit that the bank enforces on each card. Why can we use a confidence interval based on a Normal sampling distribution for the sample mean $\bar{x}$?

(b) The bank's experiment was not comparative. The increase in amount charged over last year may be explained by lurking variables rather than by the rule change. What are some plausible reasons charges might go up? Outline the design of a comparative randomized experiment to answer the bank's question.

21.41 A sampling distribution, continued. Exercise 21.39 presents 50 sample means $\bar{x}$ from 50 random samples of size 100. Using a calculator, find the mean and standard error of these 50 values. Then answer these questions.

(a) The mean of the population from which the 50 samples were drawn is $\mu = 0.5$ if the random number generator is accurate. What do you expect the mean of the distribution of $\bar{x}$'s from all possible samples to be? Is the mean of these 50 samples close to this value?

(b) The standard error of the distribution of $\bar{x}$ from samples of size $n = 100$ is supposed to be $\sigma/10$, where σ is the standard deviation of individuals in the population. Use this fact and the standard deviation you calculated for the 50 $\bar{x}$'s to estimate σ.

21.42 Plus four confidence intervals for a proportion. The large-sample confidence interval $\hat{p} \pm z^* \sqrt{\hat{p}(1-\hat{p})/n}$ for a sample proportion p is easy to calculate. It is also easy to understand because it rests directly on the approximately Normal distribution of $\hat{p}$. Unfortunately, confidence levels from this interval can be inaccurate, particularly with smaller sample sizes where there are only a few successes or a few failures. The actual confidence level is usually *less* than the confidence level you asked for in choosing the critical value z^*. That's bad. What is worse, accuracy does not consistently get better as the sample size n increases. There are "lucky" and "unlucky" combinations of the sample size n and the true population proportion p.

Fortunately, there is a simple modification that is almost magically effective in improving the accuracy of the confidence interval. We call it the "plus four" method because all you need to do is *add four imaginary observations, two successes and two failures.* With the added observations, the **plus four estimate** of p is

$$\tilde{p} = \frac{\text{number of successes in the sample} + 2}{n + 4}$$

The formula for the confidence interval is exactly as before, with the new sample size and number of successes. To practice using the plus four confidence interval, consider the following problem. Cocaine users commonly snort the powder up the nose through a rolled-up paper currency bill. Spain has a high rate of cocaine use, so it's not surprising that euro paper currency in Spain often shows traces of cocaine. Researchers collected 20 euro bills in each of several Spanish cities. In Madrid, 17 out of 20 bore traces of cocaine. The researchers note that we can't tell whether the bills had been used to snort cocaine or had been contaminated in currency-sorting machines. Use the plus four confidence interval method to estimate the proportion of all euro bills in Madrid that have traces of cocaine.

Exploring the Web

Access these exercises on the text website: macmillanlearning.com/scc10e.

What Is a Test of Significance?

In this chapter you will:

- Learn how to conduct tests of significance for proportions and means.

- Be able to interpret the results of a test of significance.

- Be able to decide if an observed difference can plausibly be attributed to chance.

A May 1, 2017, article in *Inside Higher Education* reported that first-year college students are politically more divided than ever. What was the basis for this finding?

Every year since 1985, the Higher Education Research Institute at UCLA has conducted a survey of first-year college students. The 2016 survey involved a random sample of 137,456 full-time first-year students at 184 of the nation's baccalaureate colleges and universities. An interesting finding from the 2016 survey was that the percentage of students who reported their political views as being "middle-of-the-road" was lower than it has ever been, at 42.3%. In 2015, a random sample of 141,189 first-year students from 199 colleges and universities was surveyed, and 44.9% of these students identified their political views as being "middle-of-the-road."

The *Inside Higher Education* article included a quote from Kevin Eagan, the director of the Higher Education Research Institute and the lead author of the survey reports. Eagan noted that increased political activism among first-year students seemed to intensify in the months leading up to the 2016 presidential election, and the "2016 survey points to diversity and polarity of how college freshmen perceive their place in the current political landscape."

The sample sizes for the 2015 and 2016 surveys are both quite large. Although different percentages were reported in 2015 and 2016, changes in the percentages are small. Could it be that the difference between the two samples is just due to the luck of the draw in randomly choosing the respondents?

The Reasoning of Statistical Tests of Significance

The local hot-shot playground basketball player claims to make 80% of his free throws. "Show me," you say. He shoots 20 free throws and makes eight of them. "Aha," you conclude, "if he makes 80%, he would almost never make as few as 8 of 20. So I don't believe his claim." That's the reasoning of statistical **tests of significance** at the playground level: *an outcome that is very unlikely if a claim is true is good evidence that the claim is not true.*

Statistical inference uses data from a sample to draw conclusions about a population. So, once we leave the playground, statistical tests deal with claims about a population. Statistical tests ask if sample data give good evidence *against* a claim. A statistical test says, "If we took many samples and the claim were true, we would rarely get a result like this." To get a numerical measure of how strong the sample evidence is, replace the vague term "rarely" by a probability. Here is an example of this reasoning at work.

Example 1

Is the coffee fresh?

People of taste are supposed to prefer fresh-brewed coffee to the instant variety. But, perhaps, many coffee drinkers just need their caffeine fix. A skeptic claims that coffee drinkers can't tell the difference. Let's do an experiment to test this claim.

Each of 50 subjects tastes two unmarked cups of coffee and says which he or she prefers. One cup in each pair contains instant coffee; the other, fresh-brewed coffee. The statistic that records the result of our experiment is the proportion $\hat{p}$ of the sample who say they like the fresh-brewed coffee better. We find that 36 of our 50 subjects choose the fresh coffee. That is,

$$\hat{p} = \frac{36}{50} = 0.72 = 72\%$$

To make a point, let's compare our outcome $\hat{p} = 0.72$ with another possible result. If only 28 of the 50 subjects like the fresh coffee better than instant coffee, the sample proportion is

$$\hat{p} = \frac{28}{50} = 0.56 = 56\%$$

Surely 72% is stronger evidence against the skeptic's claim than 56%. But how much stronger? Is even 72% in favor in a *sample* convincing evidence that a majority of the *population* prefer fresh coffee? Statistical tests answer these questions. Here's the answer in outline form:

- **The claim.** The skeptic claims that coffee drinkers can't tell fresh from instant so that only half will choose fresh-brewed coffee. That is, he claims that the population proportion p is only 0.5. *Suppose for the sake of argument that this claim is true.*

- **The sampling distribution (from page 489).** If the claim $p = 0.5$ were true and we tested many random samples of 50 coffee drinkers, the sample proportion $\hat{p}$ would vary from sample to sample according to (approximately) the Normal distribution with

$$\text{mean} = p = 0.5$$

and

$$\text{standard deviation} = \sqrt{\frac{p(1-p)}{n}}$$

$$= \sqrt{\frac{(0.5)(0.5)}{50}}$$

$$= 0.0707$$

Figure 22.1 displays this Normal curve.

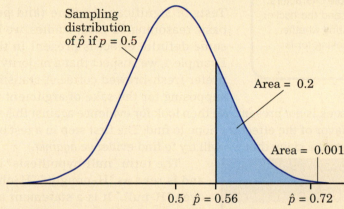

Figure 22.1 The sampling distribution of the proportion of 50 coffee drinkers who prefer fresh-brewed coffee if the truth about all coffee drinkers is that 50% prefer fresh coffee, Example 1. The shaded area is the probability that the sample proportion is 56% or greater.

- **The data.** Place the sample proportion $\hat{p}$ on the sampling distribution. You see in Figure 22.1 that $\hat{p} = 0.56$ isn't an unusual value, but that $\hat{p} = 0.72$ is unusual. We would rarely get 72% of a sample of 50 coffee drinkers preferring fresh-brewed coffee if only 50% of all coffee drinkers felt that way. So, the sample data do give evidence against the claim.

- **The probability.** We can measure the strength of the evidence against the claim by a probability. What is the probability that a sample gives a $\hat{p}$ this large or larger if the truth about the population is that $p = 0.5$? If $\hat{p} = 0.56$, this probability is the shaded area under the Normal curve in Figure 22.1. This area is 0.20. Our sample actually gave $\hat{p} = 0.72$. The probability of getting a sample outcome this large is only 0.001, an area too small to see in Figure 22.1. An outcome that would occur just by chance in 20% of all samples is *not* strong evidence against the claim. But an outcome that would happen only 1 in 1000 times *is* good evidence.

Statistics in Your World

Gotcha! A tax examiner suspects that Ripoffs, Inc., is issuing phony checks to inflate its expenses and reduce the tax it owes. To learn the truth without examining every check, she boots up her computer. The first digits of real data follow well-known patterns that do *not* give digits 0 to 9 equal probabilities. If the check amounts don't follow this pattern, she will investigate. Down the street, a hacker is probing a company's computer files. He can't read them because they are encrypted. But he may be able to locate the key to the encryption anyway—if it's the only long string that really *does* give equal probability to all possible characters. Both the tax examiner and the hacker need a method for testing whether the pattern they are looking for is present.

Be sure you understand why this evidence is convincing. There are two possible explanations of the fact that 72% of our subjects prefer fresh to instant coffee:

1. The skeptic is correct ($p = 0.5$), and by bad luck a very unlikely outcome occurred.

2. In fact, the population proportion favoring fresh coffee is greater than 0.5, so the sample outcome is about what would be expected.

We cannot be certain that Explanation 1 is untrue. Our taste test results *could* be due to chance alone. But the probability that such a result would occur by chance is so small (0.001) that we are quite confident that Explanation 2 is right.

Hypotheses and *P*-values

Tests of significance refine (and perhaps hide) this basic reasoning. In most studies, we hope to show that some definite effect is present in the population. In Example 1, we suspect that a majority of coffee drinkers prefer fresh-brewed coffee. A statistical test begins by supposing for the sake of argument that the effect we seek is *not* present. We then look for evidence against this supposition and in favor of the effect we hope to find. The first step in a test of significance is to state a claim that we will try to find evidence *against*.

The term "null hypothesis" is abbreviated H_0 and is read as "H-nought," "H-oh," and sometimes even "H-null." It is a statement about the population and so must be stated in terms of a population parameter. In Example 1, the parameter is the proportion p of all coffee drinkers who prefer fresh to instant coffee. The null hypothesis is

$$H_0: p = 0.5$$

The statement we hope or suspect is true instead of H_0 is called the **alternative hypothesis** and is abbreviated H_a. In Example 1, the alternative hypothesis is that a majority of the population favors fresh coffee. In terms of the population parameter, this is

$$H_a: p > 0.5$$

A significance test looks for evidence against the null hypothesis and in favor of the alternative hypothesis. The evidence is strong if the outcome we observe would rarely occur if the null hypothesis is true but is

Key Terms

The claim being tested in a statistical test is called the **null hypothesis (H_0).** The test is designed to assess the strength of the evidence against the null hypothesis. Usually, the null hypothesis is a statement of "no effect" or "no difference."

more probable if the alternative hypothesis is true. For example, it would be surprising to find 36 of 50 subjects favoring fresh coffee if, in fact, only half of the population feel this way. How surprising? A significance test answers this question by giving a probability: the probability of getting an outcome at least as extreme or more extreme than the actually observed outcome from what we would expect when H_0 is true. What counts as "at least as extreme or more extreme" depends on H_a as well as H_0. In the taste test, the probability we want is the probability that 36 or more of 50 subjects favor fresh coffee. If the null hypothesis $p = 0.5$ is true, this probability is very small (0.001). That's good evidence that the null hypothesis is not true.

In practice, most statistical tests are carried out by computer software that calculates the *P*-value for us. It is usual to report the *P*-value in describing the results of studies in many fields. You should, therefore, understand what *P*-values say even if you don't do statistical tests yourself, just as you should understand what "95% confidence" means even if you don't calculate your own confidence intervals.

Key Terms

The probability, computed assuming that H_0 is true, that the sample outcome would be as extreme or more extreme than the actually observed outcome is called the **P-value** of the test. The smaller the *P*-value is, the stronger is the evidence against H_0 provided by the data.

Example 2 — *Working through college*

Do college students work too much? According to a 2015 report from the Georgetown University Center on Education and the Workforce, 70% of college students in the United States have a full- or part-time job while enrolled in college. If we express 70% as a proportion, this means the claimed population proportion is $p = 0.7$ An administrator from a local college questions the accuracy of this claim. In particular, she believes the true proportion of students at her college who have full- or part-time jobs while enrolled in college is different from 0.7. The administrator is able to survey a random sample of 325 students from her college, and she finds that 238 of these students have full- or part-time jobs. The sample proportion is

$$\hat{p} = \frac{238}{325} = 0.732$$

Is this evidence that the true population proportion is something different than 0.7? This is a job for a significance test.

LightField Studios/Shutterstock

The hypotheses. The null hypothesis says that the population proportion is 0.7 ($p = 0.7$). The administrator believes that this value is incorrect, but she does not theorize ahead of time that the true value is higher than or lower than 0.7. She just believes the true population proportion is something different than 0.7, so the alternative hypothesis is just "the population proportion is not 0.7." The two hypotheses are

$$H_0: p = 0.7$$
$$H_a: p \neq 0.7$$

The sampling distribution. *If the null hypothesis is true,* the sample proportion of college students who work full- or part-time has approximately the Normal distribution with

$$\text{mean} = p = 0.7$$
$$\text{standard deviation} = \sqrt{\frac{p(1-p)}{n}}$$
$$= \sqrt{\frac{(0.7)(0.3)}{325}}$$
$$= 0.02542$$

The data. Figure 22.2 shows this sampling distribution with the administrator's sample outcome $\hat{p} = 0.732$ marked. The picture already suggests that this is not an unlikely outcome that would give strong evidence against the claim that $p = 0.7$.

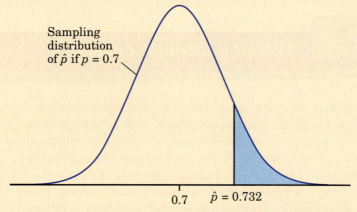

Sampling
distribution
of $\hat{p}$ if $p = 0.7$

0.7 $\hat{p} = 0.732$

Figure 22.2 The sampling distribution of the proportion of college students who report working full- or part-time, Example 2. The administrator's sample result, 0.732, is marked.

The P-value. How unlikely is an outcome as far from 0.7 as $\hat{p} = 0.732$? Because the alternative hypothesis allows p to lie on either side of 0.7, values of $\hat{p}$ far from 0.7 in either direction provide evidence against H_0 and in favor of H_a. The P-value is, therefore, the probability that the observed $\hat{p}$ lies as far from 0.7 *in either direction* as the observed $\hat{p} = 0.732$. Figure 22.3 shows this probability as area under the Normal curve. It is $P = 0.19$.

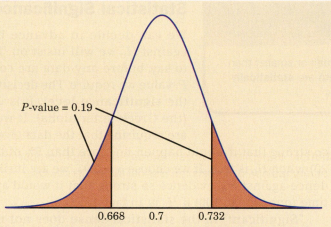

Figure 22.3 The *P*-value for testing whether the proportion of college students who work full- or part-time is different than 0.7, Example 2. This is the probability, calculated assuming the null hypothesis is true, of a sample proportion as far or farther from 0.7 as the administrator's sample result of 0.732.

The conclusion. If the true population proportion is 0.7, the probability we would obtain a sample proportion this far or farther from 0.7 is 0.19. We therefore cannot reject the claim that 70% of college students work full- or part-time while attending college.

The alternative $H_a: p > 0.5$ in Example 1 is a **one-sided alternative** because the effect we seek evidence for says that the population proportion is greater than one-half. The alternative $H_a: p \neq 0.7$ in Example 2 is a **two-sided alternative** because we ask whether the population proportion p is different from 0.7. Whether the alternative is one-sided or two-sided determines whether sample results that are extreme in one direction or in both directions count as evidence against H_0 in favor of H_a.

Now it's your turn

22.1 Working full- or part-time. Suppose the college administrator was only able to survey a random sample of 50 students. She finds that 32 of these students report having a full- or part-time job. The sample proportion is

$$\hat{p} = \frac{32}{50} = 0.64.$$

Is this evidence that the population proportion is not equal to 0.7? Formulate the hypotheses for an appropriate significance test and determine the sampling distribution of the sample proportion of students who work full- or part-time if the null hypothesis is true. As you are working through this problem, think about what you learned in Chapters 17 and 20 about the "law of large numbers." Does it make sense that your sample proportion would be different with a smaller sample size?

Statistical Significance

We can decide in advance how much evidence against H_0 we will insist on. The way to do this is to say, before any data are collected, how small a *P*-value we require. The decisive value of *P* is called the **significance level.** It is usual to write it as α (the Greek letter alpha). If we choose α = 0.05, we are requiring that the data give evidence against H_0 so strong that it would happen no more than 5% of the time (one time in 20) when H_0 is true. If we choose α = 0.01, we are insisting on stronger evidence against H_0, evidence so strong that it would appear only 1% of the time (one time in 100) if H_0 is, in fact, true.

"Significant" in the statistical sense does not mean "important." It means simply "not likely to happen just by chance." We used these words in Chapter 5 (page 97). Now we have attached a number to statistical significance to say what "not likely" means. You will often see significance at level 0.01 expressed by the statement, "The results were significant ($P < 0.01$)." Here, *P* stands for the *P*-value.

One traditional level of significance to use is 0.05. The origins of this appear to trace back to British statistician and geneticist Sir Ronald A. Fisher. Fisher once wrote that it was convenient to consider sample statistics that are two or more standard deviations away from the mean as being significant. Of course, we don't have to make use of traditional levels of significance such as 5% and 1%. The *P*-value is more informative because it allows us to assess significance at any level we choose. For example, a result with $P = 0.03$ is significant at the α = 0.05 level but not significant at the α = 0.01 level. Nonetheless, the traditional significance levels are widely accepted guidelines for "how much evidence is enough." We might say that $P < 0.10$ indicates "some evidence" against the null hypothesis, $P < 0.05$ is "moderate evidence," and $P < 0.01$ is "strong evidence." Don't take these guidelines too literally, however. We will say more about interpreting tests in Chapter 23.

Calculating *P*-values*

Finding the *P*-values we gave in Examples 1 and 2 requires doing Normal distribution calculations using Table B of Normal percentiles. That was optional reading in Chapter 13 (pages 304–306). In practice, software does the calculation for us, but here is an example that shows how to use Table B.

*This section is optional.

Example 3

Tasting coffee

The hypotheses. In Example 1, we want to test the hypotheses

$$H_0: p = 0.5$$
$$H_a: p > 0.5$$

Here, p is the proportion of the population of all coffee drinkers who prefer fresh coffee to instant coffee.

kali9/E+/Getty Images

The sampling distribution. If the null hypothesis is true, so that $p = 0.5$, we saw in Example 1 that $\hat{p}$ follows a Normal distribution with mean 0.5 and standard deviation 0.0707.

The data. A sample of 50 people found that 36 preferred fresh coffee. The sample proportion is $\hat{p} = 0.72$.

The *P*-value. The alternative hypothesis is one-sided on the high side. So, the *P*-value is the probability of getting an outcome at least as large as 0.72. Figure 22.1 displays this probability as an area under the Normal sampling distribution curve. To find any Normal curve probability, move to the standard scale. When we convert a sample statistic to a standard score when conducting a statistical test of significance, the standard score is commonly referred to as a **test statistic.** The test statistic for the outcome $\hat{p} = 0.72$ is

$$\text{standard score} = \frac{\text{observation} - \text{mean}}{\text{standard error}}$$
$$= \frac{0.72 - 0.5}{0.0707} = 3.1$$

Table B says that standard score 3.1 is the 99.9 percentile of a Normal distribution. That is, the area under a Normal curve to the left of 3.1 (in the standard scale) is 0.999. The area to the right is therefore 0.001, and that is our *P*-value.

The conclusion. The small *P*-value means that these data provide very strong evidence that a majority of the population prefers fresh coffee.

Now it's your turn

22.2 Working full- or part-time. Refer to Exercise 22.1 (page 519). The college administrator surveyed 50 students and observed that 32 of them worked full- or part-time, so the sample proportion is

$$\hat{p} = \frac{32}{50} = 0.64$$

Is this evidence that the population proportion is not equal to 0.7? For the hypotheses you formulated in Exercise 22.1, find the *P*-value based on the results of the survey of 50 students. Are the results significant at the 0.05 level?

Tests for a Population Mean*

The reasoning that leads to significance tests for hypotheses about a population mean μ follows the reasoning that leads to tests about a population proportion p. The big idea is to use the sampling distribution that the sample mean $\bar{x}$ would have if the null hypothesis were true. Locate the value of $\bar{x}$ from your data on this distribution, and see if it is unlikely. A value of $\bar{x}$ that would rarely appear if H_0 were true is evidence that H_0 is not true. The four steps are also similar to those in tests for a proportion. Here are two examples, the first one-sided and the second two-sided.

Example 4
Length of human pregnancies

Researchers have recently begun to question the actual length of a human pregnancy. According to a report published on August 6, 2013, although women are typically given a delivery date that is calculated as 280 days after the onset of their last menstrual period, only 4% of women deliver babies at 280 days. A more likely average time from ovulation to birth may be much less than 280 days. To test this theory, a random sample of 95 women with healthy pregnancies is monitored from ovulation to birth. The mean length of pregnancy is found to be $\bar{x} = 275$ days, with a standard deviation of $s = 10$ days. Is this sample result good evidence that the mean length of a healthy pregnancy for *all* women is less than 280 days?

The hypotheses. The researcher's claim is that the mean length of pregnancy is less than 280 days. That's our alternative hypothesis, the statement we seek evidence *for*. The hypotheses are

$$H_0: \mu = 280$$
$$H_a: \mu < 280$$

The sampling distribution. *If the null hypothesis is true,* the sample mean $\bar{x}$ has approximately the Normal distribution with mean $\mu = 280$ and standard error

$$\frac{s}{\sqrt{n}} = \frac{10}{\sqrt{95}} = 1.03$$

We once again use the sample standard deviation s in place of the unknown population standard deviation σ.

The data. The researcher's sample gave $\bar{x} = 275$. The standard score, or test statistic, for this outcome is

$$\text{standard score} = \frac{\text{observation} - \text{mean}}{\text{standard error}}$$
$$= \frac{275 - 280}{1.03} = -4.85$$

That is, the sample result is about 4.85 standard errors below the mean we would expect if, on the average, healthy pregnancies lasted for 280 days.

*This section is optional.

The *P*-value. Figure 22.4 locates the sample outcome −4.85 (in the standard scale) on the Normal curve that represents the sampling distribution if H_0 is true. This curve has mean 0 and standard deviation 1 because we are using the standard scale. The *P*-value for our one-sided test is the area to the left of −4.85 under the Normal curve. Figure 22.4 indicates that this area is very small. The smallest value in Table B is −3.4 and Table B says that −3.4 is the 0.03 percentile, so the area to its left is 0.0003. Because −4.85 is smaller than −3.4, we know that the area to its left is smaller than 0.0003. Thus, our *P*-value is smaller than 0.0003.

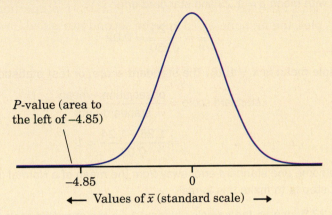

Figure 22.4 The *P*-value for a one-sided test when the standard score for the sample mean is −4.85, Example 4.

The conclusion. A *P*-value of less than 0.0003 is strong evidence that the mean length of a healthy human pregnancy is below the commonly reported length of 280 days.

Example 5 — *Time spent eating*

The Bureau of Labor Statistics conducts the American Time Use Survey (ATUS). This survey is intended to provide nationally representative estimates of how, where, and with whom Americans spend their time, with approximately 26,400 households being surveyed each year since 2003. In 2017, it was reported that Americans spend an average of 1.24 hours per day eating and drinking. A researcher at a college campus in California conducts a survey of a random sample of 150 college students and finds that the average amount of time these students report eating and drinking per day is $\bar{x} = 1.26$ hours,

PeopleImages/Getty Images

with a standard deviation of $s = 0.25$ hour. Is this evidence that the college students eat and drink, on average, a different amount of time each day when compared to the general population?

The hypotheses. The null hypothesis is "no difference" from the national mean. The alternative is two-sided because the researcher did not have a particular direction in mind before examining the data. So, the hypotheses about the unknown mean μ of the population are

$$H_0: \mu = 1.24$$
$$H_a: \mu \neq 1.24$$

The sampling distribution. *If the null hypothesis is true,* the sample mean $\bar{x}$ has approximately the Normal distribution with mean $\mu = 1.23$ and standard error

$$\frac{s}{\sqrt{n}} = \frac{0.25}{\sqrt{150}} = 0.02$$

The data. The sample mean is $\bar{x} = 1.26$. The standard score, or test statistic, for this outcome is

$$\text{standard score} = \frac{\text{observation} - \text{mean}}{\text{standard error}}$$

$$= \frac{1.26 - 1.24}{0.02} = 1$$

We know that an outcome one standard error away from the mean of a Normal distribution is not very surprising. The last step is to make this formal.

The P-value. Figure 22.5 locates the sample outcome 1 (in the standard scale) on the Normal curve that represents the sampling distribution if H_0 is true. The two-sided P-value is the probability of an outcome at least this far out *in either direction.* This is the shaded area under the curve. Table B says that a standard score of 1.0 is the 84.13 percentile of a Normal distribution. This means the

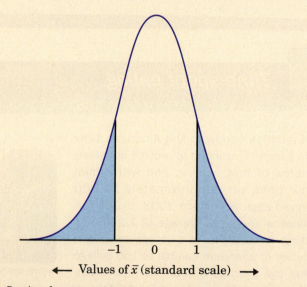

$\longleftarrow$ Values of $\bar{x}$ (standard scale) $\longrightarrow$

Figure 22.5 The P-value for a two-sided test when the standard score for the sample mean is -1, Example 5.

area to the left of 1.0 in the Normal curve is 0.8413. The area to the right is, therefore, 0.1587. The area to the right of 1.0 and to the left of -1.0 is double this, or about 0.3174. This is our approximate P-value.

The conclusion. The large P-value gives us no reason to think that the mean time spent eating and drinking per day among the California college students differs from the national average.

The test assumes that the 150 students in the sample are an SRS from the population of all college students. We should check this assumption by asking how the data were produced. If the data come from a survey conducted on one particular day, in one dining hall on campus, for example, the data are of little value for our purpose. It turns out that the researcher attempted to gather a random sample by using a computer program to randomly select names from a college directory. He then sent surveys out to those students who had been chosen to participate in this study of eating habits.

What if the researcher did not draw a random sample? You should be very cautious about inference based on nonrandom samples, such as convenience samples, or random samples with large nonresponse. Although it is possible for a convenience sample to be representative of the population and, hence, yield reliable inference, establishing this is not easy. You must be certain that the method for selecting the sample is unrelated to the quantity being measured. You must make a case that individuals are independent. You must use additional information to argue that the sample is representative. This can involve comparing other characteristics of the sample to known characteristics of the population. Is the proportion of men and women in the sample about the same as in the population? Is the racial composition of the sample about the same as in the population? What about the distribution of ages or other demographic characteristics? Even after arguing that your sample appears to be representative, the case is not as compelling as that for a truly random sample. You should proceed with caution.

The data in Example 5 do *not* establish that the mean time spent eating and drinking μ for this sample of college students is 1.24 hours. We sought evidence that μ differed from 1.24 hours and failed to find convincing evidence. That is all we can say. No doubt the mean time spent eating and drinking among the entire college population is not exactly equal to 1.24 hours. A large enough sample would give evidence of the difference, even if it is very small.

Statistics in Your World

Catching cheaters Lots of students take a long multiple-choice exam. Can the computer that scores the exam also screen for papers that are suspiciously similar? Clever people have created a measure that takes into account not just identical answers, but the popularity of those answers and the total score on the similar papers as well. The measure has close to a Normal distribution, and the computer flags pairs of papers with a measure outside ± 4 standard deviations as significant.

Now it's your turn

22.3 IQ scores. The mean IQ for the entire population in any age group is supposed to be 100. Suppose that we measure the IQ of 45 seventh-grade students in a Midwest school district and find a sample mean of 105.8 and sample standard deviation of 14.3. Treat the scores as if they were an SRS from all middle-school students in this district. Do the scores provide good evidence that the mean IQ of this population is not 100?

Chapter 22: Statistics in Summary

- A confidence interval estimates an unknown parameter. A **test of significance** assesses the evidence for some claim about the value of an unknown parameter.

- In practice, the purpose of a statistical test is to answer the question, "Could the effect we see in the sample just be an accident due to chance, or is it good evidence that the effect is really there in the population?"

- Significance tests answer this question by giving the probability that a sample effect as large as the one we see in this sample would arise just by chance. This probability is the *P*-value. A small *P*-value says that our outcome is unlikely to happen just by chance.

- To set up a test, state a **null hypothesis** that says the effect you seek is *not* present in the population. The **alternative hypothesis** says that the effect *is* present.

- The *P*-value is the probability, calculated taking the null hypothesis to be true, of an outcome as extreme in the direction specified by the alternative hypothesis as the actually observed outcome.

- A sample result is **statistically significant at the 5% level (or at the 0.05 level)** if it would occur just by chance no more than 5% of the time in repeated samples.

This chapter summary will help you evaluate the Case Study.

Link It

In this chapter, we discuss tests of significance, another type of statistical inference. The mathematics of probability, in particular the sampling distributions discussed in Chapter 18, provides the formal basis for a test of significance. The sampling distribution allows us to assess "probabilistically" the strength of evidence against a null hypothesis, through either a level of significance or a *P*-value. The goal of hypothesis testing, which is used to assess the evidence provided by data about some claim concerning a population, is different from the goal of confidence interval estimation, discussed in Chapter 21, which is used to estimate a population parameter.

Although we have applied the reasoning of tests of significance to population proportions and population means, the same reasoning applies to tests of significance for other population parameters, such as the correlation coefficient, in more advanced settings. In the next chapter, we provide more discussion of the practical interpretation of statistical tests.

Case Study Evaluated

Use what you have learned in this chapter to evaluate the Case Study. Start by reviewing the Chapter Summary. Then answer each of the following questions in complete sentences. Be sure to communicate clearly enough for any of your classmates to understand what you are saying. Could the 2015 and 2016 random samples in the Higher Education Research Institute's surveys of college freshmen differ by 44.9% versus 42.3% just by chance? Tests of significance can help answer these questions. In both cases, one finds that the P-values for the tests of whether two such random samples would differ by the amounts reported was less than 0.001.

1. Using language that can be understood by someone who knows no statistics, write a paragraph explaining what a P-value of less than 0.001 means in the context of the Higher Education Research Institute's surveys of college freshmen.

2. Are the results of the study significant at the 0.05 level? At the 0.01 level? Explain.

In this chapter you:

- Learned how to conduct tests of significance for proportions and means.
- Interpreted the results of tests of significance.
- Determined how to decide if an observed difference can plausibly be attributed to chance.

macmillan learning Online Resources

- The Snapshots Video, *Hypothesis Tests*, discusses the basic reasoning of tests of significance in the context of an example involving discrimination.

- The StatClips Video, *P-value Interpretation*, discusses the interpretation of a P-value in the context of an example of a hypothesis test for a population mean.

Check the Basics

For Exercise 22.1, see page 519; for Exercise 22.2, see page 521; and for Exercise 22.3, see page 526.

22.4 Watching the news. In 2018, the Pew Research Center published a report in which it was claimed that 44% of adults in the United States prefer watching the news on television as opposed to reading it or listening to it on the radio. A researcher believes that significantly more than 44% of adults prefer watching the news on television. To test his theory, he obtains data from a random sample of 327 adults, and it is observed that 53% of this sample prefers watching the news on television. As a first step in conducting a statistical test of significance, the researcher will write a null hypothesis. In this example, the null hypothesis should be what?

(a) $H_0: p < 0.53$

(b) $H_0: p > 0.44$

(c) $H_0: p = 0.53$

(d) $H_0: p = 0.44$

22.5 College students and the Internet. It is reported that college students spend an average of 100 minutes per day on the Internet. An educational technologist disputes this claim and believes that college students spend more than an average of 100 minutes per day on the Internet. When a random sample of data is obtained and a statistical test of significance is conducted, it is observed that the P-value is 0.01. What is the correct decision to reach based on this P-value?

(a) The data provide strong evidence that null hypothesis must be true.

(b) The data provide strong evidence that the alternative hypothesis must be true.

(c) The data provide strong evidence against the null hypothesis.

(d) The data provide strong evidence against the alternative hypothesis.

22.6 Statistical significance. If the results of a statistical test are considered to be statistically significant, what does this mean?

(a) The results are important.

(b) The results are not likely to happen just by chance.

(c) The P-value is large.

(d) The alternative hypothesis is true.

22.7 Pulse rates. Suppose that a report by a leading medical organization claims that the healthy human heart beats an average of 72 times per minute. Advances in science have led some researchers to question if the healthy human heart beats an entirely different amount of time, on average, per minute. They obtain pulse rate data from a sample of 85 healthy adults and find the average number of heart beats per minute to be 76, with a standard deviation of 13. Before conducting a statistical test of significance, this outcome needs to be converted to a standard score, or a test statistic. What would that test statistic be?

(a) 0.3

(b) −0.3

(c) 2.8

(d) −2.8

Chapter 22 Exercises

22.8 Ethnocentrism. A social psychologist reports, "In our sample, ethnocentrism was significantly higher ($P < 0.05$) among church attenders than among nonattenders." Explain to someone who knows no statistics what this means.

22.9 Students' earnings. The financial aid office of a university asks a sample of students about their employment and earnings. The report says, "For academic year earnings, a significant difference ($P = 0.028$) was found between the sexes, with men earning more on average than women. No difference ($P = 0.576$) was found between the earnings of black and white students." Explain both of these conclusions for the effects of sex and of race on mean earnings, in language understandable to someone who knows no statistics.

22.10 Diet and diabetes. Does eating more fiber reduce the blood cholesterol level of patients with diabetes? A randomized clinical trial compared normal and high-fiber diets. Here is part of the researchers' conclusion: "The high-fiber diet reduced plasma total cholesterol concentrations by 6.7 percent ($P = 0.02$), triglyceride concentrations by 10.2 percent ($P = 0.02$), and very-low-density lipoprotein cholesterol concentrations by 12.5 percent ($P = 0.01$)." A doctor who knows no statistics says that a drop of 6.7% in cholesterol isn't a lot—maybe it's just an accident due to the chance assignment of patients to the two diets. Explain in simple language how "$P = 0.02$" answers this objection.

22.11 Diet and bowel cancer. It has long been thought that eating a healthier diet reduces the risk of bowel cancer. A large study cast doubt on this advice. The subjects were 2079 people who had polyps removed from their bowels in the past six months. Such polyps may lead to cancer. The subjects were randomly assigned to a low-fat, high-fiber diet or to a control group in which subjects ate their usual diets. Did polyps reoccur during the next four years?

(a) Outline the design of this experiment.

(b) Surprisingly, the occurrence of new polyps "did not differ significantly between the two groups." Explain clearly what this finding means.

22.12 Pigs and prestige in ancient China. It appears that pigs in Stone Age China were not just a source of food. Owning pigs was also a display of wealth. Evidence for this comes from examining burial sites. If the skulls of sacrificed pigs tend to appear along with expensive ornaments, that suggests that the pigs, like the ornaments, signal the wealth and prestige of the person buried. A study of burials from around 3500 B.C. concluded that "there are striking differences in grave goods between burials with pig skulls and burials without them. . . . A test indicates that the two samples of total artifacts are significantly different at the 0.01 level." Explain clearly why "significantly different at the 0.01 level" gives good reason to think that there really is a systematic difference between burials that contain pig skulls and those that lack them.

22.13 Ancient Egypt. Settlements in Egypt before the time of the pharaohs are dated by measuring the presence of forms of carbon that decay over time. The first datings of settlements in the Nagada region used hair that had been excavated 60 years earlier. Now researchers have used newer methods and more recently excavated material. Do the dates differ? Here is the conclusion about one location: "There are two dates from Site KH6. Statistically, the two dates are not significantly different. They provide a weighted average corrected date of 3715 ± 90 B.C." Explain to someone interested in ancient Egypt but not interested in statistics what "not significantly different" means.

22.14 What's a gift worth? Do people value gifts from others more highly than they value the money it would take to buy the gift? We would like to think so because we hope that "the thought counts." A survey of 209 adults asked them to list three recent gifts and then asked, "Aside from any sentimental value, if, without the giver ever knowing, you could receive an amount of money instead of the gift, what is the minimum amount of money that would make you equally happy?" It turned out that most people would need more money than the gift cost to be equally happy. The magic words "significant $(P < 0.01)$" appear in the report of this finding.

(a) The sample consisted of students and staff in a graduate program and of "members of the general public at train stations and airports in Boston and Philadelphia." The report says this sample is "not ideal." What's wrong with the sample?

(b) In simple language, what does it mean to say that the sample thought their gifts were worth "significantly more" than their actual cost?

(c) Now be more specific: what does "significant $(P < 0.01)$" mean?

22.15 Attending church. A recent survey by a national polling firm finds that 39% of American adults say they attended religious services last week. There is reason to suspect this percentage is not accurate.

(a) Why might we expect answers to a poll about attending religious services to overstate true church attendance?

(b) You suspect strongly that the true percentage attending church in any given week is less than 39%. You plan to watch a random sample of adults and see whether or not they go to church. What are your null and alternative hypotheses? (Be sure to say in words what the population proportion p is for your study.)

22.16 Body temperature. We have all heard that 98.6 degrees Fahrenheit (or 37 degrees Celsius) is "normal body temperature." In fact, there is evidence that most people have a slightly lower body temperature. You plan to measure the body temperature of a random sample of people very accurately. You hope to show that a majority have temperatures lower than 98.6 degrees.

(a) Say clearly what the population proportion p stands for in this setting.

(b) In terms of p, what are your null and alternative hypotheses?

22.17 Unemployment. The national unemployment rate in a recent month was 5.1%. You think the rate may be different in your city, so you plan a sample survey that will ask the same questions as the Current Population Survey. To see if the local rate differs significantly from 5.1%, what hypotheses will you test?

22.18 Living on campus. A UCLA survey of college freshmen in the 2016–2017 academic year found that 74.8% of all first-year college students planned to live on campus. You wonder if this percentage is different at your school, but you have no idea whether it is higher or lower. You plan a sample survey of first-year students at your school. What hypotheses will you test to see if your school differs significantly from the UCLA survey result?

22.19 Do our athletes graduate? The National Collegiate Athletic Association (NCAA) requires colleges to report the graduation rates of their athletes. At one large university, 82% of all students who entered in 2012 graduated within six years. One hundred forty-nine of the 190 students who entered with athletic scholarships graduated. Consider these 190 as a sample of all athletes who will be admitted under present policies. Is there evidence that the percentage of athletes who graduate is less than 82%?

(a) Explain in words what the parameter p is in this setting.

(b) What are the null and alternative hypotheses H_0 and H_a?

(c) What is the numerical value of the sample proportion $\hat{p}$? The P-value is the probability of what event?

(d) The P-value is $P = 0.0968$. Explain whether or not this P-value indicates there is some reason to think that graduation rates are lower among athletes than among all students.

22.20 AP courses. In 2016, 17.6% of first-year college students responding to a national survey said that they had not taken any Advanced Placement (AP) courses in high school. Administrators at a large state university believe that more than 17.6% of their first-year students have not taken any AP courses. They find that 46 of an SRS of 200 of the university's first-year students said that they had not taken any AP courses in high school. Is the proportion of first-year students at this university who said they had not taken any AP courses significantly larger than the 2016 national value of 17.6%?

(a) Explain in words what the parameter p is in this setting.

(b) What are the null and alternative hypotheses H_0 and H_a?

(c) What is the numerical value of the sample proportion $\hat{p}$? The P-value is the probability of what event?

(d) The P-value is $P = 0.0227$. Explain carefully why this evidence should lead administrators to reject H_0.

22.21 Vote for the best face? We often judge other people by their faces. It appears that some people judge candidates for elected office by their faces. Psychologists showed head-and-shoulders photos of the two main candidates in 32 races for the U.S. Senate to many subjects (dropping subjects who recognized one of the candidates) to see which candidate was rated "more competent" based on nothing but the photos. On election day, the candidates whose faces looked more competent won 22 of the 32 contests. If faces don't influence voting, half of all races in the long run should be won by the candidate with the better face. Is there evidence that the proportion of times the candidate with the better face wins is more than 50%?

(a) Explain in words what the parameter p is in this setting.

(b) What are the null and alternative hypotheses H_0 and H_a?

(c) What is the numerical value of the sample proportion $\hat{p}$? The P-value is the probability of what event?

(d) The P-value is $P = 0.017$. Explain carefully why this is reasonably good evidence that H_0 may not be true and that H_a may be true.

22.22 Do our athletes graduate? Is the result of Exercise 22.19 statistically significant at the 10% level? At the 5% level?

22.23 AP courses. Is the result of Exercise 22.20 statistically significant at the 5% level? At the 1% level?

22.24 Vote for the best face? Is the result of Exercise 22.21 statistically significant at the 5% level? At the 1% level?

22.25 Significant at what level? Explain in plain language why a result that is significant at the 1% level must always be significant at the 5% level. If a result is significant at the 5% level, what can you say about its significance at the 1% level?

22.26 Significance means what? Asked to explain the meaning of "statistically significant at the $\alpha = 0.05$ level," a student says: "This means that the probability that the null hypothesis is true is less than 0.05." Is this explanation correct? Why or why not?

22.27 Finding a *P*-value by simulation. Is a new method of teaching reading to first-graders (Method B) more effective than the method now in use (Method A)? You design a matched pairs experiment to answer this question. You form 20 pairs of first-graders, with the two children in each pair carefully matched by IQ, socioeconomic status, and reading-readiness score. You assign at random one student from each pair to Method A. The other student in the pair is taught by Method B. At the end of first grade, all the children take a test to determine their reading skill. Assume that the higher the score on this test, the more proficient the student is at reading. Let p stand for the proportion of all possible matched pairs of children for which the child taught by Method B has the higher score. Your hypotheses are

$H_0: p = 0.5$ (no difference in effectiveness)

$H_a: p > 0.5$ (Method B is more effective)

The result of your experiment is that Method B gave the higher score in 12 of the 20 pairs, or $\hat{p} = 12/20 = 0.6$.

(a) If H_0 is true, the 20 pairs of students are 20 independent trials with probability 0.5 that Method B "wins" each trial (is the more effective method). Explain how to use Table A to simulate these 20 trials if we assume for the sake of argument that H_0 is true.

(b) Use Table A, starting at line 105, to simulate 10 repetitions of the experiment. Estimate from your simulation the probability that Method B will do better (be the more effective method) in 12 or more of the 20 pairs when H_0 is true. (Of course, 10 repetitions are not enough to estimate the probability reliably. Once you understand the idea, more repetitions are easy.)

(c) Explain why the probability you simulated in part (b) is the *P*-value for your experiment. With enough patience, you could find all the *P*-values in this chapter by doing simulations similar to this one.

22.28 Finding a *P*-value by simulation. A classic experiment to detect extra-sensory perception (ESP) uses a shuffled deck of cards containing five suits (waves, stars, circles, squares, and crosses). As the experimenter turns over each card and concentrates on it, the subject guesses the suit of the card. A subject who lacks ESP has probability 1-in-5 of being right by luck on each guess. A subject who has ESP will be right more often. Julie is right in 5 of 10 tries. (Actual experiments use much longer series of guesses so that weak ESP can be spotted. No one has ever been right half the time in a long experiment!)

(a) Give H_0 and H_a for a test to see if this result is significant evidence that Julie has ESP.

(b) Explain how to simulate the experiment if we assume for the sake of argument that H_0 is true.

(c) Simulate 20 repetitions of the experiment; begin at line 121 of Table A.

(d) The actual experimental result was 5 correct in 10 tries. What is the event whose probability is the *P*-value for this experimental result? Give an estimate of the *P*-value based on your simulation. How convincing was Julie's performance?

The following exercises concern the optional section on calculating P-values. To carry out a test, complete the steps (hypotheses, sampling distribution, data, P-value, and conclusion) illustrated in Example 3.

22.29 AP courses. Return to the study in Exercise 22.20, which found that 46 of 200 first-year students said that they had not taken any AP courses in high school. Carry out the hypothesis test described in Exercise 22.20 and compute the *P*-value. How does your value compare with the value given in Exercise 22.20(d)?

22.30 Interpreting scatterplots. In 2014, the Pew Research Centers American Trends Panel sought to better understand what Americans know about science. It was observed that among a random selection of 3278 adults, 2065 adults could correctly interpret a scatterplot. Is this good evidence that more than 60% of Americans are able to correctly interpret scatterplots?

22.31 Side effects. An experiment on the side effects of pain relievers assigned arthritis patients to one of several over-the-counter pain medications. Of the 420 patients who took one brand of pain reliever, 21 suffered some "adverse symptom."

(a) If 10% of all patients suffer adverse symptoms, what would be the sampling distribution of the proportion with adverse symptoms in a sample of 420 patients?

(b) Does the experiment provide strong evidence that fewer than 10% of patients who take this medication have adverse symptoms?

22.32 Do chemists have more girls? Some people think that chemists are more likely than other parents to have female children. (Perhaps chemists are exposed to something in their laboratories that affects the sex of their children.) The Washington State Department of Health lists the parents' occupations on birth certificates. Between 1980 and 1990, 555 children were born to fathers who were chemists. Of these births, 273 were girls. During this period, 48.8% of all births in Washington State were girls. Is there evidence, at a significance level of 0.05, that the proportion of

girls born to chemists is higher than the state proportion?

22.33 Speeding. It often appears that most drivers on the road are driving faster than the posted speed limit. Situations differ, of course, but here is one set of data. Researchers studied the behavior of drivers on a rural interstate highway in Maryland where the speed limit was 55 miles per hour. They measured speed with an electronic device hidden in the pavement and, to eliminate large trucks, considered only vehicles less than 20 feet long. They found that 5690 out of 12,931 vehicles were exceeding the speed limit. Is this good evidence, at a significance level of 0.05, that (at least in this location) fewer than half of all drivers are speeding?

The following exercises concern the optional section on tests for a population mean. To carry out a test, complete the steps illustrated in Example 4 or Example 5.

22.34 Student attitudes. The Survey of Study Habits and Attitudes (SSHA) is a psychological test that measures students' study habits and attitudes toward school. Scores range from 0 to 200. The mean score for U.S. college students is about 115, and the standard deviation is about 30. A teacher suspects that older students have better attitudes toward school. She gives the SSHA to 25 students who are at least 30 years old. Assume that scores in the population of older students are Normally distributed with standard deviation $\sigma = 30$. The teacher wants to test the hypotheses

$$H_0: \mu = 115$$
$$H_a: \mu > 115$$

(a) What is the sampling distribution of the mean score $\bar{x}$ of a sample of 25 older students if the null hypothesis is true? Sketch the density curve of this distribution. (*Hint:* Sketch a Normal curve first, then mark the axis using what you know about locating μ and σ on a Normal curve.)

(b) Suppose that the sample data give $\bar{x} = 118.6$. Mark this point on the axis of your sketch. In fact, the outcome was $\bar{x} = 125.7$. Mark this point on your sketch. Using your sketch, explain in simple language why one outcome is good evidence that the mean score of all older students is greater than 115 and why the other outcome is not.

(c) Shade the area under the curve that is the *P*-value for the sample result $\bar{x} = 125.7$.

22.35 Mice in a maze. Experiments on learning in animals sometimes measure how long it takes mice to find their way through a maze. The mean time is 19 seconds for one particular maze. A researcher thinks that a loud noise will cause the mice to complete the maze faster. She measures how long each of several mice takes to find its way through a maze with a noise as a stimulus. What are the null hypothesis H_0 and alternative hypothesis H_a?

22.36 Response time. Last year, your company's service technicians took an average of 2.5 hours to respond to trouble calls from business customers who had purchased service contracts. Do this year's data show a significantly different average response time? What null and alternative hypotheses should you test to answer this question?

22.37 Testing a random number generator. Our statistical software has a "random number generator" that is supposed to produce numbers scattered at random from 0 to 1. If this is true, the numbers generated come from a population with $\mu = 0.5$. A command to generate 100 random numbers gives outcomes with $\bar{x} = 0.536$ and $s = 0.312$. Is this good evidence that the mean of all numbers produced by this software is not 0.5?

22.38 Will they charge more? A bank wonders whether omitting the annual credit card fee for customers who charge at least $3000 in a year will increase the amount charged on its credit cards. The bank makes this offer to an SRS of 400 of its credit card customers. It then compares how much these customers charge this year with the amount that they charged last year. The mean increase in the sample is $246, and the standard deviation is $112. Is there significant evidence at the 1% level that the mean amount charged increases under the no-fee offer? State H_0 and H_a and carry out a significance test. Use significance level 0.01.

22.39 Bad weather, bad tip? People tend to be more generous after receiving good news. Are they less generous after receiving bad news? The average tip left by adult Americans is 20%. Give 20 patrons of a restaurant a message on their bill warning them that tomorrow's weather will be bad and record the tip percentage they leave. Here are the tips as a percentage of the total bill:

18.0 19.1 19.2 18.8 18.4 19.0 18.5 16.1 16.8 18.2
14.0 17.0 13.6 17.5 20.0 20.2 18.8 18.0 23.2 19.4

Suppose that tip percentages are Normal with $\sigma = 2$, and assume that the patrons in this study are a random sample of all patrons of this restaurant. Is there good evidence that the mean tip percentage for all patrons of this restaurant is less than 20 when they receive a message warning them that tomorrow's weather will be bad? State H_0 and H_a and carry out a significance test. Use significance level 0.05.

22.40 Why should the significance level matter? On June 15, 2005, an article by Lawrence K. Altman appeared in the *New York Times*. The title of the article was "Studies Rebut Earlier Report on Pledges of Virginity." The article mentioned that in two studies conducted by the Heritage Foundation, it was observed that students who took virginity pledges engaged in fewer risky sexual behaviors and had lower rates of acquiring sexually transmitted diseases. These results were at odds with the results of earlier studies that focused on the same topic, but since each study used different methods of statistical analysis, direct comparisons among the studies was difficult. One particular criticism of the new study was that the result of a statistical test at a 0.10 level of significance was reported when journals generally use a lower level of 0.05. Why might this be a concern?

Exploring the Web

Access these exercises on the text website: macmillanlearning.com/scc10e.

23

Use and Abuse of Statistical Inference

In this chapter you will:

- Learn what statistical confidence and statistical significance do and do not mean.

- Be able to identify abuses of statistical inference.

Do the foods a woman eats prior to conceiving a baby have an effect on the sex of the baby? There is some debate about this. In 2008, the *NewScientist* published an article with the headline "Breakfast cereals boost chances of conceiving boys." Other media outlets followed suit and published similar reports about the power of breakfast cereal. The reports described a research study conducted in Great Britain with a sample of 740 pregnant women. A total of 721 of these women gave a retrospective report of their typical diet in the year prior to the conception of their baby by completing a food frequency questionnaire. When the researchers analyzed data from the questionnaires, they found that of the 133 food items included on the questionnaire, breakfast cereal was the only item to be strongly associated with the sex of the baby, with those women who consumed more breakfast cereal producing a greater percentage of male infants than those women consuming less breakfast cereal.

Critics of the study were quick to point out that if there were 133 food items and possibly 133 hypothesis tests conducted, one of these tests was bound to produce a significant result just by chance alone, even if there is no impact of breakfast cereal on the sex of a baby. Why might this be the case? By the end of this chapter, you will be able to answer the question of why conducting multiple hypothesis tests in a single study can be problematic.

Using Inference Wisely

In previous chapters, we have met the two major types of statistical inference: confidence intervals and significance tests. We have, however, seen only two inference methods of each type, one designed for inference about a population proportion p and the other designed for inference about a population mean μ. There are libraries of both books and software filled with methods for inference about various parameters in various settings. The reasoning of confidence intervals and significance tests remains the same, regardless of the method. The first step in using inference wisely is to understand your data and the questions you want to answer and fit the method to its setting. Here are some tips on inference, adapted to the settings we are familiar with.

The design of the data production matters. "Where do the data come from?" remains the first question to ask in any statistical study. Any inference method is intended for use in a specific setting. For our confidence interval and test for a proportion p:

- The data must be a simple random sample (SRS) from the population of interest. When you use these methods, you are acting as if the data are an SRS. In practice, it is often not possible to actually choose an SRS from the population. Your conclusions may then be open to challenge.

- These methods are not correct for sample designs more complex than an SRS, such as stratified samples. There are other methods that fit these settings.

- There is no correct method for inference from data haphazardly collected with bias of unknown size. Fancy formulas cannot rescue badly produced data.

- Other sources of error, such as dropouts and nonresponse, are important. Remember that confidence intervals and tests use the data you collect and ignore these errors.

Example 1

The psychologist and the women's studies professor

A psychologist is interested in how our visual perception can be fooled by optical illusions. Her subjects are students in Psychology 101 at her university. Most psychologists would agree that it's safe to treat the students as an SRS of all people with normal vision. There is nothing special about being a student that changes visual perception.

A professor at the same university uses students in Women's Studies 101 to examine attitudes toward violence against women and

ContentWorks/Getty Images

reproductive rights. Students as a group are younger than the adult population as a whole. Even among young people, students as a group come from more prosperous and better-educated homes. Even among students, this university isn't typical of all campuses. Even on this campus, students in a women's studies course may have opinions that are quite different from those of students who do not take Women's Studies 101. The professor can't reasonably act as if these students are a random sample from any population of interest other than students taking Women's Studies 101 at this university during this term.

Know how confidence intervals behave. A confidence interval estimates the unknown value of a parameter and also tells us how uncertain the estimate is. All confidence intervals share these behaviors:

• The confidence level says how often the *method* catches the true parameter when sampling many times. We never know whether this specific data set gives us an interval that contains the true value of the parameter. All we can say is that "we got this result from a method that works 95% of the time." This data set might be one of the 5% that produce an interval that misses the true value of the parameter. If that risk is too high for you, use a 99% confidence interval.

Statistics in Your World

Dropping out An experiment found that weight loss is significantly more effective than exercise for reducing high cholesterol and high blood pressure. The 170 subjects were randomly assigned to a weight-loss program, an exercise program, or a control group. Only 111 of the 170 subjects completed their assigned treatment, and the analysis used data from these 111. Did the dropouts create bias? Always ask about details of the data before trusting inference.

- High confidence is not free. A 99% confidence interval will be wider than a 95% confidence interval based on the same data. To be more confident, we must have more values to be confident about. There is a trade-off between how closely we can pin down the true value of the parameter (the *precision* of the confidence interval) and how confident we are that we have captured its true value.

- Larger samples give narrower intervals. If we want high confidence *and* a narrow interval, we must take a larger sample. The width of our confidence interval for p goes down by a factor of the square root of the sample size. To cut the interval in half, we must take four times as many observations. This is typical of many types of confidence intervals.

Know what statistical significance says. Many statistical studies hope to show that some claim is true. A clinical trial compares a new drug with a standard drug because the doctors hope that the health of patients given the new drug will improve. A psychologist studying gender differences suspects that women will do better than men (on average) on a test that measures social-networking skills. The purpose of significance tests is to weigh the evidence that the data give in favor of such claims. That is, a test helps us know if we have found what we were looking for.

To do this, we ask what would happen if the claim were *not* true. That's the null hypothesis—no difference between the two drugs, no difference between women and men. A significance test answers only one question: "How strong is the evidence that the null hypothesis is not true?" A test answers this question by giving a P-value. The P-value tells us how likely data as or more extreme than ours would be if the null hypothesis were true. Data that are very unlikely and have a small P-value are good evidence that the null hypothesis is not true. We usually don't know whether the hypothesis is true for this specific population. All we can say is that "data as or more extreme than these would occur only 5% of the time if the hypothesis were true."

This kind of indirect evidence against the null hypothesis (and for the effect we hope to find) is less straightforward than a confidence interval. We will say more about tests in the next section.

Know what your methods require. Our significance test and confidence interval for a population proportion p require that the population size be much larger than the sample size. They also require that the sample size itself be reasonably large so that the sampling distribution of the sample proportion $\hat{p}$ is close to Normal. We have said little about the specifics of these requirements because the reasoning of inference is more important. Just as there are inference methods that fit stratified samples, there are methods that fit small samples and small populations. If you plan to use statistical inference in practice, you will need help from a statistician (or need to learn lots more statistics) to manage the details.

Most of us read about statistical studies more often than we actually work with data ourselves. Concentrate on the big issues, not on the details of whether the authors used exactly the right inference methods. Does the study ask the right questions? Where did the data come from? Do the results make sense? Does the study report confidence intervals so you can see both the estimated values of important parameters and how uncertain the estimates are? Does it report P-values to help convince you that findings are not just good luck?

The Woes of Significance Tests

The purpose of a significance test is usually to give evidence for the presence of some effect in the population. The effect might be a probability of heads different from one-half for a coin or a longer mean survival time for patients given a new cancer treatment. If the effect is large, it will show up in most samples—the proportion of heads among our tosses will be far from one-half, or the patients who get the new treatment will live much longer than those in the control group. Small effects, such as a probability of heads only slightly different from one-half, will often be hidden behind the chance variation in a sample. This is as it should be: big effects are easier to detect. That is, the P-value will usually be small when the population truth is far from the null hypothesis.

The "woes" of testing start with the fact that a test measures only the strength of evidence against the null hypothesis. It says nothing about how big or how important the effect we seek in the population really is. For example, our hypothesis might be "This coin is balanced." We express this hypothesis in terms of the probability p of getting a head as $H_0: p = 0.5$. No real coin is exactly balanced, so we know that this hypothesis is not exactly true. If this coin has probability $p = 0.502$ of a head, we might say that, for practical purposes, it is balanced. A statistical test doesn't think about "practical purposes." It just asks if there is evidence that p is not exactly equal to 0.5. The focus of tests on the strength of the evidence against an exact null hypothesis is the source of much confusion in using tests.

Pay particular attention to the size of the sample when you read the result of a significance test. Here's why:

- Larger samples make tests of significance more sensitive. If we toss a coin hundreds of thousands of times, a test of $H_0: p = 0.5$ will often give a very low P-value when the truth for this coin is $p = 0.502$. The test is right—it found good evidence that p really is not exactly equal to 0.5—but it has picked up a difference so small that it is of no practical interest. **A finding can be statistically significant without being practically important.**

- On the other hand, tests of significance based on small samples are often not sensitive. If you toss a coin only 10 times, a test of $H_0: p = 0.5$ will often give a large P-value even if the truth for this coin is $p = 0.7$. Again, the test is right—10 tosses are not enough to give good evidence against the null hypothesis. **Lack of significance does not mean that there is no effect, only that we do not have good evidence for an effect. Small samples often miss important effects that are really present in the population.** As cosmologist Martin Rees said, "Absence of evidence is not evidence of absence."

Example 2

Antidepressants versus a placebo

Through a Freedom of Information Act request, two psychologists obtained 47 studies used by the U.S. Food and Drug Administration for approval of the six antidepressants prescribed most widely between 1987 and 1999. Overall, the psychologists found that there was a statistically significant difference in the effects of antidepressants compared with a placebo, with antidepressants being more effective. However, the psychologists went on to report that antidepressant pills worked 18% better than placebos, a statistically significant difference, "but not meaningful for people in clinical settings."

Whatever the truth about the population, whether $p = 0.7$ or $p = 0.502$, more observations allow us to estimate p more closely. If p is not 0.5, more observations will give more evidence of this, that is, a smaller P-value. Because statistical significance depends strongly on the sample size as well as on the truth about the population, statistical significance tells us nothing about how large or how practically important an effect is. Large effects (like $p = 0.7$ when the null hypothesis is $p = 0.5$) often give data that are insignificant if we take only a small sample. Small effects (like $p = 0.502$) often give data that are highly significant if we take a large enough sample. Let's return to a favorite example to see how significance changes with sample size.

Example 3

Count Buffon's coin

The French naturalist Count Buffon (1707–1788) considered questions ranging from evolution to estimating the number "pi" and made it his goal to answer them. One question he explored was whether a "balanced" coin would come up heads half of the time when tossed. To investigate, Count Buffon tossed a coin 4040 times and got 2048 heads. His sample proportion of heads was

$$\hat{p} = \frac{2048}{4040} = 0.507$$

Is the Count's coin balanced? Suppose we seek statistical significance at level 0.05. The hypotheses are

$$H_0: p = 0.5$$
$$H_a: p \neq 0.5$$

The test of significance works by locating the sample outcome $\hat{p} = 0.507$ on the sampling distribution that describes how $\hat{p}$ would vary if the null hypothesis were true. Figure 23.1 illustrates this. It shows that the observed $\hat{p} = 0.507$ is not surprisingly far from 0.5 and, therefore, is not good evidence against the hypothesis that the true p is 0.5. The P-value, which is 0.37, just makes this precise.

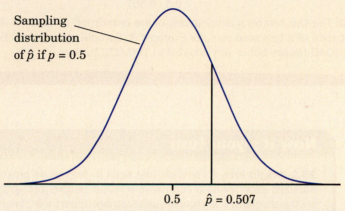

Sampling distribution of $\hat{p}$ if $p = 0.5$

0.5 $\hat{p} = 0.507$

Figure 23.1 The sampling distribution of the proportion of heads in 4040 tosses of a coin if in fact the coin is balanced, for Example 3. Sample proportion 0.507 is not an unusual outcome.

Suppose that Count Buffon got the *same result,* $\hat{p} = 0.507$, from tossing a coin 100,000 times. The sampling distribution of $\hat{p}$ when the null hypothesis is true always has mean 0.5, but its standard deviation gets smaller as the sample size n gets larger. Figure 23.2 displays the two sampling distributions, for $n = 4040$ and $n = 100,000$. The lower curve in this figure is the same Normal curve as in Figure 23.1, drawn on a scale that allows us to show the very tall and narrow curve for $n = 100,000$. Locating the sample outcome $\hat{p} = 0.507$ on the two curves, you see that the same outcome is more or less surprising depending on the size of the sample.

The P-values are $P = 0.37$ for $n = 4040$ and $P = 0.000009$ for $n = 100,000$. Imagine tossing a balanced coin 4040 times repeatedly. You will get a proportion of heads at least as far from one-half as Buffon's 0.507 in about 37% of your repetitions. If you toss a balanced coin 100,000 times repeatedly, however, you will almost never (9 times in 1 million repeats) get an outcome as or more unbalanced than this.

The outcome $\hat{p} = 0.507$ is not evidence against the hypothesis that the coin is balanced if it comes up in 4040 tosses. It is completely convincing evidence if it comes up in 100,000 tosses.

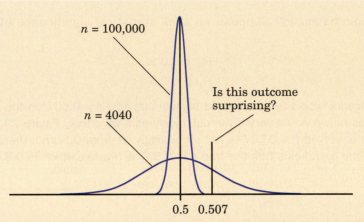

Figure 23.2 The two sampling distributions of the proportion of heads in 4040 and 100,000 tosses of a balanced coin, for Example 3. Sample proportion 0.507 is not unusual in 4040 tosses but is very unusual in 100,000 tosses.

Now it's your turn

23.1 Weight loss. A company that sells a weight-loss program conducted a randomized experiment to determine whether people lost weight after eight weeks on the program. The company researchers report that, on average, the subjects in the study lost weight and that the weight loss was statistically significant with a *P*-value of 0.013. Do you find the results convincing? If so, why? If not, what additional information would you like to have?

Beware the naked *P*-value

The *P*-value of a significance test depends strongly on the size of the sample, as well as on the truth about the population.

It is bad practice to report a naked *P*-value (a *P*-value by itself) without also giving the sample size and a statistic or statistics that describe the sample outcome.

The Advantages of Confidence Intervals

Examples 2 and 3 suggest that we should not rely on significance alone in understanding a statistical study. In Example 3, just knowing that the sample proportion was $\hat{p} = 0.507$ helps a lot. You can decide whether this deviation from one-half is large enough to interest you. Of course, $\hat{p} = 0.507$ isn't the exact truth about the coin, just the chance result of Count Buffon's tosses. So a confidence interval, whose width shows how closely we can pin down the truth about the coin, is even more helpful. Here are the 95% confidence intervals for the true probability of a head *p*, based on the two sample sizes in

Example 3. You can check that the method of Chapter 21 gives these answers.

Number of tosses	95% confidence interval
$n = 4040$	0.507 ± 0.015, or 0.492 to 0.522
$n = 100{,}000$	0.507 ± 0.003, or 0.504 to 0.510

The confidence intervals make clear what we know (with 95% confidence) about the true p. The interval for 4040 tosses includes 0.5, so we are not confident that the coin is unbalanced. For 100,000 tosses, however, we are confident that the true p lies between 0.504 and 0.510. In particular, we are confident that it is not 0.5.

> ## Give a confidence interval
>
> Confidence intervals are more informative than significance tests because they actually estimate a population parameter. They are also easier to interpret. It is good practice to give confidence intervals whenever possible.

Significance at the 5% Level Isn't Magical

The purpose of a test of significance is to describe the degree of evidence provided by the sample against the null hypothesis. The P-value does this. But how small a P-value is convincing evidence against the null hypothesis? This depends mainly on two circumstances:

- *How plausible is H_0?* If H_0 represents an assumption that the people you must convince have believed for years, strong evidence (small P) will be needed to persuade them.

- *What are the consequences of rejecting H_0?* If rejecting H_0 in favor of H_a means making an expensive changeover from one type of product packaging to another, you need strong evidence that the new packaging will boost sales.

These criteria are a bit subjective. Different people will often insist on different levels of significance. Giving the P-value allows each of us to decide individually if the evidence is sufficiently strong. But the level of significance that will satisfy us should be decided before calculating the P-value. Computing the P-value and then deciding that we are satisfied with a level of significance that is just slightly larger than this P-value is an abuse of significance testing.

Users of statistics have often emphasized standard levels of significance such as 10%, 5%, and 1%. For example, courts have tended to accept 5% as a standard in discrimination cases. This emphasis reflects the time when tables of critical values rather than computer software dominated statistical practice. The 5% level ($\alpha = 0.05$) is particularly common. **There is no sharp border between "significant" and "insignificant," only increasingly strong evidence as the P-value decreases.** There is no practical distinction between the P-values 0.049 and 0.051. It makes no sense to treat $P \leq 0.05$ as a universal rule for what is significant.

STATISTICAL CONTROVERSIES

Should Hypothesis Tests Be Banned?

In January 2015, the editors of *Basic and Applied Social Psychology* (BASP) banned "null hypothesis significance testing procedures (NHSTP)." NHSTP is a fancy way to talk about the hypothesis testing we studied in Chapter 22. In addition to the ban on NHSTP, authors submitting articles to BASP must remove all test statistics, *P*-values, and statements about statistical significance from their manuscripts prior to publication. Confidence intervals, which we studied in Chapter 21, have also been banned from BASP.

What does BASP want to see from its future authors? According to the editorial, "BASP will require strong descriptive statistics, including effect sizes" and suggest using (but will not mandate) larger sample sizes than are typical in psychological research "because as the sample size increases, descriptive statistics become increasingly stable and sampling error is less of a problem."

The editorial concludes with the editors stating that they hope other journals will join the BASP ban on NHSTP.

What is your reaction to the ban on hypothesis testing by BASP? How does the approach suggested by the editors compare with the practices that we have emphasized in this book?

Beware of Searching for Significance

Statistical significance ought to mean that you have found an effect that you were looking for. The reasoning behind statistical significance works well if you decide what effect you are seeking, design a study to search for it, and use a test of significance to weigh the evidence you get. In other settings, significance may have little meaning. Here is an example.

Example 4

Does your "sign" affect your health?

In June 2015, the *Washington Post* published an article that reported on researchers who examined the records of patients treated at the Columbia University Medical Center between 1900 and 2000. According to the article, the researchers examined the records of 1.75 million patients, and "using statistical analysis, [the researchers] combed through 1,688 different diseases and found 55 that had a correlation with birth month, including ADHD, reproductive performance, asthma, eyesight and ear infections." Believers in astrology are vindicated! There is a relationship between your horoscope sign and your health. Or is there?

wikki/Shutterstock

Before you base future decisions about your sign and your health on these findings, recall that results significant at the 5% level occur five times in 100 in the long run, even when H_0 is true. When you make dozens of tests at the 5% level, you expect a few of them to be significant by chance alone. That means that we would expect 5% of the 1688 different diseases, or about 84 diseases, to show some relationship with birth month, even when there is no association between any disease and birth month. Finding that 55 of the diseases had some relationship with birth month is actually less than we would expect if there was no relationship. The results are not surprising. Running one test and reaching the $\alpha = 0.05$ level is reasonably good evidence that you have found something. Running several dozen tests and reaching that level once or twice is not.

In Example 4, the researchers tested almost 1700 diseases and found 55 to have a relationship with birth month. Taking these results and saying that they are evidence of a relationship between your sign and your health is not appropriate. It is bad practice to confuse the roles of exploratory analysis of data (using graphs, tables, and summary statistics, like those discussed in Part II, to find suggestive patterns in data) and formal statistical inference. Finding statistical significance is not surprising if you use exploratory methods to examine many outcomes, choose the largest, and test to see if it is significantly larger than the others.

Searching data for suggestive patterns is certainly legitimate. Exploratory data analysis is an important part of statistics. But the reasoning of formal inference does not apply when your search for a striking effect in the data is successful. The remedy is clear. Once you have a hypothesis, design a study to search specifically for the effect you now think is there. If the result of this study is statistically significant, you have real evidence.

Now it's your turn

23.2 Take me out to the ball game. A researcher compared a random sample of recently divorced men in a large city with a random sample of men from the same city who had been married at least 10 years and had never been divorced. The researcher measured 122 variables on each man and compared the two samples using 122 separate tests of significance. Only the variable measuring how often the men attended Major League Baseball games with their spouse was significant at the 1% level, with the married men attending a higher proportion of games with their spouse, on average, than the divorced men did while they were married. Is this strong evidence that attendance at Major League Baseball games improves the chance that a man will remain married? Discuss.

 This might be a good place to read the "What's the verdict?" story on page 557, and to answer the questions. This is a continuation of a story from the television series "Mythbusters" that you first examined in Chapter 6.

Inference as Decision*

Tests of significance were presented in Chapter 22 as methods for assessing the strength of evidence against the null hypothesis. The assessment is made by the P-value, which is the probability computed under the assumption that the null hypothesis is true. The alternative hypothesis (the statement we seek evidence *for*) enters the test only to help us see what outcomes count against the null hypothesis. Such is the theory of tests of significance as advocated by Sir Ronald A. Fisher, and as practiced by many users of statistics.

But we have also seen signs of another way of thinking in Chapter 22. A level of significance α chosen in advance points to the outcome of the test as a *decision*. If the P-value is less than α, we reject H_0 in favor of H_a; otherwise, we fail to reject H_0. The transition from measuring the strength of evidence to making a decision is not a small step. It can be argued (and is argued by followers of Fisher) that making decisions is too grand a goal, especially in scientific inference. A decision is reached only after the evidence of many experiments is weighed, and indeed the goal of research is not "decision" but a gradually evolving understanding. Better that statistical inference should content itself with confidence intervals and tests of significance. Many users of statistics are content with such methods. It is rare (outside textbooks) to set up a level α in advance as a rule for decision making in a scientific problem. More commonly, users think of significance at level 0.05 as a description of good evidence. This is made clearer by talking about P-values, and this newer language is spreading.

Yet there are circumstances in which a decision or action is called for as the end result of inference. *Acceptance sampling* is one such circumstance. The supplier of a product (for example, potatoes to be used to make potato chips) and the consumer of the product agree that each truckload of the product shall meet certain quality standards. When a truckload arrives, the consumer chooses a sample of the product to be inspected. On the basis of the sample outcome, the consumer will either accept or reject the truckload. Fisher agreed that this is a genuine decision problem. But he insisted that acceptance sampling is completely different from scientific inference. Other eminent statisticians have argued that if "decision" is given a broad meaning, almost all problems of statistical inference can be posed as problems of making decisions in the presence of uncertainty. We are not going to venture further into the arguments over how we ought to think about inference. We do want to show how a different concept— inference as decision—changes the ways of reasoning used in tests of significance.

Tests of significance focus attention on H_0, the null hypothesis. If a decision is called for, however, there is no reason to single out H_0. There are simply two alternatives, and we must accept one and reject the other. It is convenient to call the two alternatives H_0 and H_a, but H_0 no longer has the special status

*This section is optional.

(the statement we try to find evidence against) that it had in tests of significance. In the acceptance sampling problem, we must decide between

H_0: the truckload of product meets standards
H_a: the truckload does not meet standards

on the basis of a sample of the product. There is no reason to put the burden of proof on the consumer by accepting H_0 unless we have strong evidence against it. It is equally sensible to put the burden of proof on the producer by accepting H_a unless we have strong evidence that the truckload meets standards. Producer and consumer must agree on where to place the burden of proof, but neither H_0 nor H_a has any special status.

In a decision problem, we must give a *decision rule*—a recipe based on the sample that tells us what decision to make. Decision rules are expressed in terms of sample statistics, usually the same statistics we would use in a test of significance. In fact, we have already seen that a test of significance becomes a decision rule if we reject H_0 (accept H_a) when the sample statistics is statistically significant at level α, and otherwise accept H_0 (reject H_a).

Suppose, then, that we use statistical significance at level α as our criterion for decision. And suppose that the null hypothesis H_0 is really true. Then, sample outcomes significant at level α will occur with probability α. (That's the definition of "significant at level α"; outcomes weighing strongly against H_0 occur with probability α when H_0 is really true.) But now we make a *wrong decision* in all such outcomes, by rejecting H_0 when it is really true. That is, significance level α now can be understood as the probability of a certain type of wrong decision.

Now H_a requires equal attention. Just as rejecting H_0 (accepting H_a) when H_0 is really true is an error, so is accepting H_0 (rejecting H_a) when H_a is really true. We can make two kinds of errors.

If we reject H_0 (accept H_a) when in fact H_0 is true, this is a *Type I error*. If we accept H_0 (reject H_a) when in fact H_a is true, this is a *Type II error*.

The possibilities are summed up in Figure 23.3. If H_0 is true, our decision is either correct (if we accept H_a) or is a Type I error. Only one error

		TRUTH ABOUT THE POPULATION	
		H_0 true	H_a true
DECISION BASED ON SAMPLE	Reject H_0	Type I error	Correct decision
	Accept H_0	Correct decision	Type II error

Figure 23.3 Possible outcomes of a two-action decision problem.

is possible at one time. Figure 23.4 applies these ideas to the acceptance sampling example.

So the significance level α is the probability of a Type I error. In acceptance sampling, this is the probability that a good truckload will be rejected. The probability of a Type II error is the probability that a bad truckload will be accepted. A Type I error hurts the producer, while a Type II error hurts the consumer. *Any decision rule is assessed in terms of the probabilities of the two types of error.* This is in keeping with the idea that statistical inference is based on probability. We cannot (short of inspecting the whole truckload) guarantee that good lots will never be rejected and bad lots never accepted. But by random sampling and the laws of probability, we can say what the probabilities of both kinds of errors are. Because we can find out the monetary cost of accepting bad truckloads and rejecting good ones, we can determine how much loss the producer and consumer each will suffer in the long run from wrong decisions.

Advocates of decision theory argue that the kind of "economic" thinking natural in acceptance sampling applies to all inference problems. Even a scientific researcher decides whether to announce results, or to do another experiment, or to give up research as unproductive. Wrong decisions carry costs, though these costs are not always measured in dollars. A scientist suffers by announcing a false effect, and also by failing to detect a true effect. Decision theorists maintain that the scientist should try to give numerical weights (called *utilities*) to the consequences of the two types of wrong decision. Then the scientist can choose a decision rule with the error probabilities that reflect how serious the two kinds of error are. This argument has won favor where utilities are easily expressed in money. Decision theory is widely used by business in making capital investment decisions, for example. But scientific researchers have been reluctant to take this approach to statistical inference.

To sum up, in a test of significance, we focus on a single hypothesis (H_0) and single probability (the P-value). The goal is to measure the strength of the sample evidence against H_0. If the same inference problem is thought

		TRUTH ABOUT THE TRUCKLOAD	
		Does meet standards	Does not meet standards
DECISION BASED ON SAMPLE	Reject the truckload	Type I error	Correct decision
	Accept the truckload	Correct decision	Type II error

Figure 23.4 Possible outcomes of an acceptance sampling decision.

of as a decision problem, we focus on two hypotheses and give a rule for deciding between them based on sample evidence. Therefore, we must focus on two probabilities: the probabilities of the two types of error.

Such a clear distinction between the two types of thinking is helpful for understanding. In practice, the two approaches often merge, to the dismay of partisans of one or the other. We continued to call one of the hypotheses in a decision problem H_0. In the common practice of *testing hypotheses*, we mix significance tests and decision rules as follows:

- Choose H_0 as in a test of significance.

- Think of the problem as a decision problem, so the probabilities of Type I and Type II errors are relevant.

- Type I errors are usually more serious. So choose an α (significance level), and consider only tests with probability of Type I error no greater than α.

- Among these tests, select one that makes the probability of a Type II error as small as possible. If this probability is too large, you will have to take a larger sample to reduce the chance of an error.

Testing hypotheses may seem to be a hybrid approach. It was, historically, the effective beginning of decision-oriented ideas in statistics. Hypothesis testing was developed by Jerzey Neyman* and Egon S. Pearson in the years 1928–1938. The decision theory approach came later (1940s) and grew out of the Neyman-Pearson ideas. Because decision theory in its pure form leaves you with two error probabilities and no simple rule on how to balance them, it has been used less often than tests of significance. Decision theory ideas have been applied in testing problems mainly by way of the Neyman-Pearson theory. That theory asks you to first choose α, and the influence of Fisher often has led users of hypothesis testing comfortably back to $\alpha = 0.05$ or $\alpha = 0.01$ (and also back to the warnings in this chapter about this state of affairs). Fisher, who was exceedingly argumentative, violently attacked the Neyman-Pearson decision-oriented ideas, and the argument still continues.

The reasoning of statistical inference is subtle, and the principles at issue are complex. We have (believe it or not) oversimplified the ideas of all the viewpoints mentioned and omitted several other viewpoints altogether. If you are feeling that you do not fully grasp all of the ideas of this chapter and of Chapter 22, you are in excellent company. Nonetheless, any user of statistics should make a serious effort to grasp the conflicting views on the nature of statistical inference. More than most other kinds of intellectual

---

*Neyman was born in 1894, and at the age of 85 was not only still alive but still scientifically active. In addition to developing the decision-oriented approach to testing, Neyman was also the chief architect of the theory of confidence intervals.

exercise, statistical inference can be done automatically, by recipe or by computer. These valuable shortcuts are of no worth without understanding. What Euclid said of his own science to King Ptolemy of Egypt long ago remains true of all knowledge: "There is no royal road to geometry."

Chapter 23: Statistics in Summary

■ Statistical inference is less widely applicable than exploratory analysis of data. Any inference method requires the right setting—in particular, the right design for a random sample or randomized experiment.

■ Understanding the meaning of confidence levels and statistical significance helps prevent improper conclusions.

■ Increasing the number of observations has a straightforward effect on confidence intervals: the interval gets shorter for the same level of confidence.

■ Taking more observations usually decreases the *P*-value of a test when the truth about the population stays the same, making significance tests harder to interpret than confidence intervals.

■ A finding with a small *P*-value may not be practically interesting if the sample is large, and an important truth about the population may fail to be significant if the sample is small. Avoid depending on fixed significance levels such as 5% to make decisions.

■ If a test of significance is thought of as a **decision problem,** we focus on two hypotheses, H_0 and H_a, and give a **decision rule** for deciding between them based on sample evidence. We can make two types of errors. If we reject H_0 (accept H_a) when in fact H_0 is true, this is a **Type I error.** If we accept H_0 (reject H_a) when in fact H_a is true, this is a **Type II error.**

This chapter summary will help you evaluate the Case Study.

Link It

In Chapters 21 and 22, we introduced the basic reasoning behind statistical estimation and tests of significance. We applied this reasoning to the problem of making inferences about a population proportion and a population mean. In this chapter, we provided some cautions about confidence intervals and tests of significance. Some of these echo the statements made in Chapters 1 through 6 that where the data come from matters. Some cautions are based on the behavior of confidence intervals and significance tests. These cautions will help you evaluate studies that report confidence intervals or the results of a test of significance.

Case Study Evaluated

Use what you have learned in this chapter to evaluate the Case Study that opened the chapter. Start by reviewing the Chapter Summary. Then answer each of the following questions in complete sentences. Be sure to communicate clearly enough for any of your classmates to understand what you are saying. Out of 133 food items, breakfast cereal was the only item that was significantly related to the sex of a baby, with the sample of women who consumed more breakfast cereal having a greater percentage of male babies than the sample of women who consumed less breakfast cereal.

1. Why is it not surprising that 1 food item out of 133 food items was found to be related to the sex of a baby?

2. In the original research study, the researchers mention that because of the multiplicity of testing, they only considered results to be significant if the *P*-value was less than 0.01. Why is this important?

3. Why would it be problematic to rely on retrospective reports of what the women ate prior to conceiving their babies?

4. What are some other things that would be helpful to know about the methods used in the original research study? Please explain.

In this chapter you:

- Learned what statistical confidence and statistical significance do and do not mean.
- Identified abuses of statistical inference.

macmillan learning **Online Resources**

- There is a *StatTutor* lesson on Cautions about Significance Tests.

- *LearningCurve* has good questions to check your understanding of the concepts.

Check the Basics

For Exercise 23.1, see page 542; and for Exercise 23.2, see page 545.

23.3 Finding statistical significance. If we conduct 100 hypothesis tests at the 5% level, how many of them would we expect to be statistically significant just by chance alone?

(a) 0

(b) 5

(c) 95

(d) 100

23.4 Width of a confidence interval. Suppose that the length of a confidence interval is 0.06 when the sample size is 400. To decrease the length of the confidence interval to 0.03, we should take a sample of size

(a) 100.

(b) 200.

(c) 800.

(d) 1600.

23.5 Practically significant? A new blend of gasoline costs $1.00 more per gallon than regular gasoline. Tests have shown that gas mileage with the new blend is statistically significantly higher than that of regular gasoline. The *P*-value was 0.004. A report about these results goes on to say that the new blend of gasoline increases gas mileage by 2 miles per gallon. Are these results practically significant?

(a) No, because the *P*-value is small.

(b) No, because a 2 mile-per-gallon increase is probably too small to justify an additional $1.00 per gallon.

(c) Yes, because the *P*-value is small.

(d) Yes, because a 2 mile-per-gallon increase is bigger than 0.

23.6 Practically significant? Brook conducted a study for a clothing company to determine if a rewards program caused customers to spend more. Brook found statistically significant results, with a *P*-value of 0.012. The average amount spent per order before the rewards program was $100, while the average amount spent per order after the rewards program was $223. Are these results practically significant?

(a) No, because the *P*-value is small.

(b) No, because an average increase of $123 per order is probably too small.

(c) Yes, because the *P*-value is small.

(d) Yes, because an average increase of $123 per order is more than twice the average amount spent per order before the rewards program.

23.7 Sacred significance level. A researcher decided prior to conducting research to use a 1% level of significance. The researcher finds a *P*-value of 0.027. Is it okay for the researcher to change the significance level to 5% to have statistically significant results?

(a) No, because the significance level is set beforehand and should not be changed after the *P*-value has been calculated.

(b) No, because the researcher obviously had a sample that was too small, and this is why she did not obtain significant results.

(c) Yes, because it is okay to change the significance level after the *P*-value is calculated.

(d) Yes, because the researcher should have chosen the 5% level in the first place.

Chapter 23 Exercises

23.8 A television poll. The Channel 13 news program conducts a call-in poll about the salaries of business executives. Of the 2372 callers, 1921 (or 81%) believe that business executives are vastly overpaid and that their pay should be substantially reduced. The station, following recommended practice, makes a confidence statement. They indicate that we can be 95% confident that the true proportion of citizens who believe business executives are overpaid and in need of a salary reduction is within 1.6% of the sample result. The confidence interval calculation is correct, but the conclusion is not justified. Why not?

23.9 Soccer salaries. Paris Saint-Germain, a French professional soccer team, led the 2015 ESPN/Sporting Intelligence Global Salary Survey with the largest payroll for all professional sports teams. A soccer fan looks up the salaries for all of the players on the 2015 Paris Saint-Germain team. Because the fan took a statistics course, the fan uses all of these salaries to get a 95% confidence interval for the mean salary for all 2015 Paris Saint-Germain players. This makes no sense. Why not?

23.10 How do we feel? A Gallup Poll taken in late October 2011 found that 58% of the SRS of American adults surveyed, reflecting on the day before they were surveyed, said they experienced a lot of happiness and enjoyment without a lot of stress and worry. The poll had a margin of sampling error of ±3 percentage points at 95% confidence. A news commentator at the time said the poll is surprising, given the current economic climate, in that it supports the notion that the *majority* of American adults are experiencing a lot of happiness and enjoyment without a lot of stress and worry. Why do you think the news commentator said this?

23.11 Legalizing marijuana. An October 2018 Gallup Poll found that 66% of the SRS of American adults surveyed support the legalization of marijuana in the United States. The poll had a margin of sampling error of ±4 percentage points at 95% confidence.

(a) Why do the results of this survey suggest that a *majority* of American adults support the legalization of marijuana in the United States?

(b) The survey was based on 1019 American adults. If we wanted the margin of sampling error to be ±2 percentage points at 95% confidence, how many Americans would need to be surveyed?

(c) If Gallup chose to construct a 99% confidence interval rather than a 95% confidence interval, how would the margin of sampling error change?

23.12 How far do rich parents take us? How much education children get is strongly associated with the wealth and social status of their parents. In social science jargon, this is socioeconomic status (SES). But the SES of parents has little influence on whether children who have graduated from college go on to get more education. One study looked at whether college graduates took the graduate admissions tests for business, law, and other graduate programs. The effects of the parents' SES on taking the LSAT test for law school were "both statistically insignificant and small."

(a) What does "statistically insignificant" mean?

(b) Why is it important that the effects were small in size as well as insignificant?

23.13 Searching for ESP. A researcher looking for evidence of extrasensory perception (ESP) tests 200 subjects. Only one of these subjects does significantly better ($P < 0.01$) than random guessing.

(a) Do the results of this study provide strong evidence that this person has ESP? Explain your answer.

(b) What should the researcher now do to test whether the subject has ESP?

23.14 Are the drugs really effective? A March 29, 2012, article in the *Columbus Dispatch* reported that a former researcher at a major pharmaceutical company found that many basic studies on the effectiveness of new cancer drugs appeared to be unreliable. Among the studies the former researcher reviewed was one that had been published in a reputable journal. In this published study, a cancer drug was reported as having a statistically significant positive effect on treating cancer. For purposes of this problem, assume that statistically significant means significant at level 0.05.

(a) Explain in language that is understandable to someone who knows no statistics what "statistically significant at level 0.05" means.

(b) The former researcher interviewed the lead author of the published paper. The newspaper article reported that the lead author admitted that they had repeated their experiment six times and got a significant result only once but put it in the paper because it made the best story. In light of this admission, do you think that it is accurate to claim in their published study that the findings were significant at the 0.05 level? Explain your answer. (*Note*: A statistician can show that an event that has only probability 0.05 of occurring on any given trial will occur at least once in six trials with probability about 0.26.)

23.15 Comparing bottle designs. A company compares two designs for bottles of an energy drink by placing bottles with both designs on the shelves of several markets in a large city. Checkout scanner data on more than 10,000 bottles purchased show that more shoppers bought Design A than Design B. The difference is statistically significant ($P = 0.018$). Can we conclude that consumers strongly prefer Design A? Explain your answer.

23.16 Color blindness in Africa. An anthropologist suspects that color blindness is less common in societies that live by hunting and gathering than in settled agricultural societies. He tests a number of adults in two populations in Africa, one of each type. The proportion of color-blind people is significantly lower ($P < 0.05$) in the hunter-gatherer population. What additional information would you want to help you decide whether you accept the claim about color blindness?

23.17 Blood types in Southeast Asia. One way to assess whether two human groups should be considered separate populations is to compare their distributions of blood types. An anthropologist finds significantly different ($P = 0.01$) proportions of the main human blood types (A, B, AB, O) in different tribes in central Malaysia. What other information would you want before you agree that these tribes are separate populations?

23.18 Why we seek significance. Asked why statistical significance appears so often in research reports, a student says, "Because saying that results are significant tells us that they cannot easily be explained by chance variation alone." Do you think that this statement is essentially correct? Explain your answer.

23.19 What is significance good for? Which of the following questions does a test of significance answer?

(a) Is the sample or experiment properly designed?

(b) Is the observed effect due to chance?

(c) Is the observed effect important?

23.20 What distinguishes those who have schizophrenia? Psychologists once measured 77 variables on a sample of people who had schizophrenia and a sample of people who did not have schizophrenia. They compared the two samples using 77 separate significance tests. Two of these tests were significant at the 5% level. Suppose that there is, in fact, no difference in any of the 77 variables between people who do and do not have schizophrenia in the adult population. That is, all 77 null hypotheses are true.

(a) What is the probability that one specific test shows a difference that is significant at the 5% level?

(b) Why is it not surprising that 2 of the 77 tests were significant at the 5% level?

23.21 Why are larger samples better? Statisticians prefer large samples. Describe briefly the effect of increasing the size of a sample (or the number of subjects in an experiment) on each of the following.

(a) The margin of error of a 95% confidence interval.

(b) The P-value of a test, when H_0 is false and all facts about the population remain unchanged as n increases.

23.22 Is this convincing? You are planning to test a vaccine for a virus that now has no vaccine.

Because the disease is usually not serious, you will expose 100 volunteers to the virus. After some time, you will record whether or not each volunteer has been infected.

(a) Explain how you would use these 100 volunteers in a designed experiment to test the vaccine. Include all important details of designing the experiment (but don't actually do any random allocation).

(b) You hope to show that the vaccine is more effective than a placebo. State H_0 and H_a. (Notice that this test compares *two* population proportions.)

(c) The experiment gave a *P*-value of 0.15. Explain carefully what this means.

(d) Your fellow researchers do not consider this evidence strong enough to recommend regular use of the vaccine. Do you agree?

The following two exercises are based on the optional section in this chapter.

23.23 In the courtroom. A criminal trial can be thought of as a decision problem, the two possible decisions being "guilty" and "not guilty." Moreover, in a criminal trial there is a null hypothesis in the sense of an assertion that we will continue to hold until we have strong evidence against it. Criminal trials are, therefore, similar to hypothesis testing.

(a) What are H_0 and H_a in a criminal trial? Explain your choice of H_0.

(b) Describe in words the meaning of Type I error and Type II error in this setting, and display the possible outcomes in a diagram like Figures 23.3 and 23.4.

(c) Suppose that you are a jury member. Having studied statistics, you think in terms of a significance level α, the (subjective) probability of a Type I error. What considerations would affect your personal choice of α? (For example, would the difference between a charge of murder and a charge of shoplifting affect your personal α?)

23.24 Acceptance sampling. You are a consumer of potatoes in an acceptance sampling

situation. Your acceptance sampling plan has probability 0.01 of passing a truckload of potatoes that does not meet quality standards. You might think that the truckloads that pass are almost all good. Alas, it is not so.

(a) Explain why low probabilities of error cannot ensure that truckloads that pass are mostly good. (*Hint*: What happens if your supplier ships all bad truckloads?)

(b) The paradox that most decisions can be correct (low error probabilities) and yet most truckloads that pass can be bad has important analogs in areas such as medical diagnosis. Explain why most conclusions that a patient has a rare disease can be false alarms even if the diagnostic system is correct 99% of the time.

The following exercises require carrying out the methods described in the optional sections of Chapters 21 and 22.

23.25 Liberal students. A national survey reports that the percentage of first-year college students who identify their political views as "liberal" is 31.3%. Believing that students at their college are less inclined to be "liberal," a group of educators surveys a random sample of 125 students from their institution and finds that 27 of these students identify themselves as politically "liberal." The proportion of students at this college who identified themselves as politically "liberal" is significantly lower ($P = 0.0082$) than the 31.3% of all first-year students in the country who identify themselves in this way. Use the sample proportion to construct a 95% confidence interval. Does this interval include 31.3%? Explain why you are or are not surprised by this.

23.26 Do our athletes graduate? Return to the study in Exercise 22.19 (page 530), which found that 149 of 190 athletes admitted to a large university graduated within 6 years. This did not differ significantly ($P = 0.0968$) from the 82% graduation rate for all students at the university.

(a) It may be more informative to give a 95% confidence interval for the graduation rate of athletes. Construct this confidence interval.

(b) Is the 82% graduation rate for all students included in your confidence interval? Explain why you are or are not surprised by this.

(c) Suppose that 784 of an SRS of 1000 athletes admitted to a large university graduated within 6 years. In this situation, the results would differ significantly ($P < 0.001$) from the 82% graduation rate for all students at the university. Why are these results statistically significant when the sample proportion based on the sample of size 1000 and the sample proportion based on the sample of size 190 are both 0.784?

23.27 Holiday spending. A November 2018 Gallup Poll asked a random sample of 1037 American adults "Roughly how much money do you think you personally will spend on Christmas gifts this year?" The mean holiday spending estimate in the sample was $\bar{x} = \$794$. We will treat these data as an SRS from a Normally distributed population with standard deviation $\sigma = \$150$.

(a) Give a 95% confidence interval for the mean holiday spending estimate based on these data.

(b) The mean holiday spending estimate of $\bar{x} = \$794$ included 17% of respondents who answered "No opinion." This group includes those respondents who do not celebrate Christmas and thus reported $0. Do you trust the interval you computed in part (a) as a 95% confidence interval for the mean holiday spending estimate for *all* American adults? Why or why not?

23.28 Is it significant? Over several years and many thousands of students, 85% of the high school students in a large city have passed the competency test that is one of the requirements for a diploma. Now reformers claim that a new mathematics curriculum will increase the percentage who pass. A random sample of 1000 students follow the new curriculum. The school board wants to see an improvement that is statistically significant at the 5% level before it will adopt the new program for all students. If p is the proportion of all students who would pass the exam if they followed the new curriculum, we must test

$$H_0: p = 0.85$$
$$H_a: p > 0.85$$

(a) Suppose that 868 of the 1000 students in the sample pass the test. Show that this is *not* significant at the 5% level. (Follow the method of Example 3, page 521, in Chapter 22.)

(b) Suppose that 869 of the 1000 students pass. Show that this *is* significant at the 5% level.

(c) Is there a practical difference between 868 successes in 1000 tries and 869 successes? What can you conclude about the importance of a fixed significance level?

23.29 We like confidence intervals. The previous exercise compared significance tests about the proportion p of all students who would pass a competency test, based on data showing that either 868 or 869 of an SRS of 1000 students passed. Give the 95% confidence interval for p in both cases. The intervals make clear how uncertain we are about the true value of p and how little difference there is between the two sample outcomes.

Exploring the Web

Access these exercises on the text website: macmillanlearning.com/scc10e.

WTV

What's the Verdict?

In the Discovery Channel's popular series *Mythbusters*, the hosts investigate common beliefs to determine whether there is evidence to verify them as true. In this segment, the hosts are investigating whether yawning is contagious.

https://www.discovery.com/tv-shows/mythbusters/videos/is-yawning-contagious

In Chapter 6, you examined issues regarding the design of the study. Here you will consider some of the conclusions made by the hosts of the show.

Questions

WTV23.1. The hosts reported that 29% of those who saw the yawn seed stimulus ended up yawning, and 25% of those who did not see the yawn seed stimulus yawned. One of the hosts says, "Well, it's not dramatic, but it seems like it's pretty good to me." The hosts then agree that yawning is contagious. Did the hosts perform a test of statistical significance? Explain your answer.

WTV23.2. What would be the advantages of performing a test of statistical significance in this situation?

WTV23.3. Do you agree with the *Mythbusters'* conclusion that yawning is contagious based on this evidence?

WTV23.4. What if the hosts had tested only two people—one in each group, with the yawn seed individual yawning and the no yawn seed individual not yawning? Would the results be more or less convincing?

WTV23.5. What if the hosts had tested 50,000 people and found the results of 25% and 29%? Would the results be more or less likely to be statistically significant?

WTV23.6. As mentioned in Chapter 6, *Mythbusters* recruited 50 people by posting an ad. Based on how the sample was selected, are these results likely to apply to the whole population? Why or why not?

What's the verdict? Just looking at your sample statistics and seeing a small difference is not enough to say that the results are significant. Taking sampling variability into account and using a simple random sample are important when relating the results to the whole population.

24

Two-Way Tables and the Chi-Square Test*

Case Study: Freedom of Speech and Political Beliefs

In this chapter you will:

- Interpret two-way tables.
- Test if there is a relationship between two categorical variables.
- Identify Simpson's Paradox and understand how it occurs.

First Amendment rights have come to the forefront at colleges across the nation. A 2017 Gallup/Knight Foundation survey of 3014 college students examined student views on various aspects of the first amendment as well as other issues on college campuses across the United States.

Let's look at the relationship between political party affiliation and views about freedom of speech. One question asked in the survey was, "Do you think freedom of speech is very secure, secure, threatened or very threatened in the country today?" Here is a *two-way table* that categorizes the 2990 students who have both a political party affiliation/lean and an opinion about free speech.

Freedom of speech	Democrat	Independent	Republican	Total
Very secure	183	20	98	301
Secure	1043	86	362	1491
Threatened	776	39	222	1037
Very threatened	112	14	35	161
Total	2114	159	717	2990

*This more advanced chapter is optional.

What does this table tell us about the relationship between political party affiliation and views about freedom of speech?

Two-Way Tables

Example 1

Racial differences in six-year graduation rates

One measurement of student success is the percent completing a bachelor's degree within six years. The National Center for Education Statistics publishes reports annually. The following two-way table is representative of the percentages of students completing a bachelor's degree within six years by race/ethnicity:

Race/ ethnicity	White	Black	Hispanic	Asian	American Indian/ Alaska Native	Two or more races	Total
Graduated	6963	1063	1388	792	69	175	10,450
Did not graduate	3933	1614	1163	295	110	119	7234
Total	10,896	2667	2551	1087	179	294	17,684

Rawpixel.com/Shutterstock.com

How should we evaluate the information in this table?

Graduation status (completing a bachelor's degree within six years or not) and race of students are both *categorical variables*. That is, they place individuals into categories but do not have numerical values that allow us to describe relationships by scatterplots, correlation, or regression lines. We can count the number of individuals in each category. To display relationships between two categorical variables, use a **two-way table** like the table of graduation status and race of applicants. Graduation status is the **row variable** because each row in the table describes one of the possible admission decisions for a student. Race is the **column variable** because each column describes one of the racial/ethnic groups. The entries in the table are the counts of students in each graduation status–by–race class.

How can we best grasp the information contained in this table? First, *look at the distribution of each variable separately*. The distribution of a categorical variable says how often each outcome occurred. The "Total" column at the right of the table contains the totals for each of the rows. These row totals give the distribution of graduation status for all students, for all racial/ethnic groups combined. The "Total" row at the bottom of the table gives the distribution of race for students, with both categories of graduation status ("Graduated" or "Did not graduate") combined. It is often clearer to present these distributions using percentages. We might report the distribution of race as

$$\text{percentage white} = \frac{10,896}{17,684} \times 100\% = 0.616 \times 100\% = 61.6\%$$

$$\text{percentage black} = \frac{2677}{17,684} \times 100\% = 0.151 \times 100\% = 15.1\%$$

$$\text{percentage Hispanic} = \frac{2551}{17,684} \times 100\% = 0.144 \times 100\% = 14.4\%$$

$$\text{percentage Asian} = \frac{1087}{17,684} \times 100\% = 0.061 \times 100\% = 6.1\%$$

$$\text{percentage Amer. Ind./Alaska Native} = \frac{179}{17,684} \times 100\% = 0.017 \times 100\% = 1.0\%$$

$$\text{percentage Two or more races} = \frac{294}{17,684} \times 100\% = 0.010 \times 100\% = 1.7\%$$

Note that the percentages listed above total 99.9%, not 100%. This is another example of roundoff errors, which were introduced in Chapter 10.

The two-way table contains more information than the two distributions of graduation status alone and race alone. The nature of the relationship between graduation status and race cannot be deduced from the separate distributions but requires the full table. **To describe relationships between categorical variables, calculate appropriate percentages from the counts given.**

Example 2

Racial differences in six-year graduation rates, continued

Because there are only two categories of graduation status, we can see the relationship between race and graduation status by comparing the percentages of those who completed a bachelor's degree within six years for each race:

$$\text{percentage of white students who graduated} = \frac{6963}{10,896} \times 100\% = 0.639 \times 100\% = 63.9\%$$

$$\text{percentage of black students who graduated} = \frac{1063}{2677} \times 100\% = 0.397 \times 100\% = 39.7\%$$

$$\text{percentage of Hispanic students who graduated} = \frac{1388}{2551} \times 100\% = 0.544 \times 100\% = 54.4\%$$

$$\text{percentage of Asian students who graduated} = \frac{792}{1087} \times 100\% = 0.729 \times 100\% = 72.9\%$$

$$\text{percentage of Amer. Ind./Alaska Native} = \frac{69}{179} \times 100\% = 0.385 \times 100\% = 38.5\%$$

$$\text{percentage of two or more races} = \frac{175}{294} \times 100\% = 0.595 \times 100\% = 59.5\%$$

Over 60% of the white students and more than 70% of the Asian students completed a bachelor's degree within six years, but less than 40% of black students and American Indian/Alaska Native students completed a bachelor's degree within six years.

In working with two-way tables, you must calculate lots of percentages. Here's a tip to help decide what fraction gives the percentage you want. Ask, "What group represents the total that I want a percentage of?" The count for that group is the denominator of the fraction that leads to the percentage. In Example 2, we wanted the percentage *of each racial/ethnic group* who completed a bachelor's degree within six years, so the counts of each race form the denominators.

Inference for a Two-Way Table

We often gather data and arrange them in a two-way table to see if two categorical variables are related to each other. The sample data are easy to investigate: turn them into percentages and look for an association between the row and column variables. Is the association in the sample evidence of an association between these variables in the entire population? Or could the sample association easily arise just from the luck of random sampling? This is a question for a significance test.

Example 3

Digital savviness and ability to distinguish facts in the news

Can you distinguish between fact and opinion in the news? A recent Pew Research Center study investigated the relationship between the ability to distinguish facts as facts (and opinions as opinions, but we leave that for you to investigate on your own) and various characteristics including age, education level, political awareness, and digital savviness. Respondents were American adults.

The following table of 1000 total respondents represents Pew's findings. Here are the number of respondents and percentages who were able to identify all five factual statements as fact for three levels of digital savviness:

Digital savviness	Number of respondents	Number who got all five correct	Percent correct
Very savvy	480	168	35.0
Somewhat savvy	350	70	20.0
Not savvy	170	22	12.9

Tero Vesalainen/Shutterstock

The sample percentages of respondents who identified all five factual statements to be factual are quite different. In particular, the percentage of "very savvy" respondents who got all five correct was much higher than for the "somewhat savvy" and "not savvy" groups. Are these data good evidence that there is a relationship between digital savviness and ability to identify factual statements as factual in the population of all American adults?

The test that answers this question starts with a two-way table. Here's the table for the data of Example 3:

Digital savviness	All five correct	Fewer than five correct	Total
Very savvy	168	312	480
Somewhat savvy	70	280	350
Not savvy	22	148	170
Total	260	740	1000

Our null hypothesis, as usual, says that the level of digital savviness has no effect on the ability to identify all five factual statements as fact. That is, American adults do equally well regardless of their level of digital savviness. The differences in the sample are just the result of chance. Our null hypothesis is

H_0: There is no association between the digital savviness and whether or not a person can identify all five factual statements as fact in the population of American adults.

Expressing this hypothesis in terms of population parameters can be a bit complicated, so we will be content with the verbal statement. The alternative hypothesis just says, "Yes, there is some association between digital savviness and whether or not a person can identify all five factual statements as facts in the population of American adults." The alternative doesn't specify the nature of the relationship. It doesn't say, for example, "Those who are more digitally savvy are better at identifying all five factual statements as fact."

The **expected count** in any cell of a two-way table when H_0 is true is

$$\text{expected count} = \frac{\text{row total} \times \text{column total}}{\text{table total}}$$

To test H_0, we compare the observed counts in a two-way table with the *expected counts,* the counts we would expect—except for random variation—if H_0 were true. If the observed counts are far from the expected counts, that is evidence against H_0. We can guess the expected counts for the study. In all, 260 of the 1000 respondents correctly identified all five factual statements as fact. That's an overall success rate of 26%, because 260/1000 is 26%. If the null hypothesis is true, there is no difference among the levels of digital savviness. So we expect 26% of the respondents in each group to correctly identify all five factual statements as fact.

Let's classify a person who correctly identifies all five factual statements as fact as a "success" and a person who does not as a "failure." There were 480 respondents in the very savvy group, so we expect $(0.26)(480) = 124.8$ successes and $480 - 124.8 = 355.2$ failures in this group. For the somewhat savvy group (350 respondents), we expect $(0.26)(350) = 91.0$ successes and $350 - 91 = 259.0$ failures. Finally, for the group of respondents who are not digitally savvy (170 respondents), we expect $(0.26)(170) = 44.2$ successes and $170 - 44.2 = 125.8$ failures.

If the groups had been the same size, the expected counts would be the same for each group. Fortunately, there is a rule that makes it easy to find expected counts. The expected count of successes in the very savvy group is

$$\text{expected count} = \frac{\text{row 1 total} \times \text{column 1 total}}{\text{table total}}$$
$$= \frac{(480)(260)}{1000} = 124.8$$

If the null hypothesis of no treatment differences is true, we expect 124.8 of the 480 digitally savvy respondents to succeed. That's just what we calculated based on the percentages.

Now it's your turn

24.1 Video-gaming and grades. The popularity of computer, video, online, and virtual reality games has raised concerns about their ability to negatively impact youth. Based on a recent survey, 1808 students ages 14 to 18 in Connecticut high schools were classified by their average grades and by whether they had or had not played such games. The following table summarizes the findings.

	Grade Average		
	A's and B's	C's	D's and F's
Played games	736	450	193
Never played games	205	144	80

Find the expected count of students with average grades of A's and B's who have played computer, video, online, or virtual reality games under the null hypothesis that there is no association between grades and game playing in the population of students. Assume that the sample was a random sample of students in Connecticut high schools.

The Chi-Square Test

To see if the data give evidence against the null hypothesis of "no relationship," compare the counts in the two-way table with the counts we would expect if there really were no relationship. If the observed counts are far from the expected counts, that's the evidence we were seeking. The significance test uses a statistic that measures how far apart the observed and expected counts are.

The chi-square statistic is a sum of terms, one for each cell in the table. In Example 3, 168 of the very savvy group succeeded. The expected count for this cell is 124.8. So the term in the chi-square statistic contributed by this cell is

$$\frac{(\text{observed count} - \text{expected count})^2}{\text{expected count}} = \frac{(168 - 124.8)^2}{124.8}$$
$$= \frac{1866.24}{124.8} = 14.95$$

Example 4

Digital savviness and ability to distinguish facts in the news continued

Here are the observed and expected counts for Example 3 side by side:

	Observed		Expected	
	Success	**Failure**	**Success**	**Failure**
Very savvy	168	312	124.8	355.2
Somewhat savvy	70	280	91.0	259.0
Not savvy	22	148	44.2	125.8

We can now find the chi-square statistic, adding six terms for the six contributing cells in the two-way table:

$$\chi^2 = \frac{(168 - 124.8)^2}{124.8} + \frac{(312 - 355.2)^2}{355.2}$$

$$+ \frac{(70 - 91.0)^2}{91.0} + \frac{(280 - 259.0)^2}{259.0}$$

$$+ \frac{(22 - 44.2)^2}{44.2} + \frac{(148 - 125.8)^2}{125.8}$$

$$= 14.95 + 5.25 + 4.85 + 1.70 + 11.15 + 3.92 = 41.82$$

Now it's your turn

24.2 Video-gaming and grades. The popularity of computer, video, online, and virtual reality games has raised concerns about their ability to negatively impact youth. Based on a recent survey, 1808 students ages 14 to 18 in Connecticut high schools were classified by their average grades and by whether they had or had not played such games. The following table summarizes the findings. The observed and expected counts are given side by side.

	Observed			Expected		
	A's and B's	**C's**	**D's and F's**	**A's and B's**	**C's**	**D's and F's**
Played games	736	450	193	717.7	453.1	208.2
Never played games	205	144	80	223.3	140.9	64.8

Find the chi-square statistic.

The chi-square distributions

The sampling distribution of the chi-square statistic χ^2 when the null hypothesis of no association is true is called a **chi-square distribution**.

The chi-square distributions are a family of distributions that take only nonnegative values and are skewed to the right. A specific chi-square distribution is specified by giving its **degrees of freedom**.

The **chi-square test** for a two-way table with r rows and c columns uses critical values from the chi-square distribution with $(r - 1)(c - 1)$ degrees of freedom.

Because χ^2 measures how far the observed counts are from what would be expected if H_0 were true, large values are evidence against H_0. Is $\chi^2 = 41.82$ a large value? You know the drill: compare the observed value 41.82 against the *sampling distribution* that shows how χ^2 would vary if the null hypothesis were true. This sampling distribution is *not* a Normal distribution. It is a right-skewed distribution that allows only nonnegative values because χ^2 can never be negative. Moreover, the sampling distribution is different for two-way tables of different sizes. Facts about this distribution are found in the box on the left.

Figure 24.1 shows the density curves for three members of the chi-square family of distributions. As the degrees of freedom (df) increase, the density curves become less skewed and larger values become more probable. We won't find P-values as areas under

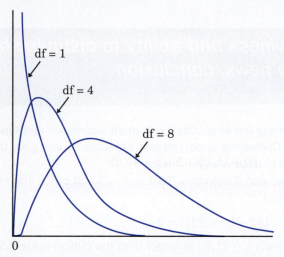

df = 1

df = 4

df = 8

0

Figure 24.1 The density curves for three members of the chi-square family of distributions. The sampling distributions of chi-square statistics belong to this family.

a chi-square curve by hand, though software can do it for us. Table 24.1 is a shortcut. It shows how large the chi-square statistic χ^2 must be in order to be significant at various levels. This isn't as good as an actual P-value, but it is often good enough. Each number of degrees of freedom has a separate row in the table. We see, for example, that a chi-square statistic with 3 degrees of freedom is significant at the 5% level if it is greater than 7.81 and is significant at the 1% level if it is greater than 11.34.

Statistics in Your World

More chi-square tests There are also chi-square tests for hypotheses more specific than "no relationship." Place people in classes by social status, wait 10 years, then classify the same people again. The row and column variables are the classes at the two times. We might test the hypothesis that there has been no change in the overall distribution of social status. Or we might ask if elevations in status are balanced by matching moves down. These hypotheses can be tested by variations in the chi-square test.

Table 24.1 To be significant at level α, a chi-square statistic must be larger than the table entry for α

	Significance Level α						
df	0.25	0.20	0.15	0.10	0.05	0.01	0.001
1	1.32	1.64	2.07	2.71	3.84	6.63	10.83
2	2.77	3.22	3.79	4.61	5.99	9.21	13.82
3	4.11	4.64	5.32	6.25	7.81	11.34	16.27
4	5.39	5.99	6.74	7.78	9.49	13.28	18.47
5	6.63	7.29	8.12	9.24	11.07	15.09	20.51
6	7.84	8.56	9.45	10.64	12.59	16.81	22.46
7	9.04	9.80	10.75	12.02	14.07	18.48	24.32
8	10.22	11.03	12.03	13.36	15.51	20.09	26.12
9	11.39	12.24	13.29	14.68	16.92	21.67	27.88

Example 5

Digital savviness and ability to distinguish facts in the news, conclusion

We have seen that higher levels of digital savviness are associated with more successes and fewer failures than lower levels of digital savviness. Comparing observed and expected counts gave the chi-square statistic $\chi^2 = 41.82$. The last step is to assess significance.

The two-way table for Example 3 has 3 rows and 2 columns. That is, $r = 3$ and $c = 2$. The chi-square statistic therefore has degrees of freedom

$$(r-1)(c-1) = (3-1)(2-1) = (2)(1) = 2$$

Look in the df = 2 row of Table 24.1. We see that $\chi^2 = 41.82$ is larger than the critical value 9.21 required for significance at the $\alpha = 0.01$ level but smaller than the critical value 13.82 for $\alpha = 0.001$. Example 3 shows a statistically significant relationship ($P < 0.01$) between digital savviness and ability to correctly identify five factual statements as facts.

The significance test says only that we have strong evidence of *some* association between digital savviness and success. We must look at the two-way table to see the nature of the relationship: more digital savviness appears to be related to a higher success rate.

Now it's your turn

24.3 Video-gaming and grades. The popularity of computer, video, online, and virtual reality games has raised concerns about their ability to negatively impact youth. Based on a recent survey, 1808 students ages 14 to 18 in Connecticut high schools were classified by their average grades and by whether they had or had not played such games. The following table summarizes the findings. The observed and expected counts are given side by side.

	Observed			Expected		
	A's and B's	C's	D's and F's	A's and B's	C's	D's and F's
Played games	736	450	193	717.7	453.1	208.2
Never played games	205	144	80	223.3	140.9	64.8

From these counts, we find that the chi-square statistic is 6.74. Does the study show that there is a statistically significant relationship between playing games and average grades? Use a significance level of 0.05.

Using the Chi-Square Test

Like our test for a population proportion, the chi-square test uses some approximations that become more accurate as we take more observations. A rough rule for when it is safe to use this test is based on the expected counts.

Example 3 easily passes this "safety check": the smallest expected cell count is 44.8, which means that no expected counts are less than 5. Here is a concluding example that outlines the examination of a two-way table.

Example 6

Do angry people have a greater incidence of heart disease?

People who get angry easily tend to have more heart disease. That's the conclusion of a study that followed a random sample of 12,986 people from three locations for about 4 years. All subjects were free of heart disease at the beginning of the study. The subjects took the Spielberger Trait Anger Scale test, which measures how prone a person is to sudden anger. Here are data for the 8474 people in the sample who had normal blood pressure. CHD stands for "coronary heart disease." This includes people who had heart attacks and those who needed medical treatment for heart disease.

pathdoc/Shutterstock

	Anger Score		
	Low	**Moderate**	**High**
Sample size	3110	4731	633
CHD count	53	110	27
CHD percent	1.7%	2.3%	4.3%

There is a clear trend: as the anger score increases, so does the percentage who suffer heart disease. Is this relationship between anger and heart disease statistically significant?

The first step is to write the data as a two-way table by adding the counts of subjects who did not suffer from heart disease. We also add the row and column totals, which we need to find the expected counts.

	Low anger	Moderate anger	High anger	Total
CHD	53	110	27	190
No CHD	3057	4621	606	8284
Total	3110	4731	633	8474

We can now follow the steps for a significance test, familiar from Chapter 22.

The hypotheses. The chi-square method tests these hypotheses:

H_0: no association between anger and CHD

H_a: some association between anger and CHD

The sampling distribution. We will see that all the expected cell counts are larger than 5, so we can safely apply the chi-square test. The two-way table of anger versus CHD has two rows and three columns. We will use critical values from the chi-square distribution with degrees of freedom $df = (2-1)(3-1) = 2$.

The data. First find the expected cell counts. For example, the expected count of high-anger people with CHD is

$$\text{expected count} = \frac{\text{row 1 total} \times \text{column 3 total}}{\text{table total}}$$

$$= \frac{(190)(633)}{8474} = 14.19$$

Here is the complete table of observed and expected counts side by side:

	Observed			Expected		
	Low	Moderate	High	Low	Moderate	High
CHD	53	110	27	69.73	106.08	14.19
No CHD	3057	4621	606	3040.27	4624.92	618.81

Looking at these counts, we see that the high-anger group has more CHD than expected and the low-anger group has less CHD than expected. This is consistent with what the percentages in Example 6 show. The chi-square statistic is

$$\chi^2 = \frac{(53-69.73)^2}{69.73} + \frac{(110-106.08)^2}{106.08} + \frac{(27-14.19)^2}{14.19}$$

$$+ \frac{(3057-3040.27)^2}{3040.27} + \frac{(4621-4624.92)^2}{4624.92} + \frac{(606-618.81)^2}{618.81}$$

$$= 4.014 + 0.145 + 11.564 + 0.092 + 0.003 + 0.265 = 16.083$$

In practice, statistical software can do all this arithmetic for you. Look at the six terms that we sum to get χ^2. Most of the total comes from just one cell: high-anger people have more CHD than expected.

Significance? Look at the df $= 2$ line of Table 24.1. The observed chi-square $\chi^2 = 16.083$ is larger than the critical value 13.82 for $\alpha = 0.001$. We have highly significant evidence ($P < 0.001$) that anger and heart disease are related. Statistical software can give the actual P-value. It is $P = 0.0003$.

The conclusion. Can we conclude that proneness to anger *causes* heart disease? This is an observational study, not an experiment. It isn't surprising to find that some lurking variables are confounded with anger. For example, people prone to anger are more likely than others to drink and smoke. The study report used advanced statistics to adjust for many differences among the three anger groups. The adjustments raised the P-value from $P = 0.0003$ to $P = 0.02$ because the lurking variables explain some of the heart disease. This is still good evidence for a relationship if a significance level of 0.05 is used. Because the study started with a random sample of people who had no CHD and followed them forward in time, and because many lurking variables were measured and accounted for, it does give some evidence for causation. The next step might be an experiment that shows anger-prone people how to change. Will this reduce their risk of heart disease?

Simpson's Paradox

As is the case with quantitative variables, the effects of lurking variables can change or even reverse relationships between two categorical variables. In the following example, we will find that sometimes a lurking variable might reverse the relationship of what we would expect to find in the data.

Example 7

Do medical helicopters save lives?

Accident victims are sometimes taken by helicopter from the accident scene to a hospital. Helicopters save time. Do they also save lives? Let's compare the percentages of accident victims who die with helicopter evacuation and with the usual transport to a hospital by road. The numbers here are hypothetical, but they illustrate a phenomenon that often appears in real data.

paulprescott/Deposit Photos

	Helicopter	Road	Total
Victim died	64	260	324
Victim survived	136	840	976
Total	200	1100	1300

We see that 32% (64 out of 200) of helicopter patients died, but only 24% (260 out of 1100) of the others died. That seems discouraging.

The explanation is that the helicopter is sent mostly to serious accidents, so that the victims transported by helicopter are more often seriously injured. They are more likely to die with or without helicopter evacuation. We will break the data down into a **three-way table** that classifies the data by the seriousness of the accident. We will present a three-way table as two or more two-way tables side by side, one for each value of the third variable. In this case there are two two-way tables, one for each method of evacuation:

Serious Accidents		
	Helicopter	Road
Died	48	60
Survived	52	40
Total	100	100

Less Serious Accidents		
	Helicopter	Road
Died	16	200
Survived	84	800
Total	100	1000

Inspect these tables to convince yourself that they describe the same 1300 accident victims as the original two-way table. For example, 200 (= 100 + 100) were moved by helicopter, and 64 (= 48 + 16) of these died.

Among victims of serious accidents, the helicopter saves 52% (52 out of 100) compared with 40% for road transport. If we look at less serious accidents, 84% of those transported by helicopter survive versus 80% of those transported by road. Both groups of victims have a higher survival rate when evacuated by helicopter.

How can it happen that the helicopter does better for both groups of victims but worse when all victims are combined? Look at the data: half the helicopter transport patients are from serious accidents, compared with only 100 of the 1100 road transport patients. So the helicopter carries patients who are more likely to die. The original two-way table did not take into account the seriousness of the accident and was therefore misleading. This is an example of *Simpson's paradox*.

Simpson's paradox is just an extreme form of the fact that observed associations can be misleading when there are lurking variables. Remember the caution from Chapter 15: *beware the lurking variable*.

Key Terms

An association or comparison that holds for all of several groups can disappear or even reverse direction when the data are combined to form a single group. This situation is called **Simpson's paradox.**

Example 8

Discrimination in mortgage lending?

Studies of applications for home mortgage loans from banks show a strong racial pattern: banks reject a higher percentage of African American and Latinx (a person of Latin American origin or descent) applicants than white applicants. Over the past few years, lawsuits have been filed against major banks in California, Maryland, and Connecticut, to name a few states. One lawsuit against a bank for discrimination in lending in the Washington, D.C., area contended that the bank rejected 17.5% of African Americans but only 3.3% of whites.

The bank replies that lurking variables explain the difference in rejection rates. African Americans have (on the average) lower incomes, poorer credit records, and less secure jobs than whites. Unlike race, these are legitimate reasons to turn down a mortgage application. It is because these lurking variables are confounded with race, the bank says, that it rejects a higher percentage of African American applicants. It is even possible, thinking of Simpson's paradox, that the bank accepts a *higher* percentage of African American applicants than of white applicants if we look at people with the same income and credit record.

Who is right? Both sides will hire statisticians to examine the effects of the lurking variables. Both sides will present statistical arguments supporting or refuting a charge of discrimination in lending. Unfortunately, there are no formal guidelines for how juries and judges are to assess statistical arguments. And juries and judges need not have any statistical expertise. Lurking variables and seeming paradoxes such as Simpson's paradox make it difficult for even experts to determine the cause of the disparity in the rejection of mortgage applications. The court will eventually decide as best it can, but the decision may not be based on the statistical arguments.

Chapter 24: Statistics in Summary

- Categorical variables classify individuals into groups. To display the relationship between two categorical variables, make a **two-way table** of counts for the groups. We describe the nature of an association between categorical variables by comparing selected percentages.

- As always, lurking variables can make an observed association misleading. In some cases, an association that holds for every level of a lurking variable disappears or changes direction when we lump all levels together. This is **Simpson's paradox.**

- The **chi-square test** tells us whether an observed association in a two-way table is statistically significant. The **chi-square statistic** compares the counts in the table with the counts we would expect if there were no association between the row and column variables. The sampling distribution is not Normal. It is a new distribution, the **chi-square distribution.**

This chapter summary will help you evaluate the Case Study.

Link It

In Chapters 14 and 15, we considered relationships between two quantitative variables. In this chapter, we use two-way tables to describe relationships between two *categorical* variables. To explore relationships between two categorical variables, make a two-way table. Examine the distribution of each row (or each column). Differences in the patterns of the distributions suggest a relationship between the two variables. No change in these patterns suggests that there is no relationship.

As in Chapters 21, 22, and 23, we use formal statistical inference to decide if any difference in the observed patterns of the distributions is simply due to chance. We compare what we would expect cell counts to be based on the distribution of each variable separately with the cell counts actually observed. The chi-square test answers the question of whether differences in these cell counts could be due to chance.

As in Chapters 14 and 15, we must be careful not to assume that the patterns we observe would continue to hold for additional data or in a broader setting. Simpson's paradox is an example of how such an assumption could mislead us. Simpson's paradox occurs when the association or comparison that holds for all of several groups reverses direction when these groups are combined into a single group.

Case Study Evaluated

Use what you have learned in this chapter to evaluate the Case Study that opened the chapter. Start by reviewing the Chapter Summary. Then answer each of the following questions in complete sentences. Be sure to communicate clearly enough for any of your classmates to understand what you are saying.

Here is the table that was presented in the Case Study at the beginning of this chapter:

Freedom of speech	Democrat	Independent	Republican	Total
Very secure	183	20	98	301
Secure	1043	86	362	1491
Threatened	776	39	222	1037
Very threatened	112	14	35	161
Total	2114	159	717	2990

1. What percentage of respondents are Independent?
2. What percentage of those who feel very secure about freedom of speech are Independent?
3. What percentage of those who feel secure about freedom of speech are Independent?
4. What percentage of those who feel threatened about freedom of speech are Independent?
5. What percentage of those who feel very threatened about freedom of speech are Independent?
6. Is there a statistically significant relationship between how a person feels about their freedom of speech and their political leaning? Assume a 5% significance level.

In this chapter you:

- Interpreted two-way tables.
- Tested if there is a relationship between two categorical variables.
- Identified Simpson's paradox and understand how it occurs.

macmillan learning **Online Resources**

■ There are several *StatTutor* lessons that will help with your understanding of some of the details of the chi-square test. There is also a *StatTutor* lesson on Simpson's paradox.

■ *LearningCurve* has good questions to check your understanding of the concepts.

Check the Basics

For Exercise 24.1, see page 564; for Exercise 24.2, see page 566; and for Exercise 24.3, see page 568.

The Pew Research Center conducts an annual Internet Project, which includes research related to social media. The following two-way table about the percent of U.S. adults who use Instagram is broken down by age and is based on data reported by Pew as of January 2018.

Use Instagram	Yes	No	Total
Age 18–29	225	127	352
Age 30–49	211	317	528
Age 50–64	114	430	544
Age 65+	52	477	529
Total	602	1351	1953

Exercises 24.4 to 24.8 are based on this table.

24.4 Instagram use. What percent of all respondents use Instagram?

(a) 11.5%

(b) 18.0%

(c) 30.8%

(d) 69.2%

24.5 Young adults who use Instagram. What percent of those age 18–29 use Instagram?

(a) 11.5%

(b) 18.0%

(c) 37.4%

(d) 63.9%

24.6 Older adults in the survey. What percent of all respondents were age 50–64?

(a) 5.8%

(b) 18.9%

(c) 21.0%

(d) 27.9%

24.7 Older adults who use Instagram. What percent of those age 50–64 use Instagram?

(a) 5.8%

(b) 18.9%

(c) 21.0%

(d) 27.9%

24.8 What does age have to do with it? Does it appear that there is a relationship between age and whether a person uses Instagram?

(a) No, age should have nothing to do with whether a person uses Instagram.

(b) No, the percentages of young adults and older adults who use Instagram are the same, which means there is no relationship.

(c) Yes, older adults do not know how to use Instagram.

(d) Yes, the percentages of young adults and older adults who use Instagram are not the same, which means there is some relationship.

Chapter 24 Exercises

24.9 Is astrology scientific? The University of Chicago's General Social Survey (GSS) is one of the most important social science sample surveys in the United States. The GSS asked a random sample of adults their opinion about whether astrology is very or somewhat scientific or not at all scientific. Is belief that astrology is scientific related to the highest level of education earned? Here is a two-way table of counts for people in the 2016 sample who had three levels of educational achievement (High School (HS) Diploma, Bachelor's Degree, Graduate or Professional Degree):

Opinion	Degree Held		
	HS diploma	Bachelor's	Graduate or professional
Not at all scientific	237	200	110
Very or somewhat scientific	170	60	38

Calculate percentages that describe the nature of the relationship between the highest level of education earned and the opinion about whether astrology is very scientific or some-what sort of scientific or not at all scientific. Give a brief summary in words.

24.10 Weight-lifting injuries. Resistance training is a popular form of conditioning aimed at enhancing sports performance and is widely used among high school, college, and professional athletes, although its use for younger athletes is controversial. A random sample of 4111 patients between the ages of 8 and 30 admitted to U.S. emergency rooms with the injury code "weight lifting" was obtained. These injuries were classified as "accidental" if caused by a dropped weight or improper equipment use. The patients were also classified into the four age categories of 8 to 13 years, 14 to 18, 19 to 22, and 23 to 30. Here is a two-way table of the results:

Age	Accidental	Not accidental	Total
8–13	295	102	397
14–18	655	916	1571
19–22	239	533	772
23–30	363	1008	1371
Total	1552	2559	4111

Calculate percentages that describe the nature of the relationship between age and whether the weight-lifting injuries were accidental or not. Give a brief summary in words.

24.11 Smoking by students and their families. How are the smoking habits of students related to the smoking habits of their close family members? Here is a two-way table from

a survey of male students in six secondary schools in Malaysia:

	Student smokes	Student does not smoke
At least one close family member smokes	115	207
No close family member smokes	25	75

Write a brief answer to the question posed, including a comparison of selected percentages.

24.12 Smoking by parents of preschoolers. How are the smoking habits of parents of preschoolers related to the father's highest degree earned? Here is a two-way table from a survey of the parents of preschoolers in Greece:

	Both parents smoke	One parent smokes	Neither parent smokes
University education	42	68	90
Intermediate education	47	69	75
High school education	183	281	273
Primary education or none	69	73	62

Write a brief answer to the question posed, including a comparison of selected percentages.

24.13 Python eggs. How is the hatching of water python eggs influenced by the temperature of the snake's nest? Researchers assigned newly laid eggs to one of three temperatures: hot, neutral (room temperature), or cold. Hot duplicates the extra warmth provided by the mother python, and cold duplicates the absence of the mother. Here are the data on the number of eggs and the number that hatched:

	Eggs	Hatched
Cold	27	16
Neutral	56	38
Hot	104	75

(a) Make a two-way table of temperature by outcome (hatched or not).

(b) Calculate the percentage of eggs in each group that hatched. The researchers anticipated that eggs would not hatch in the cold environment. Do the data support that anticipation? Assume a 5% significance level.

24.14 Firearm violence. Here are counts from a study of firearm violence reported in 2018 by the National Vital Statistics System. In particular, we will examine homicides and suicides and whether a firearm was involved.

	Firearm	No firearm	Total
Homicides	14,415	4947	19,362
Suicides	22,938	22,027	44,965

(a) Make a bar graph to compare whether a firearm was used in homicides and suicides. What does the graph suggest about homicides versus suicides?

(b) Calculate the percentage of homicides and suicides in which firearms were used. Comment on your findings.

24.15 Who earns academic degrees? How do women and men compare in the pursuit of academic degrees? The table below presents counts (in thousands), as projected by the National Center for Education Statistics, of degrees that will be earned in 2026–2027 categorized by the level of the degree and the sex of the recipient.

	Associate's	Bachelor's	Master's	Professional/Doctorate
Female	827	1207	557	109
Male	462	875	365	92
Total	1289	2082	922	191

(a) How many total people are predicted to earn a degree in the 2026–2027 academic year?

(b) How many people are predicted to earn associate's degrees?

(c) What percentage of each level of degree is predicted to be earned by women? Write

a brief description of what the data show about the relationship between sex and degree level.

24.16 Smokers rate their health. The University of Michigan Health and Retirement Study (HRS) surveys more than 22,000 Americans over the age of 50 every 2 years. A subsample of the HRS participated in an Internet-based survey that collected information on a number of topical areas, including health (physical and mental health behaviors), psychosocial items, economics (income, assets, expectations, and consumption), and retirement. Two of the questions asked were: "Would you say your health is excellent, very good, good, fair, or poor?" and "Are you currently a cigarette smoker?" Here is the two-way table that summarizes the answers on these two questions:

Health	Current Smoker	
	Yes	No
Excellent	25	484
Very good	115	1557
Good	145	1309
Fair	90	545
Poor	29	11

Describe the differences between the distributions of health status for those who are and are not current smokers with percentages, with a graph, and in words.

24.17 Totals aren't enough. Here are the row and column totals for a two-way table with two rows and two columns:

a	b	40
c	d	60
60	40	100

Find *two different* sets of counts a, b, c, and d for the body of the table that give these same totals. This shows that the relationship between two variables cannot be obtained from the two individual distributions of the variables because there is not a unique set of counts that gives the sample totals.

24.18 Airline flight delays. Here are the numbers of flights on time and delayed, for two airlines at four airports during a one-month period. Overall on-time percentages for each airline are often reported in the news. The airports that flights serve is a lurking variable that can make such reports misleading.

| | Alaska Airlines | | American Airlines | |
	On time	Delayed	On time	Delayed
Los Angeles	1410	384	2690	817
Phoenix	154	32	3725	988
San Diego	595	139	532	221
Seattle	3994	2318	3702	2299

(a) What percentage of all Alaska Airlines flights were delayed? What percentage of all American Airlines flights were delayed? These overall numbers are the numbers usually reported in the news.

(b) Now find the percentage of delayed flights for Alaska Airlines at each of the four airports. Do the same for American Airlines.

(c) American Airlines does worse at *every one* of the four airports, yet does *better* overall. That sounds impossible. Explain carefully, referring to the data, how this can happen. (The weather in Phoenix and Seattle lies behind this example of Simpson's paradox.)

24.19 Bias in the jury pool? The New Zealand Department of Justice did a study of the composition of juries in court cases. Of interest was whether Maori, the indigenous people of New Zealand, were adequately represented in jury pools. Here are the results for two districts, Rotura and Nelson, in New Zealand (similar results were found in all districts):

Rotura	Maori	Non-Maori
In jury pool	79	258
Not in jury pool	8810	23,751
Total	8889	24,009

Nelson	Maori	Non-Maori
In jury pool	1	56
Not in jury pool	1328	32,602
Total	1329	32,658

(a) Use these data to make a two-way table of race (Maori or non-Maori) versus jury pool status (In or Not in).

(b) Show that Simpson's paradox holds: a higher percentage of Maori are in the jury pool overall, but for both districts a higher percentage of non-Maori are in the jury pool.

(c) Use the data to explain the paradox in language that a judge could understand.

24.20 Field goal shooting. Here are data on field goal shooting for two members of a men's basketball team:

| | Jeremy Thomas | | Harrison Greyson | |
	Made	Missed	Made	Missed
Two-pointers	4	2	77	45
Three-pointers	3	3	1	5

(a) What percent of all field goal attempts did Jeremy Thomas make? What percent of all field goal attempts did Harrison Greyson make?

(b) Now find the percent of all two-point field goals and all three-point field goals that Jeremy made. Do the same for Harrison.

(c) Harrison had a lower percent for *both* types of field goals but had a better overall percent. That sounds impossible. Explain carefully, referring to the data, how this can happen.

24.21 Smokers rate their health. Exercise 24.16 gives the responses of a survey of 4310 Americans to questions about their health and whether they currently smoke cigarettes.

(a) Do these data satisfy our guidelines for safe use of the chi-square test?

(b) Is there a statistically significant relationship between the smoking status and opinions about health? Assume a 5% significance level.

24.22 Is astrology scientific? In Exercise 24.9, you described the relationship between belief that astrology is scientific and amount of higher education. Is the observed association between these variables statistically significant? To find out, proceed as follows:

(a) Add the row and column totals to the two-way table in Exercise 24.9 and find the expected cell counts. Which observed counts differ most from the expected counts?

(b) Find the chi-square statistic. Which cells contribute most to this statistic?

(c) What are the degrees of freedom? Use Table 24.1 to say how significant the chi-square test is. Write a brief conclusion for your study.

24.23 Smoking by students and their families. In Exercise 24.11, you saw that there is an association between smoking by close family members and smoking by high school students. The students are more likely to smoke if a close family member smokes. We want to know whether this association is statistically significant.

(a) State the hypotheses for the chi-square test. What do you think the population is?

(b) Find the expected cell counts. Write a sentence that explains in simple language what "expected counts" are.

(c) Find the chi-square statistic and its degrees of freedom. What is your conclusion about significance?

24.24 Python eggs. Exercise 24.13 presents data on the hatching of python eggs at three different temperatures. Does temperature have a significant effect on hatching? Write a clear summary of your work and your conclusion.

24.25 Stress and heart attacks. You read a newspaper article that describes a study of whether stress management can help reduce heart attacks. The 107 subjects all had reduced blood flow to the heart and so were at risk of a heart attack. They were assigned at random to three groups for the four-month study period. Subjects in the first group took a stress management program, while subjects in the second group participated in an exercise program. The remaining subjects did nothing aside from typical heart health care with their personal physicians. Over the three years following the four-month treatment programs, only 3 of the 33 subjects in

the stress management group and 7 of the 34 subjects in the exercise group had heart attacks (fatal or non-fatal) or a surgical bypass or angioplasty, while 12 of the 40 subjects who received their typical care had heart attacks (fatal or non-fatal) or a surgical bypass or angioplasty.

(a) Use the information in the news article to make a two-way table that describes the study results.

(b) What are the success rates of the three treatments in preventing cardiac events?

(c) Find the expected cell counts under the null hypothesis that there is no difference among the treatments. Verify that the expected counts meet our guideline for use of the chi-square test.

(d) Is there a significant difference among the success rates for the three treatments?

24.26 Standards for child care. Do unregulated providers of child care in their homes follow different health and safety practices in different cities? A study looked at people who regularly provided care for someone else's children in poor areas of three cities. The numbers who required medical releases from parents to allow medical care in an emergency were 42 of 73 providers in Newark, N.J.; 29 of 101 in Camden, N.J.; and 48 of 107 in South Chicago, Ill.

(a) Use the chi-square test to see if there are significant differences among the proportions of child care providers who require medical releases in the three cities. What do you conclude?

(b) How should the data be produced in order for your test to be valid? (In fact, the samples came in part from asking parents who were subjects in another study who provided their child care. The author of the study wisely did not use a statistical test. He wrote: "Application of conventional statistical procedures appropriate for random samples may produce biased and misleading results.")

Exploring the Web

Access these exercises on the text website: macmillanlearning.com/scc10e.

PART IV Review

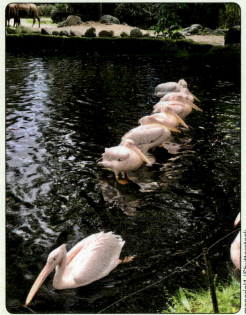

Statistical inference draws conclusions about a population on the basis of sample data and uses probability to indicate how reliable the conclusions are. A confidence interval estimates the true value of an unknown parameter. A significance test shows how strong the evidence is for some claim about a parameter. Chapters 21 and 22 present the reasoning of confidence intervals and tests and give optional details for inference about population proportions p and population means μ.

The probabilities in both confidence intervals and significance tests tell us what would happen if we used the formula for the interval or test very many times. A confidence level is the probability that the formula for a confidence interval actually produces an interval that contains the unknown parameter. A 95% confidence interval gives a correct result 95% of the time when we use it repeatedly. Figure IV.1 illustrates the reasoning using the 95% confidence interval for a population proportion p.

The Idea of a Confidence Interval

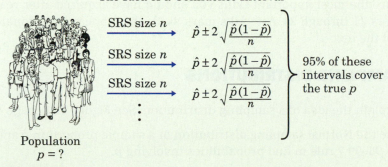

Figure IV.1

A *P*-value is the probability that the test would produce a result at least as extreme as the observed result if the null hypothesis really were true. Figure IV.2 illustrates the reasoning, placing the sample proportion $\hat{p}$ from our one sample on the Normal curve that shows how $\hat{p}$ would vary in all possible samples *if the null hypothesis were true*. A *P*-value tells us how surprising the observed outcome is if the null hypothesis were true. Very surprising outcomes (small *P*-values) are good evidence that the null hypothesis is not true.

The Idea of a Significance Test

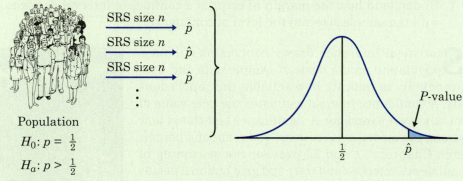

Figure IV.2

To detect sound and unsound uses of inference, you must know the basic reasoning and also be aware of some fine points and pitfalls. Chapter 23 will help. The optional sections in Chapters 21 and 22 and the optional Chapter 24 discuss a few more specific inference procedures. Chapter 24 deals with two-way tables, both for description and for inference. The descriptive part of this chapter completes Part II's discussion of relationships among variables by describing relationships among categorical variables.

PART IV SUMMARY

Here are the most important skills you should have acquired after reading Chapters 21 through 24. Asterisks mark skills that appear in optional sections of the text.

A. SAMPLING DISTRIBUTIONS

1. Explain the idea of a sampling distribution. See Figure 21.1 (page 489).

2. Use the Normal sampling distribution of a sample proportion $\hat{p}$ and the 68–95–99.7 rule to find probabilities involving $\hat{p}$.

3. *Use the Normal sampling distribution of a sample mean $\bar{x}$ to find probabilities involving $\bar{x}$.

B. CONFIDENCE INTERVALS

1. Explain the idea of a confidence interval. See Figure IV.1.

2. Explain in nontechnical language what is meant by "95% confidence" and other statements of confidence in statistical reports.

3. Use the basic formula $\hat{p} \pm 2\sqrt{\hat{p}(1-\hat{p})/n}$ to obtain an approximate 95% confidence interval for a population proportion p.

4. Understand how the margin of error of a confidence interval changes with the sample size and the level of confidence.

5. Detect major mistakes in applying inference, such as improper data production, selecting the best of many outcomes, ignoring high nonresponse, and ignoring outliers.

6. *Use the detailed formula $\hat{p} \pm z^* \sqrt{\hat{p}(1-\hat{p})/n}$ and critical values z^* for Normal distributions to obtain confidence intervals for a population proportion p.

7. *Use the formula $\bar{x} \pm z^* s/\sqrt{n}$ to obtain confidence intervals for a population mean μ.

C. SIGNIFICANCE TESTS

1. Explain the idea of a significance test. See Figure IV.2.

2. State the null and alternative hypotheses in a testing situation when the parameter in question is a population proportion p.

3. Explain in nontechnical language the meaning of the P-value when you are given its numerical value for a test.

4. Explain the meaning of "statistically significant at the 5% level" and other statements of significance. Explain why significance at a specific level such as 5% is less informative than a P-value.

5. Recognize that significance testing does not measure the size or importance of an effect.

6. Recognize and explain the effect of small and large samples on the statistical significance of an outcome.

7. *Use Table B of percentiles of Normal distributions to find the P-value for a test about a proportion p.

8. *Carry out one-sided and two-sided tests about a mean μ using the sample mean $\bar{x}$ and Table B.

D. *TWO-WAY TABLES

1. Create two-way tables for data classified by two categorical variables.

2. Use percentages to describe the relationship between any two categorical variables based on the counts in a two-way table.

3. Explain what null hypothesis the chi-square statistic tests in a specific two-way table.

4. Calculate expected cell counts, the chi-square statistic, and its degrees of freedom from a two-way table.

5. Use Table 24.1 (page 567) for chi-square distributions to assess significance. Interpret the test result in the setting of a specific two-way table.

PART IV REVIEW EXERCISES

Review exercises are short and straightforward exercises that help you solidify the basic ideas and skills in each part of this book. We have provided "hints" that indicate where you can find the relevant material for the odd-numbered problems.

IV.1. The social media divide. A January 2019 Rasmussen Reports poll asked a random sample of 1000 likely U.S. voters if they believe social media such as Facebook and Twitter make the United States a more or less divided nation. Of these likely voters, 650 said they believe social media makes the United States a more divided nation. We can assume that the sample is a simple random sample (SRS). Give a 95% confidence interval for the proportion of all likely U.S. voters who believe social media such as Facebook and Twitter make the United States a more divided nation. (*Hint*: See pages 489–492.)

IV.2. Drinking for your health. In a July 2018 Gallup Poll, a random sample of 1033 adults were asked if they believe drinking alcoholic beverages in moderation—meaning one or two drinks per day—is good for your health, makes no difference to your health, or is bad for your health. Among respondents, 165 said that they believe drinking alcoholic beverages in moderation is good for your health. Assume that the sample was an SRS. Give a 95% confidence interval for the proportion of all adults who believe drinking alcoholic beverages is good for your health.

IV.3. The social media divide. Exercise IV.1 concerns a random sample of 1000 likely U.S. voters. Suppose that (unknown to the pollsters) exactly 67% of all likely U.S. voters believe that social media makes the United States more divided as a nation. Imagine that we take very many SRSs of size 1000 from this population and record the percentage in each sample who believe that social media makes the United States more divided as a nation. Where would the middle 95% of all values of this percentage lie? (*Hint*: See pages 492–495.)

IV.4. Drinking for your health. Exercise IV.2 concerns a random sample of 1033 adults. Suppose that, in the population of all adults, exactly 14.5% would say that they believe drinking alcoholic beverages in moderation is good for your health. Imagine that we take a very large number of SRSs of size 1033. For each sample, we record the proportion $\hat{p}$ of the sample who says they believe drinking alcoholic beverages in moderation is good for your health.

(a) What is the sampling distribution that describes the values $\hat{p}$ would take in our samples?

(b) Use this distribution and the 68–95–99.7 rule to find the approximate percentage of all samples in which more than 16% of the respondents would say that they believe drinking alcoholic beverages in moderation is good for your health.

IV.5. The most honest profession. A Gallup Poll conducted from December 3 to 12, 2018, asked a random sample of 1025 adults to rate the honesty and ethical standards of people in a variety of professions. Among the respondents, 861 rated the honesty and ethical standards of nurses as very high or high. Assume that the sample was an SRS. Give a 95% confidence interval for the proportion of all adults who would rate the honesty and ethical standards of nurses as very high or high. (*Hint*: See pages 489–492.)

IV.6. Roulette. A roulette wheel has 18 red slots among its 38 slots. You observe many spins and record the number of times the ball falls in a red slot. Now you want to use these data to test whether the probability p of the ball falling in a red slot has the value that is correct for a fair roulette wheel. State the hypotheses H_0 and H_a that you will test.

IV.7. Why not? Table 11.1 (page 244) records the percentage of residents aged 65 or older in each of the 50 states. You can check that this percentage is 14% or higher in 12 of the states. So the sample proportion of states with at least 14% of elderly residents is $\hat{p} = 12/50 = 0.24$.

Explain why it does *not* make sense to go on to calculate a 95% confidence interval for the population proportion p. (*Hint*: See pages 495–498.)

IV.8. Helping welfare mothers. A study compares two groups of mothers with young children who were on welfare two years ago. One group attended a voluntary training program that was offered free of charge at a local vocational school and was advertised in the local news media. The other group did not choose to attend the training program. The study finds a significant difference ($P < 0.01$) between the proportions of the mothers in the two groups who are still on welfare. The difference is not only significant but quite large. The report says that with 95% confidence the percentage of the nonattending group still on welfare is 21% ± 4% higher than that of the group who attended the program. You are on the staff of a member of Congress who is interested in the plight of welfare mothers and who asks you about the report.

(a) Explain in simple language what "a significant difference ($P < 0.01$)" means.

(b) Explain clearly and briefly what "95% confidence" means.

(c) This study is not good evidence that requiring job training of all welfare mothers would greatly reduce the percentage who remain on welfare. Explain this to the member of Congress.

IV.9. Beating the system. Some doctors think that health plan rules restrict their ability to treat their patients effectively, so they bend the rules to help patients get reimbursed by their health plans. Here's a sentence from a study on this topic: "Physicians who agree with the statement 'Today it is necessary to game the system to provide high-quality care' reported manipulating reimbursement systems more often than those who did not agree with the statement (64.3% vs 35.7%; $P < 0.001$)."

(a) Explain to a doctor what "$P < 0.001$" means in the context of this specific study. (*Hint*: See pages 516–519.)

(b) A result that is statistically significant can still be too small to be of practical interest. How do you know this is not true here? (*Hint*: See pages 539–542.)

IV.10. Smoking in the United States. A July 2018 nationwide random survey of 1033 adults asked whether they had smoked cigarettes in the last week. Among the respondents, 165 said they had. Assume that the sample was an SRS.
(a) Give a 95% confidence interval for the proportion of all American adults who smoked in the week preceding the survey.
(b) Write a short paragraph for a news report based on the survey results.

IV.11. When shall we call you? As you might guess, telephone sample surveys get better response rates during the evening than during the weekday daytime. One study called 2304 randomly chosen telephone numbers on weekday mornings. Of these, 1313 calls were answered and only 207 resulted in interviews. Of 2454 calls on weekday evenings, 1840 were answered and 712 interviews resulted. Give two 95% confidence intervals, for the proportions of all calls that are answered on weekday mornings and on weekday evenings. Are you confident that the proportion is higher in the evening? (*Hint*: See pages 491–492.)

IV.12. Smoking in the United States. Does the survey of Exercise IV.10 provide good evidence that fewer than 18% of all American adults smoked in the week prior to the survey?
(a) State the hypotheses to be tested.
(b) If your null hypothesis is true, what is the sampling distribution of the sample proportion $\hat{p}$? Sketch this distribution.
(c) Mark the actual value of $\hat{p}$ on the curve. In your opinion, does it appear surprising enough to give good evidence against the null hypothesis?

IV.13. When shall we call you? Suppose that we know that 57% of all calls made by sample surveys on weekday mornings are answered. We make 2454 calls to randomly chosen numbers during weekday evenings. Of these, 1840 are answered. Is this good evidence that the proportion of answered calls is higher in the evening?
(a) State the hypotheses to be tested. (*Hint*: See pages 516–520.)
(b) If your null hypothesis is true, what is the sampling distribution of the sample proportion $\hat{p}$? Sketch this distribution. (*Hint*: See pages 520–521.)
(c) Mark the actual value of $\hat{p}$ on the curve. In your opinion, does it appear surprising enough to give good evidence against the null hypothesis? (*Hint*: See pages 520–521.)

IV.14. Not significant. The study cited in Exercise IV.9 looked at the factors that may affect whether doctors bend medical plan rules. Perhaps doctors who fear being prosecuted will bend the rules less often. The study report said, "Notably, greater worry about prosecution for fraud did not affect physicians' use of these tactics ($P = 0.34$)." Explain why the result $P = 0.34$ supports the conclusion that doctors' fears about potential prosecution did not affect behavior.

IV.15. Going to church. Opinion polls show that about 40% of Americans say they attended religious services in the last week. This result has stayed stable for decades. Studies of what people actually *do,* as opposed to what they *say* they do, suggest that actual church attendance is much lower. One study calculated 95% confidence intervals based on what a sample of Catholics said and then based on a sample of actual behavior. In Chicago, for example, the 95% confidence interval from the opinion poll said that between 45.7% and 51.3% of Catholics attended mass weekly. The 95% confidence interval from actual counts said that between 25.7% and 28.9% attended mass weekly.
(a) Why might we expect opinion polls on church attendance to be biased in the direction of overestimating true attendance? (*Hint*: See pages 536–539.)
(b) The poll in Chicago found that 48.5% of Catholics claimed to attend mass weekly. Why

don't we just say that "48.5% of all Catholics in Chicago claim to attend mass" instead of giving the interval 45.7% to 51.3%? (*Hint*: See pages 542.)

(c) The two results, from reported and observed behavior, are quite different. What does it mean to say that we are "95% confident" in each of the two intervals given? (*Hint*: See pages 492–495.)

The following exercises are based on the optional sections of Chapters 21 and 22.

IV.16. The social media divide. A January 2019 Rasmussen Reports poll asked a random sample of 1000 likely U.S. voters if they believe social media such as Facebook and Twitter make the United States a more or less divided nation. Of these likely voters, 650 said they believe social media makes the United States a more divided nation. Assume that the sample was an SRS. Give 90% and 99% confidence intervals for the proportion of all likely U.S. voters who believe that social media makes the United States a more divided nation. Explain briefly what important fact about confidence intervals is illustrated by comparing these two intervals and the 95% confidence interval from Exercise IV.1.

IV.17. Drinking for your health. In a July 2018 Gallup Poll a random sample of 1033 adults were asked if they believe drinking alcoholic beverages in moderation—meaning one or two drinks per day—is good for your health, makes no difference to your health, or is bad for your health. Among respondents, 289 said that they believe drinking alcoholic beverages in moderation is bad for your health. Assume that the sample was an SRS. Is this good evidence that more than one-fourth of American adults believe that drinking alcoholic beverages in moderation is bad for your health? Show the five steps of the test (hypotheses, sampling distribution, data, P-value, conclusion) clearly. Use a significance level of 0.05. (*Hint*: See pages 522–526.)

The following exercises are based on the optional material in Chapters 21, 22, and 24.

IV.18. The most honest profession. A Gallup Poll conducted from December 3 to 12, 2018, asked a random sample of 1025 adults to rate the honesty and ethical standards of people in a variety of professions. Among the respondents, 861 rated the honesty and ethical standards of nurses as very high or high. Assume that the sample was an SRS. Give a 90% confidence interval for the proportion of all adults who rate the honesty and ethical standards of nurses as very high or high. For what purpose might a 90% confidence interval be less useful than a 95% confidence interval? For what purpose might a 90% interval be more useful?

IV.19. A poll of voters. You are the polling consultant to a member of Congress. An SRS of 500 registered voters finds that 37% name "economic problems" as the most important issue facing the nation. Give a 90% confidence interval for the proportion of all voters who hold this opinion. Then explain carefully to the member of Congress what your conclusion reveals about voters' opinions. (*Hint*: See pages 495–498.)

IV.20. Smoking in the United States. Carry out the significance test called for in Exercise IV.12 in all detail. Show the five steps of the test (hypotheses, sampling distribution, data, P-value, conclusion) clearly. Use a significance level of 0.05.

IV.21. When shall we call you? Carry out the significance test called for in Exercise IV.13 in all detail. Show the five steps of the test (hypotheses, sampling distribution, data, P-value, conclusion) clearly. Use a significance level of 0.05. (*Hint*: See pages 522–526.)

IV.22. CEO pay. A study of 104 corporations found that the pay of their chief executive officers had increased an average of $\bar{x} = 6.9\%$ per year in real terms. The standard deviation of the percentage increases was $s = 17.4\%$.

(a) The 104 individual percentage increases have a right-skewed distribution. Explain why the central limit theorem says that we can, nonetheless, act as if the mean increase has a Normal distribution.

(b) Give a 95% confidence interval for the mean percentage increase in pay for all corporate CEOs.

(c) What must we know about the 104 corporations studied to justify the inference you did in part (b)?

IV.23. Water quality. An environmentalist group collects a liter of water from each of 45 random locations along a stream and measures the amount of dissolved oxygen in each specimen. The mean is 4.62 milligrams (mg) and the standard deviation is 0.92 mg. Is this strong evidence that the stream has a mean oxygen content of less than 5 mg per liter? Use a significance level of 0.05. (*Hint*: See pages 522–526.)

IV.24. Pleasant smells. Do pleasant odors help work go faster? Twenty-one subjects worked a paper-and-pencil maze wearing a mask that was either unscented or carried the smell of flowers. Each subject worked the maze three times with each mask, in random order. (This is a matched pairs design.) Here are the differences in their average times (in seconds), unscented minus scented. If the floral smell speeds work, the difference will be positive because the time with the scent will be lower.

```
−7.37  −3.14    4.10  −4.40   19.47  −10.80  −0.87
 8.70    2.94  −17.24  14.30  −24.57   16.17  −7.84
 8.60  −10.77  24.97  −4.47   11.90   −6.26   6.67
```

(a) We hope to show that work is faster on the average with the scented mask. State null and alternative hypotheses in terms of the mean difference in times μ for the population of all adults.

(b) Using a calculator, find the mean and standard deviation of the 21 observations. Did the subjects work faster with the scented mask? Is the mean improvement big enough to be important?

(c) Make a stemplot of the data (round to the nearest whole second). Are there outliers or other problems that might hinder inference?

(d) Test the hypotheses you stated in part (a). Is the improvement statistically significant? Use a significance level of 0.05.

IV.25. Sharks. Great white sharks are big and hungry. Here are the lengths in feet of 44 great whites:

```
18.7  12.3  18.6  16.4  15.7  18.3  14.6  15.8  14.9  17.6  12.1
16.4  16.7  17.8  16.2  12.6  17.8  13.8  12.2  15.2  14.7  12.4
13.2  15.8  14.3  16.6   9.4  18.2  13.2  13.6  15.3  16.1  13.5
19.1  16.2  22.8  16.8  13.6  13.2  15.7  19.7  18.7  13.2  16.8
```

(a) Make a stemplot with feet as the stems and 10ths of feet as the leaves. There are two outliers, one in each direction. These won't change $\bar{x}$ much but will increase the standard deviation s. (*Hint*: See pages 522–526.)

(b) Give a 90% confidence interval for the mean length of all great white sharks. (The interval may be too wide due to the influence of the outliers on s.) (*Hint*: See pages 501–502.)

(c) What do we need to know about these sharks in order to interpret your result in part (b)? (*Hint*: See pages 501–502.)

IV.26. Pleasant smells. Return to the data in Exercise IV.24. Give a 95% confidence interval for the mean improvement in time to solve a maze when wearing a mask with a floral scent. Are you confident that the scent does improve mean working time?

IV.27. Sharks. Return to the data in Exercise IV.25. Is there good evidence that the mean length of sharks in the population that these sharks represent is greater than 15 feet? (*Hint*: See pages 522–525.)

IV.28. Simpson's paradox. If we compare average 2018 SAT scores (Evidence-Based Reading and Writing plus Mathematics), we find that female college-bound seniors in Vermont perform better than female college-bound seniors in California. But if we look only at white students, California does better. If we look only at minority students, California again does better. That's Simpson's paradox:

the comparison reverses when we lump all students together. Explain carefully why this makes sense, using the fact that a much higher percentage of female college-bound seniors are white in Vermont than in California.

IV.29. Unhappy Health Maintenance Organization (HMO) patients. A study of complaints by HMO members compared those who filed complaints about medical treatment and those who filed nonmedical complaints with an SRS of members who did not complain that year. Here are the data on the number who stayed and the number who voluntarily left the HMO:

	No complaint	Medical complaint	Nonmedical complaint
Stayed	721	173	412
Left	22	26	28

(a) Find the row and column totals. (*Hint*: See pages 560–562.)
(b) Find the percentage of each group who left. (*Hint*: See pages 561–562.)
(c) Find the expected counts and check that you can safely use the chi-square test. (*Hint*: See pages 564, 569–571.)
(d) The chi-square statistic for this table is $\chi^2 = 31.765$. What null and alternative hypotheses does this statistic test? What are its degrees of freedom? How significant is it? What do you conclude about the relationship between complaints and leaving the HMO? (*Hint*: See pages 569–571.)

IV.30. Treating ulcers. Gastric freezing was once a recommended treatment for stomach ulcers. Use of gastric freezing stopped after experiments showed it had no effect. One randomized comparative experiment found that 28 of the 82 gastric-freezing patients improved, while 30 of the 78 patients in the placebo group improved.
(a) Outline the design of this experiment.
(b) Make a two-way table of treatment versus outcome (subject improved or not). Is there a significant relationship between treatment and outcome?
(c) Write a brief summary that includes the test result and also percentages that compare the success of the two treatments.

IV.31. College enrollment by generation. The Pew Research Center recently reported that post-millennial Hispanic youth (those born after 1996) are enrolling in college at higher rates than their peers from the previous generations. When considering Hispanic youths aged 18 to 20 who were no longer in high school and were enrolled in college last year, 55% of post-millennials were in college. Only 34% of millennials (those born 1981 to 1996) and 28% of those in Generation X (those born 1965 to 1980) were in college at a similar age. Suppose that we have 100 people from each of the three generations of Hispanic youths and that their percentages match the Pew findings.
(a) Make a two-way table of time of generation (post-millennial, millennial, Generation X) versus enrolled in college or not. (*Hint*: See pages 560–562.)
(b) It should be obvious that there is a highly significant relationship between generation and college enrollment. Why? (*Hint*: See pages 560–564.)
(c) Nonetheless, carry out the chi-square test. Use a significance level of 0.05. What do you conclude? (*Hint*: See pages 569–571.)

PART PROJECTS

Projects are longer exercises that require gathering information or producing data and that emphasize writing a short essay to describe your work. Many are suitable for teams of students.

Access these projects on the text website: **macmillanlearning.com/scc10e.**

Resolving the Controversy

Chapter 3: Should Election Polls Be Banned?

Arguments *against* public preelection polls charge that they influence voter behavior. Voters may decide to stay home if the polls predict a landslide–why bother to vote if the result is a foregone conclusion? Exit polls are particularly worrisome because, in effect, they report actual election results before the election is complete. The U.S. television networks agree not to release the results of their exit surveys in any state until the polls close in that state. If a presidential election is not too close to call, the networks may know (or think they know) the winner by midafternoon, but they forecast the vote one state at a time as the polls close across the country. Even so, a presidential election result may be known (or thought to be known) before voting ends in the western states. Some countries have laws restricting election forecasts. In France, no poll results can be published in the week before a presidential election. Canada forbids poll results in the 72 hours before federal elections. In all, some 30 countries restrict publication of election surveys.

The argument *for* preelection polls is simple: democracies should not forbid publication of information. Voters can decide for themselves how to use the information. After all, supporters of a candidate who is far behind know that fact even without polls. Restricting publication of polls just invites abuses. In France, candidates continue to take private polls (less reliable than the public polls) in the week before the election. They then leak the results to reporters in the hope of influencing press reports.

One argument *for* exit polls is that they provide a means for checking election outcomes. Discrepancies between exit polls and reported election outcomes invite investigation into the reasons for the differences. Such was the case in the 2004 U.S. presidential election. Were the exit polls flawed, or were the reported election results in error?

Chapter 4: The Harris Online Poll

The Harris Online Poll uses probability sampling and statistical methods to weight responses and uses recruitment to attempt to create a panel (sampling frame) that is as representative of the population of interest as possible. But the panel also consists of volunteers and hence suffers to some extent from voluntary response. In addition, panel members are Internet users, and it is not clear how such a panel can be representative of a larger population that includes those who do not use the Internet.

Thus, the verdict is out on whether the Harris Poll Online provides accurate information about well-defined populations such as all American adults.

Chapter 6: Is It or Isn't It a Placebo?

Should the FDA require natural remedies to meet the same standards as prescription drugs? That's hard to do in practice, because natural substances can't be patented. Drug companies spend millions of dollars on clinical trials because they can patent the drugs that prove effective. Nobody can patent an herb, so nobody has a financial incentive to pay for a clinical trial. Don't look for big changes in the regulations.

Meanwhile, it's easy to find claims that ginkgo biloba is good for (as one website says) "hearing and vision problems as well as impotence, edema, varicose veins, leg ulcers, and strokes." Common sense says you should be suspicious of claims that a substance is good for lots of possibly unrelated conditions. Statistical sense says you should be suspicious of claims not backed by comparative experiments. Many untested remedies are no doubt just placebos. Yet they may have real effects in many people—the placebo effect is strong. Just remember that the safety of these concoctions is also untested.

Chapter 7: Hope for Sale?

One issue to consider is whether bone marrow transplant (BMT) really keeps patients alive longer than standard treatments. We don't know, but the answer appears to be "probably not." The patients naturally want to try anything that might keep them alive, and some doctors are willing to offer hope not backed by good evidence. One problem was that patients would not join controlled trials that might assign them to standard treatments rather than to BMT. Results from such trials were delayed for years by the difficulty in recruiting subjects. Of the first five trials reported, four found no significant difference between BMT and standard treatments. The fifth favored BMT—but the researcher soon admitted "a serious breach of scientific honesty and integrity." The *New York Times* put it more bluntly: "he falsified data."

Another issue is "smart" compassion. Compassion seems to support making untested treatments available to dying patients. Reason responds that this opens the door to sellers of hope and delays development of treatments that really work. Compare children's cancer, where doctors agree not to offer experimental treatments outside controlled trials. Result: 60% of all children with cancer are in clinical trials, and progress in saving lives has been much faster for children than for adults. BMT for a rare cancer in children was tested immediately and found to be effective. In contrast, one of the pioneers in using BMT for breast cancer, in the light of better evidence, now says, "We deceived ourselves and we deceived our patients."

Chapter 8: Six-Year Graduation Rates, High School GPA, and Standardized Tests

We see that high school grade point average (GPA) alone appears to be a better predictor of six-year graduation rates than does SAT or ACT scores–as high school GPA increases, graduation rates increase from 26% to 47%, but as SAT or ACT score increases, graduation rates increase from 26% to only 35%. Combining high school GPA with SAT or ACT score does a better job of predicting six-year graduation rate than either by itself. At the highest level of each variable together, the predicted graduation rate is 72%.

We need to remember that students with the same high school GPA and SAT or ACT scores often perform quite differently in college, which could impact time to graduation. Motivation and study habits matter a lot. Choice of major, choice of classes, and choice of college also impact time to graduation.

However, if a college decides to go test optional, we return to the fact that high school GPA alone is a better predictor of six-year graduation rate than is SAT or ACT score alone. While the graduation rates are lower than with high school GPA and SAT or ACT combined, it might be worth the expanded diversity, which will benefit students in more ways than can be measured by time to graduation.

Chapter 12: Income Inequality

These are complicated issues, with much room for conflicting data and hidden agendas. The political left wants to reduce inequality, and the political right says the rich earn their high incomes. We want to point to just one important statistical twist. Figures 12.4 and 12.5 report "cross-sectional" data that give a snapshot of households in each year. "Longitudinal" data that follow households over time might paint a different picture. Consider a young married couple, Jamal and Tonya. As students, they work part-time, then borrow to go to graduate school. They are down in the bottom fifth. When they get out of school, their income grows quickly. By age 40, they are happily in the top fifth. In other words, many poor households are only temporarily poor.

Longitudinal studies are expensive because they must follow the same households for years. They are prone to bias because some households drop out over time. One study of income tax returns found that only 14% of the bottom fifth were still in the bottom fifth 10 years later. But very poor people don't file tax returns. Another study looked at children under 5 years old. Starting in both 1971 and 1981, it found that 60% of children who lived in households in the bottom fifth still lived in bottom-fifth households 10 years later. Many people do move from poor to rich as they grow older, but there are also many households that stay poor for years. Unfortunately, many children live in these households.

Chapter 15: Gun Control

This study is observational in nature. States cannot be randomly assigned to have particular laws, and by focusing on data from a single year, the researchers were unable to assess whether firearm mortality changed in any way after certain laws went into effect. We therefore cannot conclude that specific state laws cause low firearm mortality, and we should be very cautious in terms of suggesting that changing laws will lead to lower firearm mortality.

When Kalesan and her colleagues published their work, critics acknowledged their efforts to control for different variables (like unemployment and gun ownership), but they were quick to point out several other variables that were not taken into account. For example, what about variables such as poverty level, alcohol consumption, and mental health? In a 2016 *Washington Post* article, columnist Carolyn Johnson questioned the findings of the study by noting that "People have very different inherent risks of death by firearm simply depending on whether they live in a rural or urban area. And while gun control laws may affect death from firearms, so will other factors, such as urban environments, poverty, and gang violence." Johnson went on to include comments from David Hemenway, a researcher from Harvard, and Daniel Webster, the director of the Johns Hopkins Center for Gun Policy Research. Although both Hemenway and Webster had concerns about other methodological issues in the study, each reached a very different conclusion about the importance of the findings. Hemenway lauded the study as a step in the right direction in terms of bringing scientific evidence to the debate on gun control, whereas Webster worried about how the flaws in the study might become politicized and ultimately lead the public to become skeptical of scientific research. Like many questions of causation, this one clearly remains open.

Chapter 16: Does the CPI Overstate Inflation?

The Consumer Price Index (CPI) has an upward bias because it can't track shifts from beef to tofu and back as consumers try to get the same quality of life from whatever products are cheaper this month. This was the basis of the outside experts' criticisms of the CPI: the CPI does not track the "cost of living." Their first recommendation was that "the BLS should establish a cost of living index as its objective in measuring consumer prices." The Bureau of Labor Statistics (BLS) said it agreed in principle but that neither it nor anyone else knows how to do this in practice. It also said, "Measurement of changes in 'quality of life' may require too many subjective judgments to furnish an acceptable basis for adjusting the CPI." Nonetheless, a new kind of index that in principle comes closer to measuring changes in the cost of living was created in 2002. This new index is called the Chained CPI-U (C-CPI-U). It more closely approximates a cost-of-living index by reflecting substitution among item categories. This new index may be an improvement, but it is unlikely that the difficult problems of defining living standards and measuring changes in the cost of their attainment over time will ever be resolved completely.

Chapter 20: The State of Legalized Gambling

Opponents of gambling have good arguments against legalized gambling. Some people find betting addictive. A study by the National Opinion Research Center estimated that pathological gamblers account for as much as 15% of gambling revenue and that each such person costs the rest of us $10,000 over his lifetime for social and police work. Gambling does ruin some lives, and it does indirectly harm others.

State-run lotteries involve governments in trying to persuade their citizens to gamble. From the early days of the New York lottery, we recall billboards that said, "Support education—play the lottery." That didn't work, and the ads quickly changed to "Get rich—play the lottery." Lotteries typically pay out only about half the money bet, so they are a lousy way to get rich even when compared with the slots at the local casino. Professional gamblers and statisticians avoid them, not wanting to waste money on so bad a bargain. Poor people spend a larger proportion of their income on lotteries than do the rich and are the main players of daily numbers games. The lottery may be a voluntary tax, but it hits the poor hardest, and states spend hundreds of millions on advertising to persuade the poor to lose yet more money. Some modest suggestions from those who are concerned about state-run lotteries: states should cut out the advertising and pay out more of what is bet.

States license casinos because they pay taxes and attract tourists—and, of course, because many citizens want them. In fact, most casinos outside Las Vegas draw gamblers mainly from nearby areas. Crime is higher in counties with casinos—but lots of lurking variables may explain this association. Pathological gamblers do have high rates of arrest, but again the causal link is not clear.

The debate continues. Meanwhile, technology in the form of Internet gambling is bypassing governments and creating a new gambling economy that makes many of the old arguments outdated.

Chapter 23: Should Hypothesis Tests Be Banned?

It will probably not surprise you that the American Statistical Association (ASA) did not take kindly to the BASP band on hypothesis testing and confidence intervals. In 2016, the ASA published a formal response: https://www.amstat.org/asa/files/pdfs/P-ValueStatement.pdf

It is interesting to note that in 1999, the American Psychological Association (APA) appointed a Task Force on Statistical Inference. At that time, the task force did not want to ban hypothesis tests. The report that was produced by the task force was, in fact, a summary of good statistical practice:

- Define your population clearly.
- Describe your data production and prefer randomized methods whenever possible.

- Describe your variables and how they were measured.
- Give your sample size and explain how you decided on the sample size.
- If there were dropouts or other practical problems, mention them.
- "As soon as you have collected your data, before you compute *any* statistics, *look at your data*."
- Ask whether the results of computations make sense to you.
- Recognize that "inferring causality from non-randomized designs is a risky enterprise."

The APA task force did say, "It is hard to imagine a situation in which a dichotomous accept-reject decision is better than reporting an actual *p* value or, better still, a confidence interval. ... Always provide some effect-size estimate when reporting a *p* value." But would the task force ban hypothesis tests altogether? "Although this might eliminate some abuses, the committee thought there were enough counterexamples to justify forbearance."

Sixteen years later, BASP banned hypothesis tests and confidence intervals. The controversy is not over. We encourage you to search the Web for the most up-to-date information about this issue. A good starting point would be to view the keynote address given by Ron Wasserstein (American Statistical Association) and Allen Schirm (Mathematica Policy Research; retired) at the 2019 USCOTS meeting: https://www.causeweb.org/cause/uscots/uscots19/keynote/2.

Answers to "Now It's Your Turn" Exercises

Chapter 1

1.1 Population: all American adults, Sample: 1504 randomly selected American adults who were called in this research poll.

1.2 This is an observational study. The participants self-reported how many bottles or cans of diet soda they drank per week over a period of two years. No treatment was applied by the researchers to the participants.

Chapter 2

2.1 This is not a simple random sample. To be a simple random sample, every group of eight students has to have the same probability of selection.

2.2 Using line 116, the sample is the provinces and territories labeled 03, 10, and 06. These are Manitoba, Prince Edward Island, and Northwest Territories.

Chapter 3

3.1 3.1%

3.2 1.6%

Chapter 4

4.1 This question is clearly slanted toward a positive Yes response because the question asks the respondent to consider "escalating environmental degradation and incipient resource depletion."

4.2 Starting at line 111 in Table A, we choose the person labeled 12 as the faculty member and the person labeled 38 as the student. Answers will vary with technology.

Chapter 5

5.1

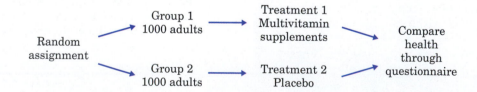

Chapter 6

6.1 There are two explanatory variables. They are the number of tweets (no tweet, single tweet, and tweet with a retweet) and the show (weekday vs. weekend). The response variable is whether or not the subject watched the show. There are $2 \times 3 = 6$ treatments as diagrammed below.

		Number of tweets		
		No tweet	Single tweet	Tweet with retweet an hour later
Show	Weekday	Treatment 1	Treatment 2	Treatment 3
	Weekend	Treatment 4	Treatment 5	Treatment 6

6.2

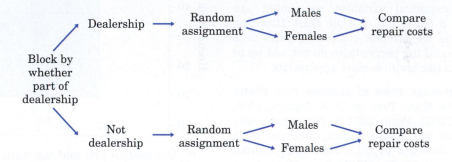

Chapter 7

7.1 Although this study exposes the students to minimal risk, it is always a good idea to seek institutional review board approval before proceeding with any study involving human subjects.

7.2 This is a complicated situation. This patient's underlying disease appears to be impairing his decision-making capacity. If his wishes are consistent during his lucid periods, this choice may be considered his real preference and followed accordingly. However, as his decision-making capacity is questionable, family members should be contacted about the procedure. Getting a surrogate decision maker involved might help determine what his real wishes are.

7.3 The first test offers anonymity. The second test offers confidentiality.

7.4 No. Observational studies can still meet all of these requirements and be deemed "ethical."

Chapter 8

8.1 The number of drivers is usually much greater between 5 and 6 P.M. (rush hour) than between 1 and 2 P.M. Thus, we would expect the number of accidents to be greater between 5 and 6 P.M. than between 1 and 2 P.M. It is therefore not surprising that the number of traffic accidents attributed to driver fatigue was greater between 5 and 6 P.M. than between 1 and 2 P.M. This is an example in which the proportion of accidents attributed to driver fatigue is a more valid measure than the actual count of accidents.

8.2 This information does not settle the debate. The results were very close (83 for the Samsung running Android and 82 for the iPhone running iOS). These ratings are unreliable because they could vary in another report outside of Consumer Reports. Their ratings are not biased because there is no evidence that these ratings systematically overestimate or underestimate the quality of these operating systems.

Chapter 9

9.1   The 2 million figure seems inaccurate. There would be 4–5 times as many people in Times Square at the New Year's Eve celebration than who walk through Times Square throughout the busiest weekday during December 2018, which is quite unreasonable.

9.2   In two years, Dejah is making $96,000, which is less than her original salary of $100,000.

Chapter 10

10.1   The "state" variable is a categorical variable. For categorical variables, we should use either a bar graph or a pie chart. However, because we are not comparing parts of a whole and the percentages do not add up to 100%, a bar graph is more appropriate.

10.2   The average price of gasoline rose pretty steadily from 1996 to 2006 (except 1998, 2001–2002). From 2007 to 2010, there was a small but stable decline, followed by a sharp increase in 2011. There was a small increase in 2012 to 2013, followed by a sharp downturn in 2014 to 2016. In 2017–2018, the average price of gasoline has increased.

Chapter 11

11.1

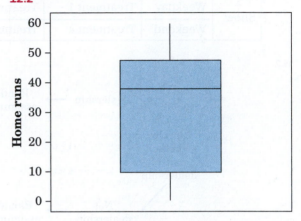

11.2   The distribution is mostly symmetric (perhaps slightly left skewed), with a center between 65 and 67 inches. The data are spread between 57 and 73 inches, with no outliers.

Chapter 12

12.1   The median is 38, the first quartile is 11, and the third quartile is 47.

12.2

The median (38) and the third quartile (47) for Ruth are slightly larger than for Bonds and Aaron. The distribution for Ruth appears more skewed (left skewed) than for Bonds and Aaron.

12.3   The mean is 32.83 and the standard deviation is 11.18. The mean (32.83) is less than the median of 34. This is consistent with the fact that the distribution of Aaron's home runs is slightly left skewed.

Chapter 13

13.1   64.2 inches and 74.2 inches.

13.2   The standard score of a height of 72 inches is 1.12.

13.3   A score of 570 or higher will be in the top 25%. 38.21% of scores will be at or below 475. 21.19% of scores will be at or above 580. 40.6% of scores are between 475 and 580.

Chapter 14

14.1 The researchers are seeking to predict IQ from brain size. Thus, brain size is the explanatory variable. The response variable is IQ. The following figure is a scatterplot of the data.

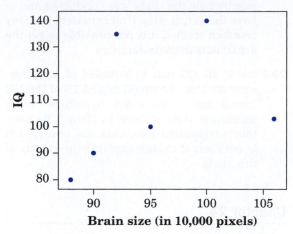

14.2 There is a weak positive association. There is no pronounced form other than evidence of a weak positive association. There are no outliers.

14.3 One might estimate the correlation to be about 0.3 or 0.4. The actual correlation is 0.38.

Chapter 15

15.1 80.13 cm.

15.2 0.0441 or 4.41%.

15.3 The observed relationship is likely due to common response. A greater number of boats registered in Florida and a greater number of manatee deaths from 1977 and 2016 can both be attributed to a growth in the Florida population over this time period.

Chapter 16

16.1 $2,193,409. This value is lower than the actual 2018 median salary of $3,627,842.

16.2 $29,388, 3.5%.

Chapter 17

17.1 As long as the coin is fair so that heads and tails are equally likely, all three sequences of 10 particular outcomes are equally likely even though the first one looks the most "random." Each sequence of 10 particular outcomes has a probability of $\left(\frac{1}{2}\right)^{10} = \frac{1}{1024}$.

17.2 A correct statement might be, "If you were to toss a coin a billion times, you could predict a nearly equal *proportion* of heads and tails."

Chapter 18

18.1 $P(\text{roll a 7}) = \dfrac{6}{36} = 0.167$;

$P(\text{roll an 11}) = \dfrac{2}{36} = 0.056$;

$P(\text{roll a 7 or roll an 11}) = \dfrac{6}{36} + \dfrac{2}{36} = \dfrac{8}{36} = 0.222$.

18.2 0.975.

Chapter 19

19.1 One option to assign digits (other assignments are possible):

0,1,2,3 = liberal

4,5,6,7 = middle-of-the-road

  8,9 = conservative

19.2 3/10 = 0.30.

Chapter 20

20.1 The average number of children under 18 in a household is 0.76.

20.2. Since Stephen Curry makes about 44% of his field-goal shots, we would expect $\dfrac{1}{.44} = 2.27$ shots before he makes his first. Starting on line 115 of Table A, the number of shots until the first made shot is 2, 1, 4, 1, 2, 5, 1, 2, 1, 2. The average of these 10 outcomes is 2.1. Thus, the estimate of the expected number of shots needed to make his first field goal is 2.1 shots, based on this simulation.

Chapter 21

21.1 0.495 to 0.525.

21.2 0.187 and 0.253.

21.3 9.89 to 10.91 minutes.

Chapter 22

22.1 $H_0: p = 0.7$ and $H_a: p \neq 0.7$. *If the null hypothesis is true*, the sample proportion has an approximately Normal distribution with mean $p = 0.7$ and $SD = 0.06481$. It does make sense that the sample proportion would be different with a smaller sample size due to sampling variability.

22.2 $\hat{p} = 0.64$. Standard score $= -0.926$, *P-value* $= 0.3682$. The large *P*-value gives no reason to think that the population proportion is different from 0.7. The results are not significant at the 0.05 level.

22.3 $H_0: \mu = 100$ and $H_a: \mu \neq 100$ *If the null hypothesis is true*, the sample mean $\bar{x}$ has approximately the Normal distribution with mean $\mu = 100$ and standard deviation 2.13. The sample mean is $\bar{x} = 105.8$. Standard score $= 2.72$, *P-value* $= 0.0070$. The small *P*-value provides evidence that that the mean IQ score of this population differs from 100.

Chapter 23

23.1 We would like to know both the sample size and the actual mean weight loss before deciding whether we find the results convincing. Better yet, we would like to know exactly how the study was conducted and to have the actual data. Unfortunately, in many research studies, it is not possible to get the actual data from researchers.

23.2 No. If all 122 null hypotheses of no difference are true, we would expect 1% of the 122 (about one) of these null hypotheses to be significant at the 1% level by chance. Because this is consistent with what was observed, it is not clear if chance explains the results of this study.

Chapter 24

24.1 717.7.

24.2 6.74.

24.3 $X^2 = 6.74$ is larger than the critical value 3.84 required for significance at the $\alpha = 0.05$ level. The study shows a significant relationship ($P < 0.05$) between playing games and average grades.

Answers to Odd-Numbered Exercises

Chapter 1

1.3 **(c)** Both (a) and (b).

1.5 **(d)** This is an experiment, and participants were randomly assigned to treatments.

1.7 **(b)** The study attempts to measure every individual in a population.

1.9 **(a)** Major League Baseball players. **(b)** Four. Age, Salary. **(c)** Age is in years. Salary is in thousands of dollars per year.

1.11 Pregnant and breast-feeding women; 21 women who returned the survey; 35%.

1.13 No. This is a sample survey in which information is gathered without imposing a treatment. The variables are gender and political party voted for in the last presidential election.

1.15 Exact descriptions of the populations may vary. **(a)** Teenagers (or "12th-graders," but from the description of the situation, the researcher would like information about all teens). **(b)** Former NFL football players. **(c)** All adults, or everyone (the radio host's question does not necessarily exclude children from consideration).

1.17 No treatment was imposed on the children being studied; the researchers simply observed the conditions under which they lived. As noted in the example, an experiment would mean exposing children to magnetic fields.

1.19 **(a)** The first is an experiment because a treatment was imposed on the subjects. The second is an observational study because no treatment was imposed on the subjects. **(b)** Because the treatments are randomly assigned, the two groups will resemble each other in age, exercise level, and other facets that may influence the HDL/LDL level. This will allow the researchers to study whether the treatment causes a change in the response.

1.21 **(a)** It is not an experiment. The amount of alcohol consumed was self-reported by the participants. No treatment was assigned by the researcher. **(b)** The population is nonagenarians (adults in their 90s). The variables are alcohol consumption, amount of exercise, and premature death status.

1.23 **(a)** This is an observational study; it could be a sample survey if we view the set of games examined as a sample from the population of all games that have been or will be played by this football team. **(b)** A sample survey would be an appropriate choice. Choose a sample of college students and ask them to rate their satisfaction. **(c)** An experiment would be best. Randomly assign students to two groups, one that has access to audio recordings of course lectures and one that does not.

Chapter 2

2.3 **(b)** It systematically favors certain outcomes.

2.5 **(c)** I write the names of all the students on similar slips of paper, put the slips of paper in a box, mix them well, and draw 10 slips from the box. The 10 names drawn are my sample.

2.7 **(b)** Angela and Lucinda.

2.9 **(a)** 466. **(b)** The online poll is a voluntary response sample, which is often biased.

2.11 Voluntary response samples are generally biased. In this case, 76% is probably higher than the true proportion of all adults who think Trump should fire Mueller.

2.13 **(a)** Post signs around the college inviting students to phone in their comments on the parking facilities or regulations. **(b)** Send interviewers out into the parking lots at a particularly busy time of day and instruct them to randomly question students using the parking lot.

2.15 Call-in polls (especially those that cost to call in), and voluntary response polls in general, tend to attract responses from those who have strong opinions on the subject, and therefore, they are often not representative of the population as a whole. A random sample of size 1000 will ideally be representative of the population as a whole. While the 300,000 callers might be an impressive number, the voluntary response of these callers is not trustworthy.

2.17 Answers will vary using technology. Using line 112 of Table A, the sample would be Armia fi, Byron, Danault, Deslauriers, Froese, Pacioretty.

2.19 **(a)** Both undergraduate and graduate students have a 1% chance of selection. **(b)** Each sample contains exactly 300 undergraduates and 100 graduate students. This is not an SRS because not all samples of size 400 are equally likely.

2.21 Eolo, Jade Mountain, Nihi Sumba.

2.23 Voluntary response samples, such as texting polls, generally misrepresent the population. Every attendee is not equally likely to text in his or her vote.

2.25 **(a)** Population: Hispanic residents of Denver. Sample: Adults interviewed at 200 mailing addresses. **(b)** Respondents may not give truthful answers to a police officer.

Chapter 3

3.3 **(c)** The percentage of all customers who purchased in the last year who would have replied they are very satisfied with the store's website.

3.5 **(d)** 2000 students.

3.7 **(b)** The result for the entire sample is more accurate because it comes from a larger sample.

3.9 45% is a statistic (related to the sample from the online poll); 52% is a parameter (related to all votes).

3.11 55 is a statistic (related to the sample of 100 phone numbers); 50.8% is a parameter (related to the population of all residential phone numbers).

3.13 **(a)** Results will vary. **(b)** The theoretical probability of choosing no females in this scenario is approximately 5%. We would expect only about 1 in 20 samples to have no women in them. Thus, it is not impossible to choose no females, but it is unlikely, so this may be enough reason to suspect discrimination.

3.15 **(a)** Population: Ontario residents. Sample: 61,239 Ontario residents interviewed. **(b)** These estimates are likely close to the truth about the entire population because the sample was large and random.

3.17 The larger sample will give less variable results. Thus, large samples are more likely to yield results close to the true value in the population as long as the bias is small.

3.19 **(a)** The value of the statistic is 61%. The parameter p in the setting is the true proportion of all American adults who are satisfied with the total cost they pay for their health care. **(b)** We are 95% confident the percent of all American adults who are satisfied with the total cost they pay for their health care is between 57% and 65%.

3.21 **(a)** 1.1%. **(b)** 0.7%. **(c)** The second snapshot of the poll more accurately reflects the views of Donato's constituents because the sample size is larger and produces a smaller margin of error. **(d)** It does change with this added information. The comment would likely produce an overestimate in the true proportion due to the paper encouraging readers to go to Donato's Facebook page and vote yes who might not otherwise have done so (or might not even be constituents). **(e)** Incorrect. The margin of error formula requires a random sample from the population of interest to be valid. The Facebook poll is a voluntary response and as a result subject to bias.

3.23 The larger sample will yield less variable results and a smaller margin of error. According to the quick method, the margin of error for the sample of size 25 will be twice as big as the margin of error for the sample of size 100 ($1/5 = 0.2$ vs. $1/10 = 0.1$).

3.25 **(a)** We are 95% confident that between 72% and 78% of all American adults think immigration is a good thing for the United States. **(b)** We have information only from a sample, not from all American adults. Although our sample results are likely to be close to the population truth, we cannot assume that they will be exactly the same. **(c)** The interval 72% to 78% is based on a procedure that gives correct results (that is, includes the true population value) 95% of the time.

3.27 The student is incorrect. The correct interpretation is that we are 90% confident that the *average* 2017 SAT score of *all* test takers of two or more races is between 800 and 1410. The emphasis should be on the average of all test takers, not individuals.

3.29 The estimated margin of error is 4.1%. This is essentially the same as the 4.1% margin of error given in Example 6.

3.31 The estimated margin of error is 0.4%.

3.33 The margin of error for 95% confidence is roughly 2.9%. Therefore, we can be 95% confident that between 60.1% and 65.9% of all American voters think marijuana should be legalized in the United States.

3.35 To reduce the margin of error to half its size, we need to quadruple the sample size. Thus, the sample size would need to be 4112.

3.37 To be more confident that the randomly varying sample statistic will land within a given margin of error of the true population parameter, we must allow a larger margin of error. So the margin of error for 95% confidence is larger than for 90% confidence.

3.39 The margin of error of ±2 percentage points belongs to the larger poll of 1993 registered voters. The margin of error for the smaller poll of 945 registered voters is close to 3%.

Chapter 4

4.3 **(b)** Random sampling error, but not other practical difficulties like undercoverage and nonresponse.

4.5 **(b)** Undercoverage.

4.7 **(a)** Because each student has a 10% chance of being selected, this is a simple random sample.

4.9 **(a)** Sampling error. **(b)** Nonsampling error. **(c)** Sampling error.

4.11 The margin of error excludes nonsampling errors such as nonresponse, poor question wording, processing errors, and response errors. The margin of error also excludes the impact of undercoverage.

4.13 This is sampling error (specifically, random sampling error). This is the only source of error accounted for in the margin of error.

4.15 Timing of the event in relation to when the polls were conducted would influence the responses. Generally, we would expect more respondents to select "generally bad" in response to the event. The results from the second poll may have also been influenced by the wording of the question and the availability of only two answer choices.

4.17 **(a)** Changing suggests there is something wrong with the Constitution. Adding suggests there is something missing. **(b)** More people will be in favor of adding to the Constitution rather than changing it.

4.19 **(a)** Yes, the question is clear; slanted toward the second option. **(b)** The question, "Do you agree or disagree?," is clear. The statement is slanted toward a desired response of agree. **(c)** Yes, the question is clear but is slanted toward a desired response of Yes. **(d)** Vague question that is prone to response errors such as lying.

4.21 *NBC News* carried out the survey via SurveyMonkey. The population was American adults. Respondents for this nonprobability survey were selected from the nearly 3 million people who take surveys on the SurveyMonkey platform each day. The sample included 6518 American adults. The response rate is unknown. The subjects were contacted via telephone (both cell phone and landline). The survey was conducted in May 2018. The exact question asked is listed in the exercise.

4.23 Closed questions are much easier to process. Answers to open questions must be read and categorized by someone. Closed questions give more precise information about the specific question asked because answers to open questions may be vague and hard to categorize. The great advantage of open questions is that not restricting responses may uncover trends you did not think of when making up the questionnaire. Closed questions can miss what is on the public's mind.

4.25 The margin of error only accounts for random sampling error. It does not include response error resulting from inaccurate or dishonest responses. In this situation, there may be some tendency for some adults who did not vote to claim that they did. That is, nonvoters may be reluctant to admit that they did not vote because they fear that they might be viewed negatively as a result.

4.27 **(a)** Answers will vary. **(b)** White 10%, Romero 20%.

4.29 Label the 500 midsize accounts from 001 to 500 and the 4400 small accounts from 0001 to 4400. On line 115, we first encounter numbers 417, 494, 322, 247, and 097 for the midsize group, then 3698, 1452, 2605, 2480, and 3716 for the small group. Results will vary based on the starting line in Table A or use of technology.

4.31 The article notes that all 50 states and the District of Columbia are represented, and it breaks down the ethnicity of the respondents. The article also notes that data were collected via both telephone and web surveys. By using an SRS from all possible telephone numbers and no web surveys, it is likely we would not get representative proportions in the sample that mirror the population of interest.

4.33 One possibility: Stratify in a manner similar to the Current Population Survey, then choose some regions from within each stratum, choose some school districts from each region, and choose a school from within each school district. Finally, choose a random sample of students from each school and contact their parents.

4.35 All dorm rooms are equally likely; each has a chance of 1/40 of being selected. To see this, note that each of the first 40 has chance 1/40 because 1 is chosen at random. But each room in the second 40 is chosen exactly when the corresponding room in the first 40 is, so each of the second 40 also has chance 1/40, and so on. This is not an SRS because the only possible samples have exactly 1 room from the first 40, 1 room from the second 40, and so on. An SRS could contain any 5 of the 200 rooms in the population. Note that this view of systematic sampling assumes that the number in the population is a multiple of the sample size.

4.37 This is a discussion question, the answers to which depend on local circumstance. Some possible answers: **(a)** "All students," or "all full-time students," or "all residential students," etc. **(b)** Obtain a list of students from the registrar's office or elsewhere. Stratification by major is probably appropriate; other strata, such as class rank, might also be used. **(c)** Mailed or e-mailed questionnaires might have high nonresponse rates. Telephone interviews exclude those without landline phones and may mean repeated calling for those who are not home. Face-to-face interviews might be more costly than your funding will allow.

4.39 All three questions are trying to address similar issues, yet subtle differences in wording result in different pictures of public opinion. Using responses to public opinion polls presents difficulties in discerning what "public opinion" really is.

Chapter 5

5.3 **(c)** An observational study.

5.5 **(a)** The size of the observed difference in longevity is not likely to be due to chance.

5.7 **(a)** Explanatory variable: frequency of digital media usage. Response: symptoms of ADHD. **(b)** This study is not an experiment because the researcher did not assign the frequency of digital media usage to the high school students. **(c)** The impact of digital media usage will be confounded with other variables such as parental involvement, learning disabilities, etc. This confounding will make it hard to isolate the effect of digital media usage on likelihood of developing ADHD.

5.9 Physiological or mental health problems may be causing the increased levels of hostility, and higher levels of the LPS-binding protein, a biomarker for leaky gut. Thus, the lurking variable of an unhappy marriage may be associated with, but not actually causing, the leaky gut.

5.11 **(a)** The subjects are the physicians, the explanatory variable is medication (aspirin or placebo), and the response variable is health, specifically whether the subjects have heart attacks. **(b)** Randomly assign the physicians to the aspirin and placebo groups, with 11,000 in each group. Compare the rate of heart attack in the aspirin and placebo group after several years. **(c)** "Significantly" means "unlikely to have occurred by chance if there were no difference between the aspirin and placebo groups."

5.13 We can never know how much of the change in attitudes was due to the explanatory variable (moving) and how much was due to the remediation those students needed. One problem with the study is that acclimating to a new environment may take time. Perhaps if we followed the students longer, we would see academic improvement.

5.15 Students registering for a course should be randomly assigned to a classroom or online version of the course. Scores on a standardized test can then be compared.

5.17 If this year is considerably different in some way from last year, we cannot compare electricity consumption over the 2 years. For example, if this summer is warmer, the customers may run their air conditioners more. The possible differences between the 2 years would confound the effects of the treatments.

5.19 **(a)** The explanatory variable is the pitch of a voice in an ad (high vs. low). **(b)** The response variable is the rating of the perceived size of the sandwich. It is only a 7-point scale that ranges from −3 to +3. **(c)** Conceivably, the researchers could have assigned students to watch an ad with no audio, thus controlling for pitch.

5.21 **(a)** Randomly assign half of the students to the high-pitch ad and the other half of the students to the low-pitch ad. After viewing the ads, compare the ratings of the perceived size of the sandwich. **(b)** High-pitch group: 4, 5, 7, 8, 9, 10, 13, 15, 16, 19; low-pitch group: 1, 2, 3, 6, 11, 12, 14, 17, 18, 20.

5.23 The difference in ratings of the perceived size of the sandwich between the low-pitch and high-pitch ads was large enough that it is unlikely due to chance alone. In other words, any observed difference in ratings between the two groups is likely due to the difference in pitch.

5.25 **(a)** Randomly assign the 50 subjects into cocoa pill and placebo groups, 25 in each group. After a period of time, compare the heart health in the two groups. **(b)** Using line 131 from Table A, the first five subjects in the cocoa pill group would be Campanella, Kaline, Palmer, Terry, and Wynn.

5.27 In a controlled scientific study, the effects of factors other than the nonphysical treatment (e.g., the placebo effect, differences in the prior health of the subjects) can be eliminated or accounted for so that the differences in improvement observed between the subjects can be attributed to the differences in treatments.

5.29 The greater risk of suicide for those with TBI could be due to the treatment the patients received for TBI rather than the brain injury itself. In addition, the records did not contain information about TBI suffered prior to 1977 that will also confound the interpretation of the results. Very large samples will yield a significant difference in treatments, even when the magnitude of the difference is small.

5.31 This was a comparative study. Randomization was used to assign the employees to only receive information about smoking-cessation programs or receive information plus financial incentives. 878 employees participated in the study. The results were statistically significant. Financial incentives for smoking cessation significantly increased the rates of smoking cessation.

Chapter 6

6.3 **(b)** A double-blind experiment.

6.5 **(a)** This was not a matched-pairs design.

6.7 **(c)** A block design.

6.9 The ratings were not blind because the researcher knew which subjects received the antidepressant and which received the placebo. If the researcher believed that the antidepressant would work, he or she might interpret subject responses differently in a way that might bias the results.

6.11 **(a)** The response variable should be "number of accidents with cars," or something similar. Ideally, one would like to randomly require some cyclists to use bright high-intensity xenon lights on both front and rear, only on front (standard light on rear), only on rear (standard light on front), and standard lights on front and rear. This might be done with volunteers. Though volunteers are not typical cyclists, a simple, completely randomized design seems promising. **(b)** The effect of bright high-intensity xenon lights may be lessened when, or if, they become common enough that people no longer notice them.

6.13 Because the subjects were severe-stroke patients and their recovery rate was compared to the average seen in all stroke patients nationwide, it is not surprising that patients with severe strokes had a slower recovery rate.

6.15 **(a)** Change in the Beck Depression Inventory score. **(b)** Randomly assign the 330 patients into the Saint-John's-wort, Zoloft, and placebo groups, with 110 in each group. Compare the change in the Beck Depression Inventory score for the three groups. **(c)** The study should be blind, and probably double-blind, to prevent expectations about effectiveness from confounding the results.

6.17 **(a)** Randomly assign the 300 children into the ibuprofen, acetaminophen, and codeine groups, with 100 in each group. Compare the change in pain relief for the three groups. **(b)** No one involved in administering the treatments or assessing their effectiveness knew which subjects were in each group. Additionally, the children did not know which group they were in. **(c)** The pain scores in Group A were so much lower than the scores in Groups B and C that a difference would not often happen by chance if acetaminophen and codeine were not more effective than ibuprofen. However, the difference between Group B and Group C could be due to chance.

6.19 **(a)** The individuals are each of the 32 New Zealand rabbits, and the response variable is the percentage of the surface covered by atherosclerotic plaques in a region of the aorta. **(b)** There are two explanatory variables (ibuprofen vs. placebo, low-cholesterol diet vs. high-cholesterol diet). There are four treatments (ibuprofen–low-cholesterol diet, ibuprofen–high-cholesterol diet, placebo–low-cholesterol diet, placebo–high-cholesterol diet). **(c)** Randomly assign the 32 rabbits into the four treatment groups, with 8 rabbits in each group. Compare the percentage of the surface covered by atherosclerotic plaques in a region of the aorta for the four groups.

6.21 The researcher will have each individual drink both the MiO enhanced water and the competing flavored water product, randomizing the order for each subject. The products will not be labeled, so the individuals will be blind to the brand. The response variable is which product the individual liked better.

6.23 **(a)** Randomly assign the 36 participants into the blue-ink regular ball pen and laptop with full-size keyboard groups, with 18 participants in each group. Compare the number of words recalled for the two groups. **(b)** Perform the experiment twice on each subject. Randomize the order (blue-ink vs. laptop with full-size keyboard) that the participants complete the task. The overall difference in number of words remembered for the participants will be analyzed.

6.25 **(a)** Block 1: Andrews 21 John 24 Cannon 25 Johnson 25; Block 2: Vaitai 27 Kelce 28 Pryor 28 Wisniewski 29; Block 3: Brown 30 Schwenke 30 Thuney 30 Mailata 32; Block 4: Karras 33 Warmack 33 Brooks 34 Peters 34; Block 5: Weathersby 35 Wynn 35 Mason 39 Waddle 42. **(b)** The exact randomization will vary with the starting line in Table A. Different methods are possible. Perhaps the simplest is to number from 1 to 4 within each block and then assign each member of Block 1 to a supplement and then assign Block 2, and so on. For example, starting at line 133, we assign 4—Johnson to treatment A, 1—Andrews to treatment B, and 3—Cannon to treatment C (so that 2—John gets treatment D). Then we carry on for Block 2, and so on (either continuing on the same line or starting over somewhere else in Table A).

6.27 As an example, we could randomly assign subjects to drink a glass of wine or not and have a snack or not. In this scenario, there would be 4 treatments since there are 4 (2×2) combinations of wine and snack. Randomly assigning enough subjects to each treatment would mitigate the impact of lurking variables such as amount of sleep the previous night or general overall health. The response variable would be the number of times the subject wakes up during the night.

6.29 Answers will vary.

6.31 The authors reach the conclusion that in general placebos have powerful clinical effects.

However, outside the setting of clinical trials, Hrobjartsson and Gotzsche conclude there is no justification for the use of placebos. Bailar finds these conclusions too sweeping. In particular, Bailar argues the evidence that placebos might contribute to pain relief may merit their continued therapeutic use when there is reason to think that a patient may benefit.

6.33 Answers will vary.

Chapter 7

7.5 **(b)** To protect the rights of human subjects (including patients) recruited to participate in research activities.

7.7 **(a)** The subjects are anonymous.

7.9 **(d)** All of the above.

7.11 **(a)** A nonscientist might raise different viewpoints and concerns from those considered by scientists. **(b)** An outsider might raise different viewpoints and concerns from those considered by insiders. **(c)** Answers will vary.

7.13 Answers will vary.

7.15 **(a)** Coercion. An overt threat of harm is intentionally presented by the investigator. **(b)** No coercion. Although the employees are concerned about potential consequences, the investigator did not intentionally present an implicit or explicit threat of harm.

7.17 Answers will vary.

7.19 They cannot be anonymous because the interviews are conducted in person in the subject's home. They are certainly kept confidential.

7.21 If there is a way for the person to be identified by the computer he or she is using, then just confidentiality is maintained.

7.23–7.39 Answers will vary.

Chapter 8

8.3 **(c)** We can use percentages to compare marijuana use by young adults in these states because we are given rates.

8.5 (c) This will be an unbiased measure of the patient's weight.

8.7 (c) Reliable.

8.9 Invalid measures: e.g., score on a written test of knowledge of sports; each student's weight. Valid measures: e.g., distance the student can run in some fixed length of time (or time required to run some fixed distance).

8.11 Compare death rates among occupants of motor vehicles involved in an accident who were wearing a restraining device with the rates for those who were not wearing a restraining device.

8.13 These numbers do make a convincing case that Turkey and the United States have a more substantial problem with obesity than Japan because the statistics cited from OECD are rates of obesity rather than counts.

8.15 These questions measure learned facts rather than general problem-solving ability.

8.17 For example, ask patients to rate their pain on a scale of 1 to 10. More detail on the rating scale might be useful, e.g., "1 is the pain caused by a splinter."

8.19 It must be shown that scores on the GATB predict future job performance. First give the GATB to a large number of job applicants for a broad range of jobs. Then, after some time, rate each applicant's actual job performance. These ratings should be objective when possible; if workers are rated by supervisors, the rating should be blind in the sense that the rater does not know the GATB score. Arranging a reliable and unbiased rating of job performance may be the hardest part of the task. Finally, examine the relationship between GATB scores and later job ratings.

8.21 Answers will vary.

8.23 (a) Every time the subject is asked, she would give the measurements 3.4 10.0 6.2 4.7 8.1 in the order in which the lengths are listed in the problem. (b) The list of reported measurements will vary, but it should have some too large and some too small. An example would be 3.4 9.0 6.2 3.7 8.1.

8.25 (a) This is biased: it is a systematic deviation, that is, an error that arises from the system used to make the measurements. (b) For example, randomly assign workers to a training program rather than let them choose whether or not to participate.

8.27 For the FBI data, local agencies may deliberately underreport crime or may simply be careless in keeping records. Victims may not report crimes to agencies.

8.29 (a) This method omits the fraction of a beat between the last beat and the end of the fixed time period. It will always result in an answer that is a multiple of 4, and the measurement could easily vary by 4 if pulse rate is constant. Imagine the case of two 15-second intervals, which could result in either 17 or 18 beats being counted, for a reported measurement of 68 or 72 beats per minute. This improves the reliability of the measurement as compared with counting the number of beats in 5 seconds, which would always be a multiple of 12. The total beats per minute could vary by 12 if pulse is consistent for each 5-second interval. (b) Counting beats for a full minute should produce results that vary by no more than a few beats per minute from one measurement to the next.

8.31 The rate of outstanding ratings is valid to measure high satisfaction. For the Apple in-ear headphone: $347/1648 = 0.211 = 21.1\%$; for the Klipsch Image S4i earphones: $69/134 = 0.515 = 51.5\%$. Looking at the rate of satisfaction, Klipsch has the better high-satisfaction rating.

8.33 Answers will vary.

Chapter 9

9.3 (c) 67% over the current figure.

9.5 (a) Your friend assumed that the amount of trash and the amount of recycling were equal to begin with.

9.7 (c) We cannot make a decision about the 15% figure because we do not know how courses are divided among teaching days (e.g., Monday-Wednesday-Friday or Tuesday-Thursday).

9.9. Doctors usually recommend aspirin, not a particular brand of aspirin. Bayer claims that doctors recommend (any brand of) aspirin more than twice as often as Tylenol.

9.11 800,000 deer/438 square miles is roughly equivalent to 1826 deer per square mile. This figure seems too high to be plausible.

9.13 15.9 cups of coffee per passenger or crew member per day is way too large to be plausible.

9.15 The percentage of flights canceled is 1%, not 10%.

9.17 **(a)** 240 beatings per hour; 5760 beatings per day; 2,102,400 beatings per year. The writer's arithmetic is wrong. It is 2.1 million, not 21 million. **(b)** The correct arithmetic puts the number of beatings per year under the total number estimated by the survey but still well above the estimated 254,000 committed by husbands or (ex-) boyfriends. However, the *Times* editorial is plausible because the National Crime Survey counts cases of violence, not incidents of violence: a woman in an abusive relationship (one case) likely gets beaten more than once a year (several incidents).

9.19 The Dow dropped by 4.60% on February 5, 2018. The Dow dropped by 22.63% on October 19, 1987.

9.21 A 100% decrease would reduce the emissions to zero. A greater than 100% decrease would result in negative emissions, which is not possible.

9.23 The chance of dying in a motor vehicle accident is 1 in 6813.

9.25 **(a)** 0.5%. **(b)** 3100 pounds. 45,000 pounds does not seem plausible because that would mean a cubic foot of soil weighs approximately 14.5 times as much as a cubic foot of water. **(c)** 50 cubic feet of soil would weigh about 3750 pounds. Adding 230 pounds of compost would give a total weight of 3980 pounds. Because 230 pounds is roughly 5.8% of 3980 pounds, the 5% organic matter conclusion is roughly correct.

9.27 Suspicious regularity: the correlations remain unchanged to three decimal places

as the sample size grows from 21 to 53 pairs over a decade. Burt's data are almost certainly fraudulent.

9.29 From 2007 to 2017, median household income increased by 3.09%. The percentage increase of top earners over the same time period was 13%.

9.31 If the statement from the *Guardian* is correct, 50% of obese people earn less than the national average income, whereas *more than 50%* of all workers (including those who are obese) earn less than the national average income. Without more information, this does not provide evidence that obese workers tend to earn less than other workers. It would be helpful to know what percent of nonobese workers earn less than the national average income.

Part I Review

I.1 **(a)** Every subset of n individuals from the population has the same probability of selection. **(b)** 95% of all possible samples will produce a confidence interval that captures the population parameter. **(c)** Sampling errors come from the act of choosing a sample. Random sampling error or the use of bad sampling methods are common types of sampling error. **(d)** The difference between the treatment groups is large enough that it would rarely occur by chance.

I.3 This Twitter poll is an example of a voluntary response sample. Voluntary response samples are, typically, biased as they tend to attract the strongest opinions and do not represent the population as a whole.

I.5 Assigning labels 01 through 30 and beginning on line 128, we choose 15, 27, 06, and 13 corresponding to Mee, Weese, Foyston, and Lehman, respectively.

Answers may vary based on the use of the applet or software.

I.7 This is undercoverage, which is a sampling error and is not accounted for in the margin of error (which only includes *random* sampling error).

I.9 Only the random sampling error in Exercise I.8 would be reduced. Doubling the sample size would reduce the variation in results from sample to sample.

I.11 (a) Non-LGBTQ Americans. (b) 2.2%. We are 95% confident the percentage of non-LGBTQ Americans who are "very" or "somewhat" uncomfortable seeing a same-sex couple holding hands lies between 28.8% and 33.2%.

I.13 Each student has a 1-in-5 (20%) chance of being interviewed. To be an SRS, every sample of size 10 from the population must have the same probability of selection. However, a sample of 10 students over age 21 is not even possible in this sampling plan. This is a stratified sample.

I.15 Use a completely randomized design, like the one outlined below, and compare time to get arrival on campus between the two groups.

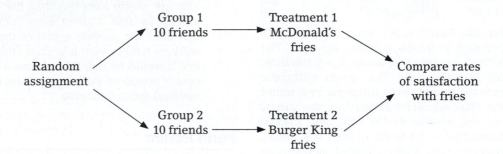

This experiment could also be done as matched pairs where all 20 friends taste both fries, and the order of the two fries is randomly assigned to each friend. Each friend then identifies which fries are preferred. Many important lurking variables would need to be controlled, including freshness and saltiness of the fries.

I.17 Subjects: 12,546 participants in seven countries (at least age 55 for men and at least age 60 for women). Explanatory variable: aspirin vs. placebo. Response variables: whether or not the participant experienced adverse (or serious adverse) cardiovascular events.

I.19 The three principles are an institutional review board, informed consent, and confidentiality. Some group must review the plan of study to protect the welfare of the patients. The subjects for the study must be informed of the procedures (specifically, they should know that they *might* receive a placebo, but of course, they should not be told that they *will* receive a placebo) and consent (typically in writing) to the

procedures. The subjects' personal information must be kept confidential.

I.21 Reliability means similar results in repeated measurements. To improve reliability, take many measurements and average them.

I.23 (a) This is an observational study: behavior is observed, but no treatment is imposed. The subjects decided whether or not to enroll in New Mexico's Medical Cannabis program. (b) Lurking variables such as other health habits, age, awareness, level of use, and so forth prevent the researchers from determining that medical cannabis *causes* a reduction in opioid use.

I.25 –6.8%.

Chapter 10

10.3 (d) Car color—categorical; miles per day—quantitative.

10.5 (c) This is due to roundoff error.

10.7 (c) Line graph.

10.9

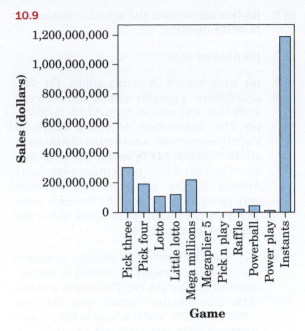

A pie chart could be used because we are displaying the distribution of a categorical variable, and we could compare parts of the whole.

10.11 **(a)** 65,320 thousand—that is, about 65.3 million. **(b)** Married is the most common marital status, followed by never married, with much smaller counts of widowed and divorced American women. **(c)** A pie chart could be used because each number in the table represents parts of the total (approximately 130,513,000 women).

10.13 The overall trend for all four generations shows the percentage in favor of same-sex marriage has increased from 2001 to 2015. For every year from 2001 to 2015, the percentage in favor of same-sex marriage is highest for millennials, followed by Generation X, baby boomers, and the Silent Generation.

10.15 **(a)** Within each year, there is both a high point and a low point. This is expected due to the growing season of oranges. **(b)** After accounting for the seasonal variation, there is a gradual positive trend in the price of oranges.

10.17 The bars were ordered generally (some exceptions) from highest to lowest credit

hours (decreasing height) for the bar graph, but the order of the categories on the x-axis does not matter in a bar graph. We could have arranged the last names alphabetically by last name, which would have produced a more random pattern in the heights of the bars.

10.19 A higher percentage of males than of females completed less than a high school degree. A higher percentage of females than of males completed some college but no degree, associate's degrees, bachelor's degrees, and advanced degrees, although the difference is fairly small in each case (1–2%). The most frequent category for both groups (male and female) is "High school degree."

10.21 The temperature will rise and fall with the changing seasons. That is, there will be a 12-month repeating pattern, rising in the summer months and falling in winter.

10.23 The official report is adjusted for the expected seasonal variation due to high school and college students as well as other seasonal employees entering the workforce for the summer.

10.25

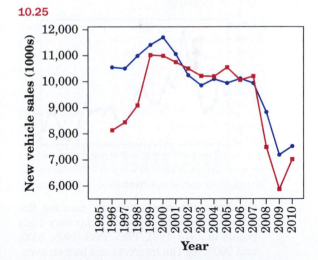

Before 2002, the number of cars was greater than the number of trucks. Overall, sales of both cars and trucks decreased from 2000 to 2007. Car and truck sales decreased sharply from 2007 to 2009, with an increase in 2010.

10.27

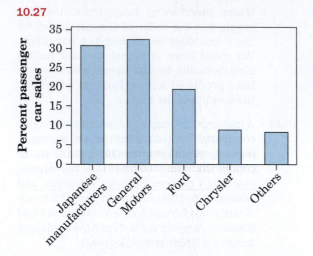

The bar graph makes it easier to see percentages (without writing them in or next to the wedges, as was done with the pie chart). It is also easier to compare sizes of bars than wedge angles.

10.29 There is overlap between the groups: young adults could have two or more of these habits and would therefore be represented twice.

10.31 **(a)** For example,

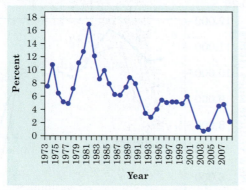

(b) From the line graph, we can see the interest rate cycle reached temporary high points in 1974, 1981, 1985, 1989, 1995, 2000, and 2007. **(c)** The interest rates had an overall peak in 1981 for the years 1973 to 2008. Since the overall peak, the interest rates have gradually decreased (ignoring the cycles).

Chapter 11

11.3 **(b)** You want to look at the distribution of a quantitative variable.

11.5 **(c)** One can recover the actual observations from the display.

11.7 **(b)** Skewed right.

11.9 **(a)** Utah has 39.7% young adults. **(b)** The distribution is roughly symmetric, centered at 35.85%, and spread from 32.5% to 39.7%. **(c)** The distribution of young adults is slightly less spread out than the distribution of older adults. **(d)** It would not be possible to identify which percentage represents Arizona in the stemplot. The stemplot only reveals the percentage for each state and does not retain information about the source of the data point.

11.11 **(a)** The distribution is roughly symmetric, although it might be viewed as slightly skewed to the right. **(b)** The center is about 15%. The smallest return was between −70% and −60%, and the largest return was between 100% and 110%. **(c)** About 23% of stocks lost money. **(d)** A histogram is preferred because the data set is large (1528 stocks).

11.13 The shape is skewed to the right. The center is 27 miles per gallon. There are five electric cars (outliers) in the data set.

11.15 The histogram is skewed to the right. There is a cluster of lower salaries and then a tail to the right toward the larger salaries. There is an outlier at 23 million.

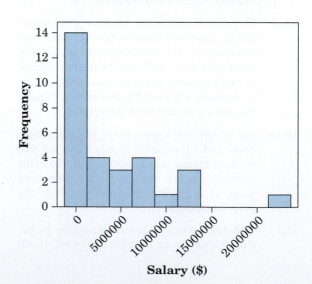

11.17 The distribution of salaries in the NBA will likely be roughly symmetric due to the salary cap.

11.19 The distribution is irregular in shape. There are two distinct groups, plus a low outlier (the veal brand with 107 calories).

11.21 Roughly symmetric; 46 is a typical year; 60 is not an outlier.

11.23 Right skewed, centered around 13 inches, with a spread of 5–42 inches. Use either a histogram or stemplot.

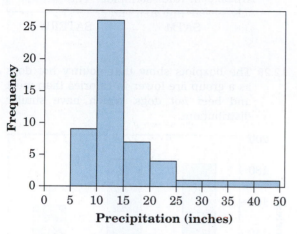

0	567788899
1	00011111111122233333344444
1	5555688
2	1223
2	5
3	2
3	8
4	2

(b) Only the variability would change. The distribution would still be skewed to the right and centered around 13 inches. The new spread with the higher value would be 5 inches to almost 50 inches.

Chapter 12

12.5 **(b)** 2.

12.7 **(c)** A boxplot.

12.9 Half of all households make more than the median ($68,145), and the other half make less.

12.11 The median wealth of these readers should be less than $240,000 (the mean) because income distributions tend to be right skewed.

12.13 **(a)** The distribution is somewhat symmetric, which means the mean will be roughly equal to the median. **(b)** The mean is $\bar{x} = 35.934\%$, and the median is $M = 35.850\%$.

12.15 We expect the distribution to be right skewed, so the mean ($4.32 million) is greater than the median ($1.8 million).

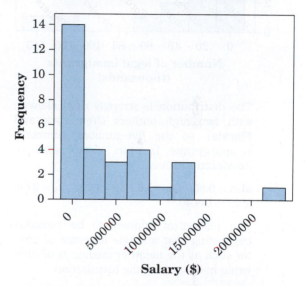

12.17 The distribution is irregular in shape. The hot dog brands fall into two groups with a gap between them, plus a low outlier. The five-number summary, in units of calories, is $\text{Min} = 107, Q_1 = 138.5, M = 153, Q_3 = 180.5$, $\text{Max} = 195$. Any numerical summary would not reveal the "gaps" in the distribution.

12.19 **(a)** The minimum is in position number 1, and the maximum is in position number 53. The median is in position 27. The first quartile is in position 13.5 (the median of the lowest 26 observations), and the third quartile is in position 40.5 (the median of observations 28 through 53). **(b)** The 40.5th number is the third quartile, the point that represents the top 25% of ages. Adding up the heights of the bars, the 40th number is in the 31 to 32.5 years class and 41st number is in the 32.5 to 34 years class. Thus, an MVP must be about 32.5 years or older to be in the top quarter.

12.21

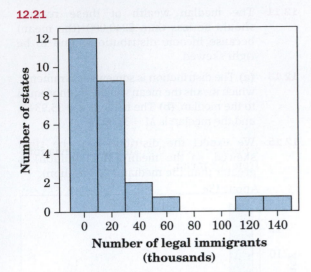

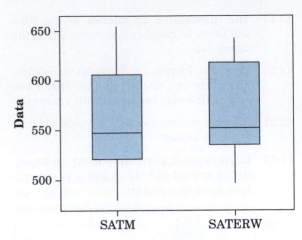

The distribution is strongly right skewed with two high outliers (New York and Florida), so the five-number summary is appropriate. In units of thousands of immigrants, this is

$Min = 0.80$, $Q_1 = 3.8$, $M = 11.0$, $Q_3 = 27.8$, $Max = 139.40$.

12.23 The histogram appears to be bimodal, suggesting that a single measure of center such as the mean or median is of little value in describing the distribution.

12.25 **(a)** The mean is more useful in this situation because of the need for a total. **(b)** The median is appropriate here. The word "typical" often refers to median.

12.27 The five-number summaries are

	Min	Q_1	M	Q_3	Max
SATM	480	521	547	606	655
SATERW	497	535	552	618	643

SAT Math scores have a slightly larger interquartile range but a much larger range than the SAT Evidence Based Reading and Writing scores. The minimum, Q_1, center (median), and Q_3 for the SAT Evidence Based Reading and Writing scores are higher than the SAT Math scores.

12.29 The boxplots show that poultry hot dogs as a group are lower in calories than meat and beef hot dogs, which have similar distributions.

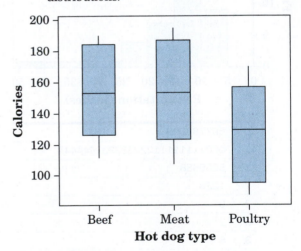

12.31 Both have mean 3; set (a) has $s = 1.095$, while set (b) has $s = 2.191$. Set (b) has a larger standard deviation and is therefore more variable.

12.33 SUV: $\bar{x} = 21.5$ mpg, $s = 3.3$ mpg; Mid-size: $\bar{x} = 37.4$ mpg, $s = 26.9$ mpg; SUVs have lower average miles per gallon than mid-size vehicles, but mid-size vehicles have more variation in miles per gallon as evidenced by the standard deviation. This agrees with the detailed comparison in Exercise 12.28.

12.35 Both sets of data have the same mean and standard deviation ($\bar{x} = 7.50$ and $s = 2.03$). However, the two distributions are quite different: set A is left skewed, while set B is roughly uniform with a high outlier.

12.37 A 5% across-the-board raise will increase the variability of the distribution. Both the distance between the quartiles and the standard deviation will increase by a factor of 5%. In other words, the new distance between the quartiles and standard deviation after the raise will be 1.05 times the original distance between the quartiles and standard deviation before the raise.

12.39 Answers will vary.

Chapter 13

13.5 **(c)** The median is the balance point in a density curve.

13.7 **(d)** Adam's pulse rate is about one-half of a standard deviation below the mean.

13.9 **(a)** Density curve with long tail to the left. **(b)** Answers vary. Rectangular density curve would be an example.

13.11 **(a)** The curve forms a 5 × 0.2 rectangle, which has area 1. **(b)** Mean and median are both 2.50. **(c)** 20% **(d)** 30%.

13.13 50%.

13.15 **(a)** 218–314 days. **(b)** Longer than 298 days. **(c)** Shorter than 250 days.

13.17 **(a)** 330–342 days. **(b)** 16%.

13.19 **(a)** The standard score for both Jessica and her mother is 2.0. **(b)** Jessica and her mother scored the same relative to each woman's age group because they have the same standard score. However, Jessica had the higher raw score, so she does have the higher absolute level of the variable measured by the test.

13.21 **(a)** 0.15%. **(b)** 64.2–74.2 inches. **(c)** 84%.

13.23 No. It has two peaks because of the two distinct subgroups (men and women).

13.25 16%.

13.27 1.39%.

13.29 57.93%.

13.31 Approximately the 76th percentile.

13.33 Approximately 0.7 standard deviations from the mean.

13.35 About 134.

Chapter 14

14.5 **(b)** Negative.

14.7 **(d)** All of the above.

14.9 **(a)** $-1 \leq r \leq 1$. **(b)** $s \geq 0$. **(c)** $-\infty \leq \bar{x} \leq \infty$ but has to fall within the min and max of a given data set.

14.11 Negative.

14.13 Moderate, roughly linear, and positive. Point A is a hot dog brand that is well below average in both calories and sodium and can be considered an outlier.

14.15 This shows a fairly strong positive association, so r should be reasonably close to 1.

14.17 The correlation decreases when A is removed from Figure 14.11 because that plot looks more linear with A. That is, if we drew a line through that scatterplot, there is a less relative scatter about that line with point A than without.

14.19 **(b)** Moderate, somewhat linear, and negative. **(c)** The pattern of the relationship in (b) does not hold for the counties that need the data to be treated with caution. The relationship appears much weaker and not negative for these counties.

14.21 **(b)** Although changing the scales makes the scatterplot look very different, it has no impact on the correlation. Changing the units of the x and y variables does not change the value of the correlation.

14.23 **(a)** It is unlikely the correlation will be about the same for the counties for which the data do not need to be treated with caution and for the counties for which the data do need to be treated with caution due to the low counts. **(b)** Counties with caution (low counts): $r = .024$; counties without caution: $r = -.469$.

14.25 The rate would decrease (the decimal point in each value would move two places to the left). The correlation between homicide and suicide rates would stay the same.

14.27 The newspaper interpreted zero correlation as implying a negative association between teaching ability and research productivity. The speaker meant that teacher rating and research productivity are not related (strictly speaking, that they are not linearly related).

14.29 **(a)** 0.8. **(b)** 0.5. **(c)** 0.2. Note that (a) will have the strongest relationship and (c) will have the weakest relationship.

14.31 **(a)** Small-cap stocks have a lower correlation with municipal bonds, so the relationship is weaker, which leads to greater diversification. **(b)** She should look for a negative correlation, although this would also mean that this investment tends to decrease when bond prices rise.

14.33 **(a)** The exact values for Alaska are 15.2 inches and 332.29 inches. **(b)** Without Alaska and Hawaii, the relationship is weakly positive and roughly linear. Knowing maximum 24-hour precipitation would be somewhat useful in predicting maximum annual precipitation for these other states, but the relationship is not very strong.

14.35 The correlation is $r = 0.413$. The correlation is greatly lowered by the one outlier. Outliers tend to have a fairly strong impact on correlation; it is much stronger here because there are so few observations.

Chapter 15

15.5 **(d)** The square of the correlation.

15.7 **(c)** As long as the patterns found in past data continue to hold true.

15.9 Inactive girls are more likely to be obese, so if hours of activity are small, we expect BMI to be high, and vice versa. 3.2%.

15.11 88.36%.

15.13 Number of manatee deaths by boats goes up by .136 when the number of boats registered in Florida goes up by 1; 88.84 manatee deaths.

15.15 **(b)** The association is negative (direction), somewhat linear (form), and moderately strong (strength). **(c)** The sign of r is negative, which agrees with our observation of the direction of the association. An r value of $-.645$ is an indication that the association is moderately strong.

15.17 **(a)**

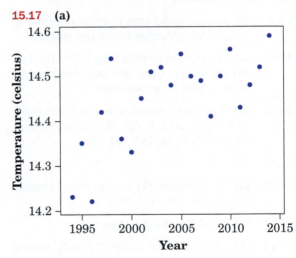

(b) 14.56 degrees Celsius. The observed average global temperature in 2014 was 14.59 degrees Celsius, so the prediction was fairly accurate. **(c)** 45.46%.

15.19 **(b)** A regression line is worthless for predicting mpg from speed because these variables do not have a straight-line relationship. For any speed, the regression line simply predicts 22.8 mpg (the average of the five fuel efficiency numbers in our data).

15.21 If the regression line has slope 0, the predicted value of y is the same (the y-intercept) regardless of the value of x. In other words, knowing x does not change the predicted value of y.

15.23 **(b)** $6000 **(c)** $y = 1000 + 300x$.

15.25 No. The sign of the slope will always match the sign of the correlation because they have to be going in the same direction.

15.27 $r = .50$.

15.29 -1091.6 deaths per 100,000 people. A negative value does not make sense. Using the regression line to predict for 150 liters per person is extrapolation because the original data ranged from 1 to 9 liters per person.

15.31 Answers will vary. If a student is not doing homework, he or she will probably not have high grades and will probably have more time for Facebook.

15.33 More income in a nation means more money to spend for everything—including health care, hospitals, medicine, etc.—which presumably should lead to better health for the people of that nation. On the other hand, if the people in a nation are reasonably healthy, resources that would otherwise be used for health care can be used to generate more wealth. Additionally, healthy people are generally more productive.

15.35 In this case, there may be a causative effect, but in the direction opposite to the one suggested: people who are overweight are more likely to be on diets and so choose low-calorie salad dressings.

15.37 $r = -0.86$ indicates a stronger straight-line relationship than $r = 0.704$ because $r = -0.86$ is closer to $r = -1$ than $r = 0.704$ is to $r = 1$.

15.39 These data can show an association between the two, but they cannot show a cause-and-effect relationship. Correlation does not imply causation.

15.41 **(a)** Eyelash Length $= 0.5 + (0.3102 \times$ Eye Width), which agrees with the equation given in Exercise 15.14. **(b)** Eye Width $= 0.394 + (2.289 \times$ Eyelash Length); 2.11 cm. **(c)** The x and y variables are reversed, which will produce least-squares regression lines with different slopes and y intercepts.

15.43 Each data set has least-squares line $y = 3 + 0.5x$.

Chapter 16

16.3 **(a)** An index number.

16.5 **(b)** A fixed market basket price index.

16.7 **(c)** Produce data needed for government policy and decisions by businesses and individuals.

16.9 **(a)** Tuition rose by a factor of 7.484 from the base period to October 2018. Equivalently, tuition costs increased by 648.4% (748.4% − 100%) over this period. **(b)** Consumer prices rose from 100 to 252.9—a 152.9% increase—which is much smaller than the increase in tuition over this same period.

16.11 Index number (1988) $= 100$; index number (2000) $= \dfrac{6,655,831}{6,939,299} \times 100 = 95.9$; index number (2013) $= \dfrac{4,137,328}{6,939,299} \times 100 = 59.6$. Between 1988 and 2000, releases decreased by 4.1%. Between 1988 and 2013, releases decreased by 40.4%.

16.13 **(a)** Because the CPI in New York is greater than the CPI in Los Angeles, prices rose faster in New York than in Los Angeles. **(b)** We do not know how prices compared in the base period.

16.15 The 2015 quantities are not relevant for a fixed market basket index. The Guru Price Index (1985 $= 100$) for 2015 is, therefore, $\dfrac{176.95}{66.45} \times 100 = 266.3$.

16.17 In terms of 2018 dollars, the 2002 trade paperback of *Rabbit, Run* cost $11.20 \times \dfrac{251.1}{179.9} = $15.63, which is more than the actual 2018 trade paperback cost of $14.50. Thus, the 2018 trade paperback of *Rabbit, Run* is a better deal.

16.19 About $53,042. $50,000 \times \dfrac{251.1}{236.7} = \$53,041.83$.

16.21 Answers will vary depending on the year. In 2014, $100 was equivalent to $100 \times \dfrac{26.8}{236.7} = \$11.32$ in 1955 dollars. Assuming the CPI continues to increase, the 1955 price of a modern microwave based on more recent CPI data will be less than $11.32, a striking decrease from the $1300 cost of a 1955 microwave.

16.23 $32,000 \times \dfrac{24.1}{14.0} = \$55,086.$

$\dfrac{100,000 - 55,100}{55,100} = 0.815 = 81.5\%.$

16.25 In terms of 2018 dollars, the 1986 WashU tuition cost $10,500 \times \dfrac{251.1}{109.6} = \$24,056$. The actual cost of $48,859 in 2018 is approximately double this value; thus, the increase indicates that the cost of WashU is going up faster than consumer prices in general.

16.27 On a percentage basis of real cost, tuition increased more from 2008 to 2018 than room and board (28.7% vs. 13.2%).

16.29 The table and plot are below. The solid line connected by dots shows that the tuition has risen fairly steadily since 1998. The dashed line connected by squares indicates that, after adjusting for inflation, the "real tuition" has also risen, though not as quickly as the actual dollar amounts. In other words, tuition is rising faster than "average" inflation: the real cost of University of Michigan tuition has increased about 62.5%.

Year	1998	2000	2002	2004	2006	2008	2010	2012	2014	2016	2018
Tuition ($)	6,098	6,513	7,411	8,202	9,723	11,037	11,837	12,994	13,486	14,402	15,262
1998 dollars	6,098	6,165	6,715	7,077	7,861	8,356	8,847	9,225	9,287	9,781	9,907

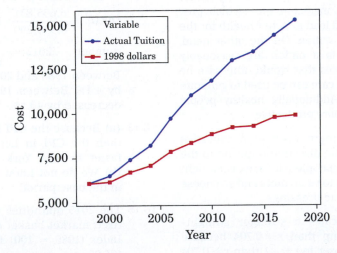

16.31 If the increase in cost is seen as paying for higher-quality cable service, it does not count toward an increase in the CPI.

16.33. **(a)** The CPI market basket is intended to represent the purchases of people living in urban areas. A person living on a cattle ranch would have a different set of needs and would make different purchases from those of an urban consumer. **(b)** Using a wood stove and no air conditioner makes Mohammed's expenditures for utilities quite different from those of a typical urban consumer. **(c)** Luis and Maria's medical expenses would be considerably higher than those for a typical consumer, so that portion of the CPI would not reflect the increase in how much they spent in the last year. **(d)** Gabriel's housing costs and spending habits would be considerably less than the typical urban consumer since he is homeless and unemployed. **(e)** Since Aliyah is employed by the military and living on a base outside the United States, her housing costs and spending habits would be different than a typical urban consumer living in the United States.

16.35 The local CPIs are based only on sample prices from the area in question. That is, they are statistics based on smaller samples than the national CPI, and smaller samples usually have greater sampling variability.

16.37 **(a)** "The 33rd percentile of the income distribution" is the income such that 33% of all workers make less than that amount. **(b)** "Real wages" means "adjusted for inflation."

16.39 By asking the same questions, the GSS can be used to track how the population is changing over time.

16.41 Answers will vary.

Part II Review

II.1 The distribution is roughly symmetrical (or slightly right skewed) with New Hampshire and Maryland as low outliers. West Virginia and Mississippi are high outliers.

II.3 Min $= 6.5$, $Q_1 = 10.4$, Med $= 11.85$, $Q_3 = 14.0$, Max $= 19.7$.

II.5 The mean for all states in Table II.1 is 12.238. Mississippi is a high outlier, so we would expect the mean to decrease. The mean without Mississippi is 11.940.

II.7 **(a)** 500. **(b)** 68%.

II.9 **(a)** Inches. **(b)** Inches. **(c)** Inches. **(d)** No units.

II.11 Dolphin: 180 kg, 1600 g. Hippo: 1400 kg, 600 g.

II.13 **(a)** Decrease because that plot looks more linear with elephants. That is, if we drew a line through that scatterplot, there is a less relative scatter about that line with elephants than without. **(b)** Increase because that plot looks less linear with dolphins, hippos, and humans. That is, if we drew a line through the scatterplot, there is more relative scatter about that line with dolphins, hippos, and humans than without.

II.15 800 g.

II.17 Roughly a straight line with r close to -1. From use, soap shrinks in size.

II.19 -56.1 g., the soap could not weigh a negative amount; prediction outside the range of the available data is risky.

II.21 $27,245 \times \dfrac{251.0}{195.3} = \$35,015.34$.

II.23 While the value of gold dropped from 1990 to 2000 (in terms of 2018 dollars), an investment in gold most certainly holds its value today.

II.25 **(a)** Roughly symmetric; the two highest and one lowest time might be considered outliers.

II.27 The mean would be higher because housing prices tend to have a right-skewed distribution. Unusually expensive houses would cause the mean to increase and be larger than the median.

II.29 **(a)** Fidelity Technology Fund because of the higher correlation. **(b)** No.

Chapter 17

17.3 **(b)** They exhibit a clear pattern in very many repetitions, although any one trial of the phenomenon is unpredictable.

17.5 **(d)** All of the above.

17.7 **(a)** It expresses an individual's judgment of how likely an outcome is.

17.9 Answers will vary.

17.11 **(a)** We expect probability 1/2. **(b)** Answers will vary. The theoretical probability is 1/3.

17.13 Answers will vary. The theoretical probability is 1/9.

17.15 (a) 0. (b) 1. (c) 0.01. (d) 0.6.

17.17 (a) Answers will vary. (b) A personal probability might take into account specific information about one's own driving habits or about the kind of traffic one usually drives in. (c) Most people believe that they are better-than-average drivers, whether or not they have evidence to support that belief.

17.19 For example, people are often eager to assume that short-run success will continue indefinitely (gamblers are reluctant to quit after winning, people justify risky behavior because it has not killed them yet, etc.). Additionally, people tend to believe that tragedies happen to others.

17.21 Answers will vary.

17.23 While on the surface it appears to be surprising, if we consider all combinations of two head football coaches throughout history, it is likely at least one pair of coaches were born at the same hospital.

17.25 Assuming free-throws are independent of each other, the 90.4% figure applies to the long run of the entire season. Short-run strings of misses are likely over the course of a season.

17.27 Assuming all at-bats are independent of each other, the 31.6% figure only applies to the long run of the season, not to short runs. Altuve is no more due for a hit than he was on his previous at-bat.

17.29 (a) The wheel is not affected by its past outcomes—it has no memory; outcomes are independent. So on any one spin, black and red remain equally likely. (b) Removing a card changes the composition of the remaining deck, so successive draws are not independent. If you hold five red cards, the deck now contains five fewer red cards, so your chance of another red decreases.

17.31 (a) Opinions will vary; one point of view is simply that airplane crashes are more sensational and interesting—perhaps because so many people die at once—so they sell more papers or get better ratings. (b) The (lower) risk of flying is not emphasized more than the risk of driving. Also, the issue of control arises: people feel more control behind the wheel of a car than sitting in a seat on an airplane, and more people survive car accidents than airplane crashes.

Chapter 18

18.3 (b) What outcomes are possible and how to assign probabilities to these outcomes.

18.5 (d) All of the above.

18.7 (d) The pattern of values of the statistic in many samples from the same population.

18.9 0.45; 0.55.

18.11 (a) 0.25. Because some outcome must occur on every trial, the sum of the probabilities for all possible outcomes must be exactly 1. (b) 0.52.

18.13 (a) 1. Because some outcome must occur on every trial, the sum of the probabilities for all possible outcomes must be exactly 1. (b) 0.96. (c) 0.22.

18.15 (a) BBB, BBG, BGB, GBB, GGB, GBG, BGG, GGG. Each has probability 1/8. (b) $3/8 = 0.375$.

18.17 (a) 1/38 since all outcomes are equally likely. (b) 18/38. (c) 10 to 9 are the correct odds. (d) 12/38.

18.19 (a) Legitimate. (b) Not legitimate. The sum of the probabilities for all possible outcomes is not 1. (c) Legitimate.

18.21 (a) 0.025. (b) 0.05.

18.23 0.2742.

18.25 (a) The sampling distribution tells us the pattern of values of $\hat{p}$ among many samples from the population. (b) 0.025. It is the (theoretical) relative frequency out of a large number of trials. (c) 0.025. (d) 0.525.

18.27 (a) 10 to 2 or equivalently 5 to 1. (b) 1/6.

Chapter 19

19.3 (a) The probabilities of the outcomes.

19.5 (b) Knowing the outcomes of one does not change the probabilities for outcomes of the other.

19.7 **(b)** Thinking carefully about the probability model for the simulation.

19.9 **(a)** Yes 0.80; No 0.20. **(b)** Yes 0–7; No 8–9. **(c)** $3/25 = 0.12$.

19.11 Let 0–2 represent Yes (that Barack Obama has done the best job in their lifetime) and 3–9 represent No. Counting the number of Yes responses out of the 100 digits, starting with line 112, 38 Americans said Barack Obama has done the best job in their lifetime. To estimate the probability that at least 35 say Yes in a random sample of 100, repeatedly sample 100 digits and count the number of samples of size 100 that have at least 35 Yes responses out of the total number of samples taken.

19.13 **(a)** 0.2. **(b)** One possibility: less than 1 hour, 0; 1 to 5 hours, 1–5; 6 to 10 hours, 6–7; more than 10 hours, 8–9.

19.15 Results vary based on the starting line in Table A or technology. Starting at line 135, the estimated probability would be $5/10 = 0.50$.

19.17 **(a)** Let 00–38 be a made three-point shot, and 39–99 be a missed three-point shot. To simulate a single three-point shot, randomly select a group of two digits from Table A, or use technology. To simulate four independent three-point shots, select a group of two digits from Table A, or use technology to represent the first shot, then another group of two digits to represent the second independent shot, then another group of two digits to represent the third independent shot, and finally one more group of two digits to represent the fourth independent shot. **(b)** $7/50 = 0.14$.

19.19 **(b)** Let 0–3 represent a pass on the first attempt, 0–4 represent a pass on the second attempt, and 0–5 a pass on the third attempt. If she passes on an attempt, begin the next repetition. In each repetition, allow for at most three attempts. **(c)** $40/50 = 0.80$.

19.21 **(a)** Let 0–1 represent die without female offspring, 2–4 have one female offspring, and 5–9 have two female offspring. **(b)** There can be between one and seven beetles after three generations (including the original

beetle). **(c)** Six descendants, five descendants, one descendant, zero descendants, six descendants. The beetle population appears to be growing because only one of these was zero. More repetitions would be advantageous.

19.23 **(a)** Let 1–6 represent the result of a single die. If 0 or 7–9 ignore, and move to the next digit. To simulate a roll of two fair die, repeat the single die process twice and add to find the total. **(b)** Starting at line 114, the player's results in 10 plays would be: Loss, Loss, Loss, Loss, Loss, Loss, Win, Loss, Loss, Win. The estimated probability would be $2/10 = 20\%$, which is a cold run at the craps table. More repetitions would be likely to converge around the true probability of winning.

19.25 **(a)** Let 00–74 represent that Matt can eliminate one of the answer choices, and 75–99 represent he cannot eliminate an answer choice. Then, let 01–33 (ignore 00 because the probability is 1/3) represent a correct guess if he can eliminate an answer choice, and 00–24 a correct guess if he cannot eliminate an answer choice. **(b)** Starting at line 124, Matt has the following results on the 10 questions: Correct, Incorrect, Incorrect, Incorrect, Incorrect, Incorrect, Incorrect, Correct, Correct, Incorrect. Matt did not pass the quiz.

19.27 Starting at line 139, the first seven triples are ignored because they are too large. Remarkably, the next two triples are both 356—so the first two people we encounter have the same birthday.

Chapter 20

20.3 **(c)** It can only be computed if the random phenomenon has numerical values.

20.5 **(a)** $0.

20.7 **(b)** Should be close to the expected value of the random phenomenon.

20.9 $0.50.

20.11 $0.95.

20.13 **(a)** The best choice is the first sequence because it is the most likely to occur. **(b)** The second sequence looks more "random" and the results mirror the actual faces of the die (four red, two green).

20.15 **(a)** $0.75. **(b)** $0.72. Mark 2 is slightly less favorable than Mark 1.

20.17 **(a)** 1.3 female offspring. **(b)** When a large population of beetles is considered, each generation of Asian stochastic beetles will contain close to 1.3 times as many females as the preceding generation. So the population will grow steadily. If the expected number of female offspring were less than 1—say, 0.80—each generation would contain only about 80% as many females as the preceding generation, and the population would decline steadily.

20.19 Use 0–1 for zero female offspring, 2–4 for one female offspring, and 5–9 for two female offspring. Choose 100 digits—one digit for each of the 100 beetles. Results vary with starting point in Table A or use of technology. The mean number of offspring for these 100 beetles should be close to the expected value of 1.3 female offspring calculated in Exercise 20.17.

20.21 **(a)** 2.44 people. **(b)** 350 households, 370 households. **(c)** 2440 people. **(d)** Right skewed. More than half of households have one or two people, while the larger family sizes are less common.

20.23 To simulate one sequence of children, read two digits from Table A, with 00–48 representing a girl and 49–99 representing a boy. Keep reading two digits at a time until finding a girl, and record the number of pairs read. Repeat the process 25 times and average the results. Results vary with starting point in Table A.

20.25 Use two digits for each simulation, with 00–19 a correct guess and 20–99 an incorrect guess. Count the number of correct answers out of 10 questions, repeat the process 20 times, and average the results. Results vary with starting point in Table A.

20.27 By the np formula, $(10)(.50) = 5$. We expect 5 heads in 10 flips of a coin. The mean number of heads in the 50 repetitions is close to 5 but not exactly 5.

Part III Review

III.1 **(a)** Results will vary among online directories. **(b)** If all are equally likely, this would be 0.3, or 30%. In some online directories, they may appear to be reasonably close to equally likely.

III.3 **(a)** .05. **(b)** A+ 00; A 01–07; A– 08–15; B+ 16–23; B 24–39; B– 40–53; C+ 54–63; C 64–76; C– 77–82; D+ 83–87; D 88–91; D– 92–94; F 95–99.

III.5 0.0256 is the theoretical probability. The estimated probability based on simulation will vary based on assignment of digits and starting line in Table A. Assigning digits like III.3 part (b) and starting at line 112, none of the 10 samples result in all four students getting a B or better. Thus, the estimated probability for that particular simulation is $0/10 = 0\%$.

III.7 3.5.

III.9 **(a)** 0.03. **(b)** For example, use two digits for each hand, with 00–91 for one pair or worse and 92–99 for two pair or better. To estimate the expected number of hands, count the number of digit pairs needed to find a number between 92 and 99. Do this many times, and find the average number of hands.

III.11 **(a)** All probabilities are between 0 and 1 and the sum of all possible outcomes is 1. **(b)** 0.196. **(c)** 0.166.

III.13 **(a)** 0.872. **(b)** 0.128.

III.15 **(a)** 0.16. **(b)** 70 to 130.

III.17 0.21.

III.19 119.5 (or about 120).

III.21 One way to simulate this is to take a single digit for the gene contributed by each plant, with, say, 0–4 (or odds) representing G and 5–9 (or even) representing Y. For each plant, take two digits; the plant will be green if both digits are 4 or less (or both are odd).

III.23 Choose one digit to determine the number of customers: 0–5 represents one customer, 6–8 represents two customers, 9 represents three customers. Then, for each customer, pick a random digit, with 0 or 1 meaning that customer bought a car and 2–9 meaning the customer did not buy a car. For example, starting at line 117, the first digit is a 3, so Bill had one customer that afternoon. The next digit is an 8, so the customer did not buy a car.

Chapter 21

21.5 **(c)** 0.420 to 0.522.

21.7 **(b)** The 95% confidence interval for a sample of size 100 will generally have a smaller margin of error than the 95% confidence interval for a sample of size 50.

21.9 **(a)** The population is the 175 residents of Tonya's dorm. **(b)** p is the proportion of all residents of the dorm who like the food. **(c)** .28.

21.11 95% confidence means we obtained this result by a method that will yield correct results approximately 95% of the time. In other words, if we were to repeatedly take samples of the same size, approximately 95% of the confidence intervals constructed from these samples would capture the true proportion of K–12 teachers who feel engaged in their work. Roughly 5% of the intervals would miss the true proportion.

21.13 **(a)** p is the proportion of all people ages 16 and older living in the United States who had visited, in person, a library or a bookmobile in the last year. **(b)** 0.455 to 0.504. **(c)** We are 95% confident that the proportion of people ages 16 and older living in the United States who had visited, in person, a library or a bookmobile in the last year is between 0.455 and 0.504.

21.15 Four times as large, $n = 4164$.

21.17 There are 15 zeros among those 200 digits, which gives $\hat{p} = 0.075$; 0.0385 to 0.1115,

which includes the true parameter value $p = 0.1$.

21.19 **(a)** Many high school students might be reluctant to admit that they have engaged in vaping, although this may be somewhat offset by some who might be inclined to brag or exaggerate. **(b)** The sample value is likely to be an underestimate of the true proportion since high school seniors will downplay vaping. **(c)** 4833.

21.21 **(a)** 0.637 to 0.683. **(b)** The margin of error for the 95% confidence interval ("19 cases out of 20") was .023 = 2.3%, which agrees with the statement "no more than 3 percentage points."

21.23 **(a)** The distribution is approximately Normal with mean 0.10 and standard deviation .015. **(b)** About 97.5%.

21.25 **(a)** $\hat{p} \pm \sqrt{\dfrac{\hat{p}(1-\hat{p})}{n}}$. **(b)** This method works (gives an interval that includes the population proportion) about 68% of the time. **(c)** 0.208 to 0.214. This is half as wide as the 95% confidence interval.

21.27 0.646 to 0.694. This interval is narrower than the 95% interval found earlier.

21.29 0.434 to 0.466, 0.430 to 0.470, 0.425 to 0.475, 0.410 to 0.490. As the confidence level increases, the confidence interval gets wider.

21.31 0.036 to 0.038.

21.33 Answers will vary.

21.35 **(a)** 6 to 30. **(b)** 16.3 to 19.7.

21.37 0.485 to 0.587.

21.39 The distribution is reasonably close to Normal (roughly symmetric and bell shaped, centered near 0.5).

21.41 **(a)** The sampling distribution of $\bar{x}$ should have mean equal to the population mean, $\mu = 0.5$. For these 50 sample means, we find $\bar{x} = 0.496$, which is close to 0.5. **(b)** The standard deviation of these 50 sample means is $s_{\bar{x}} = 0.03251$, so we estimate σ to be $10s_{\bar{x}} = 0.3251$.

21.43 Answers will vary.

Chapter 22

22.5 **(c)** The data provide strong evidence against the null hypothesis.

22.7 **(c)** 2.8.

22.9 There is good evidence that male students earn more (on the average) than do female students during the academic year. The difference in earnings in our sample was large enough that it would rarely occur in samples drawn from a population in which men's and women's average earnings are equal. Such a difference would happen in less than 3% of all samples. The average earnings of black and white students in our sample were so close together that a difference this large would not be unexpected in samples drawn from a population in which the average earnings of blacks and whites are equal. Similar differences would happen more than 50% of the time.

22.11 **(a)** Randomly assign the 2079 subjects to the low-fat/high-fiber and usual diet groups. Compare number of polyps that reoccurred in the two groups. **(b)** The difference in polyp development between the two groups was small enough that it might occur simply by chance if diet had no effect.

22.13 The differences observed between dates estimated using the old method and estimates based on the new method were small enough that they could occur simply by chance if both methods produced the same results (on the average).

22.15 **(a)** People might be embarrassed to admit that they have not attended religious services, or they might want to make a favorable impression by saying that they did. **(b)** We take p to be the proportion of American adults who attended religious services last week. $H_0: p = 0.39$ and $H_a: p < 0.39$.

22.17 With p as the local unemployment rate, our hypotheses are $H_0: p = 0.051$ and $H_a: p \neq 0.051$.

22.19 **(a)** p is the graduation rate for all athletes at this university. **(b)** $H_0: p = 0.82$ and $H_a: p < 0.82$. **(c)** 0.7842; the P-value is the

probability that $\hat{p} \leq 0.7842$ (under the assumption that $p = 0.82$). **(d)** A P-value of 0.0968 indicates that $\hat{p}$ values as extreme as 0.7842 would be somewhat rare (that is, they would occur in about 10% of all samples). This gives some reason to doubt the assumption that $p = 0.82$.

22.21 **(a)** p is the proportion of times the candidate with the better face wins. **(b)** $H_0: p = 0.50$ and $H_a: p > 0.50$. **(c)** 0.6875; the P-value is the probability that $\hat{p} \leq 0.6875$ (under the assumption that $p = 0.50$). **(d)** A P-value of 0.017 indicates that $\hat{p}$ values as extreme as 0.6875 would be unlikely, that is, they would occur in about 1.7% of all samples. This gives pretty good reason to doubt the assumption that $p = 0.50$.

22.23 Because the P-value (0.0227) is less than 5% but more than 1%, the result is significant at the 5% level but not at the 1% level.

22.25 A test is significant at the 1% level if outcomes as or more extreme than observed occur less than once in 100 times. A test is significant at the 5% level if outcomes as or more extreme than observed occur less than five in 100 times. Something that occurs less than once in 100 times also occurs less than five in 100 times, but the opposite is not necessarily true.

22.27 **(a)** Take 20 digits, and use the digits 0–4 for "Method A wins the trial" and 5–9 for "Method B wins the trial." (Or vice versa, or use even digits for Method A, etc.) **(b)** Results will vary depending on which digits represent "Method B wins." **(c)** We simulated the probability (assuming that H_0 is true) of observing results at least as extreme as those in our sample.

22.29 Our hypotheses are $H_0: p = 0.176$ and $H_a: p > 0.176$, where p is the proportion of first-year students at this university who have not taken any AP courses. If the null hypothesis is true, then the proportion of first-year students at this university who said they had not taken any AP courses from an SRS of 200 students would have (approximately) a Normal distribution with mean $p = 0.176$ and standard deviation 0.02692. Our sample had $\hat{p} = 0.230$, for which the standard score is 2.006 and a P-value 0.0227. P-value in Ex 2.20 was given as 0.0227, so the value is the same. Since the P-value is small, we reject the null hypothesis. The data do suggest that the proportion of first-year students at this university who have not taken any AP courses is larger than the national value of 17.6%.

22.31 (a) If $p = 0.1$, then the proportion suffering adverse symptoms in an SRS of 420 patients has (approximately) a Normal distribution with mean $p = 0.1$ and standard deviation 0.01464. **(b)** We test $H_0: p = 0.1$ and $H_a: p < 0.1$. Our sample had $\hat{p} = \dfrac{21}{420} = 0.05$, for which the standard score is -3.42. Compared with Table B, we see that this is very strong evidence ($P < 0.0003$) that fewer than 10% of patients suffer adverse side effects from this medication.

22.33 $H_0 : p = 0.5$ and $H_a : p < 0.5$, where p is the proportion of all drivers who are speeding. If the null hypothesis is true, then the proportion of speeders in an SRS of 12,931 drivers would have (approximately) a Normal distribution with mean $p = 0.5$ and standard deviation 0.004397. Our sample had $\hat{p} = \dfrac{5,690}{12,931} = 0.4400$, for which the standard score is -13.6. Table B tells us that $P < 0.0003$. We can conclude that fewer than half of all drivers in this location are speeding.

22.35 $H_0: \mu = 19$ seconds and $H_a: \mu < 19$ seconds, where μ is the mean time to complete the maze with the noise as a stimulus.

22.37 We test $H_0: \mu = 0.5$ versus $H_a: \mu \neq 0.5$. If H_0 were true, the sample mean $\bar{x}$ of an SRS of 100 numbers would have (approximately) a Normal distribution with mean 0.5 and standard deviation 0.0312. For $\bar{x} = 0.536$, the standard score is 1.2, for which $P = 0.2302$ using Table B. We have little reason to believe that the mean of all possible numbers produced by this software is not 0.5.

22.39 We test $H_0: \mu = 20$ versus $H_a: \mu < 20$, where μ is the mean tip percentage for all patrons of this restaurant who are given a message about bad weather. If H_0 were true, the sample

mean $\bar{x}$ of an SRS of 20 customers would have (approximately) a Normal distribution with mean 20 and standard deviation 0.4472. For $\bar{x} = 18.19$, the standard score is -4.05. Compared with Table B, we see that this is very strong evidence ($P < 0.0003$) that the mean tip percentage for all patrons of this restaurant who are given a message about bad weather is less than 20%.

22.41 Answers will vary.

Chapter 23

23.3 (b) 5.

23.5 (b) No, because a 2 mile-per-gallon increase is probably too small to justify an additional $1.00 per gallon.

23.7 (a) No, because the significance level is set beforehand and should not be changed after the P-value has been calculated.

23.9 Because the fan had the salaries for all of the players on the 2015 Paris Saint-Germain team, the mean obtained from the data is the mean salary for all 2015 Paris Saint-Germain players. There is no need to estimate the population mean with a confidence interval since it is known.

23.11 (a) Because majority means more than 50%, and the confidence interval is 62% to 70%, the results suggest a majority of American adults support the legalization of marijuana in the United States. **(b)** Four times as large, so n = 4076. **(c)** The margin of error would increase due to the higher confidence level.

23.13 (a) No. In a sample of size 200, we expect to see about two people who have a P-value of 0.01 or less. This person *might* have ESP, or he or she may simply be among the "lucky" ones we expect to see. **(b)** The researcher should repeat the procedure on this person to see whether he or she again performs well.

23.15 We might conclude that customers prefer Design A, but not strongly. Because the sample size is so large, this statistically significant difference may not be of any practical importance.

23.17 Were these random samples? How big were the samples? If the samples were very large, a significant result might not indicate an important difference.

23.19 A significance test answers only part (b). The *P*-value states how likely the observed effect (or a stronger one) is if chance alone is operating. The observed effect may be significant (very unlikely to be due to chance) and yet not of practical importance. And the calculation leading to significance *assumes* a properly designed study.

23.21 (a) The width (or margin of error) decreases, allowing more precise estimation of the parameter. (b) The *P*-value will decrease. That is, the evidence against H_0 will grow stronger.

23.23 (a) H_o: Innocent (Not Guilty) H_a: Guilty. We assume the defendant is innocent and weigh the strength of the evidence to convince the jury of guilt. Thus, H_o represents innocence. (b) A Type I error would be concluding the defendant is guilty when, in reality, he or she is not guilty. A Type II error would be failing to conclude the defendant is guilty when in reality he or she is guilty. (c) One factor would be seriousness of the crime. If the charge is murder, we would want to choose a small probability of a Type I error to minimize the chance of falsely convicting the defendant.

23.25 0.144 to 0.288. The interval does not include 31.3%. This is not surprising since the group of educators thought students at their college are less likely to identify as "liberal" than college students nationally.

23.27 (a) $784.87 *to* $803.13 (b) Because all respondents who answered no opinion were recorded as $0, and this group is likely to include people who will spend money on Christmas gifts but are unsure about the amount in addition to people who do not celebrate Christmas, the sample mean of $812 is likely to be too low. Thus, we should be skeptical of the confidence interval estimate for *all* American adults.

23.29 0.847 to 0.889; 0.848 to 0.890.

23.31 Answers will vary.

Chapter 24

24.5 (d) 63.9%.

24.7 (c) 21.0%.

24.9 The percentages, respectively, are: high school $237/407 = 58.2\%$; bachelor $200/260 = 76.9\%$; and graduate or professional $110/148 = 74.3\%$. It appears that people with more education are more likely to believe that astrology is not at all scientific.

24.11 At least one close family member smokes: $115/322 = 35.7\%$; No close family member smokes: $25/100 = 25\%$. In our sample, male students with at least one close family member who smokes are more likely to smoke than are male students with no close family member who smokes.

24.13 (a) Hatched: 16, 38, 75. Did not hatch: 11, 18, 29. (b) In order of increasing temperature, the percentage hatched are $16/27 = 59.3\%$, $38/56 = 67.9\%$, $75/104 = 72.1\%$. The percentage hatching increases with temperature; the cold water did not prevent hatching, but made it less likely.

24.15 (a) 4,484,000. (b) 1,289,000. (c) Associate's 64.2%, Bachelor's 58.0%, Master's 60.4%, Professional/Doctorate 57.1%. Women earn over 50% of associate's, bachelor's, master's, and professional/doctoral degrees. The lowest percent earned by women is professional/doctorate, and the highest percent earned by women is associate's.

24.17

30	10
30	30

40	0
20	40

24.19 (b) Overall, 0.8% of Maori and 0.6% of non-Maori were in the jury pool. However, for Rotura, the percentages are 0.9% and 1.1%, respectively; for Nelson, they are 0.1% and 0.2%. (c) Overall, 0.8% of Maori serve on juries compared to 0.6% of non-Maoris. However, in each of the two districts, non-Maoris served at a higher rate than did Maori.

24.21 (a) All expected counts are 1 or greater, and only 1 (10%) of the expected counts is less than 5, so the chi-square test is safe. (b) The chi-square statistic is 229.660, df $= 4$. This result is significant at $\alpha = 0.001$ (and much smaller). There is a statistically significant relationship between smoking status and opinions about health.

24.23 (a) The hypotheses are as follows:

H_0: There is no association between whether the student smokes and whether close family members smoke.

H_a: There is some association whether the student smokes and whether close family members smoke.

Population is male students in secondary schools in Malaysia. **(b)** The expected counts are 106.82, 215.18, 33.18, and 66.82. These are the counts we would expect—except for random variation—if H_0 were true. **(c)** The chi-square statistic is 3.951, df $= 1$. This result is significant at the 5% level, but not 1%.

24.25 (b) Stress management 90.9%, Exercise 79.4%, Usual Care 70%. **(c)** 6.79, 26.21, 6.99, 27.01, 8.22, 31.78. All expected counts are greater than 5, so the chi-square test is safe. **(d)** The chi-square statistic is 4.840, df $= 2$. This result is significant at the 10% level, but not 5%.

24.27 Answers will vary.

Part IV Review

IV.1. 0.620 to 0.680.

IV.3. 0.6403 to 0.6997.

IV.5. 0.818 to 0.862.

IV.7 We have information about all states—the whole population of interest—so we already know the value of the population proportion. Thus, there is no need to estimate the population proportion using a confidence interval.

IV.9. **(a)** If there were no difference between the two groups of doctors, results like these would be very rare (occur less than 1 in 1000 times). **(b)** The two proportions are given for comparison and are very different.

IV.11. For weekday mornings, 0.550 to 0.590; for weekday evenings, 0.733 to 0.767. The entire confidence interval for the evening proportion is considerably higher than the confidence interval for the morning proportion.

IV.13. **(a)** $H_0: p = 0.57$ and $H_a: p > 0.57$, where p is the proportion of answered calls in the evening. **(b)** If H_0 is true, then the proportion $\hat{p}$ from an SRS of 2454 would have (approximately) a Normal distribution with mean 0.57 and standard deviation 0.0100. **(c)** The observed sample proportion $\hat{p} = \dfrac{1840}{2454} = 0.7498$ lies so far out in the high tail of the sampling distribution that it is very strong evidence against the null hypothesis.

IV.15. **(a)** People are either reluctant to admit that they don't attend regularly or believe that they attend more regularly than they do. **(b)** Sample results vary from population truth; while we hope the proportion in our sample is close to the population proportion, we cannot assume the two are exactly equal. Thus, we give a range of plausible values for the population proportion via a confidence interval. **(c)** Both intervals are based on methods that work (include the true population proportion) about 95% of the time.

IV.17. Our hypotheses are $H_0: p = \dfrac{1}{4}$ and $H_a: p > \dfrac{1}{4}$, where p is the proportion of American adults who believe drinking alcoholic beverages in moderation is bad for your health. If the null hypothesis is true, then the proportion from an SRS of 1033 adults would have (approximately) a Normal distribution with $p = 1/4$ and standard deviation $\sqrt{\dfrac{\left(\frac{1}{4}\right)\left(\frac{3}{4}\right)}{1033}} = 0.0135$.

Our sample had $\hat{p} = 0.2798$, for which the standard score is 2.2 and associated p-value is 0.0139. We do have enough evidence to conclude that more than one-fourth of American adults believe drinking alcoholic beverages in moderation is bad for your health.

IV.19. 0.335 to 0.406. We are 90% confident the proportion of all voters who think economic problems are the most important issue facing the nation is between 0.335 to 0.406.

IV.21. Our hypotheses are $H_0: p = 0.57$ and $H_a: p > 0.57$, where p is the proportion of answered calls in the evening. If the null hypothesis is true, then the proportion $\hat{p}$ from an SRS of 2454 would have (approximately) a Normal distribution with mean 0.57 and standard deviation 0.0100. Sample proportion: 0.7498; Standard score: 17.99; P-value: $P < 0.0003$ based on Table B. Conclusion: We have very strong (overwhelming) evidence that more than 57% of evening calls are answered.

IV.23. We test $H_0: \mu = 5$ mg/liter and $H_a: \mu < 5$ mg/liter. If the null hypothesis were true, the sample mean $\bar{x}$ of an SRS of 45 measurements would have (approximately) a normal distribution with mean 5 mg/liter and approximate standard deviation 0.1371. Standard score: -2.77; P-value: 0.0028.

Conclusion: We have strong evidence that the stream has a mean oxygen content of less than 5 mg/liter.

IV.25. **(a)** The distribution looks reasonably symmetric; other than the low (9.4 feet) and high (22.8 feet) outliers, it appears to be fairly Normal. The mean is $\bar{x} = 15.59$ feet and $s = 2.550$ feet. **(b)** 14.96 to 16.22 feet. **(c)** We need to know what population we are examining: Were these all full-grown sharks? Were they all male? (That is, is μ the mean adult male shark length or something else?) Also, can these numbers be considered an SRS from this population?

IV.27. We test $H_0: \mu = 15$ feet and $H_a: \mu > 15$ feet. If the null hypothesis were true, the sample mean $\bar{x}$ of an SRS of 44 sharks would have (approximately) a Normal distribution with mean 15 feet and approximate standard deviation 0.3844 feet. Standard score: 1.53; P-value: 0.0668. This is significant at the 10% level, but not at the 5% level. We have some weak evidence that mean shark length is greater than 15 feet.

IV.29. **(a)** Row and column totals: Stayed 1306; Left 76; No Complaint 743; Medical Complaint 199; Nonmedical Complaint 440. **(b)** No complaint: $22/743 = 2.96\%$, Medical complaint: $26/199 = 13.07\%$, and Nonmedical Complaint: $28/440 = 6.36\%$. **(c)** 702.14, 188.06, 415.80, 40.86, 10.94, 24.20. All expected counts are greater than 5, so the chi-square test is safe. **(d)** We test

H_0: There is no relationship between a member complaining and leaving the HMO.

H_a: There is some relationship between a member complaining and leaving the HMO.

The degrees of freedom are df = 2; this X^2 value is so large (and the degrees of freedom are so small) that checking a table should not be necessary: this result is significant at $\alpha = 0.001$ (and much smaller). We have strong evidence that there is a relationship between a member complaining and leaving the HMO.

IV.31. **(a)** The two-way table is in the Minitab output below.

Rows: C5 Columns: Worksheet columns

	Post-Millennial	Millennial	Generation X	All
Enrolled	55	34	28	117
	39	39	39	
Not Enrolled	45	66	72	183
	61	61	61	
All	100	100	100	300

Cell Contents
  Count
  Expected count

Chi-Square Test

	Chi-Square	DF	P-Value
Pearson	16.898	2	0.000
Likelihood Ratio	16.823	2	0.000

(b) The percent enrolled in college varies drastically between the three generations (55% post-Millennial vs. 34% Millennial vs. 28% Generation X). **(c)** We find $X^2 = 16.898$ with df = 2, which is highly significant. There is a relationship between enrollment in college and generation.

Index

Note: Page numbers in **boldface** type indicate pages where key terms are defined.

Table A Random digits

Line								
101	19223	95034	05756	28713	96409	12531	42544	82853
102	73676	47150	99400	01927	27754	42648	82425	36290
103	45467	71709	77558	00095	32863	29485	82226	90056
104	52711	38889	93074	60227	40011	85848	48767	52573
105	95592	94007	69971	91481	60779	53791	17297	59335
106	68417	35013	15529	72765	85089	57067	50211	47487
107	82739	57890	20807	47511	81676	55300	94383	14893
108	60940	72024	17868	24943	61790	90656	87964	18883
109	36009	19365	15412	39638	85453	46816	83485	41979
110	38448	48789	18338	24697	39364	42006	76688	08708
111	81486	69487	60513	09297	00412	71238	27649	39950
112	59636	88804	04634	71197	19352	73089	84898	45785
113	62568	70206	40325	03699	71080	22553	11486	11776
114	45149	32992	75730	66280	03819	56202	02938	70915
115	61041	77684	94322	24709	73698	14526	31893	32592
116	14459	26056	31424	80371	65103	62253	50490	61181
117	38167	98532	62183	70632	23417	26185	41448	75532
118	73190	32533	04470	29669	84407	90785	65956	86382
119	95857	07118	87664	92099	58806	66979	98624	84826
120	35476	55972	39421	65850	04266	35435	43742	11937
121	71487	09984	29077	14863	61683	47052	62224	51025
122	13873	81598	95052	90908	73592	75186	87136	95761
123	54580	81507	27102	56027	55892	33063	41842	81868
124	71035	09001	43367	49497	72719	96758	27611	91596
125	96746	12149	37823	71868	18442	35119	62103	39244
126	96927	19931	36809	74192	77567	88741	48409	41903
127	43909	99477	25330	64359	40085	16925	85117	36071
128	15689	14227	06565	14374	13352	49367	81982	87209
129	36759	58984	68288	22913	18638	54303	00795	08727
130	69051	64817	87174	09517	84534	06489	87201	97245
131	05007	16632	81194	14873	04197	85576	45195	96565
132	68732	55259	84292	08796	43165	93739	31685	97150
133	45740	41807	65561	33302	07051	93623	18132	09547
134	27816	78416	18329	21337	35213	37741	04312	68508
135	66925	55658	39100	78458	11206	19876	87151	31260
136	08421	44753	77377	28744	75592	08563	79140	92454
137	53645	66812	61421	47836	12609	15373	98481	14592
138	66831	68908	40772	21558	47781	33586	79177	06928
139	55588	99404	70708	41098	43563	56934	48394	51719
140	12975	13258	13048	45144	72321	81940	00360	02428
141	96767	35964	23822	96012	94591	65194	50842	53372
142	72829	50232	97892	63408	77919	44575	24870	04178
143	88565	42628	17797	49376	61762	16953	88604	12724
144	62964	88145	83083	69453	46109	59505	69680	00900
145	19687	12633	57857	95806	09931	02150	43163	58636
146	37609	59057	66967	83401	60705	02384	90597	93600
147	54973	86278	88737	74351	47500	84552	19909	67181
148	00694	05977	19664	65441	20903	62371	22725	53340
149	71546	05233	53946	68743	72460	27601	45403	88692
150	07511	88915	41267	16853	84569	79367	32337	03316

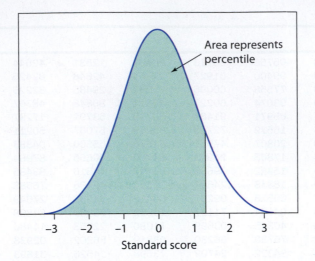

Area represents percentile

Standard score

Table B Percentiles of the Normal distributions

Standard score ⟶ Percentile		Standard score ⟶ Percentile		Standard score ⟶ Percentile	
−3.4	0.03	−1.1	13.57	1.2	88.49
−3.3	0.05	−1.0	15.87	1.3	90.32
−3.2	0.07	−0.9	18.41	1.4	91.92
−3.1	0.10	−0.8	21.19	1.5	93.32
−3.0	0.13	−0.7	24.20	1.6	94.52
−2.9	0.19	−0.6	27.42	1.7	95.54
−2.8	0.26	−0.5	30.85	1.8	96.41
−2.7	0.35	−0.4	34.46	1.9	97.13
−2.6	0.47	−0.3	38.21	2.0	97.73
−2.5	0.62	−0.2	42.07	2.1	98.21
−2.4	0.82	−0.1	46.02	2.2	98.61
−2.3	1.07	0.0	50.00	2.3	98.93
−2.2	1.39	0.1	53.98	2.4	99.18
−2.1	1.79	0.2	57.93	2.5	99.38
−2.0	2.27	0.3	61.79	2.6	99.53
−1.9	2.87	0.4	65.54	2.7	99.65
−1.8	3.59	0.5	69.15	2.8	99.74
−1.7	4.46	0.6	72.58	2.9	99.81
−1.6	5.48	0.7	75.80	3.0	99.87
−1.5	6.68	0.8	78.81	3.1	99.90
−1.4	8.08	0.9	81.59	3.2	99.93
−1.3	9.68	1.0	84.13	3.3	99.95
−1.2	11.51	1.1	86.43	3.4	99.97